International Encyclopaedia of
DAIRY TECHNOLOGY AND ANIMAL HUSBANDRY

International Encyclopaedia of
DAIRY TECHNOLOGY AND ANIMAL HUSBANDRY

Volume 2

P.K. Saraswat

ANMOL PUBLICATIONS PVT. LTD.
NEW DELHI - 110 002 (INDIA)

ANMOL PUBLICATIONS PVT. LTD.
H.O.: 4374/4B, Ansari Road, Daryaganj,
New Delhi-110 002 (India)
Ph.: 23278000, 23261597
B.O.: No. 1015, Ist Main Road, BSK IIIrd Stage
IIIrd Phase, IIIrd Block,
Bangalore - 560 085 (India)
Visit us at: www.anmolpublications.com

International Encyclopaedia of Dairy Technology and Animal Husbandry

First Edition, 2009
ISBN 978-81-261-3793-0 (Set)

PRINTED IN INDIA

Printed at Mehra Offset Press, Delhi.

Contents

Contents

[illegible]

VOL. I

VOL. II

Preface

Husbandry implies application of scientific principles to agriculture, especially to animal breeding. Animal husbandry is the looking after and breeding of animals, particularly livestock. Animal husbandry has been around for over 20,000 years. Animal husbandry today is the agricultural practice of breeding and raising livestock. Examples of animal husbandry include beekeeping, dog breeding, farming, horse breeding. Livestock are domesticated animals intentionally reared in an agricultural setting to make produce such as food or fibre, or for their labour. Livestock include pigs, cattle, goats, deer, sheep, yaks and poultry. The type of livestock reared varies worldwide and depends on factors such as climate, consumer demand, native animals, local traditions, and land type. Veterinary Science is vital to the study and protection of animal production practices, herd health and monitoring spread of widespread disease. Veterinary medicine is the application of medical, diagnostic, and therapeutic principles to companion, domestic, exotic, wildlife, and production animals. It requires the acquisition and application of scientific knowledge in multiple disciplines and uses technical skills towards disease prevention in both domestic and wild animals. Human health is protected by veterinary science working closely with many medical professionals by the careful monitoring of livestock health as well as its unique training in epidemiology and emerging zoonotic diseases worldwide. Veterinary medicine is informally as old as the human/animal bond but in recent years has expanded exponentially because of the availability of advanced diagnostic and therapeutic techniques for most species. Animals nowadays often receive advanced medical, dental, and surgical care including insulin injections, root canals, hip replacements, cataract extractions, and pacemakers. Veterinarians assist in ensuring the quality, quantity, and security of food supplies by working to maintain the health of livestock and inspecting the meat itself. Veterinary scientists are very important in chemical, biological, and pharmacological research.

This publication titled "International Encyclopedia of Dairy Technology and Animal Husbandry" deals primarily with animal husbandry, dairy technology and animal breeding. It also provides an introductory overview of animal husbandry, veterinary science and dairying in India, including an analysis of initiatives taken at centre and state levels. Issues and options for India and rest of the world with respect to veterinary science, animal health and livestock improvement are dealt in detail. The plans and projects related to intensive cattle development in India have been reflected briefly, analyzing various schemes and evaluation processes in detail. An introductory overview of global dairy science and technology has been provided for the readers. An in-depth analysis of dairy farming, cattle types and milking has been done with focus on dairy products, biosynthesis, grading and defects. An analysis of dairy microbiology, starter cultures, inhibitors and antimicrobial proteins has been undertaken in detail. Other issues which have been discussed at great length include dairy plant maintenance with focus on training, servicing, equipment and operations management. An educational approach has been followed while dealing with modern day dairy technology and milk supply in India. Appendix, glossary of terms, acronyms/abbreviations, bibliography and links have been provided to make this publication a user friendly one.

—Editors

5

Veterinary Science, Animal Health and Livestock Improvement: Issues and Options for India and Rest of the World

VETERINARIAN

A veterinarian is a physician for animals and a practitioner of veterinary medicine. The word comes from the Latin *veterinae* meaning "draught animals." The word "veterinarian" was first used in English by Thomas Browne (1605-1682).

Although veterinarians in many countries may have been awarded with doctoral degrees and receive extensive training in veterinary medical practice, there are many career fields open to those with veterinary degrees other than clinical practice. Those that do work in clinical settings often practice medicine in a specific field, such as companion animal or "pet" medicine (small animals such as dog, cat, and pocket pets), production medicine or "livestock" medicine including specialties in dairy cattle, beef cattle, swine, sheep, and poultry, equine medicine (e.g. sport, race track, show, rodeo), laboratory animal medicine, reptile medicine, or ratite medicine. Veterinarians may choose to specialize in medical disciplines such as surgery, dermatology or internal medicine, after post-graduate training and certification.

Some veterinarians pursue post-graduate training and enter research careers and have contributed many advances in many human and veterinary medical fields, including pharmacology and epidemiology. Research veterinarians were the first to isolate oncoviruses, *Salmonella* species, *Brucella* species, and various other pathogenic agents. Veterinarians were in the fore-front in the effort to suppress malaria and yellow fever in the United States, and a veterinarian was the first to note disease caused by West Nile Virus in New York zoo animals. Veterinarians determined the identity of the botulism disease-causing agent; produced an anticoagulant used to treat human heart disease; and developed surgical techniques for humans, such as hip-joint replacement, and limb and organ transplants. Like physicians, veterinarians must make serious ethical decisions about their patients' care. For example, there is ongoing debate within the profession over the ethics of performing declawing of cats and docking or cropping tails and ears, as well as "debarking" dogs and in the housing of sows in gestation crates.

Education and Regulation

Prerequisites for admission include the undergraduate studies listed under veterinary medicine and extensive veterinary and other animal-related experience (typically about 1000 or more hours combined). In the United States the average veterinary medical student has an undergraduate GPA of 3.5 and a GRE score of approximately 1350. Veterinary school lasts for four years just like human medicine programmes, with at least one year being dedicated to clinical rotations. After completion of the national board examination, some newly-accredited veterinarians choose to pursue residencies or internships in certain (usually more competitive) fields. There are some inconsistencies concerning the titles awarded upon completion of veterinary studies. US graduates are awarded either a Doctor of Veterinary Medicine (DVM) or the less common *Veterinariae Medicinae Doctoris* (VMD) degree, the latter if they are a graduate of the University of Pennsylvania School of Veterinary Medicine. In Great Britain and Ireland, a qualified veterinary surgeon holds a Bachelor's Degree (BVSc). In continental Europe and other regions adhering to the Bologna regulations of university education, the graduate is awarded a Master's Degree (MVM) that allows him/her to practice clinically. In these regions, the Doctorate (Dr. med. vet. or DVM) is a postgraduate title that requires the writing of an original scientific research dissertation. This can sometimes cause confusion when comparing the North American DVM title to the European DVM. There is some reciprocal international recognition of veterinary degrees. For example: Veterinarians graduating from AVMA (North American accredited universities), (e.g. Glasgow, Edinburgh, Royal Veterinary College, Sydney, Massey, Murdoch, Melbourne, etc.) may work in the USA after passing the NAVLE, a veterinary licensing exam taken by all American veterinarians. Graduates from these Universities are granted a BVS or BVSc degree which has been accredited in the US and Canada and is entirely equivalent to the DVM and VMD degrees. Non-AVMA accredited university graduates must also sit a week long Clinical Proficiency Examination in order to work in the USA. In the United Kingdom and some Commonwealth nations, a veterinary surgeon is an animal practitioner regulated by the Royal College of Veterinary Surgeons under the Veterinary Surgeons Act 1966. This legislation restricts the treatment of animals in the UK to qualified veterinary surgeons only, with certain specific exceptions, including physiotherapy, chiropractic, osteopathy, under the supervision of a veterinary surgeon. Various alternative medicine therapies (such as homeopathy, acupuncture, herbal medicine) can only be performed by a veterinary surgeon.

Career

In the United States veterinarians in private practice earn an average salary of $66,590 per year, while those working for the US government average $78,769 per year (2004 Bureau of Labor Statistics data). Dr. Morris reported several recent data from the American Veterinary Medical Association reports median earnings of $77,500-$98,500, for all types of private, public, and corporate veterinarians. Owning your own practice can bring in a much larger income anywhere from $200,000 up. Most veterinarians are paid based on production, rather than a straight salary, so earnings can vary based on type of practice, location of practice, and even the season of the year. The economic outlook for newly graduated veterinarians is clouded by the high debt level carried by many graduates, as the cost of veterinary medical education rises. As in other medical fields, new veterinarians tend to concentrate in urbanized areas and economic competition is limiting post-graduate opportunities in private practice. On the other hand, veterinarians are able to set-up successful new practices in established markets by providing special services such as an emergency and critical care clinics for pets and mobile veterinary clinics or by obtaining advanced training and certification in specialty fields of medicine. More than 3,800 veterinarians in the USA currently work at veterinary schools where they participate in research and teach vet students; teaching is another career path for a veterinarian.

There is some concern about the decreasing number of new veterinary graduates pursuing careers in the livestock industry. The majority of today's veterinary students grew up in urban or suburban areas, providing limited, if any, exposure to livestock medicine or farm animals prior to veterinary school. Livestock medicine, once based on serving many family farms such as those depicted in the James Herriot series, is getting increasingly specialized as farms are decreasing in number but increasing in individual size. Today's livestock veterinarian is more likely to work in a one-species discipline, perhaps as a full-time on-site veterinarian for one specific farm, than to work in the charming pastoral settings so common only one generation ago. This change in livestock medicine has brought vast improvements to the health and efficiency of food production. However, without regular exposure to this growing field of veterinary practice, students are less likely to pursue this line of profession. The concern is that as the baby-boomer generation of large animal veterinarians retires, there will not be enough young veterinarians to continue its work. Veterinary schools are aware of this issue, and most now expect a pre-veterinary background which includes large animal experience. Some veterinary schools are doing more to encourage the acceptance of students planning a career in production medicine by providing an alternate admissions process (e.g. Michigan State University's "Production Medicine Scholars Program") and specific scholarships.

Regulatory Medicine

Some veterinarians work in a field called regulatory medicine, ensuring a nation's food safety, e.g. the USDA FSIS, or work by protecting a country from imported exotic animal diseases. e.g the USDA APHIS. The emerging field of conservation medicine involves veterinarians even more directly with human health care, providing a multidisciplinary approach to medical research that also involves environmental scientists.

Government

Public health medicine is another option for veterinarians. Veterinarians in government and private laboratories provide diagnostics and testing services. Some veterinarians serve as state epidemiologists, directors of environmental health, and directors of state or city public health departments. Veterinarians are also employed by the US Agriculture Research Service, Fish and Wildlife Service, United States Environmental Protection Agency, National Library of Medicine, and National Institutes of Health. The military also employs veterinarians in a number of capacities — caring for pets on military bases, caring for military working animals, controlling various arthropod-borne diseases, or as food safety inspectors. There are several U.S Senators who are veterinarians, including Wayne Allard (R) Colorado, and John Ensign (R) Nevada.

In Popular Culture

Perhaps the best known depictions of a veterinarian at work are in the autobiographical books by James Alfred Wight, better known to his readers as James Herriot. Dr. Wight's books were also made into a famous BBC adaptation, *All Creatures Great and Small*. The most popular in mainstream media is Dr. Dolittle, which was a children's book turned into a movie in 1967 with Rex Harrison in the title role. The movie was then remade in 1998 casting Eddie Murphy as Dr. Dolittle. The original Dr. Dolittle involved an island as the main setting, whereas the remake of Dr. Dolittle has a setting in a city. The US-based cable network Animal Planet, because of its animal-based programming, features shows about veterinarians frequently. Two of its most notable shows about vets are *Emergency Vets* and *E-Vet Interns*, both set at Alameda East Veterinary Hospital in Denver, Colorado. In the hit TV-show "Grey's Anatomy", the main character (Meredith) dates a vet named

Finn Dandridge for several episodes. Their first date was interrupted when Finn received a call to birth a horse. Later, Finn helps diagnose Meredith's dog, Doc, which she shares with Derek Shepherd. Sadly, the dog has to be put to sleep. Finn becomes known as "McVet" by many of the interns at the hospital, following the show's tradition of McLabeling.

Workplace

Small animal veterinarians typically work in veterinary clinics or veterinary hospitals, or both. Large animal veterinarians often spend more time traveling to see their patients at the primary facilities which house them (zoos, farms, etc).

VETERINARY MEDICINE

Veterinary medicine is the application of medical, diagnostic, and therapeutic principles to companion, domestic, exotic, wildlife, and production animals. Veterinary science is vital to the study and protection of animal production practices, herd health and monitoring the spread of disease. It requires the acquisition and application of scientific knowledge in multiple disciplines and uses technical skills directed at disease prevention in both domestic and wild animals. Veterinary science helps safeguard human health through the careful monitoring of livestock, companion animal and wildlife health. Emerging zoonotic diseases around the globe require capabilities in epidemiology and infectious disease surveillance (by means of quantitative parasitology) and control that are particularly well-suited to veterinary science's "herd health" approach. Veterinary medicine is informally as old as the human/animal bond but in recent years has expanded exponentially because of the availability of advanced diagnostic and therapeutic techniques for most species. Animals nowadays often receive advanced medical, dental, and surgical care including insulin injections, root canals, hip replacements, cataract extractions, and pacemakers. Veterinary specialization has become more common in recent years. Currently 20 veterinary specialties are recognized by the American Veterinary Medical Association (AVMA), including anesthesiology, behaviour, dermatology, emergency and critical care, internal medicine, cardiology, oncology, neurology, radiology and surgery. In order to become a specialist, a veterinarian must complete additional training after graduation from veterinary school in the form of an internship and residency and then pass a rigorous examination. Veterinarians assist in ensuring the quality, quantity, and security of food supplies by working to maintain the health of livestock and inspecting the meat itself. Veterinary scientists occupy important positions in biological, chemical, agricultural and pharmaceutical research. In many countries, equine veterinary medicine is also a specialized field. Clinical work with horses involves mainly locomotor and orthopedic problems, digestive tract disorders (including equine colic, which is a major cause of death among domesticated horses), and respiratory tract infections and disease. Zoologic medicine, which encompasses the healthcare of zoo and wild animal populations, is another veterinary specialty that has grown in importance and sophistication in recent years as wildlife conservation has become more urgent. As in the human health field, veterinary medicine (in practice) requires a diverse group of individuals to meet the needs of patients. Veterinarians must complete four years of study in a veterinary school following 3–4 years of undergraduate pre-veterinary work. They then must sit for examination in those states in which they wish to become licensed practitioners. Veterinarians are expected to diagnose and treat disease in a variety of different species without benefit of verbal communication with their patients. In addition to veterinarians, many veterinary hospitals utilize a team of veterinary technicians and veterinary assistants to provide care for sick as well as healthy animals. Veterinary technicians are, essentially, veterinary nurses and are graduates of two or four year college-level programmes and are legally qualified to assist veterinarians in many medical procedures. Veterinary assistants are not licensed by most states, but can be well-trained through programmes offered in a variety of technical schools.

TOXICOLOGY TESTING

Toxicology testing, also known as safety testing, is conducted by pharmaceutical companies testing drugs, or by contract animal testing facilities such as Huntingdon Life Sciences and Inveresk Research International on behalf of a wide variety of customers, including the manufacturers of household and personal goods such as shampoos and domestic cleaning products. Around one million animals are used every year in Europe in toxicology tests. In the UK, one-fifth of animal experimentats are toxicologys. The tests are conducted without anesthesia, since drugs can change test results. The tests examine finished products such as pesticides, medications, food additives such as artificial sweeteners, packing materials, and air freshener, or their chemical ingredients. The substances are applied to the skin or eyes; injected intravenously, intramuscularly, or subcutaneously; inhaled either by placing a mask over the animals and restraining them, or by placing them in an inhalation chamber; or administered orally, through a tube into the stomach, or placing them in the animals' food. Doses may be given once, repeated regularly for many months, or for the lifespan of the animal.

Types of Test

LD50

Examples of toxicology tests include the LD50 test (Lethal Dose 50%), which involves administering a chemical to an animal population to determine what dose will kill 50 percent of the test subjects. The oral LD50 test has been banned in parts of Europe, and the U.S. Environmental Protection Agency (EPA) has announced that it no longer supports it. The test was phased out of the Organization for Economic Co-operation and Development's guidelines in December 2002, though the inhaled and skin-application LD50 tests remain in the guideline. The UK banned the oral LD50 in 1999, but the inhaled and skin-application tests are still used, and the oral LD50 is still used by the Ministry of Defence. Following the principles of reduction and refinement, the LD50 test is being replaced by methods such as the fixed dose procedure, that use fewer animals and aim to cause less suffering. Nonetheless, the oral LD50 is still widely used. *Nature* writes that, as of 2005, "the LD50 acute toxicity test ... still accounts for one-third of all animal [toxicity] tests worldwide."

Draize Test

The Draize test involves applying 0.5mL or 0.5g of a substance to an animal's eye or skin for four hours; the test subjects are usually albino rabbits who are conscious and held immobilized in stocks. The animals are observed for up to 14 days for signs of erythema and edema on the skin, and redness, swelling, discharge, ulceration, hemorrhaging, cloudiness, or blindness in the eye. The animals are killed after the test. Despite two decades of research into alternatives to this test, no non-animal alternatives have yet been successful, although the *low-volume eye test* is being investigated as an alternative that may reduce, but not eliminate, animal suffering. The most stringent tests are reserved for drugs and foodstuffs. For these, a number of tests are performed, lasting less than a month (acute), one to three months (subchronic), and more than three months (chronic) to test general toxicity (damage to organs), eye and skin irritancy, mutagenicity, carcinogenicity, teratogenicity, and reproductive problems.

The cost of the full complement of tests is several million dollars per substance and it may take three or four years to complete. Most toxicity tests involve testing ingredients rather than finished products. However, according to BUAV, manufacturers regard these tests as crude and believe they overestimate the toxic effects of the substances. They therefore repeat the tests using their finished products in order to obtain a less toxic label.

Others

Other tests include the:

- *90-day repeat-dose test*, administered by mouth, inhalation, or skin to 30-80 rats, rabbits, or guinea pigs;
- *90-day-repeat-dose test by mouth in non-rodents*, in which dogs are used;
- *teratogenicity test*, which tests for the effects of a substance on a fetus, and involves at least 80 mice, rats, hamsters, or rabbits;

- chronic toxicity test, which involves at least 160 rats, who are given daily doses of the substance for most of their lifespan;
- *carcinogenicity test*, another lifetime study testing for cancer in 400-500 rodents;
- *one- and two-generation reproduction toxicity test*, which involves more than 100 rats or mice;
- *test for embryonic genetic damage*, which involves 10-60 rats, hamsters, mice and their offspring;
- *toxicokinetic study*, which involves eight to ten animals to study the absorption, metabolism, distribution, and excretion of a substance.

Predictiveness and Utility

Questions have been raised about the predictiveness and overall utility of toxicology testing. *Nature* writes that most animal tests either over- or underestimate risk, or do not reflect toxicity in humans particularly well. This variability stems from trying to use the effects of high doses of chemicals in small numbers of a laboratory animal to predict the effects of low doses in large numbers of humans. Although relationships do exist, opinion is divided on how to use data on one species to make exact predictions about the level of risk in another species. Embryotoxicity tests are one example; these involve feeding chemicals to pregnant animals, then determining the effects both on the embryos and on the offspring of two subsequent generations. Horst Spielmann, a toxicologist at the Federal Institute for Risk Assessment in Berlin, told *Nature*: "Animal embryotoxicity tests are not reliably predictive for humans. When we find that cortisone is embryotoxic in all species tested except human, what are we supposed to make of them?" Addressing the criticism that toxicology testing is inaccurate, Professor Robert L. Brent wrote in *Pediatrics* that: "There is no question that [these] studies are helpful in raising concerns about the reproductive effects of drugs and chemicals, but negative animal studies do not guarantee that these agents are free from reproductive effects.."

ALTERNATIVES TO ANIMAL TESTING

Most scientists and governments say they agree that animal testing should cause as little suffering as possible, and that alternatives to animal testing need to be developed. The "three Rs",first described by Russell and Burch (1959), are guiding principles for the use of animals in research in many countries:

(i) Reduction refers to methods that enable researchers to obtain comparable levels of information from fewer animals, or to obtain more information from the same number of animals.

(ii) Refinement refers to methods that alleviate or minimize potential pain, suffering or distress, and enhance animal welfare for the animals still used.

(iii) Replacement refers to the preferred use of non-animal methods over animal methods whenever it is possible to achieve the same scientific aim.

Examples

The two major alternatives to *in vivo* animal testing are *in vitro* cell culture techniques and *in silico* computer simulation. However, some claim they are not true alternatives since simulations use data from prior animal experiments and cultured cells often require animal derived products, such as serum. Others say that they cannot replace animals completely as they are unlikely to ever provide enough information about the complex interactions of living systems. Examples of computer simulations available include models of diabetes asthma and drug absorption, though potential new medicines identified using these techniques are currently still required to be verified in animal tests before licensing. Cell culture is currently the most successful, and promising, alternative to animal use. For example, cultured cells have also been developed to create monoclonal antibodies, prior to this production required animals to undergo a procedure likely to cause pain and distress.

A third alternative now attracting considerable interest is so-called microdosing, in which the basic behaviour of drugs is assessed using human volunteers receiving doses well below those expected to produce whole-body effects.

Institutes

Institutes researching (and organizations funding) alternatives to animal testing include:

- Center for Alternatives to Animal Testing
- UCDavis Center for Animal Alternatives
- Physicians Committee for Responsible Medicine
- Dr Hadwen Trust

In October 2006 the European Centre for the Validation of Alternative Methods (ECVAM) launched an online database of toxicology non-animal alternative test methods. Categories at present include *in vitro* methods, QSAR models and a bibliographic section. Under the Framework Programmes 6 and 7, the European Commission is funding (and will be funding) a significant number of large integrated research projects aiming to develop alternatives to animal testing.

Medical Education Without Animal Experimentation

There are efforts underway in many countries to find alternatives to animal testing for education. Horst Spielmann, German director of the Central Office for Collecting and Assessing Altrnatives to Animal Experimentation (ZEBET), while describing Germany's progress in this area, told German broadcaster ARD in 2005:

Using animals in teaching curricula is already superfluous. In many countries, one can become a doctor, vet or biologist without ever having performed an experiment on an animal.

BOVINE SOMATOTROPIN

Bovine somatotropin (abbreviated bST and BST) is a protein hormone produced in the pituitary glands of cattle. It is also called bovine growth hormone, or BGH. BST can be produced synthetically, using recombinant DNA technology. The resulting product is called recombinant bovine somatotropin (rBST), recombinant bovine growth hormone (rBGH), or artificial growth hormone. It is administered to the cow by injection and used to increase milk production. Currently Monsanto is the only company that markets recombinant bovine somatotropin, under the trade name Posilac.

Physiology

A cow's pituitary gland naturally secretes BST into the bloodstream. Some of it latches on to receptors in the liver, which then produce Insulin-like Growth Factor 1 (IGF-1) which enters the blood as well. These two hormones have many different effects in the body, including increasing the breakdown of fat for energy and helping to prevent mammary cell death. The combination of increased energy from increased fat breakdown and decrease in mammary cell death is thought to be the cause of higher milk production. Studies have shown that there is no increase in the amount of BST secreted in the milk when a cow is injected with supplemental rBST. However, the studies have been conflicting about whether or not IGF-1 and IGF-2 output increases. The amount of IGF-1 and IGF-2 secreted varies greatly by stage of lactation and whether or not an animal is pregnant; most studies have shown that while these hormones are slightly elevated overall it falls within the normal range of variation.

Posilac

In 1937, the administration of BST was shown to increase the milk yield in lactating cows by preventing mammary cell death in dairy cattle. Until the 1980s, there was very limited use of the compound in agriculture as the sole source of the hormone was from bovine cadavers. During this time, the knowledge of the structure and function of the hormone increased. Monsanto developed a recombinant version of BST, brand named Posilac, in 1994, which is produced through a genetically engineered *E. coli*. A gene that codes for the sequence of amino acids that makes up BST is inserted into the DNA of the E. coli bacterium. The bacteria are then broken up and separated from the rBST and is purified to produce the injectable hormone. Growth hormones associated with injections given to dairy cows to increase milk production are known under an assortment of terms, but these terms generally refer to the Monsanto product. The Monsanto fact sheet on its proprietary product states that when injected into dairy cattle, the product can increase milk production by an average of more than 10% over the span of 300 days.

Controversy

Use of rBST is regarded as controversial due to its effects on animal and human health and the encroachment on small farmers by large corporations.

Animal Health

Two meta-analyses have been published on rBGH's effects on bovine health. Findings indicated an average increase in milk output ranging from 11%-16%, a nearly 25% increase in the risk of clinical mastitis, a 40% reduction in fertility and 55% increased risk of developing clinical signs of lameness. The same study reported a decrease in body condition score but speculated that it may have been attributable to, differences in feeding of treated (underfed) versus untreated (overfed) cows. A European Union scientific commission was asked to report on the incidence of mastitis and other disorders in dairy cows and on other aspects of the welfare of dairy cows. The commission's statement, subsequently adopted by the European Union, stated that the use of rBST substantially increased health problems with cows, including foot problems, mastitis and injection site reactions, impinged on the welfare of the animals and caused reproductive disorders. The report concluded that on the basis of the health and welfare of the animals, rBST should not be used. Health Canada prohibited the sale of rBST in 1999; the recommendations of external committees were that despite not finding a significant health risk to humans, the drug presented a threat to animal health and for this reason could not be sold in Canada.

Human Health

Concern has been expressed by various groups about the levels of two substances in milk consumed by humans - BST itself, and insulin-like growth factor 1 (IGF-1), which is increased by rBST injections. Monsanto has stated that both of these compounds are harmless given the levels found in milk and the effects of pasteurization. According to the FDA, no significant difference has been shown between milk derived from rbST treated and non rbST treated cows.

IGF-1

Monsanto's studies show use of rBST in cows increases bovine insulin-like growth factor 1 in milk, a structure that is identical in cows and humans. Monsanto states that there is no danger of consuming milk or meat from cows treated by BST, and that the only difference between milk from supplemented cattle and unsupplemented cattle is the amount of IGF-1, though even these elevated levels are similar to levels found in milk from untreated cows. Further, the amount of IGF-1 consumed in milk is negligible compared to the amount produced in the body. Opponents counter that rBGH does cause differences aside from the higher rate of IGF-1, most importantly that BST and rBST have a different chain of amino acids which can alter how a protein interacts with the immune system. A Health Canada report on rBST found no "biologically plausible" safety concerns for humans about the sale of rBST in Canada barring immune hypersensitivity that may occur in some individuals Studies have found links between serum levels of IGF-1 some medical conditions, including breast, prostate and colorectal cancer, a higher risk of diabetes and a shorter lifespan in animal studies, and has been linked to an increased number of twins born to humans.

Lawsuit Against Fox Television

Fox television affiliate WTVT/Fox13 in Tampa, Florida was sued by Steve Wilson and Jane Akre, two former anchors over the issue of reporting the harmful effects of BGH on humans. The journalists—originally, with station approval—wrote a story in 1996 that stated the human health risks of rBGH. However, the station rejected it and insisted they report a different story on rBGH with statistics supplied by Monsanto. They rewrote the story over 80 times but were eventually fired by Fox. After a five-week trial which ended August 18, 2000, a Florida state court jury unanimously determined that Fox "acted intentionally and deliberately to falsify or distort the plaintiffs' news reporting on BGH." In that decision, the jury also found that Akre's threat to blow the whistle on Fox's misconduct to the Federal Communications Commission (FCC) was the sole reason for the termination and awarded $425,000 in damages. Fox appealed and prevailed February 14, 2003, when an appeals court issued a ruling reversing the jury, accepting a defense argument that had been rejected by three other judges on at least six separate occasions. The appeals court's decision on the verdict was on the basis that FCC policies on news agencies reporting the truth did not legally require the station to report the truth in a news story, as FCC policies are not law. The story that was subsequently reported on BGH contained no statistics that may have indicated a human health risk, as these statistics (the ones found by Akre and Wilson and mentioned in their original story) were ignored. In 2004, Fox filed a $1.7 million counter-suit against Akre and Wilson for trial fees and costs. This story is featured at length in the documentaries The Corporation and Outfoxed.

Regulation

Use of the recombinant supplement has been controversial. While it is used in the United States (though not without reaction), it is banned in Canada, parts of the European Union, Australia and New Zealand.

Regulation Inside the United States

In 1993, the product was approved for use in the U.S. by the FDA, and its use began in 1994. The product is now sold in 49 states, the exception being Michigan. According to Monsanto, approximately one third of dairy cattle in the U.S. are injected with Posilac; approximately 8,000 dairy producers use the product. It is now the top selling dairy cattle pharmaceutical product in the U.S. Several scientists involved in the review of Posilac were dismissed or pressured to leave the FDA expressing concerns over the process used to approve the drug for use in dairy cows and others expressed their concerns anonymously for fear of retribution.

Enforcement

The FDA does not require special labels for products produced from cows given rbST but has charged several dairies with "misbranding" their milk as having no hormones, due to the fact that all milk contains hormones and can not be produced in such a way that it would not contain any hormones. Monsanto sued an independent dairy over their use of a label which pledged to not use artificial growth hormones. The dairy stated that their disagreement was not over the scientific evidence for the safety of Posilac (Monsanto's complaint about the label), but rather they were more interested in marketing milk than a drug. The suit was settled when the dairy agreed to add a qualifying statement to their previous label regarding the lack of difference between milk produced by Posilac-dosed cows and cows that had not received the drug. Demand for milk without using synthetic hormones has increased 500% in the US since Monsanto introduced their rbST product; organic milk is the fastest growing sector of the organic food market.

Labeling in Pennsylvania

In 2007, the U.S. state of Pennsylvania, under heavy pressure from Monsanto, adopted a regulation that would have banned the practice of labeling milk as derived from cows not treated with rBST. This prohibition was to go into effect January 1, 2008, but was delayed to February 1, 2008 in order to give interested parties more time to submit comments to the state's Department of Agriculture. This policy, had it been implemented, would have made it impossible for consumers to distinguish between milk from cows treated with rBST and milk from cows not treated with rBST. For this reason, the ban was opposed by several consumer groups, and the state reversed its position before the ban could take effect and adopted the Federal Trade Commission's recommended labeling guidelines instead.

Response from Commercial Groups

Several milk purchasers and resellers have elected not to purchase milk produced with rBGH.

(i) Safeway in the northwestern United States stopped buying from dairy farmers that use rBGH in January 2007. The two Safeway plants produce milk for all of Oregon, Southwest Washington, and parts of northern California. Safeway's plant in San Leandro, CA had already been rBGH-free for two years.

(ii) Chipotle Mexican Grill has also announced it will serve rBGH-free sour cream at its restaurants.

(iii) Kroger also announced its intent to sell only milk produced without synthetic hormones by February 2008,

(iv) Publix announced it has been rBST-free since May, 2007.

(v) Braum's has also issued a press release stating its milk is rBST-free.

(vi) Starbucks Company has as of January 2008 made all dairy in beverages rBGH free.

(vii) Wal-Mart and Sam's Club stores featured hormone-free "Great Value" brand milk, but did not label it as such in 2008.

Monsanto has responded to this trend by lobbying state governments to ban the practice of distinguishing between milk from farms pledged not to use rBST and those that do. According to the New York Times, a pro-rBST advocacy group called Afact has been most active in these lobbying efforts. Afact is made up of both producer members and allied industries and is closely affiliated with Monsanto itself; the group's acronym stands for American Farmers for the Advancement and Conservation of Technology.

Though rbST is one of Afact's main concerns, their mission is to prevent "marketers from convincing some consumers to doubt the credibility and safety assurances from of even the most respected food safety agencies and scientific oversight organizations." Thus far, a large-scale negative consumer response to Afact's legislative and regulatory efforts has kept state regulators from pushing through strictures that would ban hormone-free milk labels, though several politicians have tried, including Pennsylvania's agriculture secretary Dick Wolff, who tried to ban rBST-free milk on the grounds that it would alleviate consumer confusion. Proposed labeling changes have been floated by Afact lobbyists in New Jersey, Ohio, Indiana, Kansas, Utah, Missouri and Vermont. So far, however, this effort has been unsuccessful.

Regulation Outside the United States

In Japan, Australia, New Zealand, and Canada, rbST is not approved for use.

The European Union declared the use of rbST as safe in 1990, but in 1993, a moratorium was placed on its sale by all 27 member nations.

Canada's health board, Health Canada, refused to approve rBGH for use on Canadian dairies, citing concerns over animal health. The study they had commissioned, however, found "no biologically plausible reason for concern about human safety if rbST were to be approved for sale in Canada. The only exception to this statement is... [possible hypersensitivity]."

UNDERSTANDING REPRODUCTION AND ITS IMPLICATION IN ANIMALS

Reproduction is the biological process by which new individual organisms are produced. Reproduction is a fundamental feature of all known life; each individual organism exists as the result of reproduction. The known methods of reproduction are broadly grouped into two main types: sexual and asexual. Human reproduction belongs to sexual reproduction. In asexual reproduction, an individual can reproduce without involvement with another individual of that species. The division of a bacterial cell into two daughter cells is an example of asexual reproduction. Asexual reproduction is not, however, limited to single-celled organisms. Most plants have the ability to reproduce asexually. Sexual reproduction requires the involvement of two individuals, typically one of each sex. Normal human reproduction is a common example of sexual reproduction.

Sexual Reproduction

Sexual reproduction is a biological process by which organisms create descendants that have a combination of genetic material contributed from two (usually) different members of the species. Each of two parent organisms contributes half of the offspring's genetic makeup by creating haploid gametes. Most organisms form two different types of gametes. In these *anisogamous* species, the two sexes are referred to as male (producing sperm or microspores) and female (producing ova or megaspores). In *isogamous species* the gametes are similar or identical in form, but may have separable properties and then may be given other different names. For example, in the green alga, *Chlamydomonas reinhardtii*, there are so-called "plus" and "minus" gametes. A few types of organisms, such as ciliates, have more than two kinds of gametes. Most animals (including humans) and plants reproduce sexually. Sexually reproducing organisms have two sets of genes for every trait (called alleles). Offspring inherit one allele for each trait from each parent, thereby ensuring that offspring have a combination of the parents' genes. Having two copies of every gene, only one of which is expressed, allows deleterious alleles to be masked, an advantage believed to have led to the evolutionary development of diploidy (Otto and Goldstein).

Allogamy

Allogamy is a term used in the field of biological reproduction describing the fertilization of an ovum from one individual with the spermatozoa of another.

Autogamy

Self-fertilization (also known as autogamy) occurs in hermaphroditic organisms where the two gametes fused in fertilization come from the same individual. They are bound and all the cells merge to form one new gamete.

Mitosis and Meiosis

Mitosis and meiosis are an integral part of cell division. Mitosis occurs in somatic cells, while meiosis occurs in gametes. Mitosis The resultant number of cells in mitosis is twice the number of original cells. The number of chromosomes in the daughter cells is the same as that of the parent cell. Meiosis The resultant number of cells is four times the number of original cells. This results in cells with half the number of chromosomes present in the parent cell. A diploid cell duplicates itself, then undergoes two divisions (tetroid to diploid to haploid), in the process forming four haploid cells. This process occurs in two phases, meiosis I and meiosis II.

Same-Sex Reproduction

In recent decades, developmental biologists have been researching and developing techniques to facilitate same-sex reproduction. The obvious approaches, subject to a growing amount of activity, are female sperm and male eggs, with female sperm closer to being a reality for humans, given that Japanese scientists have already created female sperm for chickens. More recently, by altering the function of a few genes involved with imprinting, other Japanese scientists combined two mouse eggs to produce daughter mice.

Reproductive Strategies

There is a wide range of reproductive strategies employed by different species. Some animals, such as the human and Northern Gannet, do not reach sexual maturity for many years after birth and even then produce few offspring. Others reproduce quickly; but, under normal circumstances, most offspring do not survive to adulthood. For example, a rabbit (mature after 8 months) can produce 10–30 offspring per year, and a fruit fly (mature after 10–14 days) can produce up to 900 offspring per year. These two main strategies are known as K-selection (few offspring) and r-selection (many offspring). Which strategy is favoured by evolution depends on a variety of circumstances. Animals with few offspring can devote more resources to the nurturing and protection of each individual offspring, thus reducing the need for many offspring. On the other hand, animals with many offspring may devote fewer resources to each individual offspring; for these types of animals it is common for many offspring to die soon after birth, but enough individuals typically survive to maintain the population.

Other Types of Reproductive Strategies

(i) Polycyclic animals reproduce intermittently throughout their lives.

(ii) Semelparous organisms reproduce only once in their lifetime, such as annual plants. Often, they die shortly after reproduction. This is a characteristic of r-strategists.

(iii) Iteroparous organisms produce offspring in successive (e.g. annual or seasonal) cycles, such as perennial plants. Iteroparous animals survive over multiple seasons (or periodic condition changes). This is a characteristic of K-strategists.

Asexual vs. Sexual Reproduction

Organisms that reproduce through asexual reproduction tend to grow in number exponentially. However, because they rely on mutation for variations in their DNA, all members of the species have similar vulnerabilities. Organisms that reproduce sexually yield a smaller number of offspring, but the large amount of variation in their genes makes them less susceptible to disease. Many organisms can reproduce sexually as well as asexually. Aphids, slime molds, sea anemones, some species of starfish (by fragmentation), and many plants are examples. When environmental factors are favourable, asexual reproduction is employed to exploit suitable conditions for survival such as an abundant food supply, adequate shelter, favourable climate, disease, optimum pH or a proper mix of other lifestyle requirements. Populations of these organisms increase exponentially via asexual reproductive strategies to take full advantage of the rich supply resources. When food sources have been depleted, the climate becomes hostile, or individual survival is jeopardized by some other adverse change in living conditions, these organisms switch to sexual forms of reproduction. Sexual reproduction ensures a mixing of the gene pool of the species. The variations found in offspring of sexual reproduction allow some individuals to be better suited for survival and provide a mechanism for selective adaptation to occur. In addition, sexual reproduction usually results in the formation of a life stage that is able to endure the conditions that threaten the offspring of an asexual parent. Thus, seeds, spores, eggs, pupae, cysts or other "over-wintering" stages of sexual reproduction ensure the survival during unfavourable times and the organism can "wait out" adverse situations until a swing back to suitability occurs.

Life Without Reproduction

The existence of life without reproduction is the subject of some speculation. The biological study of how the origin of life led from non-reproducing elements to reproducing organisms is called abiogenesis. Whether or not there were several independent abiogenetic events, biologists believe that the last universal ancestor to all present life on earth lived about 3.5 billion years ago. Today, some scientists have speculated about the possibility of creating life non-reproductively in the laboratory. Several scientists have succeeded in producing simple viruses from entirely non-living materials. The virus is often regarded as not alive. Being nothing more than a bit of RNA or DNA in a protein capsule, they have no metabolism and can only replicate with the assistance of a hijacked cell's metabolic machinery. The production of a truly living organism (e.g. a simple bacterium) with no ancestors would be a much more complex task, but may well be possible according to current biological knowledge.

Lottery Principle

Sexual reproduction has many drawbacks, since it requires far more energy than asexual reproduction, and there is some argument about why so many species use it. George C. Williams used lottery tickets as an analogy in one explanation for the widespread use of sexual reproduction. He argued that asexual reproduction, which produces little or no genetic variety in offspring, was like buying many tickets that all have the same number, limiting the chance of "winning" - that is, surviving. Sexual reproduction, he argued, was like purchasing fewer tickets but with a greater variety of numbers and therefore a greater chance of success. The point of this analogy is that since asexual reproduction does not produce genetic variations, there is little ability to quickly adapt to a changing environment. The lottery principle is less accepted these days because of evidence that asexual reproduction is more prevalent in unstable environments, the opposite of what it predicts.

TOWARDS UNDERSTANDING REPRODUCTIVE ENDOCRINOLOGY AND INFERTILITY

Reproductive endocrinology and infertility (REI) is a surgical subspecialty of obstetrics and gynecology that addresses hormonal functioning as it pertains to reproduction. While a major focus of REI is infertility, reproductive endocrinologists also evaluate and treat hormonal dysfunctions in female and males outside of infertility. Reproductive surgeons operate on anatomical disorders that affect fertility. Reproductive endocrinologists have a specialty training in obstetrics and gynecology before they undergo subspecialty training (fellowship) in reproductive endocrinology and infertility.

Certification

In a number of countries the pathway to become a subspecialist in REI is regulated. Thus in the United States the American Board of Obstetrics and Gynecology (ABOG) sets the standards for subspecialtists to become certified. After a four-year training in Obstetrics and Gynecology a three-year approved fellowship needs to be successfully completed. To be board certified in reproductive endocrinolgy and infertility, one must first complete board certification in obstetrics and gynecology (written and oral exams) and then certify in reproductive edncrinology and infertility (written oral and thesis exams).

Societies

Reproductive endocrinologists often belong to specific medical societies, such as ASRM or ESHRE.

MODAL COURSE ANIMAL HEALTH: A CASE STUDY OF ACS DISTANCE EDUCATION

Course Structure

There are ten lessons as follows:

1. Introduction To Animal Health
2. Signs & Symptoms Of Diseases
3. Disease Classification
4. Causes & Diagnosis of Diseases
5. Treatment Of Diseases
6. Inflammation
7. Fever & Immunity
8. Tissue Repair
9. Wounds
10. Cell Changes

Each lesson culminates in an assignment which is submitted to the school, marked by the school's tutors and returned to you with any relevant suggestions, comments, and if necessary, extra reading.

Aims

- Explain common health problems affecting animals, including the circumstances under which animals contract health problems, and methods used to prevent the development of ill health
- Analyse physical indicator symptoms of ill health in animals.
- Determine the taxonomic class of animal pests and diseases.
- Explain the diagnostic characteristics of the main types of animal pathogenic microorganisms.
- Explain the methods used in the treatment of pests and diseases in farm animals.
- Explain the role of inflammation, including it's symptoms and causes, in animals.
- Explain the biological processes which affect and control the immune system in animals.
- Explain the biological processes which affect and control tissue repair in animals.
- Determine procedures for the management of wounds to animals, on a farm.
- Explain the processes involved in cellular change in animals.
- Diagnose simple health problems in farm animals.

What the Course Covers

Here are just some of the things you will be doing:

- List criteria used to assess the health status, including ill-health, of animals.
- Describe the different causes of ill-health in animals.
- Explain the methods used to prevent ill-health in animals.
- Write a standard procedure for a routine health examination of a chosen farm animal.
- Describe the symptoms of ill-health in animals.
- Compare the causes of two symptomatically similar health problems for a specified farm animal.
- Diagnose a health problem from a given set of symptoms.

- Distinguish, using labelled illustrations, between different taxonomic classes of animal pest and disease organisms.
- Describe identifying characteristics of four different disease carrying agents of specified farm animals.
- Classify commonly occuring pests and diseases of three different animals, into their taxonomic classes.
- Describe the characteristics of viruses, using illustrations and a report.
- Describe the characteristics of bacteria, using illustrations and a report.
- Describe the characteristics of protozoa, using illustrations and a report.
- Describe the characteristics of parasites, using illustrations and a report.
- Describe the characteristics of nutritional disorders, using illustrations and a report.
- Analyse the relevance of ten specified factors, to determining the health of a chosen species of farm animal.
- Describe the veterinary treatments available over the counter for on-farm use.
- Explain the vaccination programmes used to treat two different specifies of farm animal.
- Describe the applications and techniques used for dips, to control external parasites in a specified farm animal.
- List the essential items for a First Aid Kit for a specified farm animal.
- Write guidelines for general procedures to follow when nursing sick farm animals.
- List the procedures employed in quarantine, using a chosen animal as an example.
- Describe the procedures for slaughtering a diseased ruminant in order to conduct a post-mortem examination.
- Prepare an illustrated, one page report on the post-mortem procedures of a ruminant.
- Compare two different methods used to control a specified disease in farm animals.
- Identify a suitable method of control for ten different, specified pests and diseases of farm animals.

- Differentiate between at least five factors which cause inflammation in animals.
- Develop a checklist for analysing inflammation in a chosen farm animal species.
- Explain the inflammatory response in a specific case study.
- Compare the different methods used to control inflammation in animals.
- Describe the function of the immune system in animals.
- List the agents which can cause fevers in animals.
- Explain the biology of fevers in a specified case study of a farm animal species.
- Explain the methods used in treating fevers in animals.
- Explain at least five factors which influence immune response in animals.
- Explain the characteristics of the immune system in a chosen farm animal species.
- Describe the composition of tissues at three different body sites, in terms of susceptibility to different types of internal and external damage.
- Compare the characteristics of different types of tissue damage.
- List factors, in terms of both rate of, and quality of repair; which influence tissue repair.
- Explain the biological processes, which occur as damaged tissue heals in animals.
- Compare the different effects of wounding, including psychological, physiological and anatomical, to three different parts of a specified animals body.
- Explain the different biological processes which occur following wounding, including: tissue repair and infection.
- Develop a checklist for the treatment of wounds in farm animals.
- List an appropriate treatment for each of five different types of wounds to 4 different species of farm animals.
- Describe post care treatment of the wounds as discussed above.
- Determine the potential causes of wounding of farm animals.
- Develop guidelines for prevention of wounds to farm animals, based on the potential causes identified above.
- Describe the different causes of cellular change in animals.

- Explain the general processes associated with cancer at a cellular level, in animals.
- Explain the cellular processes associated with death of animal tissue.
- List the factors which influence the rate and extent of cellular change in diseased animals.
- Monitor the health condition of a farm animal over a four month period.
- Observe, and prepare a report, on the veterinarians diagnostic process/ health assessment methodology, when inspecting three different farm animals.
- Diagnose the cause of three different health problems, detected in three different genera of farm animals.
- Develop a checklist of the diagnostic indicators of common health problems, which occur in three different farm animal species.

Learn to describe common diseases in animals, how to recognise and treat these diseases and wounds. It is advisable to complete Animal Husbandry I before attempting this course.

SITUATION W.R.T. LIVESTOCK IMPORTANT IN INDIA: ISSUES & OPTIONS

In sum the livestock in India are in general of poor quality. Any improvement depends upon four factors—breeding, feeding, disease control and management. The keynote of the genetic improvement policy for cattle is the extension of the programme of cross-breeding of indigenous animals with exotic germ plasm. To begin with, this programme was taken up in a few selected areas and exotic cattle of Jersey and Friesian breeds were imported from Australia and the USA for distribution among different states. Cross-breeding programmes were also taken up with foreign collaboration, *e.g.*, Indo-Danish Project at Karnatak for Red Dane cattle, Indo-Swiss Project in Kerala for Brown Swiss cattle and Indo-German Project in Himachal Pradesh. For the intensification of this programmo, the Government of India is, presently, allocating free foreign exchange for the import of exotic cattle for distribution to State farms, agricultural universities and research institutes. A big break through in milk production is expected to be achieved through a massive cross-breeding programme with exotic germ plasm simultaneously in different areas. A Frozen Semen Bank has been set up near Bangalore for the production and supply of frozen semen to different State Governmets. Establishment of four central projects for intensive cross-breeding of cattle with imported frozen semen is also being envisaged. Feed and fodder development which is an important ingredient for cattle development, is, presently, being given close attention. A number of leguminous fodder crops like lucerne, berseem, some varieties of cow-pea and guar were being grown by our farmers even prior to the First Plan. Subsequently, giant hybrid Napier, Guinea grass, Rhodes grass and Para grass have also gained popularity in certain areas. Three regional forage stations have been set up to demonstrate improved practices of production, conservation and utilization of forage. Two of them are receiving assistance under UNDP Grass Land Fodder Development Project in the form of equipment, training, fellowships and experts. Four more such stations are being set up. These regional stations would supply seed and planting material to livestock farms, State agencies and other centres for further multiplication and distribution. The National Seeds Corporation is also engaged in the production of improved fodder seeds in large quantities to ensure steady supply of quality seeds to farmers. The State Governments are running a few fodder seed production centres to meet a part of their seed requirements. In order to supplement feed and fodder resources in the country, attention is being given to the utilization of agricultural and industrial by-products/wastes in cattle and poultry feeds. Increasing attention has been given to provide adequate health cover for livestock development. Presently, a good network of veterinary institutions cover the entire country for the control of animal diseases. A massive programme for eradication of rinderpest disease was undertaken in 1954 and the incidence of the disease has come down from about 8,000 attacks per year.to 207 attacks during 1971-72. The disease is presently under complete control. Similarly, new Cattle disease which was a courage of poultry is no longer a serious problem in the country. Other animal disease like foot and mouth diseases are being taken care of. There is a fairly strong machinery in the country for the diagnosis and investigation into animal diseases. All the vaccines required for protection of livestock are now being produced in India. The internationally reputed Indian Veterinary Research Institute at Izatnagar is undertaking research in several disciplines of animal sciences. It is providing training facilities both to students from within the country and abroad. This Institute is producing as many as 35 different biological products. Besides, the states have their disease investigation laboratories and biological product units.

Management

In livestock production, management occupies an important place as it embraces the fields of nutrition, feeding and the techniques acquired from the experience for handling livestock. This is reflected in proper housing and care of animals. The production can be considerably improved by improvement in the managemental practices. Recent researches have shown that buffaloes which do not breed during summer could be bred to obtain an even calving rate and flow of milk in case suitable modifications were made in the managemental practices. The calving interval of cattle which is normally 18-20 months in the village areas can be reduced by three months which will produce a net increase by about 15 per cent. In the case of sheep the breeding season can be telescoped to produce the lamb crop in suitable climates and environments. Punjab has started an integrated programme of Agriculture, Animal Husbandry and Dairy Development. Schemes have been drawn up especially for training the farmers and the staff. The veterinary doctor is actively associated with all training camps for farmers arranged by the Department of Agriculture so that he could give practical advice to the trainees. Advice on cultivation of good fodder, importance of artificial insemination as well as after-care formed part of the syllabi of each training programme. Fodder development was also taken up on a campaign basis. The role of animal husbandry in encouraging rotational crops and importance of animal products in the context of rural economy was emphasised at meetings and conferences. Cinema shows were held on a campaign basis to explain the advantages of cross-breeding for improving the breed of the cattle and increasing milk yield. Agencies for the development of small and marginal farmers and landless labour undertook to provide financial assistance (credit as well as subsidy) and technical know-how for those with marginal and small holdings to set up dairy, piggery and poultry units. The Dairy Development Corporation took up the onerous task of building up a marketing structure for the sale of milk and milk products through a network of milk plants and milk bars. Quality feed is provided by the Poultry Corporation. The digestibility of different feed constituents in the rumen of the cow, the buffalo and the goat are being studied with the help of fistulated animals. The differences noted are being co-related with the types of micro-organisms present in the rumen of different species. This would enable greater use of agricultural and industrial waste materials, and non-protein substances like ureas to be used for routine feeding, increasing milk production and lowering the cost of milk. A dairy animal takes 3-4 years to come into milk. Constant effort is being made to reduce the age at first calving, through better management practices, and to reduce the cost of rearing young stock by replacing milk used for feeding with equally nutritious "calf starters" prepared from cheaper materials like fishmeal, deoiled cakes etc. Balanced feeding has been shown to be important not only for maximising production but also in maintaining the composition of milk to the same level throughout the year. Cow and buffalo milk through containing the same constituents, differ mainly in their properties. This leads to difficulties in utilising established manufacturing practices for preparing products like condensed milk, sterilised milk's cheese, etc.; with buffalo milk. Methods for studying in details reasons for observed differences were standardised and the components which give rise to these differences identified and isolated. This understanding has enabled the manufacture of as good products from buffalo milk as from cow milk. A constant check is needed on quality. Rapid bacteriological and chemical tests have been devised to check hygienic quality as well as freedom from presence of extraneous substances, like water in milk, cheaper fats in ghee etc. With the help of the Hansa test, presence of as little as 3 per cent of buffalo milk can be detected in cow milk. With the same sensitivity, the freezing point of milk detects added water in milk and chromatograhic methods detect foreign fats in ghee. In selecting the equipment for storage and processing, care has to be taken for their proper design to facilitate easy cleaning and eliminating areas where bacteria thrive. The NDRI has associated with the Indian Standards Institution in evolving of desirable practices.

Case of Sheep Development

The fact remains that the wide range of environment in India extending from the Alpine climate of the Himalayas to the semi-arid areas of Deccan Plateau and desert areas of Rajasthan and Gujarat has given rise to a number of breeds and types of sheep suited to the agro-climate conditions of various areas. As mentioned earlier, the number of sheep and goats in India is fairly large—42 million in 1966. However, the productivity of most of the sheep stock is low in terms of wool and mutton. The bulk of wool produced is coarse and is used largely in the manufacture of carpets, druggets and coarse blankets. Organised efforts to improve the quality and productivity of sheep have been carried out under the successive Plans. For improved breeding rams produced at the state sheep breeding farms are being distributed to the flock-owners through the sheep and wool extension centres established in sheep rearing areas of the country. These centres also provide guidance to sheep farmers in improved methods of sheep husbandry, protection against diseases, mass drenching of sheep and other facilities. Each such centre covers, on an average, 10,000 sheep. Sheep shearing and wool grading and marketing schemes are also being implemented. Presently, seven large-scale sheep breeding farms with exotic breed of sheep are being set up in different parts of the country for supplying seed material for the cross breeding programme. An Indo-Australian sheep project has been established at Hissar with about 2,000 Corriedle sheep and the number of these sheep will go upto 7,000 in due course. About five thousand sheep of superior exotic breeds like Ram-bouillet, Polwarth and Soviet Merino, have already been imported and another 7,000 are proposed to be imported during the Fourth Plan period.

Case of Poultry Development

The poultry products are a valuable source of protein in the form of eggs and meat. Nutrition experts have laid a standard of one egg per day per head of population. The population of 500 million on this basis would require over 61 billion eggs, if it is assumed that 50 per cent of the population are potential consumers of this commodity. The total number of eggs available as against this is estimated at only 3630 millions. There is, thus a wide gap between production and requirements of eggs. This gap is likely to grow in a long run perspective ending 1975 and 1985 unless special production programmes are undertaken. As increase in egg production can be achieved both by extensive and intensive methods *i.e.,* increasing poultry population as well as yield per bird. The poultry population in the country increase from 73.5 million in 1951 to 114.3 million in 1961 *i.e.,* by about 50 per cent. According to the 1966 census poultry population increased marginally during the quinquinneum 1961-66, to 115.1 million.

Population Growth

In fact, one reason for this change in the population growth pattern may be the severe draught conditions in 1965-66 to 1966-67 which led to an acute shortage of livestock feeds. A number of poultry units closed down during 1965 and 1966. The trend as indicated by 1966 census may, therefore, be considered as a passing phase. In the past, table birds were largely from the called breeding stock and indigenous table birds upto the period of one year. Under the present practice where a larger proportion of the population is crossbreed, the male stock is sold for table before it attains the age of 6 months. It is thus likely that at the time of the enumeration, the male stock was not properly accounted for, indicating no visible increase in the poultry population. To this extent, it could be assumed that the 1966 poultry census was an under-estimate. When improved breeds are introduced on a mass scale a large incubation capacity is necessary. In the First and the Second Plan period it was possible to create adequate incubation capacity as most of the poultry equipment had to be imported. During the Third Plan period, however, adequate capacity to manufacture larage-sized incubators has been created. It would now be possible to increase poultry population rapidly.

Analylsing Yield Per Bird

Annual production of eggs per laying hen was estimated at 60 at the end of the second Five-Year Plan. This estimate was based on the performance of indigenous breeds under free range conditions. According to recent surveys conducted by IARS in a few states, egg production per bird is estimated at 89 in Andhra Pradesh, 96 in Kerala, 99 in West Bengal, 86 in Gujarat and 112 in Maharashtra. On this basis we may assume an annual yield of 75 eggs per adult bird for the country as a whole. This reflects that the development of the poultry industry during the Third Plan period has resulted in major changes in the pattern of poultry production in the country. While in the past the entire production of eggs was based on free range, introduction of intensive rearing on deep litter system resulted in a change of pattern to the extent that a sizeable proportion of the urban supply of eggs was from commercial flocks raised under intensive system of poultry rearing. Under the Plan programme, a large number of birds of exotic breeds are being introduced to replace the indigenous stock. Intensive training and distribution programmes for rearing day old chicks have been taken up thereby making it possible to introduce a large number of exotic breeds in the rural areas through their distribution. The setting up of commercial units with foreign collaboration has further intensified the introduction of high-laying strains in the country. These programmes will need to be strengthened to effect a change over of the poultry industry from free range breeding of non-descript birds to the intensive rearing of improved stock. The development of poultry in India has received considerable attention during the last two decades. A modest beginning was made in the First Five Year Plan with the launching of a pilot project with 33 extension centres. During the Second Five Year Plan, a comprehensive All-India Poultry Development Project was launched under which five regional poultry farms were established in different agro-climatic regions. These farms aimed at production and propagation in the respective regions, of acclimatised birds and training of officers and farmers. Through the assistance of TCM, a wide range of equipment and a consignment of 30,000 chicks of White Leghron and Rhode Island Red breeds were imported to provide the necessary support to the project. Besides, 300 Poultry Extension-curn-Demonstration Centres were established in different parts of the country to introduce modern poultry farming in rural areas. In the Third Year Plan, seventy-seven Intensive Projects closely linked with hatcheries, food mixing units, marketing centres and extension organizations were established and all the required inputs and services, including credit requirements were made available to the farmers. The Regional and State Poultry Farms were expanded. Two processing units, 68 feed mills and 18 marketing centres were also set up. In the private sector also, four commercial hatcheries came up for the propagation of genetically superior stock of eggs as well as meat lines, developed in other countries. Many companies have come into existence to manufacture the entire range of equipment required for the poultry industry.

Today, the country is not only self-sufficient in its requirement of equipment, but is also in a position to export the same to the developing countries. In the Fourth Plan, emphasis has been laid on the production of better breeding stock, dissemination of latest scientific farming practices and strengthening of the existing marketing structure, besides taking up intensive production work in new areas. The Regional Poultry Farms are assisting 24 State Government Farms in implementing their breeding programmes. Attempts are also being made to exploit hybrid vigour through crossing of strains in various combinations. A Random Sample Laying Test Unit has been established to assess, from time to time, the genetic gains achieved through the breeding programme. An In-Service Training Institute in Poultry has been set up in Bangalore, where training courses in breeding, nutrition and management, including marketing will be conducted each year with the help of foreign experts under EPTA Programme of the FAO. Facilities for broiler production are being expanded. Two large processing plants at Poona and Chandigarh have started functioning. Some of the State Poultry Farms have also taken up a programme of raising one or two crops of broilers by utilising their existing facilities during the off-breeding season. To ensure the farmers a reasonable price for their produce, it is proposed to encourage establishment of four egg collection and marketing centres at Calcutta, Delhi, Bombay and Madras in the cooperative sector with facilities for cold storage. An Egg Powder Plant has also been set up recently at Bombay.

The Case of Piggery Development

The pig is an important source for the supply of protein-rich meat. The indigenous pigs are un-economical for commercial exploitation because of their low growth rate, poor body weight, small litter size and late maturity. As against this, the exotic breeds like Yorkshire, Landraco and Hampshire are ideal for commercial exploitation. In order to develop pig production on systematic lines and organise it as a profitable industry, coordinated piggery development programme was initiated towards the end of the Second Plan. The salient features of this programme are: (a) setting up of pig breeding stations/farms for production of improved breeding stock for distribution especially to the weaker sections of the community who traditionally rear pigs; (b) organizing pig production in compact rural areas by distribution of breeding stock at subsidised rates and making available other inputs; and (c) setting up of bacon factories/pork processing plants for providing ready and remunerative market to the producers. Seven bacon factories have been established in different states to provide marketing facilities to the farmers for the quality exotic pigs produced by them. Around these bacon factories, piggery development has been stepped up through the supply of quality breeding stock and all health support and training of farmers in modern pig-raising.

REFERENCES

Afzal, S., R. Ahmad, T. Hussain, and G. Jilani. 1994. Developments for EM-technology to replace chemical fertilizers in Pakistan. In Third Conference on Effective Micro-organisms (EM). Kyusei Nature Farming Center. Saraburi, Thailand. pp. 45-66.

Harada, D.Y., M.J.A. Jorge, A.B. Sanches, and H. Tokeshi. 1995. Interaction between microorganisms, soil physical structure, and plant diseases. Mimeographed sheet. Universidade do São Paulo, Peracicaba, Brazil.

Higa, T. 1996. An Earth Saving Revolution. (English translation.) Sunmark Publishers, Inc. Tokyo, Japan.

Leon, Raul. 1997. Aplicación del Sistema TDF para la remediación bioenzimática de las lagunas facultativas de la Compañia Munidimar, S.A. Técnica del Futuro, S.A. San José, Costa Rica.

Lopez, M.E. 1995. Evaluación Técnica y Proyecto Mejoras Tratamiento Aguas Residuales Compañia Mundimar–Guapiles. EASA Consultores, S.A., San José, Costa Rica.

Phillips, J.M., and S.R. Phillips. 1995. The EM Application Manual for APNAN Countries. EM Technologies, Inc., Tucson, Arizona, USA

Toledo, I. 1996. Tratamiento de las aguas residuales del proceso de fabricación de puré de banano en la planta Mundimar, S.A. (Guapiles, Limon) Utilizando Microorganismos Eficientes (EM). EARTH College, Limon, Costa Rica.

Wididana, G.N. 1994. Preliminary experiment of EM technology on waste water treatment. In Third Conference on Effective Microorganisms (EM). Kyusei Nature Farming Center, Saraburi, Thailand. pp. 1-6.

Angniman, P.A. 1996. *Privatization of veterinary services within the context of structural adjustment in Mali, Cameroon and Chad.* Rome, FAO.

Ashley, S.D., Holden, S.J. & Bazeley, P.B.S. 1996. *The changing role of veterinary services: a report of a survey of chief veterinary officers' opinions.* Somerset, UK, Livestock in Development.

Babjee, A.M. 1996. *Privatization with reference to the veterinary services in Malaysia.* Paper presented at the FAO/GTZ Workshop on Systematic Improvement of the Efficiency of Public and Private Livestock Services in Asia, 22-26 April 1996, Bangkok, Thailand.

Baker, J.L. 1995. *Profile of veterinary services in New Zealand.* Rome, FAO.

Cheneau, Y. 1984. Towards new structures for the development of animal husbandry in Africa south of the Sahara. *Rev. Sci. Tech. Off. Int. Epiz.*, 3(3): 621-627.

Cheneau, Y. 1985. The organization of veterinary services in Africa. *Rev. Sci. Tech. Off. Int. Epiz.*, 5(1): 107-154.

CTA (Technical Centre for Agriculture and Rural Cooperation). 1985a. *Seminar on a Primary Animal Health Care Structure.* 24-26 October 1984, Bujumbura, Burundi.

CTA (Technical Centre for Agriculture and Rural Cooperation). 1985b. *Proceedings of the Seminar on Primary Animal Health Care in Africa.* 25-28 September 1985, Balantyre, Malawi.

de Haan, C. & Bekure, S. 1991. *Animal health services in sub-Saharan Africa: initial experiences with alternative approaches.* World Bank Technical Papers No. 134. Washington, DC, World Bank.

FAO. 1991. *Guidelines for strengthening animal health services in developing countries.* Rome.

FAO. 1994. *Expert Consultation on Application of effective herd health and production programmes to increase livestock productivity in developing countries.* 30 March-1 April 1993, Rome.

FAO. 1996. *World livestock production systems - current status, issues and trends.* FAO Animal Production and Health Paper No. 127. Rome.

FAO/RLAC. 1992. *Report of the round table on privatization of the veterinary services in Latin America and the Caribbean.* 12-14 November 1991, Santiago, Chile.

FAO/RNEA. 1994. *Study on privatization of veterinary services in the Near East region.* Report of a meeting, 28-30 June 1994, Cairo, Egypt.

Fassi-Fehri, B.E.C. & Bakkoury, M. 1995. *Profile of the veterinary services in Morocco.* Rome, FAO.

Gatongi, P.M., Scott, M.E., Ranjan, S., Gathuma, J.M., Munyna, W.K., Cheruiyot, H. & Prichard, R.K. 1993. Effects of three nematode anthelmintic treatment regimes on flock performance of sheep and goats under extensive management in semi-arid Kenya, *Vet. Parasitology,* 68: 323-336.

Holden, S., Ashley, S. & Bazeley, P. 1996. *Improving the delivery of animal health services in developing countries: a literature review.* Somerset, UK, Livestock in Development.

Howard, J.R. 1969. Costs of feedlot disease and prevention programmes. *J. Am. Vet. Med. Assoc.,* 154(10): 1166-1167.

Ilemobade, A.A. 1996. *Privatization of veterinary services within the context of structural adjustment in Zimbabwe, Namibia and Ghana.* Rome, FAO.

Jemal, A. & Hugh-Jones, M.E. 1995. Association of tsetse control with health and productivity of cattle in the Didessa Valley, western Ethiopia. *Prev. Vet. Med.,* 22: 29-40.

Leonard, D.K. 1987. The supply of veterinary services: Kenyan lessons. *Agric. Admin. & Extension,* 26: 219-236.

Leonard, D.K. 1993. Structural reform of the veterinary profession in Africa and the new institutional economics. *Dev. & Change,* 24: 227-267.

McCorkle, C.M. & Mathias, E. 1996. Paraveterinary healthcare programmes: a global overview. *In* K.-H. Zessin, ed. *Livestock production and diseases in the tropics: livestock production and human welfare.* Proceedings of the VII International Conference of Institutions of Tropical Veterinary Medicine, Vol. II, p. 544-549. 25-29 September 1994, Berlin. Feldafing, Deusche Stifung fur Internationale Entwicklung.

McInerney, J.P., Howe, K.S. & Schepers, J.A. 1992. A framework for the economical analysis of disease in farm livestock. *Prev. Vet. Med.,* 13: 137-154.

Mehraban, A.B., Barker, T. & Ward, D.E. 1996. *Community-based animal health and production in Afghanistan - a delivery system.* Paper presented at the Asian-Australian Animal Production Meeting, 13-18 October 1996, Makuhari, Japan.

Mukhebi, A.W., Kariuki, D.P., Mussukuya, E., Mullins, G., Nugumi, P.N., Thorpe, W. & Perry, B.D. 1995. Assessing the economic impact of immunization against East Coast fever: a case study in Coast Province. Kenya. *Vet. Rec.,* 137: 17-22.

OIE (International Office of Epizootics). 1995. *Recommendation of the OIE: appropriate privatization of veterinary practice in Africa.* Rabat, Morocco.

Ott, S.L., Hillberg Seistzinger, A. & Hueston, W.D. 1995. Measuring the national economic benefits of reducing livestock mortality. *Prev. Vet. Med.,* 24: 203-211.

Peagram, R.G., James, A.D., Bamharo, C., Dolan, T.T., Hove, T., Kan, G.K. & Latif, A.A. 1996. *Effects of immunization against Theileria parva on beef cattle productivity and economics of control options.* Trop. Anim. Health Prod., 28: 99-11.

Schillhorn van Veen, T.W. & de Haan, C. 1995. Trends in the organization and financing of livestock and animal health services. Prev. *Vet. Med.,* 25: 225-240.

Schreuder, B.E.C., Moll, H.A.J., Noorman, N., Halimi, C., Kroese, A.H. & Wassink, G. 1995. A benefit-cost analysis of veterinary intervention in Afghanistan based on a livestock mortality study. *Prev. Vet. Med.,* 26: 303-314.

Sidahmed, A.E. 1995. *Livestock development and rangelands management cluster: privatization of veterinary services in sub-Saharan Africa.* Rome, International Fund for Agricultural Development (IFAD).

Swallow, B.M., Mulatu, W. & Leak, S.G.A. 1995. Potential demand for a mixed public-private animal health input: evaluation of a pour-on insecticide for controlling tsetse-transmitted trypanosomiasis in Ethiopia. *Prev. Vet. Med.,* 24: 265-275.

Till, J.E. 1995. Building credibility in public studies. *Am. Sci.,* 83(Sept-Oct): 468-473.

Umali, D. & Schwartz, L. 1994. *Public and private agricultural extension: beyond traditional frontiers.* World Bank Discussion Papers No. 236, Washington, DC, World Bank.

Umali, D., Feder, G. & de Haan, C. 1992. *The balance between public and private sector activities in the delivery of livestock services.* World Bank Discussion Papers No. 63. Washington, DC, World Bank.

Wamukoya, J.P.O., Gathuma, J.W. & Mutiga, E.R. 1995. *Spontaneous private veterinary practices evolved in Kenya since* 1988. Rome, FAO.

6

Dairy Farming, Cattle Types and Milking: An In-depth Analysis

DAIRY FARMING

Dairy farming is a class of agricultural, or an animal husbandry enterprise, for long-term production of milk, which may be either processed on-site or transported to a dairy factory for processing and eventual retail sale. Most dairy farms sell the male calves born by their cows, usually for veal production, or breeding depending on quality of the Bull calf, rather than raising non-milk-producing stock. Many dairy farms also grow their own feed, typically including corn, alfalfa, and hay. This is fed directly to the cows, or is stored as silage for use during the winter season. Additional dietary supplements are added to the feed to increase quality milk production. Dairy farming has been part of agriculture for thousands of years, but historically, it was usually done on a small scale on mixed farms. Specialist scale dairy farming is only viable where either a large amount of milk is required for production of more durable dairy products such as cheese, or there is a substantial market of people with cash to buy milk, but no cows of their own. Centralized dairy farming as we understand it primarily developed around villages and cities, where residents were unable to have cows of their own due to a lack of grazing land. Near the town, farmers could make some extra money on the side by having additional animals and selling the milk in town. The dairy farmers would fill barrels with milk in the morning and bring it to market on a wagon. Before mechanization most cows were still milked by hand. At milking time they brought the vacuum pump, and the automatic milking machine. The first milking machines were an extension of the traditional milk pail. The early milker device fit on top of a regular milk pail and sat on the floor under the cow. Following each cow being milked, the bucket would be dumped into a holding tank. This developed into the Surge hanging milker. Prior to milking a cow, a large wide leather strap called a surcingle was put around the cow, across the cow's lower back. The milker device and collection tank hung underneath the cow from the strap. This innovation allowed the cow to move around naturally during the milking process rather than having to stand perfectly still over a bucket on the floor. Surge later developed a vacuum milk-return system known as the Step-Saver, to save the farmer the trouble of carrying the heavy steel buckets of milk all the way back to the storage tank in the milkhouse. The system used a very long vacuum hose coiled around a receiver cart, and connected to a vacuum-breaker device in the milkhouse. Following milking each cow, the hanging milk bucket would be dumped into the receiver cart, which filtered debris from the milk and allowed it to be slowly sucked through the long hose to the milkhouse. As the farmer milked the cows in series, the cart would be rolled further down the center aisle, the long milk hose unwrapped from the cart, and hung on hooks along the ceiling of the aisle. The next innovation in automatic milking was the milk pipeline. This uses a permanent milk-return pipe and a second vacuum pipe that encircles the barn or milking parlor above the rows of cows, with quick-seal entry ports above each cow. By eliminating the need for the milk container, the milking device shrank in size and weight to the point where it could hang under the cow, held up only by the sucking force of the milker nipples on the cow's udder. The milk is pulled up into the milk-return pipe by the vacuum system, and then flows by gravity to the milkhouse vacuum-breaker that puts the milk in the storage tank. The pipeline system greatly reduced the physical labor of milking since the farmer no longer needed to carry around huge heavy buckets of milk from each cow. The final innovation in automatic milking was the milking parlor, which streamlined the milking process to permit cows to be milked as if on an assembly line, and to reduce physical stresses on the farmer by putting the cows on a platform slightly above the person milking the cows to eliminate having to constantly bend over. Many older and smaller farms still have tie-stall or stanchion barns, but a growing number now have parlors. Newer innovations include automatic take-off systems, which remove the milker from the cow when the milk flow reaches a preset level, computer to measure te production of each animal while it is milking, and computer chips that identify cows individually when they walk into a parlor so their feed intake and milk output can be monitored. These last three are becoming more common because of their value on large farms where it is hard to monitor each cow individually.

History of Milk Preservation Methods

Keeping milk cool helps preserve it. When windmills and well pumps were invented, one of its first uses on the farm besides providing water for animals was for cooling milk, to extend the storage life before being transported to the town market. The naturally cold underground water would be continuously pumped into a tub or other containers of milk set in the tub to cool after milking. This method of milk cooling was extremely popular before the arrival of electricity and refrigeration.

When refrigeration first arrived, the equipment was fairly small and did not have the ability to rapidly cool the large volume of milk that was entering the storage tank in a short period of time. This problem was resolved through the development of the ice bank. This is a double-walled tank design where water and cooling coils fill the space underneath and around the milk tank above. All day long, the small compressor and cooling system slowly draws heat out of the water, while a second pump continuously circulates the water around the coils. Ice eventually builds up around the coils, until it reaches a thickness of about three inches surrounding each pipe, and the cooling system shuts off. When the milking operation starts only the milk agitator and the water circulation pump blowing water across the ice and the steel walls of the tank are needed to rapidly reduce the incoming milk to a temperature below 40 degrees. But because the ice is not permitted to build up until it touches the milk storage tank, the milk does not get cold enough to also freeze. This cooling method worked well for smaller dairies up to about 40 cows, but for large numbers of animals a better system was needed to rapidly cool the incoming warm milk. This is usually done using a device known as a plate chiller, which is a heat exchanger. Alternating stainless steel plates cause the milk to flow in a thin sheet across the plates, while cold water is circulated in a thin sheet on the other side of the plates. Flattening out the milk flow permits quick. even cooling for all the milk, compared to a round tube where the center core does not cool as rapidly as the walls. The plate chiller has high cooling demands, and for many farms this involves a step back into the past, back to the days of windmills and milk-can cooling, except now a large volume of naturally cold underground water is continuously streamed through the plate chiller to quickly bring down the milk down to the temperature of the underground water at about 50 degrees F. The water is usually not just dumped back into the ground again, but reused for washing and other purposes. But the milk still is not as cold as it needs to be, so the milk storage tank is still used to do further cooling, to bring the milk down to 40 degrees. But with the development of high-power 3-phase electrical service, ice-bank chillers are typically no longer used. Instead the milk storage tank is a direct-cooling system with cooling coils embedded in the walls of the tank, that quickly pull the heat out and dump it across a large array of possibly several different high horsepower compressors and condensing units. Once the milk has achieved 40 degrees F after milking is finished, only one or two cooling units need to run occasionally to maintain the correct temperature.

Milking Operation

Original Hand Milking Processes

Until the late 1800s, the milking of the cow was done by hand. In the United States, several large dairy operations existed in some northeastern states and in the west, that involved as many as several hundred cows, but an individual milker could not be expected to milk more than a dozen cows a day. Smaller operations predominated.

Milking took place indoors in a barn with the cattle tied by the neck with ropes or held in place by stanchions. Feeding could occur simultaneously with milking in the barn, although most dairy cattle were pastured during the day between milkings. Such examples of this method of dairy farming are difficult to locate, but some are preserved as a historic site for a glimpse into the days gone by. One such instance that is open for public tours is at Point Reyes National Seashore.

With the availability of electric power and suction milking machines, the production levels that were possible in stanchion barns increased but the scale of the operations continued to be limited by the labor intensive nature of the milking process. Attaching and removing milking machines involved repeated heavy lifting of the machinery and its contents several times per cow and the pouring of the milk into milk cans. As a result, it was rare to find single-farmer operations of more than 50 head of cattle.

Modern Milking Parlor Operations

Farmers use any number of styles of milking parlors to milk dairy cattle. Many older farms have stanchion or tie-stall facilities, where the milkers are brough to the cows and the milker bends down to apply the milking machine to the cow. More modern farms use recessed parlors, where the milker stands in a recess such that his arms are at the level of the cow's udder. Recessed parlors can be herringbone, where the cows stand in two angled rows either side of the recess and the milker accesses the udder from the side, parallel, where the cows stand side-by-side and the milker accesses the udder from the rear or, more recently, rotary (or carousel), where the cows are on a raised circular platform, facing the center of the circle, and the platform rotates while the milker stands in one place and accesses the udder from the rear. There are many other styles of milking parlors which are less common. In herringbone and parallel parlors, the milker generally milks one row at a time. The milker will move a row of cows from the holding yard into the milking parlor, and milk each cow in that row. Once all or most of the milking machines have been removed from the milked row, the milker releases the cows to their feed. A new group of cows is then loaded into the now vacant side and the process repeats until all cows are milked. Depending on the size of the milking parlor, which normally is the bottleneck, these rows of cows can range from four to sixty at a time. In rotary parlors, The cows are loaded one at a time onto the platform as it slowly rotates. The milker stands near the entry to the parlor and puts the cups on the cows as they move past. By the time the platform has completed almost a full rotation, another milker or a machine removes the cups and the cow steps backwards off the platform and then walks to her feed. Milking machines are held in place automatically by a vacuum system that draws the ambient air pressure down to 15 to 21 pounds of vacuum. The vacuum is also used to lift milk vertically through small diameter hoses, into the receiving can. A milk lift pump draws the milk from the receiving can through large diameter stainless steel piping, through the plate cooler, then into a refrigerated bulk tank. Milk is extracted from the cow's udder by flexible rubber sheaths known as liners or inflations that are surrounded by a rigid air chamber. A pulsating flow of ambient air and vacuum is applied to the inflation's air chamber during the milking process. When ambient air is allowed to enter the chamber, the vacuum inside the inflation causes the inflation to collapse around the cow's teat, squeezing the milk out of teat in a similar fashion as a baby calf's mouth massaging the teat. When the vacuum is reapplied in the chamber the flexible rubber inflation relaxes and opens up, preparing for the next squeezing cycle. It takes the average cow three to five minutes to give her milk. Some cows are faster or slower. Slow-milking cows may take up to fifteen minutes to let down all their milk. Milking speed is only minorly related to the quantity of milk the cow produces - milking speed is a separate factor from milk quantity; milk quantity is not determinative of milking speed. Because most milkers milk cattle in groups, the milker can only process a group of cows at the speed of the slowest-milking cow. For this reason, many farmers will cull slow-milking cows. The extracted milk passes through a strainer and plate heat exchangers before entering the tank, where it can be stored safely for a few days at approximately 3°C or around 42°F. At pre-arranged times, a milk truck arrives and pumps the milk from the tank for transport to a dairy factory where it will be pasteurized and processed into many products.

Animal Waste from Large Dairies

As measured in phosphorus, the waste output of 5,000 cows roughly equals a municipality of 70,000 people. In the U.S., dairy operations with more than 1,000 cows meet the EPA definition of a CAFO (Concentrated Animal Feeding Operation), and are subject to EPA regulations. For example, in the San Joaquin Valley of California a number of dairies have been established on a very large scale. Each dairy consists of several modern milking parlor set-ups operated as a single enterprise. Each milking parlor is surrounded by a set of 3 or 4 loafing barns housing 1,500 or 2,000 cattle. Some of the larger dairies have planned 10 or more series of loafing barns and milking parlors in this arrangement, so that the total operation may include as many as 15,000 or 20,000 cows. The milking process for these dairies is similar to a smaller dairy with a single milking parlor but repeated several times. The size and concentration of cattle creates major environmental issues associated with manure handling and disposal, which requires substantial areas of cropland (a ratio of 5 or 6 cows to the acre, or several thousand acres for dairies of this size) for manure spreading and dispersion, or several-acre methane digesters. Air pollution from methane gas associated with manure management also is a major concern. As a result, proposals to develop dairies of this size can be controversial and provoke substantial opposition from environmentalists including the Sierra Club and local activists. The potential impact of large dairies was demonstrated when a massive manure spill occurred on a 5,000-cow dairy in Upstate New York, contaminating a 20-mile stretch of the Black River, and killing 375,000 fish. On Aug. 10, 2005, a manure storage lagoon collapsed releasing several million gallons of manure into the Black River. Subsequently the New York Department of Environmental Conservation mandated a settlement package of $2.2 million against the dairy.

Use of Hormones and Antibiotics

Approximately 22% of diary cows in the US are injected with recombinant growth hormones known as recombinant BST or rBGH to maintain slightly higher milk production. The use of rBST is regarded as controversial due to its effects on animal and possibly human health. The European Union, Japan, Australia and Canada have banned rBSTdue to such concerns.

Management of the Dairy Herd

Modern dairy farmers use milking machines and sophisticated plumbing systems to harvest and store the milk from the cows, which are usually milked twice or thrice daily. During the warm months, in the northern hemisphere, cows may be allowed to graze in their pastures, both day and night, and are brought into the barn only to be milked. Many barns also incorporate tunnel ventilation into the architecture of the barn structure. This ventilation system is highly efficient and involves opening both ends of the structure allowing cool air to blow through the building. Farmers with this type of structure keep cows inside during the summer months to prevent sunburn and damage to udders. During the winter months, especially in northern climates, the cows may spend the majority of their time inside the barn, which is warmed by their collective body heat. Even in winter, the heat produced by the cattle requires the barns to be ventilated for cooling purposes. Many modern facilities, and particularly those in tropical areas, keep all animals inside at all times to facilitate herd management. Housing the cow can be either loose housed or stalls (called cow cubicles in UK). In the southern hemisphere milking animals are more likely to spend most of their lives outside on pasture. There is little research available on dimensions required for cow stalls, and much housing can be out of date, however increasingly companies are making farmers aware of the benefits, in terms of animal welfare, health and milk production. The production of milk requires that the cow be in lactation, which is a result of the cow having given birth to a calf. The cycle of insemination, pregnancy, parturition, and lactation, followed by a "dry" period before insemination can recur, requires a period of 12 to 16 months for each cow. Dairy operations therefore included both the production of milk and the production of calves. Bull calves are either castrated and raised as steers for beef production or raised for veal. As the size of herds has increased, the conditions in which large numbers of veal calves are raised, fed and marketed on larger dairies also have provoked controversy among animal rights activists.

Analysis Dairy Farming in the World

In the United States, the top four dairy states are, in order by total milk production, California, Wisconsin, New York, and Pennsylvania. Dairy farming is also an important industry in Florida, Minnesota, Ohio and Vermont. In Pennsylvania, the dairy industry is the number one industry in the state. Pennsylvania is home to 8,500 farms and 555,000 dairy cows. Milk produced in Pennsylvania yields about US$1.5 million in farm income every year, and is sold to various states up and down the east coast. The world's largest exporter of dairy products is New Zealand. Japan is the world's largest importer of dairy products. There follows two lists of countries by milk production (MT = million tonnes).

TABLE 1

World Production Not Including Countries in the European Union

Rank	*Country*	*Production (MT / yr)[a]*
1.	India	96.1
2.	United States	67.2
3.	Russia	32.8
4.	Brazil	23.3
5.	China	16.8
6.	New Zealand	14.6
7.	Australia	10.6
8.	Mexico	9.8
9.	Turkey	9.5
10.	Japan	8.4
11.	Canada	8.0
12.	Argentina	8.0
13.	Switzerland	3.9
14.	South Africa	2.6
15.	South Korea	2.4
16.	Norway	1.6

Notes: (a) Source, unless otherwise noted: 2005 OECD Agricultural Outlook Tables, 1970-2014, OECD, 2003.

The EU is the largest milk producer in the world, with 143.7 million tonnes in 2003. This data, encompassing the present 25 member countries, can be further broken down into the production of the original 15 member countries, with 122 million tonnes, and the new 10 mainly former Eastern European countries with 21.7 million tonnes.

TABLE 2

Milk Production Data for EU Countries

Rank	*Country*	*Production (MT / yr)[a]*
1.	Germany	28.5
2.	France	24.6
3.	United Kingdom	15.0
4.	Poland	11.9
5.	Netherlands	11.0
6.	Italy	10.8
7.	Spain	6.6
8.	Ireland	5.4
9.	Denmark	4.7
10.	Sweden	3.2
11.	Austria	3.2
12.	Belgium	3.1
13.	Czech Republic	2.7
14.	Finland	2.5
15.	Hungary	1.9
16.	Portugal	1.9
17.	Lithuania	1.8

Notes: (a) Source unless otherwise noted: Production of cow's milk and milk deliveries to dairies, European union by country, MDC Datum, 2003, <http://www.mdcdatum.org.uk/MilkSupply/euproduction.html>. Retrieved on 29 October 2007.

Dairy Competition

Most milk-consuming countries have a local dairy farming industry, and most producing countries maintain significant subsidies and trade barriers to protect domestic producers from foreign competition. In large countries, dairy farming tends to be geographically clustered in regions with abundant natural water supplies (both for feed crops and for cattle) and relatively inexpensive land (even under the most generous subsidy regimes, dairy farms have poor return on capital). New Zealand, the fourth largest dairy producing country, does not apply any subsidies to dairy production. The milking of cows was traditionally a labor-intensive operation and still is in less developed countries. Small farms need several people to milk and care for only a few dozen cows, though for many farms these employees have traditionally been the children of the farm family, giving rise to the term "family farm". Advances in technology have mostly led to the radical redefinition of "family farms" in industrialized countries such as the United States. With farms of hundreds of cows producing large volumes of milk, the larger and more efficient dairy farms are more able to weather severe changes in milk price and operate profitably, while "traditional" very small farms generally do not have the equity or cashflow to do so. The common public perception of large corporate farms supplanting smaller ones is generally a misconception, as many small family farms expand to take advantage of economies of scale, and incorporate the business to limit the legal liabilities of the owners and simplify such things as tax management. Before large scale mechanization arrived in the 1950s, keeping a dozen milk cows for the sale of milk was profitable. Now most dairies must have more than one hundred cows being milked at a time in order to be profitable, with other cows and heifers waiting to be "freshened" to join the milking herd. In New Zealand the average herd size, depending on the region, is about 600 cows. [dubious – discuss] Herd size in the US varies between 1,200 on the West Coast and Southwest, where large farms are commonplace, to roughly 50 in the Northeast, where land-base is a significant limiting factor to herd size.[dubious – discuss] The average herd size in the U.S. is about one hundred cows per farm.

TOTAL MIXED RATION (TMR)

It is a method of feeding cattle. The term total mixed ration (TMR) may be defined as, "The practice of weighing and blending all feedstuffs into a complete ration which provides adequate nourishment to meet the needs of dairy cows." Each bite consumed contains the required level of nutrients (energy, protein, minerals and vitamins) needed by the cow. The advantages are the following:

- All forages, grains, protein supplements, minerals and vitamins are thoroughly mixed. Therefore, the cow can do very little sorting for individual ration ingredients.
- Completely blended feeds, coupled with grouping the cows, permits greater flexibility in feeding exact amounts of nutrients (energy, protein, etc.) to more accurately feed cows for their particular stage of lactation and level of milk yield.
- Grain mixtures can be liberally fed to grouped high producers without over-feeding the late-lactation or lower-producing cows, resulting in more efficient use of feeds.
- The dairy farmer has more control over the feeding programme.

Disadvantages:

Cows should be grouped by production levels.

- Grouping cows is not feasible in small herds (less than 50 cows).

- If not grouped according to production, cows in late lactation are likely to get too fat.

Special equipment is needed:

- The equipment must have the capability to thoroughly blend the feed ingredients.
- The mixer-wagon, preferably mobile, must be capable of accurately weighing each ingredient.

In Sum:

- Complete rations feature the blended approach all forages, concentrates, protein supplements, minerals and vitamins are mixed together and offered as a single feed.
- Complete ration system can save labour and reduce overall feeding cost.
- It is extremely important to keep the mixture exactly the same day after day and to make big changes gradually.
- Early detection of problems with the ration system is possible by observing the bulk tank milk level after each milking.
- Forage analysis is necessary and should include dry matter, crude protein, acid detergent fiber, neutral detergent fiber, calcium and phosphorus.
- TMR can be used effectively by many dairy farmers, but it is not a substitute for good management. In fact, the intensity of management may be increased. Most of all, management skills and competency of the dairy farmer is critical to make this system work effectively.

MIXER-WAGON

A mixer-wagon, or diet feeder, is a specialist agricultural machine used for accurately weighing, mixing and distributing Total Mixed Ration (TMR) for ruminant farm animals, in particular cattle and most commonly, dairy cattle. Trailed mixer-wagons vary in size from 10 m^2 to more than 45 m^2. Some self-propelled mixer-wagons may be bigger than this. Displacement varies according to ration dry matter. More water means more weight. With dry (45% DM (dry matter) rations, a 14m^2 mixer-wagon such as the one pictured, may contain 3 tonnes fully loaded, or enough for about 60 Holstein cows. A mixer-wagon commonly consists of:

- a trailed chassis, for coupling to a tractor power unit, and fitted with one or more usually braked axles, and fitted with a road-legal lighting system.

- a mixing body, attached to the chassis by four weighing sensors, one at each corner. There are three main types of mixing body:
  - — paddle, whereby a central axial shaft turns a series of paddles which rotate the contents and provide front to back mixing.
  - — vertical auger, with one or two augers used to move the contents from top to bottom.
  - — horizontal auger, containing between one and five augers, used to circulate the contents from front to back and bottom to top.
- a digital weighing computer working through the above mentioned weighing sensors. Such a computer can typically memorise 9 or more different rations and 99 or more ingredients.

The system can weigh in all the ingredients of a ration in any chosen order and weigh out according to chosen quantities corresponding to the needs of different groups of animals. As each ingredient is loaded, a visual and audible signal alerts the operator when the required amount is reached. The flashing light and "beep" system repeats faster and faster until the ingredient is complete, when the signal becomes continuous for 2 seconds. The computer then shifts to the next ingredient and displays the product name and weight to be loaded. Loading is generally done using a high capacity tractor loader, or a telescopic handler.

- the mixing paddle, rotors or augers are connected to the tractor pto through a reducer system, provided by either a planetary gearbox, and / or a step-down pulley and chain system.
- a set of stationery knives, against which long fibre may be chopped by the forcing action of the mixing rotor.
- a hydraulic door to seal the ration in during mixing, thus permitting the use of liquid feeds such as molasses.
- an unloading system, consisting of a simple hydraulically adjustable chute, up to a hydraulic powered conveyor belt.
- Self-propelled mixer-wagons are mounted on a lorry chassis or may be specialist self-loading machines.

DAIRY CATTLE

Dairy cattle, generally of the species *Bos taurus*, are domesticated animals bred to produce large quantities of milk. For general information on milk production see dairy farming. A young dairy animal is known as a calf. A female calf which has not given birth to a calf and is less than thirty months old is called a heifer. When the heifer is seven months pregnant or has reached the stage in pregnancy where the udder starts to swell, it is known as a springer. After more than thirty months old, a female dairy animal is known as a cow. The process of birthing a calf is known as calving or parturition. A male dairy animal is called a bull at any stage of life, unless castrated, in which case it is known as a steer until it is four years old, then it is called an ox. "Ox" is also the term for any bovine trained for draft work; this is usually a steer. A dairy animal's mother is known as its dam. Similarly, a dairy animal's father is known as its sire.

Historically, there was less distinction between dairy cattle and beef cattle than is the case now, with animals of the same species often being used for both meat and milk production. Dairy cattle are now specialized animals, and most of them belong to breeds which have been bred specifically to give large volumes of milk. This milk is made into various products, including cheese, yogurt, butter, ghee, cottage cheese, whey, and ice cream, and is consumed around the world.

Dairy Farms

Dairy cattle may be found in herds on farms where dairy farmers own, manage, care for, and collect milk from them. These herds range in size from small farms of fewer than five cows to large conglomerates of 25,000 cows or more. The average dairy farmer in the United States owns about one hundred cows and is about 55 years old.

Life of Dairy Cattle

Pure bred heifer calves are usually reared as herd replacements, and as such are of great value to the dairy farmer. Sales of crossbred heifers and bull calves are subject to the demand for beef animals within transport range of the farm. Surplus calves are generally sold at two weeks of age and bulls may fetch a premium over heifers due to their size and potential. Calves may be sold for veal, or for one of several types of beef production, depending on available local crops and markets. Such bull calves may be castrated if turnout onto pastures is envisaged, in order to render the animals less aggressive. Purebred bulls from elite cows may be put into progeny testing schemes to find out whether they might become superior sires for breeding. Such animals may become extremely valuable (20,000 euros or more for Holsteins). The aim of most farms is to separate the calf from its dam within 24 hours of birth. Contrary to popular perception, this early separation eases the stress on cow and calf as "bonding" is prevented. A restless cow bellowing for her lost calf can be avoided by earlier separation; the calf, in turn, may be more easily taught to drink milk from a bucket, as its "sucking reflex" strengthens with time spent suckling the cow. The dam's first milk, called colostrum, is rich in antibodies and is required for newborn calves to protect them from infection. A calf must drink two quarts (2 L) of colostrum within twelve hours of birth, as it has no immune system of its own for the first two weeks of life. The antibodies are directly assimilated into the bloodstream by the calf's digestive system, a phenomenon which shuts down permanently from 12 to 24 hours after birth. The colostrum changes into milk suitable for commercial use within three or four days after calving. Most young stock then subsist on commercial milk replacer, a feed based on dried milk powder and reconstituted using hot water, until old enough to start consuming solid foods at 3 to 4 weeks old. Milk replacer is cheaper than milk from the cow. A day old calf can only drink around 2 litres of milk per day, whereas the average Holstein cow will produce 30 litres or more. Even at weaning, at 8 weeks old, a calf will only be consuming around 6 litres per day.

The Bull

A select few high genetic potential pedigree bulls sired by elite bulls out of elite cows, will be reared for breeding purposes. These bulls will generally have excellent production indices, based on the results of their progeny testing. They will also demonstrate superior type conformation (for the breed), as measured by their daughters udder quality, feet and leg quality etc., compared with the general cow population. Herd bulls, are bulls kept on the farm to provide natural breeding for the herd. A bull may service up to 50 or 60 cows during a continuous breeding period. Any more and the semen will become too diluted, leading to cows "returning to service". A herd bull may only stay for one season since over two years old their temperament becomes too unpredictable. More recently, since the 1950s, artificial insemination,or AI, has become practically ubiquitous. Through AI, fewer than a thousand elite bulls can serve as sires for an entire world generation of cows. Although conception is dependent upon effective herd management and heat detection which increases the time the dairy farmer must spend with the cows, a few factors have prompted farmers to use AI nearly exclusively. The foremost is the high quality of cows produced through AI. AI also limits the need for farmers to maintain their own bulls, which contributes to safety, as bulls can be dangerous animals to keep on the farm. Some dairy farms however, still use herd bulls, as it is difficult to cover the entire year's breeding programme using AI, and there is rarely a need to breed the entire herd with quality purebred bulls.

The Cow

Dairy heifers are of great value to their breeders, as they will become the next generation of dairy cows. As a cow cannot produce milk until after calving (giving birth), most farmers will begin breeding heifers as soon as they are fit, at about fourteen months of age for Holsteins. A cow's gestation period is about nine months (279 days long), so most heifers give birth and become cows at about two years of age. A cow will produce large amounts of milk over its lifetime. Certain breeds produce more milk than others; however, different breeds produce within a range of around 4,000 to over 10,000 kg of milk per annum. The average for dairy cows in the US in 2005 was 8,800 kg (19,576 pounds). Production levels peak at around 40 to 60 days after calving. The cow is then bred. Production declines steadily afterwards, until, at about 305 days after calving, the cow is 'dried off', and milking ceases. About sixty days later, one year after the birth of her previous calf, a cow will calve again. High production cows are more difficult to breed at a one year interval. Many farms take the view that 13 or even 14 month cycles are more appropriate for this type of cow. Dairy cows will continue to be productive members of the herd for many lactations. 10 or more lactations are not uncommon. The chances of problems arising which may lead to a cow being culled are however, high; the average herd life of US Holsteins is today fewer than 3 lactations. This is unfortunate as it requires more expensive herd replacements to be reared or purchased. Over 90% of all cows are culled for 4 main reasons:

(i) *Infertility*—failure to conceive and therefore perpetuate production. Cows are at their most fertile between 60 and 80 days after calving. Cows remaining "open" (not in calf) after this period become increasingly difficult to breed. An important influence is the condition of the previous calving. If this was difficult and required assistance, or if the cow was overly fat or thin, or the cow came down with milk fever, or failed to cleanse (expel the afterbirth), this could lead to infertility. Metritis, or infection of the uterus, is a common consequence. This can be cured with intra-uterine antibiotic treatments. Other common consequences are luteal cysts. These can be cured with injections of synthetic prostaglandins.

(ii) *Mastitis*—persistent mammary gland infection, leading to high somatic cell counts and loss of production. Mastitis is recognized by a reddening and swelling of the infected quarter of the udder and the presence of whitish clots in the milk or even pus in serious cases. The cow may run a high temperature and serious untreated cases can be fatal. Antibiotics and anti-inflammatory treatments may be indicated. Congestion and swelling may hinder evacuation of the infection. The speed of the cow's immune system response is important in serious cases. The infection stimulates white blood cell concentration in the infected quarter. Sometimes infectious pathogens may not be entirely eliminated and set up a chronic infection. This leads to semi-permanent high white cell counts which mean the milk from such cows is unmarketable. Treatment is impossible although injection of long-acting antibiotics at drying off has some success.

(iii) *Lameness*—persistent foot infection or leg problems causing infertility and poor feed intakes leading to loss of production. High feed levels of highly digestible carbohydrate cause acidic conditions in the cow's rumen. This leads indirectly to laminitis and subsequent lameness, leaving the cow vulnerable to other foot infections and problems. This is almost entirely preventable with proper feeding techniques. Laminitis vastly increases horn production leading to abnormal foot growth. If this is not trimmed back regularly, the cow becomes susceptible to foot abscesses and ulcers. Infections such as inter-digital dermatitis cause pain and swelling. This can be cured by antibiotic treatments. Cows in pain from foot problems will not stand or walk often and this seriously lowers feed intake levels. A healthy 700 kg Holstein may eat 50 kg of feed per day and require over 100 litres of water.

(iv) *Production*—some animals, despite high genetic potential, simply fail to produce economic levels of milk to justify their feed costs. Feed costs may not be justified by production levels below around 12 to 15 litres of milk per day. A good cow may begin lactation at around 20 litres, peak at 40+ litres and be dried off at calving - 45 days. If production falls too much too early she may be sold as uneconomic. Heifers calving for the first time are unknown quantities. At calving they might be too highly strung and panic during milking, or simply have only two or three functioning quarters. Herd life is strongly correlated with production levels. Lower production cows live longer than high production cows, but are arguably less profitable all the same. Cull cows are sent for slaughter. Their meat is of relatively low value and is generally used for processed meat such as hamburger beef.

Embryo Transfer

More recently, certain practices have been developed to enable the multiplication of progeny from elite cows. Such cows are given hormone treatments to produce multiple embryos. These are then 'flushed'. 7-12 embryos are consequently removed from these donor cows and transferred into other cows who serve as surrogate mothers. The result will be between 3 and 6 calves instead of the normal single, or rarely, twins. This process is called embryo transfer.

Breeds

In the United States, dairy cattle are divided into six major breeds. These are the: Holstein-Friesian, Brown Swiss, Guernsey, Ayrshire, Jersey, Milking Shorthorn. Many other breeds are used nearly exclusively for dairy, or for both dairy and beef purposes.

HOLSTEIN CATTLE

The Holstein is a breed of dairy cow known today as the world's highest production dairy animal. Originating in Europe, Holsteins were developed in what is now the Netherlands and more specifically in the two northern provices of North Holland and Friesland (not from Holstein, Germany). The original animals were the regional cattle of the Batavians and Friesians, two tribes who settled in the coastal Rhine region around 2,000 years ago. The Dutch breeders bred and oversaw the evolution of the breed with the aim of obtaining animals which would make best use of grass, the area's most abundant resource. The result, over the centuries, was an efficient, high-producing black-and-white dairy cow. With the growth of the new world, markets began to develop for milk in America, and dairy breeders turned to Holland for their livestock. After about 8,800 Holsteins had been imported, disease problems in Europe led to the cessation of imports. In Europe, the breed is used for milk in the North, meat in the South - Since 1945, European development has led to cattle production becoming increasingly regionalised. Over 60% of the cattle herd and under 50% of the useable agricultural area, but over 80% of dairy production, is to be found to the north of a line joining Bordeaux and Venice. This change led to the need for specialised animals for dairy (and beef) production. Until this time, milk and beef had been produced from dual-purpose animals, and the leading breeds, national derivatives of the *Dutch Friesian*, had become very different animals than their American counterparts. It was the obvious choice to import superior production animals to cross with the European black and whites. For this reason, in modern usage of the word *Holstein* is used to describe North American stock and its use in Europe. *Friesian*, denotes animals of a traditional European ancestry. Crosses between the two are described by the term *Holstein-Friesian*.

Physical Characteristics

Holsteins, easily recognized by their distinctive markings and outstanding milk production, are large, stylish animals with color patterns of black and white. Size: A healthy calf weighs 40 kg or more at birth. A mature Holstein cow weighs 600 to 1,000 kg and stands 130 cm tall at the shoulder. Holstein heifers can be bred at 18 months of age, when they weigh around 500 kg. Generally, breeders aim for Holstein heifers to calve for the first time between 23 and 26 months of age. Gestation period is approximately nine months.

History

300 BC: Pastoral nomads from Central Asia arrived with their cattle in the river Ems / middle arm of the Rhine area. 100 BC: A displaced group of people from Hesse, migrated with their cattle to the shores of the North Sea near the Friesians, occupying the island of Batavia, between the Rhine, Maas and Waal. Historical

records suggest these cattle were black; and that the Friesian cattle at this time were "pure white and light coloured". Crossbreeding may have led to the foundation of the present Holstein-Friesian breed, as the cattle of these two tribes from then on appear identical in historical records The portion of the country bordering on the North Sea was called Frisia, situated within the provinces of North Holland, Friesland and Groningen and in Germany to the river Ems. The people were known for their care and breeding of cattle. The Friesians, preferring pastoral pursuits to warfare, paid a tax of ox hides and ox horns to the Roman government, whereas the Batavians furnished soldiers and officers to the Roman army; these fought successfully in the various Roman wars. The Friesians were thus able to breed the same strain of cattle unadulterated for two thousand years, except from accidental circumstances. 1282: Floods produced the Zuider Zee, separating the cattle breeders into two groups. The western group occupied West Friesland, now part of North Holland; the eastern, the present provinces of Friesland and Groningen. The rich Polder land in the Netherlands is unsurpassed for the production of grass, cattle and dairy products. Between the thirteenth and sixteenth centuries, the production of butter and cheese was enormous, and history tells of the existence of remarkably heavy meat cattle, weighing from twenty-six hundred to three thousand pounds. The aim was to produce as much milk and beef as possible from the same animal, and selection, breeding and feeding have been carried out with huge success. Inbreeding was not tolerated, and (distinct) families never arose, although differences in soil in different localities produced different sizes and variations.

The Case of United Kingdom

17th to 18th C.: Dutch cattle appear to have been imported into the British Isles, becoming influential in the formation of several breeds in England and Scotland. The eminent Prof. Low recorded: "the Dutch breed was especially established in the district of Holderness, on the north side of the Humber; northward through the plains of Yorkshire. The finest dairy cattle in England...", of Holderness in 1840 still retained the distinct traces of their Dutch origin. Further north in the Tees area, imports likewise took place of continental cattle from Holland, Holstein or the countries on the Elbe. He added: "Of the precise extent of these early importations we are imperfectly informed, but that they exercised a great influence on the native stock appears from this circumstance, that the breed formed by the mixture became familiarly known as the Dutch or Holstein breed". Holstein Friesians were found throughout the rich lowlands of France, Belgium, Holland and the western provinces of Germany. The breed did not become established in Great Britain at the time, nor did it invade the islands of Jersey or of Guernsey, where laws existed against imports for breeding purposes from the continent. Late 19th C and early 20th C: around 2,000 in calf heifers and several bulls and cows were imported from Canada together with several shiploads of store cattle. This was followed up with the importation of almost 200 animals in the years following WW2. This included a gift of 3 yearling bulls from Canadian breeders to help establish the breed. The pure Holstein Breed Society was started in 1946 while the British Friesian Cattle Society had already existed for some time. The breed developed slowly up to the 1970's after which there was an explosion in its popularity and following which a considerable number of important imports took place. More recently, the two Societies merged in 1999 establishing Holstein UK as we know it today.

Numbers

Records on 1 April 2005 from NUTS1 (Nomenclature for Units of Territorial Statistics level 1) show Holstein influence appearing in 61% of all 3.47 million dairy cattle in the UK:

- Holstein Friesian (Friesian with more than 12.5% and less than 87.5% of Holstein blood): 1 765 000 (51%)
- Friesian (more than 87.5% Friesian blood): 1 079 000 (31%)
- Holstein (more than 87.5% of Holstein blood): 254 000 (7%)
- Holstein Friesian Cross (any of the above crossed with other breeds): 101 000 (3%)
- Other dairy breeds: 278 000 (7%)

The above statistics are for all dairy animals possessing passports at the time of the survey, i.e. including young stock. DEFRA lists just over 2 million adult dairy cattle in the UK

Holstein in this instance, and indeed in all modern discussion, refers to animals traced from North American bloodlines, while *Friesian* refers to indigenous European black and white cattle. Criteria for inclusion in the Supplementary Register (i.e. not pure breed) of the Holstein UK herd book are as follows:

CLASS A: A typical representative of the Holstein or Friesian breed, as to type, size and constitution, with no obvious signs of cross breeding, or be proved from its breeding records to contain between 50% and 74.9% Holstein genes or Friesian genes. If the breeding records show that one parent is of a breed other than Holstein Friesian or Holstein or Friesian then such parent must be a purebred animal fully registered in a Herd Book of a dairy-breed society recognised by the Society.

CLASS B: For a calf by a bull registered or dual registered in the Herd Book or in the Supplementary Register and out of a foundation cow or heifer registered in Class A or B of the Supplementary Register and containing between 75% and 87.4% Holstein genes or Friesian genes.

For inclusion in the Pure (Holstein or Friesian) herd book, a heifer or bull calf from a cow or heifer in Class B of the Supplementary Register and by a bull registered or dual registered in the Herd Book or the Supplementary Register, and containing 87.5% or more Holstein genes or Friesian genes will be eligible to have its entry registered in the Herd Book.

Production

The breed currently averages 7655 litres/year throughout 3.2 lactations with pedigree animals averaging 8125 litres/year over an average of 3.43 lactations. By calculation, lifetime production therefore stands at around 26,000 litres.

The Case of United States

American breeders began to become interested in Holstein- Friesian cattle around the 1830s. Black and White cattle were introduced into the US from 1621 to 1664. The eastern part of the State of New York was the Dutch colony of New Netherlands, where many Dutch farmers settled along the Hudson and Mohawk river valleys. They probably brought cattle with them from their native land and crossed them with cattle purchased in the colony. For many years afterwards, the cattle here were called Dutch cattle and were renowned for their milking qualities. The first recorded imports were more than one hundred years later, consisting of six cows and two bulls. These were sent in 1795 by the Holland Land Company, which then owned large tracts in the State of New York, to their agent, Mr. John Lincklaen of Cazenovia. A settler described them thus; "the cows were of the size of oxen, their colors clear black and white in large patches; very handsome". In 1810 a bull and two cows were imported by the Hon. William Jarvis for his farm at Wethersfield, Vt. About the year 1825 another importation was made by Herman Le Roy, a part of which were sent into the Genesee valley. The rest were kept near New York City. Still later an importation was made into the State of Delaware. No records were kept of the descendants of these cattle. Their blood was mingled and lost in that of the native cattle. The first permanent introduction of this breed was due to the perseverance of Hon. Winthrop W. Chenery, of Belmont, Mass. The animals of his first two importations, and their offspring were destroyed by the government in Massachusetts because of a contagious disease. He made a third importation in 1861. This was followed in 1867 by an importation for the Hon. Gerrit S. Miller, of Peterboro, N. Y., made by his brother, Dudley Miller, who had been attending the noted agricultural school at Eldena, Prussia, where this breed was highly regarded. These two importations, by Hon. William A. Russell, of Lawrence, Mass.; and three animals from East Friesland, imported by Gen. William S. Tilton of the National Military Asylum, Togus, Me., formed the nucleus of the Holstein Herd Book. After about 8,800 Holsteins had been imported, a cattle disease broke out in Europe and importation ceased.

In the late 1800's there was enough interest among Holstein breeders to form associations to record pedigrees and maintain herd books. These associations merged in 1885, to found the Holstein-Friesian Association of America. In 1994 the name was changed to Holstein Association USA, Inc.

Production

Recorded cows in the USA produced 22,347 pounds (10,158 kg) of milk at 3.64% fat and 3.05% protein in the 2005 DHIA (National Dairy Herd Information Association). Total lifetime productivity can be inferred from the average lifetime of US cows. This has been decreasing regularly in recent years and now stands at around 2.75 lactations, which when multiplied by average lactation yield above gives around 28,000 kg of milk. The considerable advantage, compared to the UK, for example, can be explained by several factors:

(i) Use of milk production hormone, bST. A study in February 1999 determined that "response to bST over a 305-day lactation equaled 894 kg of milk, 27 kg of fat, and 31 kg of protein". Monsanto estimates a figure of about 1.5 million out of 9 million dairy cows being treated with bST, or about 17% of cows nationally.

(ii) greater use of 3x a day milking. In a study performed in Florida between 1984 and 1992 using 4293 Holstein lactation records from 8 herds, 48% of cows were milked 3 times a day. The practice was responsible for an extra 17.3% milk, 12.3% fat and 8.8% protein. Three Times-A-Day Milking has become a common milking frequency in recent years. Twice-A-Day Milking is the most common milking schedule of dairy cattle. In Europe, Australia, and New Zealand, milking 2x at 10 to 14-hour intervals is common.

(iii) higher cow potential (100% Holstein herds). European Friesian types traditionally had lower production performance than their North American Holstein counterparts. Despite Holsteinization over the last 50 years, a large genetic trace of these cattle is still present.

(iv) greater use of Total Mixed Ration (TMR) feeding systems. Total mixed rations (TMR) continue to expand in use on dairy farms. A 1993 Hoard's Dairy survey reported that 29.2% of surveyed U.S. dairy farms had adopted this system of feeding dairy cows. A 1991 Illinois dairy survey found 26% of Illinois dairy farmers utilized TMR rations with 300 kg more milk per cow compared to other feeding systems. The American type of operation (North and South America) is characterised by large loose-housing operations, total mixed ration feeding (TMR) and relatively many employees. However, dairy farms in Northeast US and parts of Canada differ from the typical American operation. There you find many smaller family farms with either loose-housing or stanchion barns. These operations are quite similar to the European type, which is characterised by relatively small operations where each cow is fed and treated individually.

Genetics

The Golden Age of Holstein breeding occurred during the last 50 years, greatly helped latterly by embryo transfer techniques, which permitted a huge multiplication of bulls entering progeny testing out of elite, bull-mother cows.

Osborndale Ivanhoe b. 1952, brought stature, angularity, good udder conformation and feet and leg conformation, but his daughters lacked depth and strength. His descendants included:

(i) *Round Oak Rag Apple Elevation* b. 1965, another top class bull. "Father" to over 70,000 Holstein cattle, with descendants numbering over 5 million, Elevation was named Bull of the Century by Holstein International Association in 1999. *Elevation* was the result of a cross of *Tidy Burke Elevation* being used on one of the best ever *Ivanhoe* daughters, *Round Oak Ivanhoe Eve*. He was unsurpassed at the time for type and production.

(ii) *Penstate Ivanhoe Star* b. 1963, sired daughters with similar stature and dairiness as the *Ivanhoes* but with higher production. He also notably sired *Carlin-M Ivanhoe Bell*, the great production bull of the 80s, known also for good udders, feet and legs.

(iii) *Hilltop Apollo Ivanhoe* b. 1960 ,sire of *Whittier Farms Apollo Rocket* b. 1967, the highest milk production bull of the 70s; and *Wayne Spring Fond Apollo* b. 1970, the first bull ever to have a milk transmission index of over 2,000 M *and* have a positive type index. *Wayne* had a very famous daughter, *To-Mar Wayne Hay*, who was dam of the great *To-Mar Blackstar* b. 1983.

Pawnee Farm Arlinda Chief b. 1962, was descended from *ABC Reflection Sovreign* b. 1946, had a huge effect on modern Holsteins. He sired powerful, high production cows with high fat and protein rates although udder conformation was a weak point. He also gave rise to other famous bulls such as:

(i) *SWD Valiant* b. 1973, who in turn sired such popular bull sires as *Applenotch Valiant Calypso* b. 1980, *Mel-Est Valiant Iros Melvin* b. 1980, *Tesk-Holm Valiant Rockie* b. 1981, *Exranco Thor* b. 1981 *Hanoverhill Inspiration* b.

1981 and *Utag Valiant Fancy Paul* b.1982. Valiant was the sire of the famous cow *Walkup Valiant Lu Ella*, the dam of *Rothrock Tradition Leadman* b. 1985.

(ii) *Glendell Arlinda Chief* b. 1968, one of the top production sires of the 70s. He was noted as being sire of *Arlinda Rotate* b. 1975, who in turn sired *Arlinda Melwood* b. 1982,*Juniper Rotate Jed* b. 1987, and *To-Mar Wister* b.1985.

(iii) *Milu Betty Ivanhoe Chief* b. 1969, sire of *Cal-Clark Board Chairman b. 1976*. *Chairman* sired over 50,000 daughters and was a good correcting bull,as well as providing an alternative source of genes than the narrow base then in use. Another influential bull bringing in "outside" genes was *Whittier-Farms Ned Boy* b. 1979. *Ned Boy* sired *Singing-Brook NB Mascot* b. 1986, a massive production bull of the 90s.

(iv) *Walkway Chief Mark* b. 1979. He sired *Bushy-Park Chesapeake*, *Donnandale Skychief* b. 1986,*Mark CJ Gilbrook Grand* b. 1986,*Lutz-Meadows Mark Malloy* b. 1986, *Paradise-R Roebuck*b. 1986 and *Highlight Mr Mark Cinder* b. 1986. *Mark* was a great udder improver who was unfortunately poor for feet and leg conformation. Many good bulls were bred from his daughters, especially sons of *Blackstar*.

Elevation sired:

(i) *Sweet Haven Tradition* b. 1974, one of the best udder improvers who also improved feet and legs. As well as being the sire of *Rothrock Tradition Leadman* b. 1985, he also sired *Bis-May Tradition Cleitus* b. 1981, and *United Nick* b. 1985.

(ii) *Rockalli Son of Bova* b. 1974. This milk yield improver was the first ever bull born from embryo transfer. He sired *Osdell-Endeavour Bova Cubby* b. 1985.

(iii) *Hanoverhill Starbuck* b. 1979. Arguably Canada's ticket to the Holstein big time, *Starbuck* was not only sire to many of the world's most influential cows, he was sire of *Madawaska Aerostar* (*Startmore Rudolph*, *Oliveholme Aeroline* and *Maughlin Storm* and their descendants), *A Ronnybrook Prelude* (*Duncan Progress*, *Carol Prelude Mtoto* and *Comestar Outside*), *Hanoverhill Raider* (*Comestar Lee*) , *Marcrest Encore*(*Regancrest Dundee and Indianhead Encounter*) and *Duregal Astre Starbuck* (*Roybrook Leduc Dusty*, *STBVQ Rubens* and *Tcet Lyster*). In the US he was leading sire of All-Americans for 7 consecutive years (1987-1993). Starbuck facts:

(a) Weight : 1173 kg (2580 lb)

(b) Size : 187 cm (73 1/2 in.)

(c) 36,976 daughters classified in Canada : 70 % GP & Better (326 Ex., 6812 VG et 18 963 GP)

(d) More than 200,000 daughters around the world

(e) 209 proven sons and 406 proven grandsons

(f) 130 All-Canadian and 82 All-American nominations

(g) 685 000 doses of semen sold in 45 countries

Chief and *Elevation* accounted for nearly 25% of genes segregating Holsteins in 1990, so heavy was the use of their genes. This led to concerns regarding inbreeding and narrowing of the genetic base.

Other important sires:

- *Burkgov Inka de Kol* b. 1948
- *Whirlhill Kingpin* b. 1959
- *No-Na-Me Fond Matt* b. 1960
- *Sunnyside Standout-Twin* b. 1962
- *Roybrook Telstar* b. 1963
- *Paclamar Bootmaker* b. 1963
- *Paclamar Astronaut* b. 1964
- *Wapa Arlinda Conductor* b. 1970
- *A Puget-Sound Sheik* b. 1972
- *Marshfield Elevation Tony* b. 1972
- *Hanover-hill Triple-Threat-Red*b. 1972
- *Straight-Pine Elevation Pete* b. 1973
- *Glenafton Enhancer* b. 1977
- *Thonyma Secret* b. 1980
- *To-Mar Blackstar* b. 1983; the leading sire of cows over 50,000 lb of milk.
- *Emprise Bell Elton" b. 1983*
- *Southwind Bell of Bar-lee* b. 1984
- *Maizefield Bellwood* b. 1989
- *MJR Blackstar Emory* b. 1989 said to be the best son of Blackstar.

Cloning

STARBUCK II, clone of the famous CIAQ sire Hanoverhill Starbuck, was born on 7 September 2000 in Saint-Hyacinthe. The clone is a result of the combined efforts of CIAQ, L'Alliance Boviteq inc. and the Faculté de médecine vétérinaire de l'Université de Montréal. The cloned calf was born 21 years and 5 months after Starbuck's own birth date. The calf weighed 54.2 kg at birth and showed the same vital signs as calves produced from regular A.I. or E. T. Starbuck II is derived from frozen fibroblast cells, recovered one month before the death of Starbuck. Huge controversy in the Uk in january 2007 linked Cloning Company Smiddiehill and Humphreston Farm owned by father and son team Micheal and Oliver Eaton with a calf that was cloned from a cow in Canada. Despite their futile efforts to block the farm from view of the press, news cameras broadcast this as breaking news among many of the countries top news stations. Since then there have been rumors that this calf had been put down to protect the owners from invasions of the press 27

British Friesian Cattle

While interest in increasing production through indexing and lifetime profit scores saw a huge increase in Holstein bloodlines in the UK, proponents of the traditional British Friesian did not see things that way and maintain that these criteria do not reflect the true profitability or the production of the Friesian cow. Friesian breeders say that modern conditions in the U.K., similar to the 50's through to the 80's, with low milk price and the need for extensive, low-cost systems for many farmers, may ultimately cause producers to re-examine the attributes of the British Friesian. This animal came to dominate the U.K. dairy cow population during these years, with exports of stock and semen to many countries throughout the world. Although the idea of "dual-purpose" animals has arguably become outmoded, the fact remains that the Friesian is eminently suitable for many farms, particularly where grazing is a main feature of the system. Proponents argue that Friesians last for more lactations through more robust conformation, thus spreading depreciation costs. There is the added advantage of income from the male calf, which can be placed into barley beef systems (finishing from eleven months) or steers taken on to finish at two years, on a cheap system of grass and silage. Very respectable grades can be obtained, commensurate with beef breeds, thereby providing extra income for the farm. Such extensive, low cost systems may infer lower veterinary costs, through good fertility, resistance to lameness and a tendency to higher protein percentage and, therefore, higher milk price. An 800 kg Holstein has a higher daily maintenance energy requirement than the 650kg Friesian. Friesians have also been disadvantaged through the comparison of their type to a Holstein base. It is suggested that a separate "index" be composed to greater reflect the aspects of maintenance for body weight, protein percentage, longevity and calf value. National Milk Records

(NMR) figures suggest that highest yields are achieved between the fifth and seventh lactations; if so, this is particularly so for Friesians, with a greater lift for mature cows, and sustained over more lactations. However, production index only takes the first five lactations into account. British Friesian breeding has certainly not stood still and, through studied evaluation, substantial gains in yield have been achieved without the loss of type.

Friesians were imported into the east coast ports of England and Scotland, from the lush pastures of North Holland, during the 1800s until live cattle importations were stopped in 1892, as a precaution against endemic foot and mouth disease on the Continent. So few in number were they, that they were not included in the 1908 census. However, in 1909, the Society was formed as the British Holstein Cattle Society, soon to be changed to British Holstein Friesian Society and, by 1918, to the British Friesian Cattle Society. The Livestock Journal of 1900 referred to both the "exceptionally good" and "remarkably inferior" Dutch cattle. The Dutch cow was also considered to require more quality fodder and need more looking after than some English cattle that could easily be out-wintered. It is interesting to note that, in an era of agricultural depression, Breed Societies had flourished as a valuable export trade developed for traditional British Breeds of cattle. At the end of 1912, the herd book noted 1,000 males and 6,000 females, the stock which originally formed the foundation of the breed in England and Scotland. Entry from then until 1921, when grading up was introduced, was by pedigree only. No other Friesian cattle were imported until the official importation of 1914, which included several near descendants of the renowned dairy bull *Ceres 4497 F.R.S.* These cattle were successful in establishing the Friesian as an eminent long living dairy breed in Britain. This role was continued in the 1922 importation from South Africa through *Terling Marthus* and *Terling Collona*, who were also near descendants of *Ceres 4497*. The 1936 importation from Holland introduced a more dual purpose type of animal, the Dutch having moved away from the Ceres line in the meantime. The 1950 importation has a lesser influence on the breed today than the previous importations, although various *Adema* sons were used successfully in some herds. The Friesian enjoyed great expansion in the 1950s, through to the 80s, until the Holsteinization of the national herd in the 1990s; a trend which is being questioned by some commercial dairy farmers in the harsh dairying climate that prevails today, with the need to exploit grazing potential to the full. Friesian semen is once again being exported to countries with grass based systems of milk production. The modern Friesian is pre-eminently a grazing animal, well able to sustain itself over many lactations, on both low lying and upland grassland, being developed by selective breeding over the last 100 years. Some outstanding examples of the breed have 12 to 15 lactations to their credit, emphasising their inherent natural fecundity. In response to demand, protein percentages have been raised across the breed and herd protein levels of 3.4% to 3.5% are not uncommon. Whilst the British Friesian is first and foremost a dairy breed, giving high lifetime yields of quality milk from home produced feed, by a happy co-incidence surplus male animals are highly regarded, as producers of high quality lean meat, whether crossed with a beef breed or not. Beef cross heifers have long been sought after as the ideal suckler cow replacement. Although understanding the need to change the Society's name to include the word Holstein in 1988, British Friesian enthusiasts are less than happy now that the word Friesian has been removed from the Society's name. With the history of the breed spanning 100 years, the British Friesian cow is continuing to prove her worth. The general robustness and proven fertility provide an ideal black and white cross for Holstein breeders seeking these attributes. The disposal of male black and white calves continues to receive media attention, and would appear to be a waste of a valuable resource. One of the great strengths of the British Friesian is the ability of the male calf to finish and grade satisfactorily, either in intensive systems, or as steers, extensively. This latter system may become increasingly popular due to the prohibitive increase in grain prices. The robustness of the British Friesian and its suitability to grazing and forage systems is well known. The breed:-

- Calves more frequently.
- Calves more often in their lifetime.
- Needs less replacements.
- Provides valuable male calves.
- Has lower cell counts
- Has higher fat and protein percent.

Red and White Holsteins

The expression of red colour replacing the black in Holsteins is a function of a recessive gene. The genetic combinations possible with simple dominance can be expressed by a diagram called a Punnett square. One parent's alleles are listed across the top and the other parent's alleles are listed down the left side. The interior squares represent possible offspring, in the ratio of their statistical probability. In the case of red colour, *R* represents the *dominant* non red (or standard black and white) coloured allele and *r* the *recessive* red coloured allele. If both parents are non red coloured, and heterozygous (Rr) i.e. red *carriers* but outwardly normal black and white, the Punnett square for their offspring.

From this, we can see that if two outwardly black and white carriers are bred together, the chances of the calf being red are 1 in 4, or 25%. In the *RR* and *Rr* cases, the offspring will be black and white due to the dominant *R*. Only in the *rr* case is there expression of the recessive red coloured phenotype.

If a red bull, *rr* is used on a carrier cow, *Rr*, then the Punnett square will give: There is a 50% chance that the calf will be red. Only if a red bull and a red cow are bred will the resulting calf be sure to be red. If a red bull is used on a non-carrier cow, i.e. the black and white majority of animals. There is no chance of a red calf being born. There is, however, a 100% certainty that the calf will be a red carrier, *Rr*.

History

13th C: Early records show that cattle of "broken" colours entered the Netherlands from Central Europe. Most foundation animals in the US were imported between 1869 and 1885. A group of early breeders decreed that animals of any colour other than black and white would not be accepted in the herd book, and that the breed would be known as Holsteins. There were objections, saying that quality and not colour should be the aim, and that the cattle should be called *Dutch*, rather than *Holsteins*. Only a small number of carriers were identified over the hundred-year span from the early importations until they were accepted into the Canadian and American herd books in 1969 and 1970 respectively. Most of the early accounts of red calves being born to black and white parents were never documented. A few stories of "reds" born to elite parents persist over time, as there is a tendency to credit the ancestor with the highest (closest) relationship to a red-carrier animal as the one that transmitted the trait, whereas sometimes it is the other parental line that has passed it on, even though the ancestor responsible may have entered the pedigree several generations earlier. In 1952 there was a sire in an artificial insemination (AI) unit in the US that was a carrier of red coat colour. Although the AI unit reported the condition and advised breeders as to its mode of inheritance, almost a third of the breeding unit's Holstein inseminations that year were to that red-carrier bull. That year, American AI units had used 67 red factor bulls that had sired 8250 registered progeny. However, in spite of this, any change to the colour marking rules was rejected. The Red and White Dairy Cattle Association (RWDCA) began registry procedures in 1964 in the United States. Its first members were Milking Shorthorn breeders, who wanted a dairy registry for their cattle they had bred in prior years, including some red and white Holsteins. The name was changed to the Red and White Dairy Cattle Association in 1966. When Milking Shorthorn breeders were looking for potential outcrosses to improve milk production, red and white Holsteins came into the picture, since the red colour factor is the same for both breeds. The RWDCA had adopted an "open herd book" policy, and the Red and White Holstein became the major player. The red trait was thus able to survive the attempts to eradicate it that came from all sides of the Holstein industry. It was inevitable that even when a red calf was killed or sent to a grade herd, the herd owner rarely did anything to remove the dam from his herd and only hoped that she would not have another red calf. Many red calves, born in both countries prior to the

1970s, were quietly disposed of, with a view to preserving the acceptance of their elite pedigrees. Also, thousands of Holsteins were imported from Canada each year, and many were carriers. More than 14,000 Holsteins were exported to the United States in 1964 and again in 1965. This was at a time when both countries were debating the "red question." While the United States was trying to eliminate the red trait, the Canadian imports simply counter-balanced the US effort to reduce its incidence. Canada's number one *red carrier* sire in the 1940s was *A B C Reflection Sovereign*. His sons and grandsons in the 1950s and 60s spread the red gene throughout Canada and increased its frequency in the United States. Three of the biggest names siring Red and Whites in the United States were *Rosafe Citation R*, *Roeland Reflection Sovereign*, and *Chambric A B C*. The red trait was readily available in Canadian Holstein genetics. Early on, there was criticism of the policy of the Canadian AI units to remove bulls found to carry red. A number of superior bulls were slaughtered or exported. The studs were simply supporting the Canadian policy to prevent the intensification of the red recessive in the breed. The phrase *carries the red factor* had to be included in the description, and excessive promotion of unproven red factor bulls was discouraged. They later added the aim of permitting intelligent breeders to utilize any red carrier sire that had an outstanding proof for production and type. It became obvious that AI was the primary way of finding out which bulls were red carriers.

Prior to AI, few red carrier sires were uncovered because their service was limited to one or a few herds. Such herds often had no carrier females, and there was only a twenty-five percent chance that a carrier bull mated to a carrier female would produce a red calf. If a red and white calf was dropped, it was often concealed and quietly removed from the herd. In 1964, the Netherlands Herd Book Society indicated a breakdown of 71% Black and White Friesian and 28% Red and Whites. A herd book that accepted Red and Whites had already been established in the United States. A separate herd book for Canadian Red and Whites was then established, following which Red and Whites became acceptable to the major Canadian (export) markets. The sales ring began to establish interest in the new breed. The US Holstein-Friesian Association and its membership worked diligently from its early days until 1970 to eliminate the red trait from the registered population. However, once the door was open, red and whites began to appear in some of the more elite herds. The rush to get the best of Canadian breeding even prior to the opening of the herd book brought red calves to many dairymen who had never even seen one. Canadian Red and Whites became eligible for registration in the herd book on July 1, 1969. This was done through an alternate registry. Red and Whites were to be listed with the suffix –RED and Black and Whites with ineligible markings would be registered with the suffix –ALT. Both groups and their progeny would be listed only in the Alternate book and the suffixes had to be part of the name. In the Canadian herd books, all "Alt" and "Red" animals were listed in the regular herd book in registration number order and were identified with an *A* in front of their number. The *Alternates* were separate in name only. The *A* in front of the registration number was discontinued in 1976 and the "Alt" suffix was dropped in 1980, but *–Red* was continued. It did not bar the registration of animals whose hair turned from red to black. The US Holstein Association decided not to have a separate herd book for red and whites and *off-colour* animals. The suffixes of "Red" and –OC would be used, and numbering would be consecutive. The first red and white Holsteins were recorded with an *R* in front of their number. There were 212 males and 1191 females recorded in the initial group of *red* registrations. Red and Whites registered in the Canadian herd book numbered 281 in 1969 and 243 in 1970. An American Breeders Service (ABS) ad in the Canadian Holstein Journal in 1974 on *Hanover-Hill Triple Threat* mentioned one of several colour variants that were not *true red*. Its existence was undoubtedly common knowledge among breeders in both countries, but up until that time it had not been mentioned in print. Calves were born red and white and registered as such, but over the first 6 months of age turned black or mostly black with some reddish hairs down the backline, around the muzzle and at the poll. The hair coat colour change became known as *Black/Red* (B/R) and sometimes as *Telstar/Red,* since the condition appeared in calves sired by *Roybrook Telstar*. *Telstar* was the sire of *Triple Threat*, but nothing about this had hitherto been in print about *Telstar* who was by then over 10 years old. *Black/Reds* were often discriminated against when sold and were barred from Red and White-sponsored shows. In 1984 Holstein Canada considered recoding B/R bulls that had always been coded simply as *red carriers*, a designation that was not acceptable to all buyers.

The breed agreed to change after checking with other breed associations and with the AI Industry. In 1987, Holstein Canada and the Canadian A I Industry modified their coding procedures to distinguish between *Black/Red* and *true red* colour patterns for bulls. Holstein Canada dropped the suffix *Red* as a part of the name in 1990, but continue to carry it as part of the *birth date and other codes* field. The easiest pathway to trace when looking at the migration of the red trait in Canada is to work back through the ancestry of ABC's sire, *Montvic Rag Apple Sovereign*. Sovereign was sired by *Emperor of Mount Victoria-RC*. In Canada the sire *Agro-Acres Marquis Ned-RC* was born in 1964 and was extremely popular in Canada, siring a number of well known Red and White females that were imported by US breeders. Among them were *Blue-Haven Rose Ned-Red*. A popular red-carrier sire was *Roybrook Telstar*. *Hanover-Hill Triple Threat* had a huge influence on Holstein breeding in both the United States and Canada during the 1970s and 80s. He was considered a good mate for the daughters of *Elevation*, *Bootmaker*, and other highly-rated sires from both countries, despite the fact that he carried the Black/ Red trait. *Glenafton Enhancer* was dominant in the 1980s as a sire of sons for AI units. He was a son of *Roybrook Starlite*. Canadian headliner herds such as *Romandale*, *Rosafe*, *Springbank*, *Rockwood*, *Spring Farm*, and *Glenafton*, with their many sales and interchange of breeding stock, were certain to have the red trait whether they wanted it or not. Heavy use of *Montvic Rag Apple Sovereign* and *A B C Reflection Sovereign* and their sons and grandsons, along with the *Palmyra* line, guaranteed it. Some early RWDCA sires, including those with *Duallyn* and *Hayssen* prefixes, made their way to Canada. However, the two US bulls that had the most influence on Canadian breeding of Red and Whites were *Citation R Maple* and *Hanover Hill Triple Threat*. The Red and White breed has come a long way since it was founded. Great strides have been made in both production and type, and the best of them can go head-to-head with their black and white counterparts. However, the breed does need more herds that are committed to breeding cattle that will produce consistent results in future generations.

THE CASE OF UBRE BLANCA

Ubre Blanca (c. 1972 - 1985) was the name given to a cow in Cuba known for its prodigious milk production. The cow, along with the "cordon de la Habana" coffee plantations, the Voisin pasture system, and the microjet irrigation system, symbolizes Fidel Castro's efforts to modernize Cuba's agricultural economy. The Spanish phrase *ubre blanca* translates to the English phrase *white udder*.

Ubre Blanca produced 109.5 liters (241 pounds) of milk on a single day in January 1982 – more than four times a typical cow's production. The cow also produced 24,268.9 liters of milk (about 55,090 pounds at 2.27 pounds per liter) in 305 days (one lactation period) ending in February 1982. Both feats were recognized by Guinness World Records as world records. The cow was a cross between a Holstein bull and a zebu. The current annual production record is 75,275 pounds, set by LA-Foster Blackstar Lucy in 1998 at the LaFoster Dairy in Cleveland, North Carolina.

Castro referred to Ubre Blanca's prodigious output in speeches as evidence of communism's superior breeding skills, and the cow's achievements were often printed in Cuba's government-controlled newspapers. To many Cubans, Ubre Blanca evokes memories of the era before the so-called "Special Period" – the economic collapse that followed the demise of the Soviet Union, Cuba's main benefactor, beginning in 1989. Cuba's cattle herd diminished from 10 million head in the 1980s to less than half of that today, most starving to death for lack of feed.

In 1985, Ubre Blanca was euthanized at about the age of 13 (exact age unknown). The cow's death was commemorated by Communist Party newspaper *Granma* with a full obituary and eulogy. Taxidermists stuffed Ubre Blanca and put the body in a climate-controlled glass case at the entrance to the National Cattle Health Center 10 miles outside Havana, where it still remains. Ubre Blanca was honored by her hometown of Nueva Gerona, which erected a marble statue in memory of the cow. Since the cow's death, Cuban scientists have unsuccessfully attempted to clone Ubre Blanca using frozen tissue samples.

MANUAL MILKING MACHINE: A CASE STUDY FROM INDIAN STOCK

Raghava Gowda hails from Pallathadka, Murulya village in the Sulya taluka of South Canara district in Karnataka. A schoolteacher by profession, he is 52-years-old and is brimming with ideas that would help solve various problems faced by members of his community. His father was also a schoolteacher. He acknowledges the support provided by his family throughout the process of innovation. He attributes his success to willingness to work hard, kindness and truthfulness to his parents. These virtues according to him have contributed a lot to his growth as an individual. Raghava is respected and held in high esteem, and his innovations are by now legendary in his village.

Genesis

Finding skilled labour for milking a small herd of cows is a problem often faced by a small-time farmer. Adding to this is the fact that milking by hand is not considered healthy or hygienic anymore. But milking using a machine is a luxury which only a large farm or dairy house can boast of, calling for a huge investment in power supply and machinery. All of these set school teacher-farmer, Raghava Gowda, thinking very hard about developing an alternate means of mechanized milking which would be affordable to all farmers.

A keen observer and a fast learner, Gowda observed the working of the Gutter spray pump, used for spraying pesticides. He came up with the idea of using PVC teat cups and a plastic pipe on the Gutter pump. Thus he developed the first machine and began experimenting on his own farm. But for this the teat cup had to be moulded according to the size of the teat and he had to heat the PVC pipe. But milking using this device proved to be quite painful for the cow as an excessive vacuum was created. To solve this problem, a vacuum container was adapted. With this addition, milking could be done from four teats. Experimenting with it further he switched over to a foot press. Then he fitted the vacuum pump on a four legged fabricated frame which later was replaced by a three-legged frame to provide stability. A vacuum level gauge was provided to know the level of the vacuum generated. To reduce the strain on the operator he then tried out a gear and wheel vacuum pump set-up. A stainless steel can, lid and stainless steel junctions for the teat adaptors were also introduced. Further trials resulted in reducing the milking effort by adopting alternate pulsing for each set of two teats and by reducing stroke and diameter of vacuum pump. It took him four years of hard work and 15 models to finally arrive at the successful milking machine, which costs less than Rs.7000/- and at last Raghava is satisfied that the machine is problem free.

Raghava has developed an elaborate, refined milking machine that can milk cows and buffaloes using a set of reciprocating vacuum pumps with a vacuum gauge, a suction assembly unit and an air bubble free well gasketed milk canister to receive the milk. The suction assembly has two sub-assemblies with a set of nipples and stainless steel plate on one side and transparent conduit pipes and a regulator valve on the other.

Each of the sub-assemblies are taken apart for cleaning before and after every milking operation.The udder and teats of the cow are also washed with cold water and wiped using cotton cloth. The hand lever attached to the pumping unit is cranked till a vacuum of 200-250 inches is created in the suction–nipple unit. This is attached to the udders and the milk gets deposited in the receiving canister via the transparent conduit pipes. When air bubbles flow along with the milk, the operation is stopped and on gently pressing the top of the teats, the teats get released. The control valve and vacuum pressure gauge, located on the main pumping unit, control the suction circuit and the milk receiving canister via the transparent conduit pumps. The four nipple suction configuration can alternate pulsations between two sets for the operation for facilitating milk flow and reducing milking time.

Good publicity in the press helped and soon his phone was ringing off the hook, with frantic and urgent enquiries about the milking machine. This is now widely used and appreciated by small-scale diary farmers in his area. As of date he has sold 170 machines and has another ten orders in hand from states like Andhra Pradesh and Tamil Nadu also.

He has got good responses from universities, banks & other government agencies. The University of Agricultural Sciences, Dharwad, had given him a certificate during the Krishimela. The Syndicate bank, Hiriyadka, also has given a certificate in appreciation of his innovation. But so far he has not been approached with any business enquiries or for licensing the technology.

Advantage

The advantages of this milking machine are manifold. It is easy to operate, costs low, saves time as it milks 1.5 litre to 2 litres per minute. It is also very hygienic and energy-conserving as electricity is not required. All the milk from the udder can be removed. The machine is also easily adaptable and gives a suckling feeling to the cow and avoids pain in the udder as well as leakage of milk.

A Life-Long Innovator

Raghava had started innovating at the tender age of ten when he fashioned a Spray Gun out of bamboo and a noodle-press which was followed a couple of years later by a bamboo pump. His resume of innovations also include a rotating toy using heat and light, a Gobar gas plant using plastic and PVC spares. Improving his innovations has always been a priority for him as he devised a method to improve the productivity of gobar gas using two drums and he also ran an engine using the gas. He has also won an award for creating an artificial Queen Bee and his interview was broadcast on AIR, Mangalore.

Experiments at his Farm

He has developed a sprayer which can machine spray areca trees. This sprayer uses a gutter pump with a control mechanism in hand and costs about Rs.350. It is easy to use, assemble and dismantle and he has been using it for the last 12 years. He claims the method is cost effective and efficient and that the sprayer can spray in a 360° direction and to a distance of 20ft that can cover around 20 areca plants.

He has developed a fodder cutter that slices even hard fodder material like coconut and Areca Palms easily. He claims that the present fodder cutter he is using is unique and more efficient and safer to handle and that he developed it six years back. Now he has developed a new model with some modifications and is selling it at a cost of Rs.750.

Raghava Gowda also practises many alternative as well as resource saving technologies as part of his daily life. He uses solar energy to light the lamps in his house. For this he has installed two solar panels on the rooftop and charges 24 volts batteries. He practices multi-crop farming and has 15acres of well irrigated farm with lust greenery. His farm consists of a variety of plans of both horticultural and medicinal importance and he claims that for the past 14 years he hasn't used any chemical fertilizers in his farm. He uses the waste slurry as manure for the plants in his farm. An advocate of rainwater harvesting, he effectively uses the run-off water to charge his own bore wells.

Future Plans and Dreams...

Raghava's aim is to make his milking machine available to dairy farmers at a cheaper rate. But he is quite sure that he does not want to sit back and enjoy the success of his "Milk Master", but would prefer to dedicate his time to developing the following innovations which are currently in a conceptual form- easy methods of bee-keeping, a gas water heater and an automatic water dispenser for a cattle shed.

SHAREMILKING

Sharemilking is a form of sharefarming applied to the dairy industry. The application of this model of farming is particularly common in New Zealand. Typically sharemilkers own their own cows, and will often take the herd with them when shifting between properties. The model is not exploitative, and over time, sharemilkers often slowly buy out the landholder, or alternatively use it as a method to save for their own property. This practice is useful for dairymen anywhere, who do not wish the burdens of owning their own land and allows them to focus their investment in livestock and equipment. The sharemilking also provides income for former dairy farms that have given up their herds, by providing them with an income from rental of fields, pastures and barns.

APPENDIX

LIST OF BREEDS OF CATTLE

The following is a list of breeds of cattle. Over 800 breeds of cattle are recognized worldwide, adapted both for local climate and for specialized uses. Breeds fall into two main types, regarded as either two closely related species, or two subspecies of one species. *Bos indicus* (or *Bos taurus indicus*) cattle, also called zebu, are adapted to hot climates. *Bos taurus* (or *Bos taurus taurus*) are the typical cattle of Europe, north-eastern Asia, and parts of Africa – they are referred to in this list as "taurine" cattle, and many are adapted to cooler climates. *Taurus/indicus* hybrids are widely bred in many warmer regions, combining useful characteristics of both the ancestral types. In some parts of the world further species of cattle are found (both as wild and domesticated animals), and some of these are related so closely to taurine and indicus cattle that interspecies hybrids have been bred. Examples include the Dwarf Lulu cattle of the mountains of Nepal with yak blood, the Beefalo of North America with bison genetics, the Selembu breed of India and Bhutan with gayal genetics. The Madura breed of Indonesia may have banteng in its parentage. The Dzo of Nepal is an infertile cattle-yak crossing which is bred for agricultural work. Like the mule which is also infertile, they have to be continually bred from the parent species.

Multipurpose Breeds

These breeds are bred for multiple purposes.

Taurine Cattle (Bos taurus)

- Abondance (*Dairy/beef*)
- Adamawa (*Dairy/beef/draught*)

- Ala Tau (*Dairy / beef*)
- Albanian (*Dairy / draught*)
- Anatolian Black (*Dairy / draught*)
- Armorican (*Dairy / beef*)
- Arouquesa Cattle (*Beef / draught*)
- Aure et Saint-Girons (*Dairy / beef / draught*)
- Barrosã Cattle (*Beef / draught*)
- Belgian Blue (*Dairy / beef*)
- Blonde d'Aquitaine (*Dairy / beef*)
- Braunvieh (*Dairy / beef*)
- British Friesian (*Dairy / beef*)
- Buša cattle (*Dairy / beef / draught*)
- Canadienne (*Dairy / beef*)
- Carinthian Blondvieh (*Dairy / beef / draught*)
- Charolais (*Beef*)
- Droughtmaster (*Dairy / beef*)
- Dexter (*Dairy / beef*)
- Eastern Finncattle (*Dairy / beef*)
- Ennstal Mountain Pied Cattle (*Beef / draught*)
- Fleckvieh (*Dairy / draught*)
- French Simmental (*Dairy / beef*)
- Gelbvieh (*Dairy / beef / draught*)
- German Red Pied (German Rotbunte) (*Dairy / beef*)
- Gloucester (*Dairy / beef / draught*)
- Harz Red mountain cattle (*Dairy / beef / draught*)
- Herens (*Dairy / beef / sport*)
- Hinterwald Cattle (*Dairy / beef*)
- Holando-Argentino (*Dairy / beef*)
- Hybridmaster (*Dairy / beef*)
- Irish Moiled (*Dairy / beef*)
- Jutland cattle (*Dairy / beef*)

- Kostroma Cattle (*Dairy / beef*)
- Kurgan (*Dairy / beef*)
- Lourdais (*Dairy / beef / draught*)
- Meuse-Rhine-Issel (*Dairy / beef*)
- Milking Devon (*Dairy / beef / draught*)
- Montbeliard Cattle (*Dairy / beef*)
- Murboden Cattle (*Dairy / beef / draught*)
- Normande Cattle (*Dairy / beef*)
- Northern Finncattle (*Dairy / beef*)
- Pembroke cattle (*Dairy / beef*)
- Pie Rouge des Plaines or French Lowland Red Pied (*Dairy / beef*)
- Piedmontese (*Dairy / beef*)
- Pinzgau Cattle (*Dairy / beef*)
- Randall (*Dairy / beef / draught*)
- Red Poll (*Dairy / beef*)
- Reina (*Dairy / beef*)
- Salers (*Beef / draught*)
- Shorthorn (*Dairy / beef*)
- Simmental (*Dairy / beef / draught*)
- South Devon (*Dairy / beef*)
- Tarentaise (*Dairy / beef*)
- Tyrolese Grey Cattle (*Dairy / beef*)
- Vorderwald Cattle (*Dairy / beef*)
- Welsh Black (*Dairy / beef*)
- Western Finncattle (*Dairy / beef*)
- Western Red Polled (*Dairy / Beef*)

***Zebu* (Bos indicus)**

- Gir (*Dairy / beef*)
- Guzerat and Kankrej (*Beef / draught*)
- Tharparkar (*Dairy / draught*)

Hybrids

- Madura cattle: Banteng/Zebu hybrid (Beef/draught/dairy/racing)
- Nguni: taurine/zebu hybrid (*Dairy / beef*)
- Selembu: cattle/gayal hybrid (*Dairy / beef*)

Dairy Breeds

These breeds are primarily used to produce milk.

Taurine Cattle (Bos taurus)

- Abigar
- Agerolese
- Alderney
- Angeln
- Aulie-Ata
- Australian Friesian Sahiwal
- Australian Milking Zebu
- Ayrshire
- Belgian Red Cattle
- Blaarkop
- Black Pied Dairy Cattle
- Brown Swiss
- Chinese black pied
- Danish Red and variants
- Dutch Belted (Lakenvelder)
- Évolène Cattle
- Holstein (Friesian)
- German Black Pied Cattle
- Guernsey
- Icelandic
- Illawarra cattle
- Jersey
- Kerry cattle

- Lineback
- Milking Shorthorn
- Murnau-Werdenfels Cattle
- Red Holstein
- Swedish Red Cattle
- Tux Cattle

***Zebu* (Bos indicus)**

- Butana and Kenana
- Red Sindhi
- Sahiwal
- Vechoor cow

Beef-producing Breeds

This is a list of cattle breeds which are primarily used for beef production.

***Taurine Cattle* (Bos taurus)**

- Aberdeen Angus
- Adaptaur
- Africangus
- Amerifax
- Angus
- Aubrac Cattle
- Asturian Mountain
- Australian Lowline
- Barzona
- Beef Shorthorn
- Beefmaker
- Belgian Blue
- Belmont Red
- Belted Galloway
- Balancer
- Bazadais

- Black Baldy
- Black Hereford
- Blonde d'Aquitaine
- Blue Albion
- Blue Grey
- Bonsmara
- Braford
- British White
- BueLingo
- Charolais
- Chianina
- Droughtmaster
- Galloway
- German Angus Cattle
- Glan Cattle
- Greyman Cattle
- Hereford
- Highland Cattle
- Hungarian Grey Cattle
- Limousin
- Lincoln Red
- Longhorn
- Luing
- Maine-Anjou cattle
- Murray Grey
- Mandalong Special
- Original Braunvieh
- North Devon cattle (Devon, Devon Ruby, Red Ruby)
- Poll Devon
- Polled Hereford
- Pustertal Pied Cattle

- Queensland Miniature Borans
- Red Angus
- Romagnola
- Senepol
- Simford
- Square Meater
- South Devon
- Sussex cattle
- Tasmanian Grey
- Texas Longhorn
- Traditional Hereford
- Vaynol cattle
- Wagyu
- Wangus
- Whitebred Shorthorn
- White Park

Zebu (*Bos indicus*)

- Boran
- Brahman
- Indo-Brazilian
- Nelore
- White Fulani

Hybrids

- Afrikaner: taurine/zebu hybrid
- Australian Braford hybrid
- Australian Brangus hybrid
- Australian Charbray hybrid
- Beefalo: cattle/American Bison hybrid
- Beefmaster: taurine/zebu hybrid
- Brangus: taurine/zebu hybrid
- Canchim: taurine/zebu hybrid

- Red Fulani: taurine/zebu hybrid
- Sanga: taurine/zebu hybrid (A group of breeds)
- Santa Gertrudis cattle: taurine/zebu hybrid
- Simbrah: taurine/zebu hybrid
- Symons Type: taurine/zebu hybrid
- Tswana: taurine/zebu hybrid
- Tuli: taurine/zebu hybrid
- Tulim: taurine/zebu hybrid

Draught Breeds

These breeds are primarily used as beasts of burden.

*Taurine Cattle (***Bos taurus***)*

- Aosta cattle
- Sayaguesa Cattle

*Zebu (***Bos indicus***)*

- Abyssinian Highland Zebu
- Abyssinian Shorthorned Zebu

Hybrids

- Batangas: taurine/zebu hybrid

Other Purposes

These breeds were generated for other purposes than those above.

- African Boran Cattle (small) (*beef/pets*)

*Taurine Cattle (***Bos taurus***)*

- Camargue cattle (*Sport*)
- Corriente (*Sport*)
- Fighting Cattle (*Sport*)
- Heck Cattle (*Scientific experiment*)
- Queensland Miniature Boran (*Beef/pets*)

Hybrids

- Ankole-Watusi: taurine/zebu hybrid (*Status symbol/currency*)
- Cattle/wisent hybrid (*Scientific experiment*)

Feral Cattle

These breeds evolved from domestic cattle that took to the wild again

- Aleutian wild cattle
- Chillingham Cattle
- Enderby Island Cattle
- Ushuaia Wild Cattle

Uncategorised Breeds

- Aceh
- Achham
- Aden
- Afghan
- Alberes
- Alambadi
- Albanian Dwarf
- Alberes
- Albese
- Alentejana
- Alentejana
- Aliad Dinka
- Alistana-Sanabresa
- Allmogekor
- Alur
- American Angus
- American Beef Friesian
- American Brown Swiss
- American
- American White Park
- Amrit Mahal zebu
- Andalusian Black
- Andalusian Blond
- Andalusian Grey

- Angoni
- Ankina
- Apulian Podolian
- Aracena
- Arado
- Argentine Crillo
- Argentine Friesian
- Arsi
- Asturian Valley
- Australian Grey
- Australian Shorthorn
- Australian White
- Austrian Simmental
- Austrian Yellow
- Avetonou
- Avilena
- Avilena-Black Iberian
- Aweil Dinka
- Azaouak
- Azebuado
- Azerbaijan Zebu
- Azores
- Bernese
- Bohus Polled
- Caracu
- Chateaubriand
- Chiangus
- Chinese Central Plains Yellow: taurine/zebu hybrid
- Chinese Northern Yellow
- Commercial
- Danish Black-Pied

- Danish Jersey
- Doela
- Dwarf Lulu: taurine/zebu/yak hybrid
- Eastern Red Polled
- Estonian Native
- Estonian Red
- Finnish Ayrshire
- Finnish Holstein-Friesian
- Fjällnära
- Iberian cattle
- Lampurger
- Latvian Blue
- Latvian Brown
- Lithuanian Black-and-White
- Lithuanian Light Grey
- Lithuanian Red
- Lithuanian White-backed
- Lohani zebu
- Marchigiana
- Miniature
- N'Dama
- Norwegian Red
- Pantaneiro cattle
- Polish Black-and-White
- Red Brangus: taurine/zebu hybrid
- Ringamåla
- Schwyz
- Sided Tröder and Nordland
- Swedish Friesian
- Swedish Mountain
- Swedish Red-and-White

- Swedish Red Polled
- Telemark
- Väne
- Western Fjord

REFERENCES

Atkinson, D.; Watson, C.A. Animal Science. Penicuik, [Scotland]: British Society of Animal Science. Dec 1996. v. 63 (pt.3) p. 353 361.

Bergen, J.A.M. van Studies in Environmental Sciences. Amsterdam ; New York, Elsevier Scientific Publishing Co. 1995. (65B) p. 1123 1126.

Chandler, P. Feedstuffs. Carol Stream, Ill.: Miller Publishing Company. June 12, 1995. v. 67 (24) p. 11 12, 17.

Chen, S.; Cothren, G.M.; DeRamus, H.A.; Langlinais, S.; Huner, J.V.; Malone, R.F. Animal waste and the land water interface. Boca Raton: Lewis Publishers, c1995. p. 197 204.

Dewes, T. J agric sci. Cambridge: Cambridge University Press. Dec 1996. v. 127 (pt.4) p. 501 509.

Frost, J.P. Agricultural Engineer. Silsoe: Institution of Agricultural Engineers. Winter 1995. v. 50 (4) p. 2 4.

Ghaly, A.E. Bioresour technol. Oxford, U.K.: Elsevier Science Limited. Oct 1, 1996. v. 58 (1) p. 61 72. Reaves, R.P.; DuBowy, P.J.; Jones, D.D.; Sutton, A.L. Clean water, clean environment, 21st century team agriculture, working to protect water resources conference proceedings, March 5 8, 1995, Kansas City, Missouri /. St. Joseph, MI: ASAE, c1995. v. 2 p. 179 182.

Henry, G.M.; DeLorenzo, M.A.; Beede, D.K.; Van Horn, H.H.; Moss, C.B.; Boggess, W.G. J dairy sci. Champaign, Ill.: American Dairy Science Association. Mar 1995. v. 78 (3) p. 693 703.

Holloway, M.P.; Bottcher, A.B.; Nordstedt, R.A.; Campbell, K.L. Applied Engineering in Agriculture. St. Joseph, MI: American Society of Agricultural Engineers, 1985 . Mar 1996. v. 12 (2) p. 197 202.

Insam, H.; Amor, K., Renner, M.; Crepaz, C. Microb ecol. New York, N.Y.: Springer Verlag New York Inc. 1996. v. 31 (1) p. 77 87.

Liang, B.C.; Gregorich, E.G.; Schnitzer, M.; Schulten, H.R. Soil Sci Soc Am j. [Madison, Wis.] Soil Science Society of America. Nov/Dec 1996. v. 60 (6) p. 1758 1763.

Liang, B.C.; Gregorich, E.G.; Schnitzer, M.; Voroney, R.P. Biol fertil soils. Berlin ; a Secaucus, N.J.: Springer International, 1985 . 1996. v. 21 (1/2) p. 10 16.

Morse, D.; Guthrie, J.C.; Mutters, R. Journal of Dairy Science. Champaign, Ill.: American Dairy Science Association. Jan 1996. v. 79 (1) p. 149 153.

Murwira, H.K. Trop agric. St. Augustine, Trinidad: The University of the West Indies Press. Oct 1995. v. 72 (4) p. 269 273.

Stante, L.; Cellamare, C.M.; Malaspina, F.; Bortone, G.; Tilche, A. Water Resources. Oxford, U.K.: Elsevier Science Ltd. June 1997. v. 31 (6) p. 1317 1324.

Sweeten, J.M.; Wolfe, M.L.; Chasteen, E.S.; Sanderson, M.; Auvermann, B.A.; Alston, G.D. Animal waste and the land water interface /. Boca Raton: Lewis Publishers, c1995. p. 99 106.

Topp, E.; Tessier, L.; Gregorich, E.G. Can j soil sci. Ottawa: Agricultural Institute of Canada, 1957 . Aug 1996. v. 76 (3) p. 403 409.

Wang, E.; Sparling, E. Am J altern agric. Greenbelt, MD: Henry A. Wallace Institute for Alternative Agriculture. Fall 1995. v. 10 (4) p. 167 172.

7

Dairy Products, Biosynthesis, Grading and Defects: An Analysis

DAIRY PRODUCT

Dairy products are generally defined as foodstuffs produced from milk. They are usually high-energy-yielding food products. A production plant for such processing is called a dairy or a dairy factory. Raw milk for processing generally comes from cows, but occasionally from other mammals such as goats, sheep, water buffalo, yaks, or horses. Dairy products are commonly found in European, Middle Eastern and Indian cuisine, whereas they are almost unknown in East Asian cuisine.

Type of Dairy Products

*Milk, after optional homogenization, pasteurization, in several grades of bacteria *Streptococcus lactis* and *Leuconostoc citrovorum*

- Crème fraîche, slightly fermented cream
- Smetana, Central and Eastern European variety of sour cream
- Clotted cream, thick spoonable cream made by heating
  - o Cultured buttermilk, fermented concentrated (water removed) milk using the same bacteria as sour cream
  - o Milk powder (or powdered milk), produced by removing the water from milk
- Whole milk & buttermilk
- Skim milk
- Cream

- High milk-fat & nutritional powders (for infant formulas)
- Cultured and confectionery powders
    - o Condensed milk, milk which has been concentrated by evaporation, often with sugar added for longer life in an opened can
    - o Evaporated milk, (less concentrated than condensed) milk without added sugar
    - o Ricotta cheese, milk heated and reduced in volume, known in Indian cuisine as Khoa
    - o Infant formula, dried milk powder with specific additives for feeding human infants
    - o Baked milk, a variety of boiled milk that has been particularly popular in Russia
- Butter, mostly milk fat, produced by churning cream
    - o Buttermilk, the liquid left over after producing butter from cream, often dried as livestock food
    - o Ghee, clarified butter, by gentle heating of butter and removal of the solid matter
    - o Anhydrous milkfat
- Cheese, produced by coagulating milk, separating from whey and letting it ripen, generally with bacteria and sometimes also with certain molds
    - o Curds, the soft curdled part of milk (or skim milk) used to make cheese (or casein)
    - o Whey, the liquid drained from curds and used for further processing or as a livestock food
    - o Cottage cheese
    - o Quark
    - o Cream cheese, produced by the addition of cream to milk and then curdled to form a rich curd or cheese made from skim milk with cream added to the curd
    - o Fromage frais
- Casein
    - o Caseinates
    - o Milk protein concentrates and isonates

  - Whey protein concentrates and isonates
  - Hydrolysates
  - Mineral concentrates
- Yogurt, milk fermented by *Streptococcus salivarius ssp. thermophilus* and *Lactobacillus delbrueckii ssp. bulgaricus* sometimes with additional bacteria, such as *Lactobacillus acidophilus*
  - Ayran
  - Lassi
- Gelato, slowly frozen milk and water
- Ice cream, slowly frozen cream and emulsifying additives
  - Ice milk
  - Frozen custard
  - Frozen yogurt, yogurt with emulsifiers that is frozen
- Other
  - Kumis/Airag, slightly fermented mares' milk popular in Central Asia
  - Viili
  - Kajmak
  - Kephir
  - Filmjölk
  - Piimä
  - Vla
  - Dulce de leche
  - Uloo kao patha laeen ka

Criticism

Dairy may cause health issues for individuals with lactose intolerance and milk allergies. Vegans and some vegetarians avoid dairy products due to a variety of ethical, dietary, environmental, political, and religious concerns.

Eggs as Dairy?

Eggs are sometimes categorized as dairy, defining dairy as "food that is produced by animals (other than meat)" rather than as milk specifically. For example, the Open Directory Project at one point listed cooking eggs as a subcategory of cooking dairy products. Defining dairy as limited to milk products, however, is more common.

BULK TANK

In dairy farming a bulk milk cooling tank is a large storage tank for cooling and holding milk at a cold temperature until it can be picked up by a milk hauler. The bulk milk cooling tank is an important milk farm equipment. It is usually made of stainless steel and used every day to store the raw milk on the farm in good condition. It must be cleaned after each milk collection. The milk cooling tank can be the property of the farmer or being rented to the farmer by the dairy plant.

Bulk Tank Types

Raw milk producers have a choice of either open (from 150 to 3000 litres) or closed (from 1000 to 10000 litres) tanks with direct expansion or with ice builder. The cost can vary considerably, depending on manufacturing norms and whether a new or second hand tank is purchased. Direct Expansion cooling involves placing the milk in direct contact with the cooled evaporator unit. This is the most commonly used method of cooling in countries with a large dairy industry, such as Australia. On the other hand, cooling using an Ice Builder is more valuable when dairy products need to be quickly cooled, and involves the use of ice water as a cooling medium which is pumped (usually via copper tubes) between the evaporator unit and the milk, rather than direct contact with the evaporator unit. The tank capacity and type will depend on herd size, calving pattern, frequency of milk collection, required milk quality, energy and water availability and future plans for development. In direct expansion tanks, milk is cooled by refrigerant plates, which are in direct contact with the outer surfaces of the tank walls. In ice bank tanks, milk is cooled when chilled water, from an ice bank, is sprayed against the outer surfaces of the tank walls. Milk silos (10,000 litres and plus) are suitable for the very large producer. These are designed to be installed outside and adjacent to the dairy, all controls and the milk outlet pipe being situated in the dairy.

Bulk Tanks Manufacturing Norms

Norms define among other criteria : insulation, milk agitation, cooling power required, variations in milk quantity measurement, calibration, ... Some are more demanding than others. ISO ISO standard 5708 published en 1983. Northern American standard 3A 13-10 updated in 2003. European Union standard EN 13732 published in 2003

Milk Cooling Tank Description

A milk cooling tank, also known as a bulk tank or milk cooler, consists of an inner and an outer tank, both made of high quality stainless steel. For a direct expansion tank, attached to the inner tank is a system of plates and pipes through which the refrigerant fluid/gas flows. The refrigerant withdraws heat from the tanks content (milk in this case). Direct expansion milk cooling tank come with a generator set with a condensing unit which circulates the refrigerant and conveys the withdrawn heat to the air. The space between the outer tank and the inner tank is isolated with polyurethane foam. In case of a power failure with an outside temperature of 30°C, the content of the tank will warm up only 1°C in 24 hours. To facilitate an adequate and rapid cooling of the entire content of a tank, every tank is equipped with at least one agitator. Stirring the milk ensures that all milk inside the tank is of the same temperature and that the milk stays homogeneous. On top of every closed milk cooling tank is a manhole of about 40 centimetres diameter. This enables thorough cleaning and inspection of the inner tank if necessary. The manhole is covered by a lid and sealed watertight with a rubber ring. Also on top are 2 or 3 small inlets. One is covered with an air-vent, the other(s) can be used to pump milk into the tank. A milk cooling tank usually stands on 4, 6, or 8 adjustable legs. The built-in tilt of the inner tank ensures that even the last drop of milk will eventually flow to the outlet. At the bottom, every milk cooling tank has a threaded outlet, usually including a valve. All tanks have a thermometer, allowing for immediate inspection of the inner temperature. Most tanks include an automatic cleaning system. Using hot and cold water, an acid and/or alkaline cleaning fluid, a pump and a spray lance will clean the inner tank, ensuring an hygienic inner environment each time the tank is emptied. Almost every tank has a control box. It manages the cooling process by use of a thermostat. The user can turn the system on and off, allow for extra and immediate stirring, start the cleaning routine, and reset the entire system in case of a failure. New and bigger milk cooling tanks are now being equipped with monitoring and alarm systems. These systems guard temperature of the milk inside the tank, check the functioning of the agitator, the cooling unit and temperature of the cleaning water. In case of malfunctioning of any of these functions, the alarm will activate. The monitoring system will also keep a record of the temperature and of all malfunctions for a given period.

Milk Pre-Cooling

For energy savings and quality reasons it is advisable to pre-cool with the milk before it enters the tank using a plate or a tube cooler (shell and tube heat exchanger) supplied with chilled water from the well water, the ice builder or the condensing unit. The quicker milk is cooled after leaving the cow the better. This system achieves most of the cooling before the milk enters the tank, so that chilled milk, rather than warm milk, is being added to the already cooled milk in the tank.

Cooling Temperature

Generic temperature for milk storage is 3 to 4°C. For raw milk cheese manufacturing, it would be advisable to keep the milk at 12°C, as milk characteristics will be kept in a better state. The milk cooling tank is usually not completely filled at once. A 2 milking tank is designed to cool 50% of its capacity at once. A 4 milking tank is designed to cool 25% of its capacity at once, and a 6 milking tank is designed to cool 16.7% of its capacity at once. The cooling performance depends on the number of milking it takes to completely fill the tank, the ambient temperature and the cooling time.

Milk Cooling Tank Cleaning System

Automatic bulk tank cleaners are available for closed tanks and are normally activated by the milk collection truck driver after each milk collection. Hot wash involves the following stages : - pre-rinsing with cold water, - pre-rinsing with hot water to heat up the surfaces of the tank, - circulating a suitably formulated detergent steriliser solution at 50°C for ten minutes, - rinsing with cold water (in some cases chlorinated), - final rinsing with cold drinkable water. Tanks washed with hot detergent sterilisers should be treated with milkstone remover once per month or more often if required, in accordance with manufacturer's instructions. Care should always be taken when handling chemicals. The correct health and safety procedures need to be followed.

Operating Costs

Substantial reductions in running costs can be made when an ice builder is used in conjunction with off-peak electricity. Pre-cooling milk using a plate or a tube cooler supplied with mains or well water can also reduce costs and add to the cooling capacity of the tank. Bulk tank condenser units, which are not an integral part of the tank, should be fitted in an adjacent, suitable and well ventilated place. If at all possible, condenser units should not be fitted on a south facing wall. They should be installed in a way which allows them to draw in and discharge adequate quantities of air for efficient operation. Bulk tank should be easily accessible by large bulk collection tankers and positioned so that the tanker approaches can be kept clean and free from cow traffic at all times. Although tanks have been calibrated when first installed, bulk tank miscalibration is not uncommon and in some cases it can result in significant loss of income. Milk tanks calibrated on the low side, can cheat raw milk producers by up to 22 litres on each shipment. It is therefore advisable to re-calibrate a bulk tank.

Bulk Tank Outlet Standards

Swedish outlet (SMS 1145), German outlet (DIN 11851), English RJT (BS 4825), IDF (ISO 2853), tri-clamp (ISO 2852), Danish outlet (DS 722), ... can be found, not to mention different diameters. They vary from country to country. Non standard outlets are often painful for the milk collection process as the operator needs to adapt to each different standard/diameter.

Bulk Milk Cooling Tank Maintenance Suggestions

Check the agitator motor for grease leaks, noisy operation, worn shaft shields and bearings. Replace leaking agitator motor seals. Tighten bolts holding motor mounting brackets. Replace worn shaft shields and bearings. Check the timer to be sure that it will start the agitation process and advance to the "off" position. Replace timer if not ok. Check thermometer accuracy, should read 0°C when in ice water. Be sure it is not sticking. Replace thermometer if faulty. Check milk tank outlet valve for leaks and cleanness. Replace valve "O" ring, if leaking. Check the running time of the cooler. It should cool to 10°C within one hour of first milking and cool to 4°C and hold that temperature after the second hour. The blend temperature of the second, third and fourth milking should remain under 10°C. If running time is too long check and clean condenser coil. Check refrigerant. Check for foam, churned milk, and frozen milk on the milk surface. The presence of foam indicates air leaks in the milking system or excessive agitation of the milk; churned milk (clumps of fat floating on the surface) usually is caused by excessive agitation and slow cooling of the milk. Check refrigerant; and frozen milk on the surface or as layers of ice on the bottom of the tank. The freezing of milk can be avoided by turning on the refrigeration when the milk level reaches the level of the agitator blades and setting the tank thermostat so that the milk is cooled to 4°C. Check condenser coil for dirt or dust. Air must flow freely through the coil and exit into the atmosphere. If condensor coil is dirty turn disconnect switch to "off" position, brush wash with milk detergent solution, rinse with tap water from the fan side out and allow to drain for three hours before restarting. Check refrigeration unit and tubing for signs of leaks (grease spots). Call to the attention of refrigeration service man, any air or refrigeration leaks. Check refrigerant sight glass after unit has been operating 15 minutes refrigerant should be clear without any indication of foaming. If refrigerant is foaming, have refrigerant added by serviceman. Check condensor fan motors. Call to the attention of refrigeration service man any malfunction of the condensor fan motor

Other Usage of Milk Cooling Tanks

Milk cooling tanks are also used to heat or cool a fluid or simply to keep it isolated and warm/cold. Because of the hygienical finishing of the inner and outer side of the tanks, almost any fluid can be stored : water, fruit juices, honey, wine, beer, ink, paint, cosmetics, aromatic food-additives, bacterial cultures, cleansers, oil, blood, ...

YOGHURT

Yoghurt is a dairy product produced by bacterial fermentation of milk. Fermentation of the milk sugar (lactose) produces lactic acid, which acts on milk protein to give yoghurt its texture and its characteristic tang. Soy yogurt, a dairy-yogurt alternative, is made from soy milk.

There is evidence of cultured milk products being produced as food for at least 4,500 years. The earliest yoghurts were probably spontaneously fermented by wild bacteria living on the goat skin bags carried by the Bulgars (or Hunno-Bulgars), a nomadic people who began migrating into Europe in the second century AD and eventually settled in the Balkans at the end of the seventh century. Today, many different countries claim yoghurt as their own, yet there is no clear evidence as to where it was first discovered. The use of yoghurt by mediaeval Turks is recorded in the books *Diwan Lughat al-Turk* by Mahmud Kashgari and *Kutadgu Bilig* by Yusuf Has Hajib written in the eleventh century. In both texts the word "yoghurt" is mentioned in different sections and its use by nomadic Turks is described. The first account of a European encounter with yoghurt occurs in French clinical history: Francis I suffered from a severe diarrhea which no French doctor could cure. His ally Suleiman the Magnificent sent a doctor, who allegedly cured the patient with yoghurt. Until the 1900s, yoghurt was a staple in diets of the South Asian, Central Asian, Western Asian, South Eastern European and Central European regions. The Russian biologist Ilya Ilyich Mechnikov had an unproven hypothesis that regular consumption of yoghurt was responsible for the unusually long lifespans of Bulgarian peasants. Believing *Lactobacillus* to be essential for good health, Mechnikov worked to popularise yoghurt as a foodstuff throughout Europe. It fell to a Sephardic Jewish entrepreneur named Isaac Carasso to industrialise the production of yoghurt. In 1919, Carasso, who was from Salonika, started a small yoghurt business in Barcelona and named the business Danone ("little Daniel") after his son. Carasso emigrated to the United States during World War II and set up a business in New York City under an Americanised version of the name: *Dannon*. Yoghurt with added fruit jam was invented to protect yoghurt from decay. It was patented in 1933 by the Radlická Mlékárna dairy in Prague, and introduced to the United States in 1947, by Dannon. Yogurt's popularity in the United States was enhanced in the 1950's and 60's when it was presented as a health food.

Culture

Yoghurt is made by introducing specific bacteria strains into milk, which is subsequently fermented under controlled temperatures and environmental conditions (inside a bioreactor), especially in industrial production. The bacteria ingest natural milk sugars and release lactic acid as a waste product. The increased acidity causes milk proteins to tangle into a solid mass (curd) in a process called denaturation. The increased acidity (pH=4–5) also prevents the proliferation of potentially pathogenic bacteria. In the U.S., to be named yoghurt, the product must contain the bacteria strains *Streptococcus salivarius subsp. thermophilus* and *Lactobacillus delbrueckii subsp. bulgaricus*. Often these two are co-cultured with other lactic acid bacteria for taste or health effects. These include *L. acidophilus*, *L. casei* and *Bifidobacterium* species. In most countries, a product may be called yoghurt only if live bacteria are present in the final product. In the U.S., non-pasteurised yoghurt can be marketed as "live" or containing "live active culture". A small amount of live yoghurt can be used to inoculate a new batch of yoghurt, as the bacteria reproduce and multiply during fermentation. Pasteurised products, which have no living bacteria, are called fermented milk (drink).

Benefits

Yoghurt has nutritional benefits beyond those of milk: people who are moderately lactose-intolerant can enjoy yoghurt without ill effects, because the lactose in the milk precursor is converted to lactic acid by the bacterial culture. The reduction of lactose bypasses the affected individuals' need to process the milk sugar themselves. Yoghurt also has medical uses, in particular for a variety of gastrointestinal conditions, and in preventing antibiotic-associated diarrhea. One study suggests that eating yoghurt containing *L. acidophilus* helps prevent vulvovaginal candidiasis, though the evidence is not conclusive.

Presentation

To offset its natural sourness, yoghurt can be sold sweetened, flavored, or in containers with fruit or fruit jam on the bottom. If the fruit has been stirred into the yoghurt before purchase, it is commonly referred to as Swiss-style. Most yoghurts in the United States have added pectin or gelatin. Some specialty yoghurts have a layer of fermented fat at the top. Fruit jam is used instead of raw fruit pieces in fruit yoghurts to allow storage for weeks."Strained" yoghurt is the concentrated residue (described as a sort of "yoghurt cheese") produced by filtering plain yoghurt that is without flavorings, gelatin, pectin, or other additives through a paper or cloth filter, and allowing water and whey to drain away.

Varieties

- Strained yoghurts, which include Greek Yoghurt (yiaourti), Dahi and Bulgarian Yoghurt are types of yoghurt which are strained through a cloth or paper filter, traditionally made of muslin, to remove the whey, giving a much thicker consistency, and a distinctive, slightly tangy taste. Some types are boiled in open vats first, so that the liquid content is reduced. The popular East Indian dessert, Mishti Dahi, is a variation of traditional Dahi, offers a thicker, more custard-like consistency, and is usually sweeter than western yoghurts.

- Dadiah, or Dadih, is a traditional West Sumatran yoghurt made from water buffalo milk. It is fermented in bamboo tubes.
- Labneh yoghurt of Lebanon is a thickened yoghurt used for sandwiches. Olive oil, cucumber slices, olives, and various green herbs may be added. It can be thickened further and rolled into balls, preserved in olive oil, and fermented for a few more weeks. It is sometimes used with onions, meat, and nuts as a stuffing for a variety of Lebanese pies or Kebbeh balls.
- Tarator/cacýk is a popular cold soup made from yoghurt, popular during summertime in Bulgaria, Macedonia, and Turkey. It is made with Ayran, cucumbers, dill, salt, olive oil, and optionally garlic and ground walnuts in Bulgaria, and generally without walnuts in Turkey.
- Rahmjoghurt, a creamy yoghurt with much higher milkfat content (10%) than most yoghurts offered in English-speaking countries (Rahm is German for cream), is available in Germany and other countries.
- Caspian Sea Yoghurt is believed to have been introduced into Japan in 1986 by researchers returning from a trip to the Caucasus region in Georgia. This variety, called *Matsoni*, is started with *Lactococcus lactis* subsp. *cremoris* and *Acetobacter orientalis* species and has a unique, viscous, honey-like texture. It is milder in taste than other varieties of yoghurts. Ideally, Caspian Sea yoghurt is made at home because it requires neither special equipment nor unobtainable culture. It can be made at room temperature (20–30°C) in 10 to 15 hours. In Japan, freeze-dried starter cultures are sold in department stores and online, although many people obtain starter cultures from friends.
- Jameed is yoghurt which is salted and dried to preserve it. It is popular in Palestine and Jordan.

Drinks

- Ayran is a yoghurt-based, salty drink popular in Turkey, Azerbaijan, Bulgaria, Macedonia, Kazakhstan and Kyrgyzstan. It is made by mixing yoghurt with water and adding salt. The same drink is known as *tan* in Armenia, "Laban Ayran" in Syria and Lebanon, "Shenina" in Jordan, "Moru" in South India, and "Laban Arbil" in Iraq. A similar drink, doogh, is popular in the Middle East between Lebanon and Afghanistan; it differs from ayran by the addition of herbs, usually mint, and is carbonated, usually with seltzer water. In the United States, yoghurt-based beverages are often marketed under names like "yoghurt smoothie" or "drinkable yoghurt". They are also popular in Ecuador where the primary form of yoghurt is "bebida de yogurt", which literally means *drink of yoghurt*.

- Lassi is a yogurt-based beverage originally from the Indian subcontinent that is usually slightly salty or sweet. Much like a smoothie, the sweet version is typically flavored with coconut, rosewater, lemon, mango or other fruit juice. Salty lassi is usually flavored with ground, roasted cumin and chili peppers.
- Yop a fruity French yoghurt coming from the Yoplait Dairy Company, is popular in France, Canada and the UK.
- Kefir is a fermented milk drink originating in the Caucasus. A related Central Asian Turco-Mongolian drink made from mare's milk is called kumis, or airag in Mongolia. Some American dairies have offered a drink called "kefir" for many years with fruit flavours but without carbonation or alcohol. As of 2002, names like "drinkable yoghurt" and "yoghurt smoothie" have been introduced.

Homemade

Yoghurt is customarily made in domestic environments in regions where yoghurt has an important place in traditional cuisine. It can be made from a small amount of store-bought, plain, live culture yoghurt by adding milk and heating at a constant, but not boiling, temperature. Special yoghurt-making machines assist in small-batch yoghurt-making. In 2005, Mireille Guiliano released her best selling book, *French Women Don't Get Fat*, in which she touts yogurt as her secret weight loss weapon. In her book, she campaigned for Americans to discover the benefits and pleasures of homemade plain, non-fat yogurt, as opposed to the sugar and corn syrup-laden 'imposters' found in most U.S. supermarkets. Her book was the first contemporary weight loss plan to center around making homemade yogurt.

Etymology and Spelling

The word "yoghurt" comes from the Turkish *yoðurt*. The word is derived from the adjective *yoðun*, which means "dense" and "thick", or from the verb *yoðurmak*, meaning "to knead". Originally, the verb may have meant "to make dense", which is how yoghurt is made. The letter ð was traditionally rendered as "gh" in transliterations of Turkish, which used to be written in a variant of the Arabic alphabet until the introduction of the Latin alphabet in 1928.

In older Turkish the letter denoted a voiced velar fricative /c/, but this sound is elided between back vowels in modern Turkish, in which the word is pronounced. Some eastern dialects retain the consonant in this position, and Turks in the

Balkans pronounce the word with a hard /g/. In English, there are several variations of the spelling of the word. In the United States, yogurt is the usual spelling and yoghurt a minor variant. In the United Kingdom, yoghurt and yogurt are both current, *yoghurt* being more common, and yoghourt is an uncommon alternative. Canada uses mostly yogurt and yogourt, the latter being particularly common in bilingual packaging, as it is also the spelling in Canadian French; in Australia and New Zealand yoghurt prevails. Whatever the spelling, the word is pronounced with either a short "o" or a long "o" in the UK and New Zealand, and with a long "o" in North America, Ireland and Australia.

MILK PRODUCTION

Milk is the source of nutrients and immunological protection for the young cow. The gestation period for the female cow is 9 months. Shortly before calving, milk is secreted into the udder in preparation for the new born. At parturition, fluid from the mammary gland known as colostrum is secreted. This yellowish coloured, salty liquid has a very high serum protein content and provides antibodies to help protect the newborn until its own immune system is established. Within 72 hours, the composition of colostrum returns to that of fresh milk, allowing to be used in the food supply. The period of lactation, or milk production, then continues for an average of 305 days, producing 7000 kg of milk. This is quite a large amount considering the calf only needs about 1000 kg for growth. Within the lactation, the highest yield is 2-3 months post- parturition, yielding 40-50 L/day. Within the milking lifetime, a cow reaches a peak in production about her third lactation, but can be kept in production for 5-6 lactations if the yield is still good. About 1-2 months after calving, the cow begins to come into heat again. She is usually inseminated about 3 months after calving so as to come into a yearly calving cycle. Heifers are normally first inseminated at 15 months so she's 2 when the first calf is born. About 60 days before the next calving, the cow is dried off. There is no milking during this stage for two reasons:

1. milk has tapered off because of maternal needs of the fetus
2. udder needs time to prepare for the next milking cycle

The life of a female cow can be summerized as follows:

Age	
0	Calf born
15 mos	Heifer inseminated for first calf
24 mos	First calf born - starts milking
27 mos	Inseminated for second calf

34 mos	Dried off
36 mos	Second calf born - starts milking

Cycle repeats for 5-6 lactations

Effects of Milk Handling on Quality and Hygiene

Cleanliness

The environment of production has a great effect on the quality of milk produced. From the food science perspective, the production of the highest quality milk should be the goal. However, this is sometimes not the greatest concern of those involved in milk production. Hygienic quality assessment tests include sensory tests, dye reduction tests for microbial activity, total bacterial count (standard plate count), sediment, titratable acidity, somatic cell count, antibiotic residues, and added water.

The two common dye reduction tests are methylene blue and resazurin. These are both synthetic compounds which accept electrons and change colour as a result of this reduction. As part of natural metabolism, active microorganisms transfer electrons, and thus rate at which dyes added to milk are reduced is an indication of the level of microbial activity. Methylene blue turns from blue to colorless, while resazurin turns from blue to violet to pink to colourless. The reduction time is inversely correlated to bacterial numbers. However, different species react differently. Mesophilics are favoured over psychrotrophs, but psychrotrophic organisms tend to be more numerous and active in cooled milk.

Temperature

Milk production and distribution in the tropical regions of the world is more challenging due to the requirements for low-temperature for milk stability. Consider the following chart illustraing the numbers of bacteria per millilitre of milk after 24 hours:

5°C	2,600
10°C	11,600
12.7°C	18,800
15.5°C	180,000
20°C	450,000

Traditionally, this has been overcome in tropical countries by stabilizing milk through means other than refrigeration, including immediate consumption of warm milk after milking, by boiling milk, or by conversion into more stable products such as fermented milks.

Mastitis and Antibiotics

Mastitis is a bacterial and yeast infection of the udder. Milk from mastitic cows is termed abnormal. Its SNF, especially lactose, content is decreased, while Na and Cl levels are increased, often giving mastitic milk a salty flavour. The presence of mastitis is also accompanied by increases in bacterial numbers, including the possibility of human pathogens, and by a dramatic increase in somatic cells. These are comprised of leukocytes (white blood cells) and epithelial cells from the udder lining. Increased somatic cell counts are therefore indicative of the presence of mastitis. Once the infection reaches the level known as "clinical' mastitis, pus can be observed in the teat canal just prior to milking, but at sub-clinical levels, the presence of mastitis is not obvious.

Somatic Cell Count (000's / ml)	*Daily Milk Yield (kg)*	*1st Lactation*	*Older Lactations*
0-17	23.1	29.3	
18-34	23.0	28.7	
35-70	22.6	28.0	
71-140	22.4	27.4	
141-282	22.1	27.0	
282-565	21.9	26.3	
566-1130	21.4	25.4	
1131-2262	20.7	24.6	
2263-4525	20.0	23.6	
>4526	19.0	22.5	

Antibiotics are frequently used to control mastitis in dairy cattle. However, the presence of antibiotic residues in milk is very problematic, for at least three reasons. In the production of fermented milks, antibiotic residues can slow or destroy the growth of the fermentation bacteria. From a human health point of view, some people are allergic to specific antibiotics, and their presence in food consumed can have severe consequences. Also, frequent exposure to low level antibiotics can cause microorganisms to become resistant to them, through mutation, so that they are ineffective when needed to fight a human infection. For these reasons, it is extremely important that milk from cows being treated with antibiotics is withheld from the milk supply.

The withdrawal time after final treatment for various antibiotics is shown below:

Amoxcillin 60 hrs.

Cloxacillin 48 hrs.
Erythromicin 36 hrs.
Novobiocin 72 hrs.
Penicillin 84 hrs.
Sulfadimethozine 60 hrs.
Sulfabromomethozine 96 hrs.
Sulfaethoxypyridozine 72 hrs.

Anti-Microbial Systems in Raw Milk

There exists in milk a number of natural anti-microbial defense mechanisms. These include:

- lysozyme - an enzyme that hydrolyses glycosidic bonds in gram positive cell walls. However, its effect as a bacteriostatic mechanism in milk is probably negligible.
- lactoferrin - an iron binding protein that sequesters iron from microorganisms, thus taking away one of their growth factors. Its effect as a bacteriostatic mechanism in milk is also probably negligible.
- lactoperoxidase - an enzyme naturally present in raw milk that catalyzes the conversion of hydrogen peroxide to water. When hydrogen peroxide and thiocyanate are added to raw milk, the thiocyanate is oxidized by the enzyme/ hydrogen peroxide complex producing bacteriostatic compounds that inhibit Gram negative bacteria, E. coli , Salmonella spp , and streptococci. This technique is being used in many parts of the world, especially where refrigeration for raw milk is not readily available, as a means of increasing the shelf life of raw milk.

Milk Biosynthesis

Milk is synthesized in the mammary gland. Within the mammary gland is the milk producing unit, the alveolus. It contains a single layer of epithelial secretory cells surrounding a central storage area called the lumen, which is connected to a duct system. The secretory cells are, in turn, surrounded by a layer of myoepithelial cells and blood capillaries.

The raw materials for milk production are transported via the bloodstream to the secretory cells. It takes 400-800 L of blood to deliver components for 1 L of milk.

- Proteins: building blocks are amino acids in the blood. Casein micelles, or small aggregates thereof, may begin aggregation in Golgi vesicles within the secretory cell.

- Lipids:
  - o C4-C14 fatty acids are synthesized in the cells
  - o C16 and greater fatty acids are preformed as a result of rumen hydrogenation and are transported directly in the blood
- Lactose: milk is in osmotic equilibrium with the blood and is controlled by lactose, K, Na, Cl; lactose synthesis regulates the volume of milk secreted

The milk components are synthesized within the cells, mainly by the endoplasmic reticulum (ER) and its attached ribosomes. The energy for the ER is supplied by the mitochondria. The components are then passed along to the Golgi apparatus, which is responsible for their eventual movement out of the cell in the form of vesicles. Both vesicles containing aqueous non-fat components, as well as liquid droplets (synthesized by the ER) must pass through the cytoplasm and the apical plasma membrane to be deposited in the lumen. It is thought that the milk fat globule membrane is comprised of the apical plasma membrane of the secretory cell.

Milking stimuli, such as a sucking calf, a warm wash cloth, the regime of parlour etc., causes the release of a hormone called oxytocin. Oxytocin is relased from the pituitary gland, below the brain, to begin the process of milk let-down. As a result of this hormone stimulation, the muscles begin to compress the alveoli, causing a pressure in the udder known as letdown reflex, and the milk components stored in the lumen are released into the duct system. The milk is forced down into the teat cistern from which it is milked. The let-down reflex fades as the oxytocin is degraded, within 4-7 minutes. It is very difficult to milk after this time.

The importance of milk grading lies in the fact that dairy products are only as good as the raw materials from which they were made. It is important that dairy personnel have a knowledge of sensory perception and evaluation techniques. The identification of off-flavours and desirable flavours, as well as knowledge of their likely cause, should enable the production of high quality milk, and subsequently, high quality dairy products.

MILK GRADING

An understanding of the principles of sensory evaluation are neccessary for grading. All five primary senses are used in the sensory evaluation of dairy products: sight, taste, smell, touch and sound. The greatest emphasis, however, is placed on taste and smell.

The Sense of Taste

Taste buds, or receptors, are chiefly on the upper surface of the tongue, but may also be present in the cheek and soft palates of young people. These buds, about 900 in number, must make contact with the flavouring agent before a taste sensation occurs. Saliva, of course, is essential in aiding this contact. There are four different types of nerve endings on the tongue which detect the four basic "mouth" flavours - sweet, salt, sour, and bitter. Samples must, therefore, be spread around in the mouth in order to make positive flavour identification. In addition to these basic tastes, the mouth also allows us to get such reactions as coolness, warmth, sweetness, astringency, etc.

The Sense of Smell

We are much more perceptive to the sense of smell than we are to taste. For instance, it is possible for an odouriferous material such as mercaptain to be detected in 20 billion parts of air. The centres of olfaction are located chiefly in the uppermost part of the nasal cavity. To be detectable by smell, a substance must dissolve at body temperature and be soluble in fat solvents.

Note: The sense of both taste and smell may become fatigued during steady use. A good judge does not try to examine more than one sample per minute. Rinsing the mouth with water between samples may help to restore sensitivity.

Milk Grading Techniques

Temperature should be between 60-70° F (15.5-21° C) so that any odour present may be detected readily by sniffing the container. Also, we want a temperature rise when taking the sample into the mouth; this serves to volatize any notable constituents.

Noting the odour by placing the nose directly over the container immediately after shaking and taking a full "whiff" of air. Any off odour present may be noted.

Need to make sure we have a representative sample; mixing and agitation are important.

Agitation leaves a thin film of milk on the inner surface which tends to evaporate giving off odour if present.

During sampling, take a generous sip, roll about the mouth, note flavour sensation, and expectorate. Swallowing milk is a poor practice.

Can enhance the after-taste by drawing a breath of fresh air slowly through the mouth and then exhale slowly through the nose. With this practice, even faint odours can be noted.

Milk has a flavour defect if it has an odour, a foretaste or an aftertaste, or does not leave the mouth in a clean, sweet, pleasant condition after tasting.

Characterization of Flavour Defects - ADSA

Lipolytic or Hydrolytic Rancidity

Rancidity arises from the hydrolysis of milkfat by an enzyme called the lipoprotein lipase (LPL). The flavour is due to the short chain fatty acids produced, particularly butyric acid. LPL can be indigenous or bacterial. It is active at the fat/water interface but is ineffective unless the fat globule membrane is damaged or weakened. This may occur through agitation, and/or foaming, and pumping. For this reason, homogenized milk is subject to rapid lipolysis unless lipase is destroyed by heating first; the enzyme (protein) is denatured at 55-60° C. Therefore, always homogenize milk immediately before or after pasteurization and avoid mixing new and homogenized milk because it leads to rapid rancidity.

Some cows can produce spontaneous lipolysis from reacting to something indigenous to the milk. Late lactation, mastitis, hay and grain ratio diets (more so than fresh forage or silage), and low yielding cows are more suseptible.

Lipolysis can be detected by measuring the acid degree value which determines the presence of free fatty acids. Lipolytic or hydrolytic rancidity is distinct from oxidative rancidity, but frequently in other fat industries, rancid is used to mean oxidative rancidity; in dairy, rancidity means lipolysis.

Characterized: soapy, blue-cheese like aroma, slightly bitter, foul, pronounced aftertaste, does not clear up readily

Oxidation

Milk fat oxidation is catalysed by copper and certain other metals with oxygen and air. This leads to an autooxidation reaction consisting of initiation, propagation, termination.

RH — R + H initiation - free radical

R + O2 —— RO2 propagation

RO2 + RH — ROOH + R

R + R — R2 termination

R + RO2 — RO2R

It is usually initiated in the phospholipid of the fat globule membrane. Propagation then occurs in triglycerides, primarily double bonds of unsaturated fatty acids. During propagation, peroxide derivatives of fatty acids accumulate. These undergo further reactions to form carbonyls, of which some, like aldehydes and ketones, have strong flavours. Dry feed, late lactation, added copper or other metals, lack of vit E (tocopherol) or selenium (natural antioxidates) in the diet all lead to spontaneous oxidation. It can be a real problem especially in winter. Exposure to metals during processing can also contribute.

Characterized: metallic, wet cardboard, oily, tallowy, chalky; mouth usually perceives a puckery or astringent feel

Sunlight

Often confused with oxidized, this defect is caused by UV-rays from sunlight or flourescent lighting catalyzing oxidation in unprotected milk. Photo-oxidation activates riboflavin which is responsible for catalyzing the conversion of methionine to methanal. It is, therefore, a protein reaction rather than a lipid reaction. However, the end product flavour notes are similar but tends to diminish after storage of several days.

Characterized: burnt-protein or burnt-feathers-like, "medicinal"-like flavour

Cooked

This defect is a function of the time-temperature of heating and especially the presence of any "burn-on" action of heat on certain proteins, particulary whey proteins. Whey proteins are a source of sulfide bonds which form sulfhydryl groups that contribute to the flavour. The defect is most obvious immediately after heating but dissipates within 1 or 2 days.

Characterized: slightly cooked or nutty-like to scorched or caramelized

Transmitted Flavours

Cows are particulary bad for transmitting flavours through milk and milk is equally as susceptible to pick-up of off flavours in storage. Feed flavours and green grass can be problems so it is necessary to remove cows from feed 2-4 hrs before milking. Weeds, garlic/onion, and dandelions can tranfer flavours to the milk and even subsequent products such as butter. Barny flavours can be picked up in the milk if there is poor ventilation and the barn is not properly cleared and cows breathe the air. These flavours are volatile so can be driven off through vacuum de-aeration.

Characterization: hay/silage, cowy/barny

Microbial

There are many flavour defects of dairy products that may be caused by bacteria, yeasts, or moulds. In raw milk the high acid/sour flavour is caused by the growth of lactic acid bacteria which ferment lactose. It is less common today due to change in raw milk microflora. In both raw or processed milk, fruity flavours may arise due to psychrotrophs such as *Pseudomonas fragi*. Bitter or putrid flavours are caused by psychrotrophic bacteria which produce protease. It is the proteolytic action of protease that usually causes spoilage in milk. Malty flavours are caused by *S.lactis* var. *maltigenes* and is characterized by a corn flakes type flavour. Although more of a tactile defect, ropy milk is also caused by bacteria, specifically those which produce exopolysaccharides.

Miscellaneous Defects

- astringent
- chalky
- chemical/medicinal - disease - associated or adulteration
- flat - adulteration (water)
- foreign
- salty - disease associated
- bitter - adulteration

Milk flavour is graded on a score of one to 10. Some flavour defects, even if only slightly present, can decrease the score drastically. The following are suggested flavour scores for milk with designated intensities of flavour defects.

Flavour Criticisms	*Intensity of Defect*		
	Slight	*Definite*	*Pronounced*
Astringent	8	7	5
Barny	7	5	3
Bitter	7	5	3
Cooked	9	8	6
Cowy	6	4	1
Feed	9	7	5
Flat	9	8	7
Foreign	5	3	0
Garlic/onion	5	3	1
High acid	3	1	0
Bacterial	5	3	0
Lacks Freshness	7	5	3
Malty	7	5	3
Oxidized	7	5	3
Rancid	7	5	3
Salty	8	6	4
Unclean	7	5	3

GRAIN MILK

Grain milk is a milk substitute made from fermented grain or from flour. Grain milk can be made from oats, spelt, rice, rye, einkorn wheat or quinoa. Grain milk looks very similar to cow's milk. It has a lower protein content and a higher carbohydrate content than cow's milk. Just as cow's milk is often fortified with Vitamin D, which it naturally lacks, grain milks may have calcium and some vitamins (especially cobalamin) added to them. Cobalamin or vitamin B12 is produced exclusively by microorganisms. Higher plants and animals are unable to produce it. Grain milk is low in saturated fat and contains no lactose, which is beneficial for those who are lactose intolerant. Grain milk also lacks milk protein, making it suitable for vegans and people with milk allergies. Flavored grain milk can come in plain, vanilla, chocolate or a variety of other flavors. Like unflavored grain milk, it is often available with added nutrients. There are also grain milk cream and desserts available

Soured Milk

Soured milk is a food product, distinguished from spoiled milk, and is a general term for milk that has acquired a tart taste, either through the addition of an acid, such as lemon juice or vinegar, or through bacterial fermentation. The acid causes milk to coagulate and form a thicker consistency. Soured milk that is produced by bacterial fermentation is more specifically called fermented milk or cultured milk. Soured milk that is produced by the addition of an acid, with or without the addition of microbial organisms, is more specifically called acidified milk.In the US, the acids that may be used in the manufacture of acidified milk are acetic acid (commonly found in vinegar), adipic acid, citric acid (commonly found in lemon juice), fumaric acid, glucono-delta-lactone, hydrochloric acid, lactic acid, malic acid, phosphoric acid, succinic acid, and tartaric acid.

In recipes, soured milk created by the addition of an acid or by bacterial fermentation can often be used interchangeably. For example, 1 cup of cultured buttermilk, a soured milk produced by bacterial fermentation, can be substituted with 1 tablespoon of lemon juice or vinegar plus enough milk to make 1 cup. The chemically soured milk can be used after standing for 5 minutes.

Sour Milk

Currently, sour milk commonly means (pasteurized) milk that has spoiled and is "in an unpalatable state". However, the meaning of sour milk can be ambiguous since it has been used in different contexts in different time periods. Since the 1970s, sour milk has been used both for chemically and biologically soured milk. Some older recipes use sour milk, but today it is unclear which dairy product was "sour milk". Pasteurized milk was not available in the US until the early 1900s. For example, the South Jersey Milk Company created a poster in 1930 entitled "What is pasteurized milk?" to educate the American public about the safety of its newly pasteurized milk.

Thus, older recipes that use sour milk may have been written prior to the availability of pasteurized milk. Before pasteurized milk became widely available, sour milk may have been unpasteurized milk that had naturally acquired a sour taste through bacterial fermentation at room temperature. At least until the 1920s, there was a clear distinction between sour milk and buttermilk, where buttermilk was the sour tasting thin liquid leftover from making butter. Today, in North America, either cultured buttermilk, also commonly known as buttermilk but not the same product as the aforementioned buttermilk, or milk soured by the addition of lemon juice or vinegar is often used when sour milk is needed in a recipe.

UDDER CREAM

Udder cream is a topical cream developed specifically for application to the chapped udders of cows. It cures the chafing irritation caused by excessive milking and other skin disorders. Udder cream may also be used by people as a moisturizer to treat over-dry skin due to its restorative effect on skin moisture. It is even sold in "purse-size tins" for this purpose. According to Bill Kennedy Sr. the CEO of Redex Industries Inc., the makers of 'Udderly Smooth Udder Cream', 90% of the company's products are used by humans and only 10% of the products are used for cows. Udderly Smooth is available in Hand Cream, Foot Cream, Hand & Body Lotion and now also Chamois Cream. It is also now available in the UK - distributed by Glandford Ltd who distribute to various retailers across the UK including London's Garden Pharmacy. Oncology Certain oral and infusion chemotherapy drugs can affect the skin on the hands and feet (Hand-Foot Syndrome or Desquamative Syndrome) or Palmar plantar erythrodysesthesia. This side effect can be severe enough to discontinue therapy. Some people using these chemotherapy drugs are finding relief from using Udderly Smooth on their hands and feet'

CREAM SEPARATOR

A cream separator is device invented by Martin Wiberg to separate cream from milk. Before milk-cream separators, cream was separated from milk by letting milk sit until the fat floated to the top and could then be skimmed off by hand. The separator was first made by Gustaf de Laval in 1877. With it, it was possible to separate the cream from the milk. When it spins, the milk, which is heaviest, is pushed outward against the walls and the cream, which is lighter, is collected in the middle. Gustaf de Laval's construction made it possible to start the largest separator factory in the world, Alfa Laval AB. The milk-separator became a big advance in industry in Sweden. Within the first decade in the 1900s there were over twenty separator manufacturers in Stockholm. Separators (albeit slightly modified)are also used on ships to purify oil.

ICE CREAM MIX CALCULATOR

Ice Cream

Over the past 18 months or so numerous enquiries about ice cream have been directed to the Dairy Science and Food Technology website. These have been mainly related to ice cream mix calculations and to faults with the final product and are probably a reflection of increasing numbers of people becoming involved in ice cream manufacture.

Production of Balanced Ice Cream Mixes

Manufacturers produce ice cream to meet the requirements of consumers as interpreted by the retailers or those selling directly to consumers. In many situations the ice cream manufacturer will have a final product specification to meet. One element of this specification is related to the composition of the final product including legal requirements e.g. for fat and milk protein. In the UK ice cream must contain a minimum of 5% fat and a minimum of 2.5% milk protein.

A compositional specification will typically specify the fat, milk solids not fat (MSNF), sugar, emulsifier and stabiliser concentrations in the final product. In this situation the manufacturer will select ingredients that can supply the above components, blend these and then process to produce a finished ice cream. Calculations will be required to ensure that the ingredients are correctly formulated to meet the final product specification. In other situations the manufacturer may want to produce a new or improved product and will have to devise their own product compositional specification. This specification must be developed so that the fat, sugar and MSNF components are balanced.

Fat/Sugar Balance

Fat adds certain taste and other qualities to food. In ice cream fat is balanced by sugar and there is a relationship between fat concentration and sugar concentration.

The relationship between fat and sugar concentrations is also influenced by the type of freezer used (Table 1).

TABLE 1
Approximate Fat and Sugar Contents for Ice Cream Mixes Using Various Types of Freezer

Freezer Type	*Fat concentration (% w/w)*	*Sugar concentration (% w/w)*
Vertical	6	12
Vertical	7	12.5
Vertical	8	13
Horizontal	9	13.5
Horizontal	10	14
Continuous	10	14
Continuous	12	15

From Rothwell (1985).

Note the fat/sugar ratios here are approximate and that this information does need to be modified to meet consumer requirements. The market has also changed since 1985 and some of the more popular types of dairy ice cream may contain 15 % fat. However, from a product development perspective the information in table 1 provides a good start.

Guinard *et al.* (1996) studying the influence of fat and sugar in a vanilla ice cream used university students to taste and rate on a nine-point hedonic scale the texture and mouth feel, flavour (taste and odour), and overall degree of liking for nine samples of vanilla ice cream varying in sugar (8.94 to 18.81%, wt/wt) and fat (8.73 to 19.30%, w/w) concentration.

The hedonic ratings differed significantly among samples, and the best-liked sample for texture and mouth feel, flavour, and overall degree of liking contained 13.54 % sugar and 14.99 % fat. Response surface methodology, in simple terms a three dimensional graphical analysis of overall degree of liking versus sugar and fat concentrations, was used to relate hedonic ratings to sugar and fat percentages in the ice cream. Dome-shaped response surfaces,e.g. figure 1, were obtained for all three degree of liking parameters, and optimal sugar and fat, respectively, were 13.16 % and 14.02 % for degree of liking of texture and mouth feel, 14.07 % and 15.35 % for degree of liking of flavour, and 14.30 % and 14.77 % for overall degree of liking.

The results from figure 1 show clearly that particular sugar concentrations are required to obtain maximum overall degree of liking ratings for particular fat concentrations and show the value of this simple research tool.

Some research has been published in the optimal fat and sugar concentrations required for maximum acceptability of ice cream, table 2. Surprisingly, perhaps,there appears to be agreement in that fat concentrations around 14% and sugar concentration of around 15% give the most acceptable ice cream.

TABLE 2

Optimal Fat and Sugar Levels Required for Maximum Acceptability of ice-cream

Country	*Fat (% w / w)*	*Sugar (% w / w)*	*Reference*
Egypt	14	15	Salam *et al* . (1981)
US	14.3	14.77	Guinard *et al.* (1996)

MSNF/ Water Balance

Ice cream is a complex colloidal system and includes ice crystals in a concentrated unfrozen aqueous phase. This aqueous phase contains a concentrated lactose solution.

When the MSNF concentration is optimal the ice cream has a smooth body, a characteristic ice cream taste and will have a satisfactory shelf life.

At high MSNF levels the lactose concentration in the unfrozen aqueous phase may be so high that the lactose comes out of solution and that crystals of lactose grow. This may result in 'sandiness' in the final product.

At low MSNF there is a tendency for the ice cream to taste 'watery' and to lack characteristic ice-cream flavour. Growth of ice crystals may also be promoted.

Rothwell (1985) has illustrated how to calculate the maximum, note not necessarily the optimal, concentration of MSNF; all the solids apart from MSNF are summed and subtracted from 100. The product is then divided by 7 to give the maximum acceptable value. MSNF absorb about 6 times their own weight of water. As an example a mix containing 8% fat, 13% sugar, 1% stabiliser/emulsifier should have a maximum MSNF of

$$= \frac{100\text{-(fat + emulsifier + stabiliser)}}{7}$$

$$= \frac{100 - (8 + 13 + 1)}{7}$$

$$= 11.14\ \%\ \text{MSNF.}$$

You can check your calculation using the calculator provided.

Organic ice-cream makers using cream and or butter to raise the fat content of on farm produced ice cream but who do not use skim milk powder or concentrated milk as a source of additional MSNF will find it difficult to meet ideal MSNF recommendations. To some extent the use of organic stabilisers can help overcome some of the problems that can arise with low MSNF ice cream during storage.

PRINCIPLES OF ICE CREAM MIX CALCULATION

There are several methods that can be used to determine the quantities of ingredients required to meet a target ice cream mix formulation. However, ingredient costs are also important and there is a balance between cost and quality that often must be considered. One of the simpler methods for performing mix calculations is known as the serum point method and can be learned quite quickly. Mixes compositions can also be calculated using linear programming and other more advanced mathematical techniques.

Descriptions of this method can be found in several text books including Hyde and Rothwell (1973) and Marshall and Arbuckle (2000). While these books use the older imperial measures e.g. pounds they are easy to follow. A particularly clearly written explanation, for a limited number of mixes, is given in the booklet by Rothwell (1985). Professor Douglas Goff, University of Guelph in Canada has an excellent website dealing with ice cream and has a section on worked examples of mix calculations.

The serum point method is based on the principle that the quantities of MSNF and fat contributed by 'milk' of any composition can be subtracted from the entire quantity of fat and MSNF required in a mix, leaving the remainder to be supplied by concentrated sources of MSNF or fat.

There is a logical approach to solving problems using this method and the following description has been adapted from that given by Marshall and Arbuckle (2000).

1. List the fat, MSNF, sugar, emulsifier, and stabiliser concentrations, usually as percentages, that are required in the ice cream mix.
2. The quantities of single source ingredients required for 100 kg of mix are then calculated next. For example if 15 % sugar is required, then:

   quantity of ingredient, in this case sugar = 15 kg (15/100*100).
3. Next the weight of serum in the mix (serum is water and MSNF or milk serum solids) and is obtained by subtracting the weights of all of the other ingredients from 100 kg of total mix.

Serum = 100 – (fat + sweetener + emulsifier +stabiliser+ other ingredients)

• To calculate the quantity of concentrated milk needed, it is necessary to know the quantity of serum solids and the quantity of serum in 1 kg of concentrated milk as well as quantities of the same components in the mix. The formula for the quantity of concentrated milk is

Quantity of concentrated milk = (MSNF needed) – serum of mix x 0.09

(MSNF/kg concentrated milk) – serum/kg concentrated milk) x 0.09

Note the figure 0.09 represents the approximate % MSNF of skim milk and represents 9% MSNF in milk, if the actual value has been determined by analysis then this should be used. Concentrated milk is a general term for milk powder, condensed or evaporated milks and other sources of concentrated milk solids.

5. If the concentrated milk also supplies sugar or fat, these contributions must be calculated:

   Fat contribution = (quantity of concentrated milk) x (% fat)

   Sugar contribution = (quantity of conc. milk) x (% sugar)

6. The quantity of fat required in the mix is calculated from the milk and cream, or milk and cream sources, by subtracting the quantity of fat in the concentrated milk from the total quantity of fat needed in the mix:

   Fat (milk and cream) = fat (mix) – fat (concentrated milk)

7. The quantity of sugar that must be added to the mix is calculated by subtracting the quantity of sugar in the concentrated milk from the total quantity needed in the mix:

   Sugar (needed) = sugar (total) – sugar (concentrated milk)

8. If there is no fat or sugar in the concentrated milk, steps 6 and 7 are not required.

9. The quantity of milk and cream required are calculated by subtracting the total of all other ingredients from the 100 kg of mix:

   Milk and cream = (100) – total quantities of other ingredients)

10. The quantity of cream needed is then calculated as follows:

    Cream = (fat needed – [{milk and cream needed x (% fat in milk)}

    (quantity of fat/kg cream) – (quantity of fat/kg milk)

11. Calculate the quantity of milk needed by subtracting the quantity of cream from the total quantity of milk and cream.
12. Check and confirm that the total weight of all ingredients equals 100 kg.
13. The calculations should be verified by preparing a table listing all the ingredients, their weights and their MSNF, fat and sugar contributions. These should be added and compared with target values to ensure that the calculation has been undertaken correctly.

Dairy Science and Food Technology Ice Cream Mix Calculator

The websites and text books below along with the general instructions above provide a good introduction to the principles of ice-cream mix calculation.

You can use the Dairy Science and Food Technology Ice-Cream Mix calculator to check your calculations. This uses linear programming to solve the mix calculation variables.

8

Dairy Microbiology, Starter Cultures, Inhibitors and Antimicrobial Proteins: A Detailed Description

DAIRY MICROBIOLOGY

Microorganisms

Microorganisms are living organisms that are individually too small to see with the naked eye. The unit of measurement used for microorganisms is the micrometer (μ m); 1 μ m = 0.001 millimeter; 1 nanometer (nm) = 0.001 μ m. Microorganisms are found everywhere (ubiquitous) and are essential to many of our planets life processes. With regards to the food industry, they can cause spoilage, prevent spoilage through fermentation, or can be the cause of human illness.

There are several classes of microorganisms, of which bacteria and fungi (yeasts and moulds) will be discussed in some detail. Another type of microorganism, the bacterial viruses or bacteriophage, will be examined in a later section.

Bacteria

Bacteria are relatively simple single-celled organisms. One method of classification is by shape or morphology:

- *Cocci:*
  - — spherical shape
  - — 0.4 - 1.5 μ m

  Examples: staphylococci - form grape-like clusters; streptococci - form bead-like chains

- *Rods:*

  — 0.25 - 1.0 μ m width by 0.5 - 6.0 μ m long

  Examples: bacilli - straight rod; spirilla - spiral rod

There exists a bacterial system of taxonomy, or classification system, that is internationally recognized with family, genera and species divisions based on genetics.

Some bacteria have the ability to form resting cells known as endospores. The spore forms in times of environmental stress, such as lack of nutrients and moisture needed for growth, and thus is a survival strategy. Spores have no metabolism and can withstand adverse conditions such as heat, disinfectants, and ultraviolet light. When the environment becomes favourable, the spore germinates and giving rise to a single vegetative bacterial cell. Some examples of spore-formers important to the food industry are members of *Bacillus* and *Clostridium* generas.

Bacteria reproduce asexually by fission or simple division of the cell and its contents. The doubling time, or generation time, can be as short as 20-20 min. Since each cell grows and divides at the same rate as the parent cell, this could under favourable conditions translate to an increase from one to 10 million cells in 11 hours! However, bacterial growth in reality is limited by lack of nutrients, accumulation of toxins and metabolic wastes, unfavourable temperatures and dessication. The maximum number of bacteria is approximately 1 X 10e9 CFU/g or ml.

Note: Bacterial populations are expressed as colony forming units (CFU) per gram or millilitre.

Bacterial growth generally proceeds through a series of phases:

- Lag phase: time for microorganisms to become accustomed to their new environment. There is little or no growth during this phase.
- Log phase: bacteria logarithmic, or exponential, growth begins; the rate of multiplication is the most rapid and constant.
- Stationary phase: the rate of multiplication slows down due to lack of nutrients and build-up of toxins. At the same time, bacteria are constantly dying so the numbers actually remain constant.
- Death phase: cell numbers decrease as growth stops and existing cells die off.

The shape of the curve varies with temperature, nutrient supply, and other growth factors. This exponential death curve is also used in modeling the heating destruction of microorganisms.

Yeasts

Yeasts are members of a higher group of microorganisms called fungi . They are single-cell organisms of spherical, elliptical or cylindrical shape. Their size varies greatly but are generally larger than bacterial cells. Yeasts may be divided into two groups according to their method of reproduction:

1. budding: called Fungi Imperfecti or false yeasts
2. budding and spore formation: called Ascomycetes or true yeasts

Unlike bacterial spores, yeast form spores as a method of reproduction.

Moulds

Moulds are filamentous, multi-celled fungi with an average size larger than both bacteria and yeasts (10 X 40 μ m). Each filament is referred to as a hypha. The mass of hyphae that can quickly spread over a food substrate is called the mycelium. Moulds may reproduce either asexually or sexually, sometimes both within the same species.

Asexual Reproduction

- fragmentation - hyphae separate into individual cells called arthropsores
- spore production - formed in the tip of a fruiting hyphae, called conidia, or in swollen structures called sporangium

Sexual Reproduction: sexual spores are produced by nuclear fission in times of unfavourable conditions to ensure survival.

Microbial Growth

There are a number of factors that affect the survival and growth of microorganisms in food. The parameters that are inherent to the food, or intrinsic factors, include the following:

- nutrient content
- moisture content
- pH
- available oxygen
- biological structures
- antimicrobial constituents

Nutrient Requirements: While the nutrient requirements are quite organism specific, the microorganisms of importance in foods require the following:

- water
- energy source
- carbon/nitrogen source
- vitamins
- minerals

Milk and dairy products are generally very rich in nutrients which provides an ideal growth environment for many microorganisms.

Moisture Content: All microorganisms require water but the amount necessary for growth varies between species. The amount of water that is available in food is expressed in terms of water activity (aw), where the aw of pure water is 1.0. Each microorganism has a maximum, optimum, and minimum aw for growth and survival. Generally bacteria dominate in foods with high aw (minimum approximately 0.90 aw) while yeasts and moulds, which require less moisture, dominate in low aw foods (minimum 0.70 aw). The water activity of fluid milk is approximately 0.98 aw.

pH: Most microorganisms have approximately a neutral pH optimum (pH 6-7.5). Yeasts are able to grow in a more acid environment compared to bacteria. Moulds can grow over a wide pH range but prefer only slightly acid conditions. Milk has a pH of 6.6 which is ideal for the growth of many microorganisms.

Available Oxygen: Microorganisms can be classified according to their oxygen requirements necessary for growth and survival:

- Obligate Aerobes: oxygen required
- Facultative: grow in the presence or absence of oxygen
- Microaerophilic: grow best at very low levels of oxygen
- Aerotolerant Anaerobes: oxygen not required for growth but not harmful if present
- Obligate Anaerobes: grow only in complete absence of oxygen; if present it can be lethal

Biological Structures: Physical barriers such as skin, rinds, feathers, etc. have provided protection to plants and animals against the invasion of microorganisms. Milk, however, is a fluid product with no barriers to the spreading of microorganisms throughout the product.

Antimicrobial Constituents: As part of the natural protection against microorganisms, many foods have antimicrobial factors. Milk has several nonimmunological proteins which inhibit the growth and metabolism of many microorganisms including the following most common:

1. lactoperoxidase
2. lactoferrin
3. lysozyme
4. xanthine

More information on these antimicrobials can be found in a chapter on dairy microbiology and safety written by Vasavada and Cousin.

Where the intrinsic factors are related to the food properties, the extrinsic factors are related to the storage environment. These would include temperature, relative humidity, and gases that surround the food.

Temperature: As a group, microorganisms are capable of growth over an extremely wide temperature range. However, in any particular environment, the types and numbers of microorganisms will depend greatly on the temperature. According to temperature, microorganisms can be placed into one of three broad groups:

- Psychrotrophs: optimum growth temperatures 20 to 30° capable of growth at temperatures less than 7° C. Psychrotrophic organisms are specifically important in the spoilage of refrigerated dairy products.
- Mesophiles: optimum growth temperatures 30 to 40° C; do not grow at refrigeration temperatures
- Thermophiles: optimum growth between 55 and 65° C

It is important to note that for each group, the growth rate increases as the temperature increases only up to an optimum, afterwhich it rapidly declines.

Detection and Enumeration of Microorganisms

There are several methods for detection and enumeration of microorganisms in food. The method that is used depends on the purpose of the testing.

Direct Enumeration

Using direct microscopic counts (DMC), Coulter counter etc. allows a rapid estimation of all viable and nonviable cells. Identification through staining and observation of morphology also possible with DMC.

Viable Enumeration

The use of standard plate counts, most probable number (MPN), membrane filtration, plate loop methos, spiral plating etc., allows the estimation of only viable cells. As with direct enumeration, these methods can be used in the food industry to enumerate fermentation, spoilage, pathogenic, and indicator organisms.

Metabolic Activity Measurement

An estimation of metabolic activity of the total cell population is possible using dye reduction tests such as resazurin or methylene blue dye reduction, acid production, electrical impedence etc. The level of bacterial activity can be used to assess the keeping quality and freshness of milk. Toxin levels can also be measured, indicating the presence of toxin producing pathogens.

Cellular Constituents Measurement

Using the luciferase test to measure ATP is one example of the rapid and sensitive tests available that will indicate the presence of even one pathogenic bacterial cell.

Isolation of microorganisms is an important preliminary step in the identification of most food spoilage and pathogenic organisms. This can be done using a simple streak plate method.

Microorganisms in Milk

Milk is sterile at secretion in the udder but is contaminated by bacteria even before it leaves the udder. Except in the case of mastisis, the bacteria at this point are harmless and few in number. Further infection of the milk by microorganisms can take place during milking, handling, storage, and other pre-processing activities.

Lactic acid bacteria: this group of bacteria are able to ferment lactose to lactic acid. They are normally present in the milk and are also used as starter cultures in the production of cultured dairy products such as yogurt. Note: many lactic acid bacteria have recently been reclassified; the older names will appear in brackets as you will still find the older names used for convenience sake in a lot of literature. Some examples in milk are:

- lactococci
    - *L. delbrueckii* subsp. *lactis* (*Streptococcus lactis*)
    - *Lactococcus lactis* subsp. *cremoris* (*Streptococcus cremoris*)
- lactobacilli
    - *Lactobacillus casei*
    - *L.delbrueckii* subsp. *lactis* (*L. lactis*)

  - o *L. delbrueckii* subsp. *bulgaricus* (*Lactobacillus bulgaricus*)
- *Leuconostoc*

Coliforms: coliforms are facultative anaerobes with an optimum growth at 37° C. Coliforms are indicator organisms; they are closely associated with the presence of pathogens but not necessarily pathogenic themselves. They also can cause rapid spoilage of milk because they are able to ferment lactose with the production of acid and gas, and are able to degrade milk proteins. They are killed by HTST treatment, therefore, their presence after treatment is indicative of contamination.*Escherichia coli* is an example belonging to this group.

Significance of Microorganisms in Milk

- Information on the microbial content of milk can be used to judge its sanitary quality and the conditions of production
- If permitted to multiply, bacteria in milk can cause spoilage of the product
- Milk is potentially susceptible to contamination with pathogenic microorganisms. Precautions must be taken to minimize this possibility and to destroy pathogens that may gain entrance
- Certain microorganisms produce chemical changes that are desirable in the production of dairy products such as cheese, yogurt.

Spoilage Microorganisms in Milk

The microbial quality of raw milk is crucial for the production of quality dairy foods. Spoilage is a term used to describe the deterioration of a foods' texture, colour, odour or flavour to the point where it is unappetizing or unsuitable for human consumption. Microbial spoilage of food often involves the degradation of protein, carbohydrates, and fats by the microorganisms or their enzymes.

In milk, the microorganisms that are principally involved in spoilage are psychrotrophic organisms. Most psychrotrophs are destroyed by pasteurization temperatures, however, some like *Pseudomonas fluorescens, Pseudomonas fragi* can produce proteolytic and lipolytic extracellular enzymes which are heat stable and capable of causing spoilage.

Some species and strains of *Bacillus, Clostridium, Cornebacterium, Arthrobacter, Lactobacillus, Microbacterium, Micrococcus* , and *Streptococcus* can survive pasteurization and grow at refrigeration temperatures which can cause spoilage problems.

Pathogenic Microorganisms in Milk

Hygienic milk production practices, proper handling and storage of milk, and mandatory pasteurization has decreased the threat of milkborne diseases such as tuberculosis, brucellosis, and typhoid fever. There have been a number of foodborne illnesses resulting from the ingestion of raw milk, or dairy products made with milk that was not properly pasteurized or was poorly handled causing post-processing contamination. The following bacterial pathogens are still of concern today in raw milk and other dairy products:

- *Bacillus cereus*
- *Listeria monocytogenes*
- *Yersinia enterocolitica*
- *Salmonella* spp.
- *Escherichia coli* O157:H7
- *Campylobacter jejuni*

It should also be noted that moulds, mainly of species of *Aspergillus* , *Fusarium* , and *Penicillium* can grow in milk and dairy products. If the conditions permit, these moulds may produce mycotoxins which can be a health hazard.

HACCP

Raw and end-products may be tested for the presence, level, or absence of microorganisms. Traditionally these practices were used to reduce manufacturing defects in dairy products and ensure compliance with specifications and regulations, however, they have many drawbacks:

1. destructive and time consuming
2. slow response
3. small sample size
4. delays in the release of the food

In the 1960's, the Pillsbury Company, the U.S. Army, and NASA introduced a system for assuring pathogen-free foods for the space programme. This system, called Hazard Analysis and Critical Control Points (HACCP), is a focus on critical food safety areas as part of total quality programmes. It involves a critical examination of the entire food manufacturing process to determine every step where there is a possibility of physical, chemical, or microbiological contamination of the food which would render it unsafe or unacceptable for human consumption. These identified points are the critical control points (CCP). There are seven prinicples to HACCP:

1. analyze hazards
2. determine CCPs
3. establish critical limits
4. establish monitoring procedures
5. establish deviation procedures
6. establish verification procedures
7. establish record keeping procedures

Before these principles can be put into place, a prerequisite programme and preliminary setup is necessary.

Prerequisite Program

- premise control
- receiving and storage control
- equipment performance and maintenance control
- personnel training
- sanitation
- recall procedure

Preliminary Setup

- assemble team
- describe the product
- identify intended use
- construct flow diagram and plant schematic
- verify the diagram on-site

Food Safety Enhancement Program-FSEP is The Canadian Food Inspection Agency's HACCP initiative. There is extensive information at their Web site regarding FSEP, including implementation manuals, HACCP curriculum guidelines, and generic models.

Starter Cultures

Starter cultures are those microorganisms that are used in the production of cultured dairy products such as yogurt and cheese. The natural microflora of the milk is either inefficient, uncontrollable, and unpredictable, or is destroyed altogether by the heat treatments given to the milk. A starter culture can provide particular characteristics in a more controlled and predictable fermentation. The primary function of lactic starters is the production of lactic acid from lactose. Other functions of starter cultures may include the following:

- flavour, aroma, and alcohol production
- proteolytic and lipolytic activities
- inhibition of undesirable organisms

There are two groups of lactic starter cultures:

1. simple or defined: single strain, or more than one in which the number is known
2. mixed or compound: more than one strain each providing its own specific characteristics

Starter cultures may be categorized as mesophilic or thermophilic:

Mesophilic

- *Lactococcus lactis* subsp. *cremoris*
- *L. delbrueckii* subsp. *lactis*
- *L. lactis* subsp. *lactis* biovar *diacetylactis*
- *Leuconostoc mesenteroides* subsp. *cremoris*

Thermophilic

- *Streptococcus salivarius* subsp. *thermophilus* (*S.thermophilus*)
- *Lactobacillus delbrueckii* subsp. *bulgaricus*
- *L. delbrueckii* subsp. *lactis*
- *L. casei*
- *L. helveticus*
- *L. plantarum*

Mixtures of mesophilic and thermophilic microorganisms can also be used as in the production of some cheeses.

Bacteriophage

Bacteriophages are viruses that require bacteria host cells for growth and reproduction. Initially, the bacteriophage attaches itself to the bacteria cell wall and injects nuclear substance into the cell. Inside the cell, the nuclear substance produces shells, or phage coats, for the new bacteriophage which are quickly filled with nucleic acid. The bacterial cell ruptures and dies as the new bacteriophage are released.

Bacteriophages are ubiquitous but generally enter the milk processing plant with the farm milk. They can be inactivated heat treatments of 30 min at 63 to 88° C, or by the use of chemical disinfectants.

Bacteriophages are of most concern in cheese making. They attack and destroy most of the lactic acid bacteria which prevents normal ripening known as slow or dead vat.

Starter Culture Preparation

Commercial manufacturers provide starter cultures in lyophilized (freeze-dryed), frozen or spray-dried forms. The dairy product manufacturers need to inoculate the culture into milk or other suitable substrate. There are a number of steps necessary for the propagation of starter culture ready for production:

1. Commercial culture
2. Mother culture - first inoculation; all cultures will originate from this preparation
3. Intermediate culture - in preparation of larger volumes of prepared starter
4. Bulk starter culture - this stage is used in dairy product production

MICROBIOLOGY OF STARTER CULTURES

The cultures of microorganisms used in the manufacture of cheese and other fermented milk products are known as starters. While these consist mainly of bacteria, cultures containing yeasts and fungi are also used.

Role of Starters in Dairy Fermentations

Starters play a vital part in the manufacture of these products; they produce the lactic acid that influences important quality characteristics such as texture, moisture content, freedom from pathogenic microorganisms and their toxins, and taste. The rate of acid production is critical in the manufacture of certain products, e.g. Cheddar cheese. Depending on the product, especially in mechanised cheese production units, starters may also be required to produce acid at a consistently fast rate through the manufacturing period each day and every day. The negative redox potential created by starter growth in cheese also aids in preservation and the development of flavour in Cheddar and similar cheeses. Additionally antibiotic substances, now referred to as bacteriocins, produced by starters, e.g. nisin may also have a role in preservation (Mullan, 1986).

Ecology of Starter Bacteria

Most starters in use to today have originated from lactic acid bacteria originally present as part of the contaminating microflora of milk. These bacteria have probably originated from vegetation in the case of lactococci (Sandine *et al.*, 1972) or the intestinal tract in the case of *Bifidobacterium* spp., enterococci and *Lactobacillus acidophilus*.

Modern starter cultures have developed from the practice of retaining small quantities of whey or cream from the successful manufacture of a fermented product on a previous day and using this as the inoculum or starter for the preceding day's production. This practice has been called various names but the term 'back-slopping' is used widely.

Classification of Starter Bacteria

The bacteria used in the manufacture of fermented dairy products are generally lactic acid bacteria (LAB); however, *Propionibacterium shermanii* and *Bifidobacterium* spp. which are not lactic acid bacteria, although *Bifidobacterium* species do produce lactic acid, are also used. In addition, other bacteria including *Brevibacterium linens*, responsible for the flavour of Limburger cheese; and moulds (*Penicillium* species) are used in the manufacture of Camembert, Roquefort and Stilton cheeses.

Enterococcus

These organisms are Gram-positive, catalase negative cocci that tend to form chains of varying length. They are normal inhabitants of the intestinal tract of man and other animals and are often used in microbiology as indicators of faecal contamination; some species of the genus are pathogens. Apart from their ability to grow at 45°C, at pH 9.6, in high concentrations of salt, in high concentrations of bile salts, their general heat tolerance and their insensitivity to a range of antimicrobial agents they are superficially similar to lactococci. The biochemical identification key developed by Manero and Blanch (1999) is particularly helpful in identifing enterococci.

There are concerns about enterococci in foods partly because some are pathogens. However, it is their ability to exchange antibiotic resistance genes, particularly for glycopeptide antibiotics (vancomycin and teicoplanin), that perhaps raises most concern. Vancomycin is one of only a small number of antibiotics that may be effective against methicillin-resistant *Staphylococcus aureus* (MRSA).

Prior to recent taxonomic research, the *Enterococcus* species used as starters were classified as faecal streptococci and Group D streptococci.The posts in the forum on this group may also be of interest to readers.

Leuconostoc

Leuconostoc species are important flavour producers in some fermented dairy products. There is general agreement that two species, *Leuconostoc mesenteroides* subsp. *cremoris* and *Lecon. lactis* are important in starter cultures. Unlike lactococci, leuconostocs grow on Rogosa agar and are hetrofermentative producing carbon dioxide from glucose and usually fructose. While the carbon dioxide production is undesirable in Cheddar cheese gas production is desirable in some varieties e.g. Emmental.

On microscopic examination, leuconostocs generally appear as Gram-positive cocci similar in size and shape (occur in pairs and in usually short chains) to lactococci. However, small rods can often be found and since leuconostocs grow on Rogosa agar, there can be a tendency to assume that these cultures are contaminated, with lactobacilli for example. Unlike lactococci, leuconostocs do not produce ammonia from arginine and produce the D isomer of lactic acid. With some exceptions leuconostocs only grow weakly in milk, and are not capable of reducing litmus before coagulation in litmus milk medium.

Isolation and identification of leuconostocs in starters is time consuming and laborious and the author has found that the use of Rogosa agar to obtain initial isolates helpful. Carbohydrate fermentation and identification of the lactic acid isomer are useful elements in an identification protocol.

Streptococcus

Str. thermophilus is the only species of this genus found in starter cultures. This streptococcus is classified as a thermophile growing at 45°C, and higher, and is widely used in the manufacture of yoghurt and in Mozzarella, and in some other cheeses. More recently, probably since the mid-1990s-it has been used widely in the manufacture of Cheddar cheese. It is a component, along with lactococci, in some DVI/DVS cultures where it produces acid rapidly during scalding and may confer an additional measure of bacteriophage (phage) protection. Its incorporation in Cheddar-cultures also has the advantage of increasing the profitability of DVI/DVS cultures to culture-suppliers.

The slide shown in plate 1 was obtained by Gram-staining a yoghurt preparation; the cocci are cells of *Str. thermophilus* and the rods are cells of *Lb. delbrueckii* subsp. *bulgaricus*. Like lactococci and many leuconostocs, strains of *Str. thermophilus* are catalase-negative; cocci shaped and occur in pairs and chains. Generally, most strains produce long chains. L-lactic acid only is produced and carbon dioxide is not produced from glucose. Some strains produce urease and have the potential to produce CO_2 from urea. Since *Str. thermophilus* can grow in the regeneration section of pasteurisers high levels can occasionally occur in cheese. Urease-producing strains have the potential to cause openness in cheese (Mullan, 2000).

Strains differ in their ability to utilise galactose. Use of non-galactose fermenting strains will result in high levels of this reducing sugar in products. Since galactose and other reducing sugars react with amino acids in the Maillard reaction it is usual to only select galactose-utilising strains to reduce the probability of undesirable colour changes occurring in heated products.

Str. salivarius, a streptococcus commonly found in saliva, has been shown by DNA: DNA hybridization studies to be similar to *Str. thermophilus*. Because of this, for some years *Str. thermophilus* was classified as a subspecies of Str. salivarius. However, it is now accepted that *Str. thermophilus* while similar, is sufficiently distinct to justify species designation.

Str. thermophilus is sensitive to low levels of salts and to high osmotic strength media. M17 medium (Terazghi and Sandine, 1975) widely used in studies with lactococci is not an ideal medium for the growth of some strains unless modified to reduce its osmotic strength by the reduction of its glycerophosphate content.

Lactobacillus

This genus consists of a large group of Gram positive, catalase negative, rod-shaped bacteria. Some species are homofermentative while others are hetrofermentative. While some species produce mainly L-lactate from glucose, others produce D-lactate. Since some strains exhibit significant racemase activity, a racemase is an isomerase enzyme, D/L lactic acid is also produced. Strains may also exhibit coccoid morphology and this can lead, as discussed previously, to confusion with leuconostocs and perhaps even lactococci.

Lactobacilli are used as starters in the manufacture of yoghurt, and Mozzarella cheese. They are also used as starter adjuncts to promote faster ripening of Cheddar and similar cheeses, to reduce the incidence of bitterness and as probiotics in yoghurt type products. Note that there are several postings concerning the control of bitterness in Cheddar and Gouda cheeses in the discussion area. Note registration is required to obtain a username and password for entry to this part of the website.

Lb. delbrueckii subsp. *bulgaricus* is widely used along with *Str. thermophilus* as a starter in yoghurt manufacture. This subspecies is homofermentative, produces almost 2% w/v lactic acid in milk, has an optimum temperature of 42° and grows at temperatures of 45°C and higher. It will not grow in low concentrations of salt and is sensitive to bile salts.

Lb. acidophilus, which is normally present in the intestine, is generally not used as a starter; it is widely used as a probiotic. This bacterium, is homofermentative, producing high concentrations of D-lactic acid in milk, has an optimum temperature of 37°C, and is relatively tolerant of oxygen, compared with Bifidobacterium species that are frequently used in conjunction with this organism. Little growth occurs at temperatures less than 20°C and most strains show no growth at 15°C. Because *Lb. acidophilus* produces D-lactate there have been some concerns about its use in infant nutrition. This aspect will be discussed further in the probiotics section.

Lb. casei is also a normal inhabitant of the small intestine and is resistant to bile. It is used as a probiotic although it is found in some starter cultures and is commonly one of a number of non-starter lactic acid bacteria (NSLAB) found in Cheddar cheese. L-lactate is the main isomer of lactose produced although some strains produce small concentrations of D-lactate due to weak racemase activity. Rogosa agar is widely used as a general isolation medium for lactobacilli. Further information on enumeration is given in the article on probiotic bacteria.

Lb.helveticus is frequently used along with other thermophilic lactic acid bacteria in the manufacture of a range of fermented milk products including Emmental cheese, Mozzarella and yoghurt. One advantage of including this species along with *Lb. delbrueckii* subsp. *bulgaricus* is that *Lb.helveticus* utilises galactose and this can be useful if products free of reducing sugars are required. Since many strains have been shown to possess proline-iminopeptidase-like activity, *Lb.helveticus* has been used to produce modified 'Cheddar-type cheese' with some of the 'sweetness' characteristics of Swiss cheeses like Emmental. More recently, designated strains have been used as starter adjuncts to reduce bitterness in a range of cheeses, to improve flavour and/or to accelerate ripening. Bitterness is reduced due to peptidase action on starter-derived hydrophobic peptides. The species is homofermentative and produces high concentrations of D/L lactic acid in milk. Many strains grow at 45°C although lower temperatures 42-43°C generally give higher recoveries when enumerated using selective media such as Rogosa or modified MRS agars. Most strains show no or little growth at 15 °C (some atypical strains may take several weeks to grow at 15 °C or below). See the discussion area for further discussion on bitterness.

Lactococcus

Originally, the bacteria in this group were classified as members of the genus *Streptococcus*. They were differentiated from other streptococci, some of which are pathogens, by their specific reaction with Group N antiserum and by their tolerance to temperature, salt and dyes (Jones, 1978). It is now known that serotyping lactic LAB has limited value in species differentiation; strains of the same species may react with different sera and some strains may exhibit no group antigen (Schleifer and Kilpper-Balz, 1987).

Lactic Streptococci

The lactic group of the genus *Streptococcus* originally included the species *Str. lactis* and *Str. cremoris* and a subspecies of *Str. lactis*, *Str. lactis* subsp. *diacetylactis* (Deibel and Seeley, 1974). However, even in the 1970s workers were suggesting that *Str. lactis* strains might be variants of *Str. diacetylactis* that were unable to ferment citric acid, since citrate permease – negative strains of *Str. diacetylactis* had been described (Lawrence, Thomas and Terzaghi, 1976).

Bacteria in this group were designated as the lactic streptococci. The designation 'lactic' was used by Sherman (1937) for mainly historical reasons, including the use of the term by Lister (1878) to describe a bacterium that we now know as *Lc. lactis* subsp. *lactis*.

Evidence was also accumulating to suggest that *Str. cremoris*, *Str. lactis* and *Str. diacetylactis* formed a phenotypically and genotypically continuous spectrum of variation. The three so-called species were known to be similar in DNA base-composition (Knittel, 1965) indicating that they are closely related. Genetic relationships between the genus *Streptococcus* and the genera *Lactobacillus* and *Leuconostoc* were becoming apparent (London, 1976).

Differentiation of Lactic Streptococci to Species Level

Schemes for the differentiation of lactic streptococci to the species or subspecies level were developed by several groups (Sandine, Elliker and Anderson, 1959; Reiter and Møller-Madsen, 1963; Reddy et. al., 1972; Diebel and Seeley, 1974; Jones, 1978) and are generally still relevant for lactococci. The interpeptide bridge structure that was proposed by Schleifer and Kandler (1972), namely l-Lysine-D-isoasparagine has been confirmed (Schleiferr and Kilpper-Balz, 1987).

Rationale for Assigning Lactic Streptococci to Lactococcus Genus

The lactic streptococci have been assigned to the *Lactococcus* genus because of molecular and chemotaxonomic studies. This work has been reviewed. Essentially the extent of DNA: DNA and DNA-rRNA hybridization, similarity between profiles produced by restriction mapping of chromosomal DNA and the nucleotide sequence of the 16S and 32S RNAs have formed the basis for the creation of the genus. Antisera against purified superoxide dismutase have been used to demonstrate similarity between lactococci but not streptococci or enterococci.

Because citrate utilization is plasmid DNA coded and not coded on the cell chromosome, citrate utilization and the associated diacetyl production are variable traits and the former *Str. diacetylactis* has not been given subspecies status. This genus, which has only recently been established, contains one important starter species *Lactococcus lactis*. This species consists of Gram positive, catalase negative, homofermentative cocci that normally form short to long chains. L-lactic acid only is formed from lactose. *Lactococcus lactis* is divided into two subspecies *Lc. lactis* subsp. *lactis* and *Lc. lactis* subsp. *cremoris* and a biovariant of *Lc. lactis* subsp. *lactis*, designated variously as either *Lc. lactis* subsp. *lactis* biovar. *diacetylactis* or Cit+. *Lc. lactis* subsp. *lactis*. The latter is identical to *Lc. lactis* subsp. *lactis* apart from its ability to utilise citrate in the presence of carbohydrate.

Differentiation of Lactococci to Species Level

Lactococci can be differentiated to the species or biovariant level using the scheme developed for lactic streptococci-see above. Note that lactococci will not grow on Rogosa agar (Bille *et al.*, 1992). Differential, but not selective, media are available and can be useful for quality control and strain isolation purposes. The medium, Reddys' Differential Agar, developed by Reddy *et al.* (1972) is still of value. This medium contains the differential ingredients lactose, calcium citrate, L-arginine and the pH indicator bromocresol purple. This indicator gives yellow and blue/purple colours under acid and alkaline conditions respectively. *Lc. lactis* subsp. Cremoris gives yellow colonies due to acid production from lactose. *Lc. lactis* subsp. lactis while producing acid also produces ammonia from arginine. The ammonia neutralises the acid and eventually produces an alkaline reaction that results in blue/purple coloured colonies. *Lc. lactis* subsp. *lactis* biovar. *diacetylactis* also gives a blue/purple colony. Unlike *Lc. lactis* subsp. *lactis*, however, *Lc. lactis* subsp. *lactis* biovar. *diacetylactis* exhibits zones of clearing around colonies because of citrate utilization. Because some strains of *Lc. lactis* subsp. *lactis* possess only weak arginase activity streaking techniques on an improved version of this medium may be helpful in identifying these strains (Mullan and Walker, 1979).

Types of Starter Culture

Several authors have produced definitions of starters including Lawrence *et al.* (1976). The latter has classified the three main types of starter used commercially in Australia, New Zealand, the UK and North America as follows:

(a) Single-strain starters: single strains of *Str. cremoris* and less commonly *Str. lactis*. These have been used in pairs in some factories in New Zealand and in Scotland but also singly in Australia.

(b) Multi-strain starters: defined mixtures of three or more single strains of *Str. cremoris* and/or *Str. lactis*. Leuconostoc and *Str. diacetylactis* strains may also be used. Multiple-strain starters are frequently referred to as mixed-strain starters in the United States of America.

(c) Mixed-strain starters: mixtures of strains of *Str. cremoris*, *Str. lactis*, *Str. diacetylactis* and leuconostocs. The identity of the component strains is frequently unknown to the user and their composition may vary on subculture.

This classification is now of limited value for şeveral reasons. This scheme was developed before the combined use of lactococci and *Str. thermophilus* as starters for Cheddar and similar cheeses became common. Incidentally the so called stabilised cultures, used in the manufacture of some soft cheeses contain blends of lactococci and *Str. thermophilus*. In practice it now usual to define starters as either defined, meaning that the strain and species of the component strains are known or as undefined cultures. Included in this latter grouping are the artisanal cultures widely used in some European countries, particularly in Italy.

Some Observations on Artisanal Cultures

Artisanal cultures are of significant scientific and technological interest. Natural whey starters, despite their unpredictable performance, are still used extensively, for example, in the manufacture of Mozzarella cheese using milk obtained from water buffaloes (*Bubalus arnee*) in Southern Italy. Water-buffalo whey starters are derived from the whey of a previously successful batch of cheese and are generally stored at ambient temperature for 24 h prior to use. Relatively little research has been undertaken on these natural starters but they are known to contain leuconstocs, lactobacilli, lactococci and frequently streptococci

During a study to characterise the bacteriophage sensitivity of strains isolated from whey starters we (Aprea *et al.*, 2005) observed large numbers of lactic acid bacteria containing inclusion bodies e.g. fig. 1. Inclusion bodies are often found in bacteria grown under certain conditions and may be composed of the biopolymers poly-ß-hydroxybutrate, polyphosphate, sulphur, lipid or polysaccharide.

Large inclusion bodies may also be mistaken for endospores. Using specific staining techniques it is possible to distinguish and identify the inclusion bodies present.

Using Neisser staining, inclusions appeared purple/black-characteristic of polyphosphate (polyP). The inclusions were subsequently confirmed as polyP by their unique yellow fluorescence, when stained with 4, 6-diamidino-2-phenylindole dihydrochloride (DAPI). Because of the potential role of polyP in intracellular pH control, where it may assist in regulation of pH in a low pH environment, the presence of this polymer in starter bacteria may be an environmental adaptation to the typical high acidity storage conditions of natural starters.

Purple/black inclusions of polyphosphate observed under Neisser stain

A, *Lb. paracasei* 4B-10,

B, *Lb. fermentum* 5BL6.

Yellow fluorescence of polyphosphate under DAPI staining.

C, *Lb. paracasei* 4B-10,

D, *Lb. fermentum* 5BL6.

The detrimental effects of storing starter cultures under low pH and at elevated temperatures on their subsequent growth and acid-producing potential are well documented.

Since natural whey starters are subject to variable storage temperatures under high acid conditions for extended periods and generally function satisfactorily, it would appear that they have some resistance to the detrimental effects of high acidity. It is possible that the polyP metabolism of these starters may contribute, at least in part, to this resistance. Obviously further study is required to confirm this hypothesis. Interestingly, the BBC in Northern Ireland picked up on some of this work.

For more information on the role of polyphosphate in environmental microorganisms read the contribution from Dr Alan Mullan of Queen's University Questor Centre.

Practical Use of Starter Cultures

Cheesemakers generally use starters prepared in two main forms;bulk starter or starter concentrates.

This section will discuss bulk starter manufacture in detail. In particular how the starter medium is sterilised, protected from contamination during cooling, the importance of head space control, how the starter vessel is inoculated aseptically and how phage-contamination can be avoided during incubation. Mention will also be made of systems for maintaining internal and external pH control and how common problems, as investigated by the author, can be solved.

INHABITORS IN MILK

Growth and acid production by starter cultures may be inhibited by bacterial viruses, bacteriophages, or added substances including antibiotics, sterilant and detergent residues, or free fatty acids produced by or as a result of the growth of microorganisms, and natural often called indigenous antimicrobial proteins. These inhibitors include-

- Antibiotics
- Nisin
- Sterilant and detergent residues
- Free fatty acids
- Indigenous antimicrobial proteins

Antibiotics

Milk should not contain antibiotic residues. Milk production is regulated by the Dairy Products (Hygiene) Regulations 1995. These regulations include the standards for raw milk. Prior to 1990 milk was deemed to be contaminated if an antibiotic concentration of > 0.01 international units (iu) /ml was present, the standard has now been increased to 0.006 iu/ml. Manufacturers buying milk from producers impose stringent financial penalties on farmers producing contaminated milk and have procedures to exclude this from the food chain. Despite legislation and financial penalties, there is evidence to suggest that residues occasionally still cause problems. In a survey of the causes of slow acid production by cheese starters in the

UK (Boyle and Mullan, 2000, unpublished results) found that, some 28 % of respondents attributed slow acid problems to antibiotics.

Antibiotics gain entry to milk because of mastitis treatment; mastitis means inflammation of the udder. Although this term includes all inflammatory conditions of the udder, it isdefined here as a bacterial infection of the udder. The common causative organisms of mastitis in the UK are *Str. agalactiae*, *Str. dysgalactiae, coagulase-negative* staphylococci *and Staphylococcus aureus.*

The antibiotics used in veterinary medicine belong to six major groups:-

Aminoglycosides e.g. gentamicin

Penicillins and cephalosporins (ß-Lactams) e.g. cloxacillin

Macrolides-e.g. erythromycin

Quinolones and fluroquinolones see World Health Organization

Sulphonamides e.g. trimethoprim

Tetracyclines e.g. tetracycline

In the recent past, penicillin G was one of the most commonly used antibiotics. While penicillin G still has value, particularly in treating streptococcal infections, the emergence of penicillin resistant strains, the finding of new, penicillin G insensitive causal agents of mastitis and the continual search for, and introduction of, new therapeutic agents has resulted in the use of a large number of antibiotics. Penicillin

G, a member of the ß-lactam group of antibiotics is rapidly inactivated by ß-lactamase, an enzyme produced by staphylococci and other bacteria. However, ß-Lactams can be chemically modified to produce semi-synthetic penicillins that are resistant to most ß- lactamases e.g. cloxacillin.

Antibiotics are frequently administered to infected animals in a solution or suspension that is infused into the infected quarter or quarters of the udder. The antibiotic preparations available in the United Kingdom for intramammary infusion include penicillin G, erythromycin, ampicillin, cloxacillin, streptomycin, aureomycin, neomycin, and novobiocin. The sensitivity of some lactococci to antibiotics is given in table 2. Depending upon formulation, quick or slow release of antibiotic may be obtained. Withholding times, during which the milk will contain the antimicrobial agent and must not be added to the uncontaminated milk of the remainder of the herd, are given on the manufacturer's instructions.

The level and duration of antibiotic diffusion into milk depends upon several factors including the particular antibiotic, its concentration and method of preparation (aqueous solution, suspending medium). The method of preparation markedly influences retention and can affect adhesion of the antibiotic to equipment and pipelines.

The amount of antibiotic excreted into milk may vary from eight to 80%; usually it averages about 50%. It is therefore difficult to know the exact amount of antibiotic residue in milk at different milkings after treatment. Generally, the concentration of antibiotic in milk decreases rapidly with successive milkings, usually at an exponential rate.

The extent to which starter inactivity is related to residual antibiotics in milk is not precisely known but as discussed earlier, antibiotics are still thought to be responsible for acidification problems.

Meanwell (1962) made an interesting observation that is still relevant to those factories manufacturing their own bulk starter. Low levels of penicillin in the bulk starter milk had a more pronounced affect on acid production when the culture grown in that milk was used in cheese manufacture compared with acid production by a normal starter inoculated into cheese milk containing a high level of penicillin. For this reason, it is imperative that milk used for starter production is antibiotic free. This is one of the many reasons why so many companies use commercial frozen or freeze-dried starter cultures for direct vat inoculation rather than manufacture their own bulk starter.

TABLE 1
Sensitivity of Starter Cultures to Antibiotics

	Lactococcus lactissubsp. cremoris		*Lactococcus lactis*		*Mixed ormulti-strain*	
	Partial inhibition	*Marked inhibition (3)*	*Partial inhibition (3)*	*Marked*	*Partial inhibition (3)*	*Marked inhibition (3)*
Penicillin*	0.05-0.13	0.21-0.3	0.09-0.15	0.26-0.3	0.1-0.25	0.27-0.3
Tetracycline	0.11-0.16	0.3-0.4	0.09-0.21	0.28-0.65	0.09-0.20	0.29-0.35
Streptomycin	0.52-0.84	1.9-2.0	0.35-0.71	1.9-3.0	0.4-0.7	1.6-3.0
Erythromycin	—	2.0[(1)]	—	2.0[(1)]	0.05-0.10	—
Chloramphenicol	—	5.0[(1)]	—	5.0[(1)]	0.02-0.80	—
Chlortetracycline	0.015[(1)]	0.075[(1)]	—	5.0[(1)]	0.02-0.80	—
Neomycin	—	5.0[(1)]	5.0[(1)]	30.0[(1)]	2.5-3.5	—
Polymyxin B*	50[(1)]	300[(1)(2)]	300[(2)]	—	—	—
Ampicillin	—	2.0[(1)]	—	2.0[(1)]	—	—
Novobiocin	—	5.0[(1)]	—	5.0[(1)]	—	—
Cloxacillin	1.16-2.05	2.2-4.6	1.6-2.5	3.9-5.0	1.0-2.2	3.0-4.5
Bacitracin	—	—	—	—	0.3-0.5	2.0-3.0

* Concentration expressed in International units ml^{-1}. Concentration of other antibiotics in mg ml^{-1}. [(1)] Determined using an agar diffusion method. [(2)] Markedly strain dependent. [(3)] Ranging from a reduction in acid production of 80% to complete cessation of acid production except for (1). Table taken from Haverbeck *et al.* (1983).

The effects of residues other than penicillin on starter activity during cheese manufacture has not been well documented. One of the difficulties is many current tests for antibiotics have been originally designed to detect penicillin and quality assurance tests using these methods may not detect other antibiotics or else give a false low indication of their concentration. In addition, it is possible that milk may contain several different antibiotics, and that they operate synergistically to inhibit starter growth although present in low concentration.

In the past, some manufacturers of cheese and yoghurt added penicillinase to milk intended for starter manufacture. This treatment will inactivate the natural penicillins but not synthetic penicillins, for example, cloxacillin or non ß-Lactams.

On farm, testing of milk tanks combined with individual tanker load sampling is being used to detect and prevent contaminated milk from entering the supply chain. This practice combined with the extensive bulking of milk at modern dairies will normally prevent major acidification problems due to antibiotic contamination.

Production of Nisin

Raw milk contains lactococci. If favourable conditions for growth occur then high numbers will result. Some *Lc. lactis* strains produce the antibiotic nisin. Nisin is a broad spectrum antibiotic and if produced it will inhibit some starter cultures. Providing there is adequate control of temperature during the production, storage and distribution of milk nisin production should not be a problem.

Sterilant and Detergent Residues

Sterilant and detergent residues may inhibit the growth of starter bacteria. The minimum concentration required for inhibition varies with the different anti-microbial agents and between different strains of starter bacteria. Residues gain entry to milk at the (a) farm, (b) during transport to the factory and (c) the factory due to careless use of sterilants or detergents, incomplete draining or inadequate rinsing of equipment.

The concentration of sterilants required to inhibit the lactococci and other LAB have been studied. In general, the concentration of these compounds in properly produced milk should not markedly inhibit starters. However, the situation with quaternary ammonium compounds is less clear. These compounds are stable in milk and can be difficult to rinse off surfaces.

Nevertheless, quality assurance staff in some plants has ascribed some instances of slow acid production to sterilant and detergent residues. Boyle and Mullan (2000) found that some 17 % of cheese factories attributed acid production problems to these residues.

The inhibitory effects of sterilant and detergent residues are prevented by the correct and ethical use of these materials. Proper use includes the use of the chemical at the correct concentration and adequate rinsing and draining. Their presence is mitigated by dilution with uncontaminated milk.

Free Fatty Acids

Free fatty acids are present at low concentration in freshly drawn milk. Their concentration may increase due to the activity of milk lipase or microbially produced lipases. *Pseudomonas* sps. if permitted to grow in refrigerated milk will produce lipases and high concentrations of. free fatty acids. However, such milk normally contains a total bacterial count in the region of 1 x 10^7 CFU/ml.

Fatty acids are inhibitory to lactococci and in particular to *Lc. lactis* subsp. *cremoris*. However, relatively high levels of fatty acids are required; 0.1% butyric, decanonic, hexanoic and oleic acid were required for the inhibition of *Lc. lactis* subsp. *cremoris*. Such high concentrations of free fatty acids do not normally occur in modern hygienically produced milk that has been held at correct storage temperatures.

Natural Indigenous Antimicrobial Proteins

The ability of raw milk to inhibit the growth of many bacterial species has been known for many years and one of the earliest reports was authored by Hesse in 1894. Jones and his co-workers around 1920 termed the heat labile inhibitors in milk as 'lactenins'. The early work has been comprehensively reviewed by Reiter and MØller-Madsen (1963) and their paper in the Journal of Dairy Research is recommended reading. Recent work has shown that these inhibitors include:

The lactoperoxidase-thiocyanate-hydrogen peroxide (LP) system

- Immunoglobins
- Lysozyme
- Lactoferrin
- Vitamin binding proteins.

The remainder of this section will deal in summary form with the two inhibitors that affect cheese starters, namely the LP system and immunoglobulins. For more information on these inhibitors please see the section on the exploitation of antimicrobial proteins. Note that milk can also contain bioactive peptides some of which may be antimicrobial.

The LP System

Lactoperoxidase is a basic glycoprotein that contains a heme group. It has been well characterised and will be discussed further in the application section of this website. As with peroxidases generally lactoperoxidase catalyses reactions in which hydrogen peroxide is reduced and a suitable electron donor is oxidised. A wide variety of electron donors can be oxidised and some can be used to assay the enzyme.

Bovine milk, but not human milk, contains a high concentration of lactoperoxidase (ca. 30 μg/ml). It constitutes about 1% of the total serum proteins. This enzyme is synthesised in the mammary gland and is a normal constituent of milk. Wright and Cramer (1957) first showed that the inhibition of a particular strain of *Lc. lactis* subsp. *cremoris* (*Str. cremoris* 972) was associated with lactoperoxidase. Later workers demonstrated that hydrogen peroxide and thiocyanate were required in addition to lactoperoxidase for inhibition of acid production to occur.

It is now known that hydrogen peroxide forms a complex with lactoperoxidase; this complex oxidises thiocyanate (SCN^{-} -) to sulphate, carbon dioxide, ammonia, and water via an unstable intermediate oxidation product that is inhibitory to some lactococci, and other LAB bacteria, but kills other bacteria including some pathogens.

It appears likely that the antimicrobial effects are due to oxyacids of thiocyanate e.g. $OSCN^{-}$. The inhibitor(s) is/are heat labile and are inactivated by sulphur containing reducing agents e.g. cysteine, glutathione. Catalase, which breaks down hydrogen peroxide and is present in milk, does not inhibit the lactoperoxidase system unless deliberately added to milk, to give high un-physiological levels.

The mechanism whereby the LP system inhibits or kills bacteria has been well studied. In their study of the mechanism of the inhibition of lactococci Oram and Reiter (1966) found that hexokinase, the enzyme which catalyses the first step in the glycolytic pathway, was completely inhibited. Glucose 6 — phosphate dehydrogenase and aldolase were partly inhibited while phospohexokinase was only slightly affected. Lactococci resistant to the lactoperoxidase system possess an enzyme, which in the presence of $NADH_2$ reduces the intermediate oxidation product and renders it non-inhibitory. Other studies have confirmed that the LP system oxidises essential SH-groups in vital metabolic enzymes and depletes reduced nicotinamide adenine nucleotides,

The thiocyanate content of milk depends largely on the feeding regime and varies from 0.02 mM (during the winter months) to 0.25 mM (during the summer months). Increased levels of SCN^{-} can result from the detoxification of cyanide present in clover or from glucosides contained in certain plants, e.g. SCN^{-} is produced because of the action of rhodanese on glucosides.

Hydrogen peroxide, the third component of the system, is supplied metabolically by the lactococci. In fact, H_2O_2 production by starter cultures is one of the factors that may limit biomass in the production of starter concentrates.

Lactoperoxidase is a relatively heat resistant enzyme; a treatment of 70°C for 20 minutes reduces enzyme levels by 50%, total inactivation can be obtained by heating to 80°C for 5 minutes. Since these treatments are markedly in excess of that achieved in HTST pasteurization (72°C -15s) it should be apparent that lactoperoxidase would normally be present in high concentration in pasteurised milk e.g. cheese milk

The quantitative importance of the lactoperoxidase system as a cause of slow acid production in the dairy industry is not known. Starters for cheese manufacture, for example, are required to produce acid at a fast rate in pasteurised milk. Since pasteurised milk contains lactoperoxidase starters resistant to the LP system must be used. However, resistant starters can give rise to sensitive variants and since starters are generally maintained in high-heat-treated milk, sensitive variants will not be inhibited (no lactoperoxidase present, so there is no pressure to stop the growth of insensitive variants) during normal culturing and may reach high levels. It is not therefore surprising that starters continuously propagated in such media, although originally resistant to the lactoperoxidase system, have been reported to become susceptible to the inhibitors present in raw milk. Stadhouders (1961) has shown that most strains of starters are mixtures of lactoperoxidase sensitive and insensitive bacteria.

There have been been several attempts to commercial the LP system. Applications include an anticaries toothpaste, the preservation of milk in developing countries and the manufacture of calf milk replacers containing the Lp system.

Immunoglobulins

It has been known for many years that certain bacteria, including lactococci, agglutinate in raw milk. Agglutination is caused by relatively specific antibodies that occur in the globulin protein fraction in milk serum, particularly in the euglobulin and pseudoglobulin fractions. These antibodies, which normally originate from the blood of the cow, are always present in milk but occur at high concentration in the colostrum.

Agglutins inhibit acid production in raw and pasteurised whole or skim-milk. There is evidence of two types of antibody response. Response 1 is where sensitive bacteria are attached to fat globules and the second is where antibodies in the skim fraction cause bacteria to 'stick' together, form clumps and sediment to the bottom of the vat. In raw or pasteurised whole milk, the immunogloblins inhibit acid production by facilitating the removal of sensitive bacteria to the cream layer and/or causing sensitive strains to agglutinate, form clumps and sediment to the bottom of the vat. Inhibition may also occur with some strains in skim milk. In this medium, the agglutins cause sensitive strains to form clumps that settle on the bottom of the container and localise acid production.

Starter bacteria are only inhibited in milk by agglutins if the bacteria can rise with the cream or fall to the bottom of the vat. Agglutins are not normally a problem in the manufacture of hard cheeses, but see the comments below concerning long ripening times, since the rennet coagulum immobilises the bacteria in the curd and prevents their movement to the surface with the cream or to the bottom of the vat. Agglutins, however, cause well documented problems in the manufacture of Cottage cheese (a skim-milk cheese). Sensitive starters may exhibit slow acid development, 'sludge' on the bottom of the vat; shattered texture and mealy body in the curd; or complete failure to make cheese. This defect may be avoided by careful selection of cultures and by using simple quality assurance tests to eliminate agglutinin sensitive starters. Essentially test starters are inoculated into tall tubes of pasteurised skim-milk containing a pH or redox indicator, the control is milk treated with rennet. Starters sensitive to agglutination will exhibit localised acid production/redox change at the bottom of the tube. Control swill show homogenous acid/redox change throughout.

The problems with agglutinins have been well documented and it is generally accepted that because of rennet addition these inhibitors do not cause problems in hard cheese manufacture. However, this is perhaps, at best, an oversimplification and may not apply to situations in which extended ripening times, in excess of 1 hour, are used. Under these conditions, there may be significant time for agglutinin sensitive starters to display problems.

Agglutinins are inactivated by heat treatment or homogenization, unlike lactoperoxidase, sometimes known as Lactenin L2, which survives a heat treatment of 71°C for 20 minutes; immunoglublins are inactivated by this treatment. HTST pasteurization (72°C - 15s) partly inactivates the imunogloblins; 50-75% of the agglutating activity is destroyed by this heat treatment. Sterilised milk or bulk starter milk does not contain active agglutinins; the agglutinins have been denatured by the heat treatment.

MAJOR ANTIMICROBIAL PROTEINS MILK

Milk provides the newborn (neonate) with nutrients and an array of antimicrobial factors. These are believed to help protect the neonates from infection until their own immune system has developed.

This section of the dairy science website reviews the properties and potential nutritional and industrial significance of the major antimicrobial system of milk, with particular reference to the lactoperoxidase system.

The major antimicrobial proteins of milk are lysozyme, lactoferrin (Lf), lactoperoxidase (LP), and the immunoglobulins. These have been discussed in summary form in the section on Inhibitors in Milk. In some species minor proteins including a folate binding protein have been characterised. While these 'minor' factors are not discussed here they can be important in certain situations. Goat milk can have significant levels of folate binding protein and it may therefore be advisable to supplement infant formulas with folic acid in addition to the normal 'humanising' compositional adjustments when caprine milk is used in human neonate nutrition.

- Lysozyme
- Lactoferrin
- Lactoperoxidase system
- Assay of lactoperoxidase activity
- Calf milk replacers
- Shelf life of lactoperoxidase-containing milk replacers
- Improved human milk replacers

Lysozyme

Lysozyme (E.C. 3.2.1.17) is a relatively small basic protein and is classified as a 1,4-ß -N-acetylmuramidase.

Lysozyme is present in secretions such as saliva, egg white, milk and blood. Egg white lysozyme and human milk lysozyme are similar proteins (Dubois *et al.*, 1982). Lysozyme cleaves the glycosidic bond between N-acetylmuramic acid and N-acetyglucosamine in bacterial peptidoglycans, which constitute the major part of the bacterial cell wall of gram-positive bacteria.

The susceptibility of different bacteria to lysozyme depends on a number of factors including the accessibility of the substrate and the ionic environment. Gram-positive bacteria are generally more susceptible because they have a much simpler cell wall consisting to a major extent (90%) of peptidoglycan. Staphylococci, on the other hand, are more resistant. This is probably due to the presence of teichoic acid in their cell walls. In gram-negative bacteria the peptidoglycan concentration in the cell wall is much lower (5-10%) and it is protected by an outer layer of lipopolysaccharide, which prevents lysozyme from reaching its substrate. If this barrier is disrupted, which in *vitro* can be achieved by various chemical and physical treatments, many gram-negative bacteria become sensitive to lysozyme.

Because egg-white lysozyme will lyse *Clostridium tyrobutyricum,* lysozyme is used to prevent blowing in cheeses of low lactic acid content e.g. brined-cheeses such as Gouda and Grana Padano.

Lysozymes are generally heat-stable in an acid medium. Bovine milk lysozyme losses only 43% of its activity after heating for 20 min at 100°C at pH 4.0 (Eitenmiller *et al.*, 1975). Human milk lysozyme, under the same conditions, has been reported to loose 85% of its activity (Parry *et al.*, 1969). At an alkaline pH, i.e. pH 9.0, bovine milk lysozyme is inactivated completely within a few minutes at 100°C.

The biological role of lysozyme is still not understood completely. Lysozyme may have an antibacterial role in milk, serum and avian eggs.

While bovine milk normally contains very low levels of lysozyme, i.e. 0.1 μg/ml, mastitic milk, however, contains higher concentrations (1-2 μg/ml). Human milk is much richer in lysozyme and contains on average 100 μg/ml.

It has also been suggested that lysozyme may have an indirect effect on the defence systems as an immunomodulator through the stimulation of the immune system by break down products of the hydrolysis of peptidoglycan (Jolles, 1976). This hypothesis has been supported by the findings that feeding infants lysozyme-enriched formulas results in an increased level of secretory IgA in faeces (Lodinové and Jouja, 1977).

Lf may also be responsible for some false positive results for antibiotics in milk. However, such milk is likely to be of high somatic cell count.

Lactoferrin

There are several closely related iron binding proteins in animals and biological secretions. An iron chelating protein, Lf occurs in many exocrine secretions of mammals. It is closely related to transferrin, a soluble glycoprotein that binds two iron ions (III) per molecule. Lf is also related to an iron chelating protein found in hen egg white, ovotransferrin or conalbumin.

All three iron-containing proteins have similar structures and biological properties. They consist of a single polypeptide chain of molecular weight 80,000 daltons to which one or two carbohydrate chains are attached. The molecules have two symmetrical parts, which each has one site for binding of a ferric iron ion. For each iron ion that is bound one bicarbonate ion is also bound. Lf has been found in milk of some 12 species (Masson, 1970). In humans it also has been shown to be present in most exocrine secretions, including bronchial mucus, tears, saliva, nasal secretions, cervical mucus and gastric juice (Masson, 1970).

In human milk Lf is one of the major whey proteins, constituting 10-30% of the total protein content. Nagasawa *et al.* (1972) have reported levels of Lf in human colostrums and milk of 4.9 and 1.6 mg/ml respectively. In bovine milk the levels are substantially lower, i.e. about 1 mg/ml and 0.2 mg/ml in colostrums and milk, respectively. However, the concentration of lactoferrin in bovine milk increases significantly with inflammatory conditions of the mammary glands, i.e. mastitis.

In the native state Lf in only partially saturated with iron. In human milk, which contains 0.3 to 0.5 μg/ml Fe, only 2-4% of this is bound to Lf. A major part of the Fe in milk is found in the fat phase, mainly as a component of xanthine oxidase. The rest is present in the form of low molecular complexes.

A number of investigations have been carried out to elucidate the biological role of Lf. Although a number of activities have been demonstrated *in vitro*, the function of Lf *in* vivo is not well understood. However, two main roles for LF *in vivo* have been postulated, its antibacterial effect and role in the iron metabolism.

The antibacterial effect of Lf in *vitro* is well established (Bullen & Armstrong, 1979). The antibacterial mechanism has been assigned to the ability of capacity to chelate iron and thereby depriving bacteria of the iron that is essential for growth. This is supported by the fact that Lf saturated with iron does not demonstrate an antibacterial effect. To what extent the antibacterial effect of Lf also is active *in vivo* is less clear. While there is limited evidence of antibacterial effect in humans there is indirect evidence, which suggests that Lf has an important role *in vivo*. Iron is an essential nutrient for both mammalian cells as well as many bacteria. It has been postulated that Lf supplies a source of necessary iron that can be utilized by the host but is not available for bacteria. Growth of the latter are therefore inhibited due to lack of essential iron. However, some bacteria have evolved mechanisms for obtaining iron that has been 'bound' in this way.

Lf may also be responsible for some false positive results for antibiotics in milk. However, such milk is likely to be of high somatic cell count.

Lactoperoxidase System

LP the protein has been discussed previously in the introductory section on inhibitors in milk.

LP (EC 1.11.1.7) is a basic(it is negatively charged below its isoelectric point) glycoprotein that has a molecular weight of 78,000 and contains one heme group. The enzyme has an iron content between 0.068-0.071% and a carbohydrate content ranging from 9.9 - 10.2% (Carlström, 1969). The identity and type of binding of the heme moiety has been investigated. Initially it was proposed that the heme group was covalently bound to the protein moiety via an ester or amide bond (Hultqvist & Morrison, 1963). Slevers (1979) has however shown that no covalent heme-protein bond exists and that the prosthetic group is protoheme IX. Peroxidases catalyse reactions in which hydrogen peroxide is reduced and an appropriate electron donor is subsequently oxidised. A wide variety of organic and inorganic substances can serve as electron donors, but substrate specificity varies between various peroxidases.

LP itself has no antibacterial effect but in combination with certain co-factors, SCN^- and H_2O_2, forms a potent antimicrobial system. The mechanism of this inhibition is largely understood as a result of studies of the inhibition of acid production by lactococci in cheese manufacture.

The antibacterial mechanism is caused by oxidation of vital SH-groups by OSCN- / O_2SCN- (Thomas & Aune, 1978) in vital metabolic enzymes, e.g. hexokinase and/or depletion of reduced nicotinamide adenine nucleotides. A wide variety of bacteria are influenced by the LP-system. With some bacteria the effect is reversible, i.e. the bacteria are inhibited for a certain period of time, for other bacteria the effect is bactericidal. Many gram-positive bacteria such as lactococci and lactobacilli are inhibited while many gram-negative bacteria such as *Escherichia coli*, *Pseudomonas* spp, *Salmonella* spp are killed (Reiter *et al.*, 1976; Björck *et al.*, 1975).

Mammalian cells do not appear to be adversely affected by the LP-system. The LP-system did not have any mutagenic effects when tested against several mutagen-sensitive strains of *Salmonella typhimurium* (White *et al.*, 1983). The activity of the LP system *in vivo* is not fully known. In bovine milk, LP is always present in excess (Björck *et al.*, 1975). Buffalo's milk (Härnulv & Kandasamy, 1982), ewe's milk (Medina *et al.*, (1989) and goat's milk (Zapico *et al.*, 1990) generally also contain high levels of LP. The factors limiting the activity of the LP system in milk from these species are the low concentrations of thiocyanate and hydrogen peroxide present. Some thiocyanate is normally present in milk although the level is variable, i.e. between 1 to 15 mg/l. There is some evidence of *in vivo* production of hydrogen peroxide by lactic acid bacteria (Marschall, 1983; Mullan, Extrand and Waterhouse, unpublished information).

Most investigations of the antibacterial activity of the LP system have been carried out *in vitro*. In these experiments LP, thiocyanate and hydrogen peroxide are added to a suitable medium. In milk, the system can be "activated" by addition of thiocyanate and hydrogen peroxide or an appropriate hydrogen peroxide source. There are several methods for determiningthe concentration and antimicrobial activity of the LP system.

REFERENCES

Abshier, George S. "Labor Requirements for Selected Crops in Maryland." Master's thesis, University of Maryland, 1942.

Acker, Keith G. "Some Curing and Storing Procedures Affecting the Aging of Hams." Master's thesis, University of Maryland, 1935.

Akeley, Richard Warren. "Relationship between the Price and Quality of Canned Vegetables Purchased in Maryland." Master's thesis, University of Maryland, 1941.

Algire, G. W. "The Life History of the Oyster, Oystrea Virginica Gmelin, with Special Emphasis on the Spat Fall." Master's thesis, University of Maryland, 1931.

Anderson, Earl Jennings. "The Association of Certain Chemical and Histological Characters with Susceptibility in Strawberry Roots to Black Root Rot as Influenced by Soil Treatment." PhD diss., University of Maryland, 1937.

Anderson, Otto Watson. "Cow Testing Associations in Maryland." Master's thesis, University of Maryland, 1924.

Anzulovic, J. Venceslav. "A Study of Escherichia Coli in Ice Cream." Master's thesis, University of Maryland, 1931.

Asadorian, Ara A. "Population Trends in Maryland from 1880 to 1930." Master's thesis, University of Maryland, 1938.

Atkin, Maurice D. "An Analysis of the Factors of Price, Supply, and Consumption in the Washington, D.C. Milk Market, with Special Reference to Maryland Producers." Master's thesis, University of Maryland, 1941.

Bailey, Clyde. H. "The Respiration of Shelled Corn." PhD diss., University of Maryland, 1921.

Bailey, Wallace K. "Some Growth Studies on Okra with Special Reference to Time and Rate of Development of Fruit Buds." Master's thesis, University of Maryland, 1933.

Baker, Henry H. "Studies on the Root Growth of Willow Cuttings at Controlled Temperatures." Master's thesis, University of Maryland, 1931.

Baldwin, David H. "Comparison of the Vitamin A Potency between Eggs of Leghorn and Barred Rock Hens." Master's thesis, University of Maryland, 1938.

Baldwin, Willis Harford. "Part I. Chemical Studies on Menhadin Fish Oil. Part II. Isomerism of 9, 10-Dihydroxystearic Acid." PhD diss., University of Maryland, 1942.

Bartilson, Thomas H. "The Lard Problem and the Swine Industry." Master's thesis, University of Maryland, 1941.

Bartlett, John Bruen. "The Role of Organic Matter in Base Exchange." Master's thesis, University of Maryland, 1936.

Bartram, M. Thomas. "The Detection and Significance of the Escherichia — Aerobacter Group of Bacteria in Milk." PhD diss., University of Maryland, 1936.

———. "The Effect of Pasteurizing Temperatures on (a) Brucella Abortus and (B) Brucella Abortus Agglutinins in Milk." Master's thesis, University of Maryland, 1931.

Basehore, W. J. "The Organization and Business Analysis of 265 Farms on the Eastern Shore of Maryland." Master's thesis, University of Maryland, 1932.

Beall, Harry S. "A Suggested Curriculum for a Rural High School with Special Reference to Sherwood High School at Sandy Spring, Montgomery County, Maryland." Master's thesis, George Washington University, 1940.

Beaven, George Francis. "A Study of the Flora of the Pocomoke River Swamp on the Eastern Shore of Maryland." Master's thesis, Duke University, 1936.

Beavens, E. Arthur. "The Significance of Organisms of the Escherichia Aerobacter Group in Raw and Pasteurized Milk." Master's thesis, University of Maryland, 1929.

Beck, Frances F. "The Fate of Erythritol and Erythritan in the Animal Body." Master's thesis, University of Maryland, 1936.

Becker, Julia Ernestine. "The Role of Milk Powder in Nutrition." Master's thesis, Johns Hopkins University, 1928.

Bell, W. E. "An Economic Study of Maryland Livestock Farms in 1934 and 1935." Master's thesis, University of Maryland, 1936.

Bellows, John Morton. "Embryo-Sac Development in a Triploid Tulip." PhD diss., University of Maryland, 1940.

Bennett, Benjamin Hugh. "Production, Distribution and Consumption of Whole Milk in the City of Baltimore." Master's thesis, University of Maryland, 1927.

Berry, Evelyn Lindsey. "A Study of Withdrawals from a Small Rural High School." Master's thesis, College of William and Mary, 1941.

Berry, M. H. "Comparison of Feeding Practices in the Rearing of Dairy Calves." Master's thesis, University of Maryland, 1929.

Besley, A. K. "The Bacterial Content of Powdered Milk." Master's thesis, University of Maryland, 1924.

Besley, H. Elmer. "The Development of an Electric Dairy Utensil Sterilizer." Master's thesis, University of Maryland, 1931.

Bewley, John Patterson. "A Comparison of Available Potassium Tests with the Exchangeable Potassium of Some Maryland Soils." Master's thesis, University of Maryland, 1939.

Binford, Raymond. "The Germ-Cells and the Process of Fertilization in the Crab Menippe Mercenaria." PhD diss., Johns Hopkins University, 1912.

Bingham, Harold Clyde. "Visual Perception of the Chick." PhD diss., Johns Hopkins University, 1922.

Blackmore, John. "An Economic Study of the Potato Enterprise on 103 Farms in Garrett County, Maryland." Master's thesis, University of Maryland, 1938.

Blackwell, Jefferson Davis. "The Organization and Supervision of Vocational Education in Maryland County High Schools." PhD diss., Johns Hopkins University, 1929.

Blaisdell, Dorothy Jane. "Anatomy and Morphology of the Embryo and Seedling of Cyperus Esculentus L." Master's thesis, University of Maryland, 1933.

Blandford, Josephine M. "The Effect of Castration Upon the Palatability of Market Lamb." Master's thesis, University of Maryland, 1929.

Blood, Pearle. "A Profile Study of Erosion Surfaces in Pennsylvania and Maryland." Master's thesis, Columbia University, 1924.

Blunt, Forrest Percival. "The Development of the Public (White) High School in the Counties of Maryland from 1865 to 1930." Master's thesis, University of Maryland, 1932.

Bond, Ridgely B. "The Effects of Colloids on Earth and Boneblack Filtration in Sugar Refining." Master's thesis, University of Maryland, 1935.

Boswell, Victor R. "A Study of Some Environmental Factors Influencing the Shooting to Seed of Wintered-over Cabbage." PhD diss., University of Maryland, 1926.

Bouis, George E. "Roadside Markets in Maryland." Master's thesis, University of Maryland, 1926.

Bowers, J. L. "A Greenhouse Study of the Influence of Different Fertilizers on the Production of the Alaska Peas Grown on Four Different Soils." Master's thesis, University of Maryland, 1939.

Bowman, J. J. "Studies in the Prevention of Decay in Orange Fruits." Master's thesis, University of Maryland, 1929.

Boyer, W. C. "Some Effects of Freezing on the Chemical, Physical, and Nutritional Properties of Milk." Master's thesis, University of Maryland, 1932.

Brambel, Charles Eugene. "Allantochorionic Differentiations of the Pig Studied Morphologically and Histochemically." PhD diss., Johns Hopkins University, 1931.

Breyer, Frank Gottlob. "The Fixation of Atmospheric Nitrogen, Particularly in the Manufacture of Nitric Acid and Nitrates." Master's thesis, Johns Hopkins University, 1910.

Bristow, Rosa L. St. Clair. "Participation of Parents in the Development of Home Economics Programmes in Four Maryland High Schools." Master's thesis, University of Maryland, 1937.

Brittingham, Otto Washington. "Free Association Test and Vascular Variations." Master's thesis, Johns Hopkins University, 1924.

Brittingham, William Henry. "The Nature and Extent of Variation in Kentucky Bluegrass as Criteria for Type of Seed Formation and Its Practical Implications." PhD diss., University of Maryland, 1942.

Bromley, Walter Davis. "The Effect of Overrun, Temperature, and Composition on the Dipping Losses of Ice Cream." Master's thesis, University of Maryland, 1926.

Brookens, Perley Floyd. "Foreign Competition in the Dairy Industry in the United States." PhD diss., University of Maryland, 1927.

Brown, Elizabeth B. "Bacterial Studies of Certain Defrosted Vegetables." Master's thesis, University of Maryland, 1932.

Brown, Russel Guy. "A Comparative Study of the Aerobic and Anaerobic Respiration in the Storage Organs of Different Plants." PhD diss., University of Maryland, 1934.

Browne, Edward L. "The Milk Sheds of Washington, Baltimore and Philadelphia." Master's thesis, University of Maryland, 1924.

Brownlee, Donald S. "Bacteriological Studies on the Defrosting of Frozen Eggs and the Differentiation of Human and Avian Escherichia Coli." Master's thesis, University of Maryland, 1938.

Burdette, Roger F. "A Study of Roadside Markets in Maryland." Master's thesis, University of Maryland, 1934.

Burroughs, John A. "A Statistical Study of the Commercial Aspects of Sweet Potato Production." Master's thesis, University of Maryland, 1926.

Call, Tracey Gillette. "A Study of the Official Vegetable Drug Powder Descriptions Together with a Simplified Method for Identification of the Powders." Master's thesis, University of Maryland, 1944.

Campbell, Berry. "The Comparative Myology of the Hippopotamus, Pig, and Tapir." PhD diss., Johns Hopkins University, 1935.

Campbell, Marjorie H. "An Appraisal of a Study of Soil and Its Conservation in an Elementary City School." Master's thesis, University of Maryland, 1940.

Canter, Francis D. "A Study of the Present Systems of Co-Operative Marketing of Dairy Products in the United States." Master's thesis, University of Maryland, 1923.

Carr, C. Jelleff. "The Metabolism of the Sugar Alcohols and Their Anhydrides." PhD diss., University of Maryland, 1937.

Carroll, Floyd D. "Salt Penetration in Hams During Curing and Aging." Master's thesis, University of Maryland, 1939.

Carter, Edward P. "Studies on the Heating of High Moisture Wheat in Storage, with Special Reference to the Action of Microorganisms." Master's thesis, University of Maryland, 1939.

Cash, Bernice Balch. "Comparison of the Maryland Pure Food Laws Concerning Institutional Sanitation, Canneries Cold Storage, and Baked Goods with Similar Laws in California, Colorado, Connecticut, Kentucky, Illinois, Michigan, Minnesota, Ohio, Pennsylvania, and Wisconsin." Master's thesis, University of Maryland, 1934.

Chandler, Frederick Barker. "Mineral Nutrition of the Genus Brassica with Particular Reference to Boron." PhD diss., University of Maryland, 1939.

Chandler, Robert F. "Studies on the Potassium Nutrition of the Apple and Peach. The Effect of Potassium Fertilizer on the Potassium Content of Soil and Tree, Growth and Yield of Tree, and Keeping Quality of Fruit. The Absorption, Distribution and Seasonal Movement of Potassium in Young Apple, Trees, and the Effect of Potassium Fertilizer on Potassium and Nitrogen Content and Growth of Tree." PhD diss., University of Maryland, 1934.

Chase, Spencer. "An Economic Survey of Home Orchards in Counties near Baltimore, Maryland and Washington, D.C." Master's thesis, University of Maryland, 1937.

Chen, Chunjen C. "Internal Parasites of Agricultural Seeds." Master's thesis, University of Maryland, 1920.

Christensen, Hilde M. "Haploids in Twin Seedlings of Capsicum Annum L." Master's thesis, University of Maryland, 1942.

Clare, Irwin Charles. "The Qualitative Detection of Fish Scrap, Animal Tankage, Cyanamid in Fertilizers." Master's thesis, University of Maryland, 1924.

Cochran, Doris M. "The Skeletal Musculature of the Blue Crab, Callinectes Sapidus Rathbun." PhD diss., University of Maryland, 1933.

Cocke, Louise. "Studies on the Wilt of Sweet Peas." Master's thesis, University of Maryland, 1931.

Coddington, James W. "Farm Management Study of 266 Farms on the Eastern Shore of Maryland, 1931." Master's thesis, University of Maryland, 1933.

Coggeshall, Mary. "Influence of Acetic, Propionic, Normal Butyric and Sulphuric Acids and Potassium Acetate on Elongation of Primary Roots of Seedlings of White Lupine." PhD diss., Johns Hopkins University, 1931.

Conrad, Carl Marcus. "A Biochemical and Physiological Study of the Insoluble Pectic Materials in Vegetables and Fruits." PhD diss., University of Maryland, 1925.

Cordner, H. B. "A Study of Certain Factors Affecting the Set of Fruit in Henderson Bush Lima Bean." PhD diss., University of Maryland, 1932.

Cotter, Joseph W. "The Rate of Oxidation of Fat from Natural and Remade Cream." Master's thesis, University of Maryland, 1942.

Cotton, Cornelia Marie. "Diagnositc Tests Including Hemotology in Swine Brucellosis and Capsule Formation in Brucella." PhD diss., University of Maryland, 1943.

Cotton, John. "Organized Camping in Maryland." Master's thesis, University of Maryland, 1941.

Cowgill, William Henry. "Studies on the Germination, Early Seedling Growth and Nutrition of Cinchona." PhD diss., University of Maryland, 1944.

Cox, Carroll Eastburn. "The Host-Parasite Relationship in Susceptible Cantaloups, Resistant Cantaloups, and Cucumbers Inoculated with Fusarium Bulbigenum (Cke. And Mass.) Var. Niveum Wr. F. 2." PhD diss., University of Maryland, 1943.

Crane, Julian Coburn. "Growth and Fruitfulness of the Blakemore Strawberry Especially in Relation to Plant Thinning." PhD diss., University of Maryland, 1943.

Craven, Avery Odelle. "Soil Exhaustion as a Factor in the Agricultural History of Virginia and Maryland, 1606-1860." PhD diss., University of Chicago, 1924.

Cron, Lawrence E. "The Marketing of Eggs in Maryland, with Special Reference to the Baltimore Market." Master's thesis, University of Maryland, 1939.

Cronin, L. Eugene. "A Histological Study of the Development of the Ovary and Accessory Reproductive Organs of the Blue Crab, Callinectes Sapidus Rathbun." Master's thesis, University of Maryland, 1942.

Crosthwait, Samuel L. "An Attempt to Determine Factors Associated with the Resisitance of the Harlequin Bug to Certain Contact Insecticides." Master's thesis, University of Maryland, 1933.

Crow, Jane Hanes. "Consumer Practices in Buying Canned Fruits and Vegetables in College Park, Maryland." Master's thesis, University of Maryland, 1938.

Culton, Thomas Grover. "Variations in the Riboflavin Content of Feedstuffs and Their Relation to Chick Growth." Master's thesis, University of Maryland, 1940.

Cwalina, Gustav Edward. "A Phytochemical Study of Ipomea Pes-Caprae (L.) Sweet." PhD diss., University of Maryland, 1937.

Daly, Rex F. "An Economic Study of the Production and Marketing of Potatoes in Garrett County, Maryland." Master's thesis, University of Maryland, 1939.

D'Ambrogi, G. D. "Some Factors Affecting the Accuracy of the Babcock Test on Composite Samples of Milk." Master's thesis, University of Maryland, 1937.

Darkis, F. R. "The Solvent Action of Salts on Two Phase of a Soil." Master's thesis, University of Maryland, 1924.

Darrow, George M. "Experimental Studies on Growth and Development in Strawberry." PhD diss., Johns Hopkins University, 1927.

Davies, Thomas J. "The Broiler Industry in Maryland." Master's thesis, University of Maryland, 1942.

Dawson, Roy C. "Some Factors Influencing the Germicidal Action of Hypochlorites against Mastitis Streptococci." Master's thesis, University of Maryland, 1939.

Day, Frank D. "Systems of Practice Teaching and Methods of Supervising Practice Teachers in Secondary Vocational Agriculture." Master's thesis, University of Maryland, 1926.

Degman, Elliott Sanford. "The Influence of Nitrogen Fertilizers on the Shipping and Keeping Qualities of Fruits." PhD diss., University of Maryland, 1931.

Dengler, Felton Samuel. "I. The Detection and Determination of Minute Quantities of Glycerine. II. The Volumes of Weight-Normal Cane Sugar Solutions at Different Temperatures." PhD diss., Johns Hopkins University, 1912.

Derr, David E. "An Economic Study of 176 Dairy Farms in Maryland." Master's thesis, University of Maryland, 1935.

Deupree, Robert Gaston. "The Wholesale Market for Fresh Fruits and Vegetables in Baltimore." PhD diss., Johns Hopkins University, 1937.

DeVault, Samuel H. "The Wheat Industry in Maryland." PhD diss., Massachusetts State College at Amherst, 1931.

Diggs, Eleanor. "North Atlantic Coast Fisheries Dispute, 1783-1910." Master's thesis, Johns Hopkins University, 1918.

Dillard, Louisa Gardner. "Changes in Wheat Production in Canada." Master's thesis, University of Maryland, 1943.

Dillman, A. C. "The Daily Growth of the Flax Seed with Special Reference to the Time at Which the Oil Is Laid Down." Master's thesis, University of Maryland, 1927.

Ditman, Lewis Polster. "The Biology and Control of the Corn Earworm." Master's thesis, University of Maryland, 1929.

———. "Studies on the Corn Earworm." PhD diss., University of Maryland, 1931.

Dixon, Paul J. "Evaluation Criteria for Effective 4-H Local Leadership." Master's thesis, University of Maryland, 1939.

Downey, Mylo S. "The Essentials of a Handbook for Local 4-H Club Leaders." Master's thesis, University of Maryland, 1940.

Dunnigan, A. P. "Bacteria Isolated from Milk Drawn with Aseptic Precautions." Master's thesis, University of Maryland, 1932.

Dynes, Isabel. "An Investigation to Determine the Cake Making Qualities of Various Strains of Maryland Wheat Grown in Different Locations in the State." Master's thesis, University of Maryland, 1931.

Eaton, Norwood A. "Insecticidal Values of Pyrethrum Extracts." Master's thesis, University of Maryland, 1928.

Eaton, Orson Northrup. "An Anatomical and Chemical Study of Inbred and Crossbred Strains of Guinea Pigs." PhD diss., University of Maryland, 1935.

Edgeworth, Clyde Baltzer. "A Community Survey of Opinion on Consumer Education, Baltimore, Maryland." Master's thesis, University of Maryland, 1939.

Edmond, J. B. "Sex Expression in Cucumbers." PhD diss., University of Maryland, 1933.

Edwards, Thomas Isaac. "Relations of Germinating Soybeans to Temperature and Length of Incubation Time." PhD diss., Johns Hopkins University, 1932.

Engle, Ruth B. "The Effect of Castration Upon the Palatability of Market Lamb." Master's thesis, University of Maryland, 1927.

Eppley, Geary. "Hay and Pasture Grasses for Maryland. Adaptability Studies." Master's thesis, University of Maryland, 1926.

Espino, Rafael B. "Some Aspects of the Salt Requirements of Young Rice Plants." PhD diss., Johns Hopkins University, 1919.

Everhart, Herbert W. "The Production and Marketing of Peanuts." Master's thesis, University of Maryland, 1942.

Eyssell, Henry Otto. "I. The Detection of Mannite in Alkaline Solutions of Copper Sulphate. Combustion of Mannite by Alkaline Solutions of Potassium Permanganate in the Presence of Copper Sulphate. II. A Determination of the Volumes of Weight-Normal Solutions of Cane Sugar at 150, 200, 250, and 300." PhD diss., Johns Hopkins University, 1912.

Ezekiel, Walter Naphtali. "Fruit Rotting Sclerotinias." Master's thesis, University of Maryland, 1921.

———. "Fruit-Rotting Sclerotinias. The American Brown-Rot Fungi." PhD diss., University of Maryland, 1924.

Faber, Jack. "Differences in Bacterial Count of Raw and Pasteurized Milk Plated on Extract Agar with Different Ph Values - Range 6.2; 6.4; 6.6; 6.8; and 7.0." Master's thesis, University of Maryland, 1927.

Faber, John E. "Measurement, Production and Preservation of the Hemolytic Activity of Guineas Pig Complement." PhD diss., University of Maryland, 1937.

Fahey, Daniel Cox, Jr. "The Response of Bent Grasses to the Soil Reaction." Master's thesis, University of Maryland, 1929.

Fancher, George H. "A Comparison of Diatomaceous Earths." Master's thesis, University of Maryland, 1926.

Farley, Horace B. "A Study of Spinach Varieties with Special Reference to Their Canning Qualities." Master's thesis, University of Maryland, 1928.

Fawcett, H. S. "The Temperature Relations of Growth in Certain Parasitic Fungi." PhD diss., Johns Hopkins University, 1918.

Fennell, Richard Adams. "The Relation between Temperatures, Salts, Hydrogen Ion Concentration and Frequency of Ingestion of Food by Amoeba Proteus." PhD diss., Johns Hopkins University, 1936.

Fields, John Newton. "The Effects of Feeds and Weeds on the Flavors and Odors of Milk." Master's thesis, University of Maryland, 1924.

Fisher, George Clyde. "Seed Development in the Genus Peperomia." PhD diss., Johns Hopkins University, 1913.

Fisher, Paul Lewis. "The Effects of Deficiencies of Phosphorus on Tomato Plants at Different Stages of Growth." Master's thesis, University of Maryland, 1930.

Fisher, Raymond Anderson. "The Colorimetric Determination of the Forms of Inorganic Phosphorus in Soils." PhD diss., University of Maryland, 1935.

———. "A Study of Lime-Phosphorus Potassium Fertility Relations in Grassland Soils." Master's thesis, University of Maryland, 1932.

Fisher, V. E. "An Experimental Study of the Effects of Tobacco Smoking on Certain Psycho-Physical Functions." PhD diss., Johns Hopkins University, 1926.

Flenner, A. "Adsorption from Solution by Hydrous Alumina." Master's thesis, University of Maryland, 1924.

Flesch, Lawrence E. "Postwar Expenditure Plans of Maryland Farmers for Improving Farm Equipment and Buildings." Master's thesis, University of Maryland, 1945.

Fletcher, Lewis Arrowood. "A Study of the Factors Influencing the Red Color on Apples." PhD diss., University of Maryland, 1930.

Florestano, Herbert Joseph. "Studies on Oral Health as Reflected in the Saliva, with Special Reference to the Local and Systemic Use of Citrus Fruits, Oral Aciduric Microorganisms, Diastatic Activity and Ph." PhD diss., University of Maryland, 1940.

Flynn, John E. "Studies of Pathogenic Fungus Toxins." Master's thesis, University of Maryland, 1924.

Fotoz, Mustafa. "A Statistical Analysis of Tobacco Prices in the United States with Reference to Maryland Tobacco, U.S. Type 32." Master's thesis, Columbia University, 1940.

Foulks, James Grigsby. "An Analysis of the Source of Melanophores in Regenerating Feathers." PhD diss., Johns Hopkins University, 1942.

Frank, Paul S. "Is Our National, State, and Community Investment in Vocational Agricultural Education a Profitable One?" Master's thesis, University of Maryland, 1926.

Frazer, Mary Washington. "A Comparison of the Characteristics of Excherichia Coli of Chicken Origin with Those of Escherichia Coli of Human Origin." Master's thesis, University of Maryland, 1939.

Frazier, William Allen. "A Study of Some Factors Associated with the Occurrence of Cracks in the Tomato Fruit." PhD diss., University of Maryland, 1933.

———. "Physical and Chemical Measurements of Winter Hardiness in Certain Varieties of Spinach." Master's thesis, University of Maryland, 1931.

Frischnecht, Carl Oliver. "Amount of Breast Meat and Grade of Carcass in Relation to Body Measurements in Purebred Cornish, New Hampshires, Barred Plymouth Rocks, and Their Crosses." PhD diss., University of Maryland, 1945.

Frush, Harriet Louise. "Sugar Acetates, Halogeno-Acetates, and Orthoesters in Relation to the Walden Inversion." PhD diss., University of Maryland, 1941.

Gahan, James B. "Laboratory and Field Tests of the Solubility and Killing Effectiveness of Paradichlorobenzene in Solution in Various Oils." Master's thesis, University of Maryland, 1932.

Galbreath, Paul M. "The Production and Use of Rye in Maryland." Master's thesis, University of Maryland, 1940.

Gammon, Nathan, Jr. "A Base Exchange Study on Some Maryland Soils." Master's thesis, University of Maryland, 1939.

Gates, Philip W. "A Vocational Survey of the Poolesville Community as an Index to Curriculum Content in Vocational Agriculture." Master's thesis, University of Maryland, 1926.

Geiser, Samuel Wood. "Sex-Ratios and Spermatogenesis in the Top-Minnow, Gambusia Affinis." PhD diss., Johns Hopkins University, 1922.

Gericke, W. F. "Some Relations of Maintained Temperatures to Germination and the Early Growth Rate of Wheat in Nutrient Solutions." PhD diss., Johns Hopkins University, 1921.

Gienger, Guy W. "A Comparative Study of Maryland Agriculture." Master's thesis, University of Maryland, 1936.

Gifford, George E. "A Study of the Marks of a Certain Group of High School Graduates in Maryland and Their Relationships to Marks Received in Colleges and Normal Schools." Master's thesis, University of Maryland, 1933.

Glazener, Edward W. "Feed Utilization and Rate of Feathering as Related to Body-Size in the Domestic Fowl." Master's thesis, University of Maryland, 1945.

Glickman, Shirley M. "Stability of Vitamin C in Citrus Fruit Juices." Master's thesis, University of Maryland, 1940.

Godfrey, Albert B. "The Inheritance of Rate of Feathering and Plumage Color in the Jersey Black Giant and the S.C. White Leghorn." Master's thesis, University of Maryland, 1932.

———. "A Study of the Inheritance of Rate of Growth and Rate of Feathering in the Domestic Fowl." PhD diss., University of Maryland, 1942.

Golden, Lex B. "Methods for Determining Total Exchange Capacity and Exchangeable Bases of Maryland Soils." Master's thesis, University of Maryland, 1940.

Goldsborough, George H. "Farmers' Mutual Fire Insurance in Maryland." Master's thesis, University of Maryland, 1940.

Goldstein, Samuel W. "A Phytochemical Study of Phytolacca Americana." Master's thesis, University of Maryland, 1931.

Goss, Donald M. "Variations in the Total Sugar Content of Individual Hopeland Sweet Corn Ears." Master's thesis, University of Maryland, 1934.

Gould, Clarence P. "Economic History of Maryland 1720-1765. Land." PhD diss., Johns Hopkins University, 1911.

Graham, Castillo. "Biology and Control of the Plum Curculio (Conotrachelus Numphar Herbst) with Special Reference to Certain Phenological Data." PhD diss., University of Maryland, 1937.

———. "A Study of the Life History and Seasonal Activities of the Plum Curculio, Conotrachelus Nenuphor Herst in Maryland." Master's thesis, University of Maryland, 1930.

Graham, James G. "Seasonal Attachment of Marine Organisms in the Solomons Island Region of the Chesapeake Bay." Master's thesis, University of Maryland, 1939.

Grau, Frederick Vahlcamp. "Factors Affecting Pasture Quality; an Inventory of Soils, Vegetation, and Management of Maryland Permanent Pastures." PhD diss., University of Maryland, 1935.

———. "The Use of Chemicals in the Control of Turf Weeds." Master's thesis, University of Maryland, 1933.

Gray, Ellen H. "Certain Ecological and Life History Aspects of Uca Minax, the Red-Jointed Fiddler Crab, as Found in the Vicinity of Solomons Island, Maryland." Master's thesis, University of Maryland, 1942.

Greenfield, Albert Norman. "Income Tax Principles of Farm Record Keeping." Master's thesis, University of Maryland, 1943.

Greenwood, Grace-Louise. "Developing Inflorescences of Some Maryland Grasses in Relation to Their Taxonomic Characters." Master's thesis, University of Maryland, 1938.

Griggs, William Holland. "Some Factors Associated with the Fruitfulness of the Delicious Apple." PhD diss., University of Maryland, 1943.

Grober, Samuel. "The Botanical, Erosion Control, and Economic Significance of White Poplar in Maryland." PhD diss., University of Maryland, 1942.

Groschke, Albert C. "Effects of Borderline Levels of Vitamin A in the Hen's Ration on Hatchability, Livability and Chick Growth." Master's thesis, University of Maryland, 1940.

Grove, Elmer William. "Some Responses of the Howard 17 Strawberry Plant to Applications of Nitrogen and Moisture in the Non-Fruiting and Fruiting Year." PhD diss., University of Maryland, 1935.

Gwynn, Thomas S., Jr. "The Selectivity of Five Rural High Schools of Prince George's County, Maryland." Master's thesis, University of Maryland, 1938.

Haasis, Ferdinand Wead. "Some Relations of Temperature and Length of Germination Period to the Percentage of Germination of Seeds, Especially of Pitch Pine and Rice." PhD diss., Johns Hopkins University, 1928.

Hackendorf, Arthur C. "The Organization and Business Analysis of 270 Farms in the Piedmont Plateau Region of Maryland." Master's thesis, University of Maryland, 1931.

Haenni, Edward Otto. "A New Method for the Determination of Cholesterol and Its Application to the Estimation of the Egg Content of Alimentary Pastes." PhD diss., University of Maryland, 1940.

Hale, Roger F. "Wheat Storage on Maryland Farms." Master's thesis, University of Maryland, 1926.

Haller, M. H. "A Study of the Effect of Certain Factors on the Size and Composition of Apples and the Effect of Fruiting on Bud Differentiation." PhD diss., University of Maryland, 1931.

———. "The Relation of Leaf Area to the Growth and Composition of Apples." Master's thesis, University of Maryland, 1926.

Hallock, Frances Adelia. "The Relationship of Garrya. The Development of the Flowers and Seeds of Garrya and Its Bearing on the Phylogenetic Position of the Genus." PhD diss., Johns Hopkins University, 1926.

Hambright, Willam Anderson. "Infective Adsortion and Its Effects on the Maryland Agricultural Experiment Station Herd." Master's thesis, University of Maryland, 1927.

Hamburger, Pinhas. "Tobacco Taxation in the United States." PhD diss., Johns Hopkins University, 1932.

Hamilton, Arthur B. "Economic Efficiency of the Farm Layout." Master's thesis, University of Maryland, 1931.

Hamilton, Howard Laverne. "I. A Study of the Physiological Properties of Melanophores with Special Reference to Their Role in Feather Coloration. Ii. Influence of Adrenal and Sex Hormones on the Differentiation of Melanophores in the Chick." PhD diss., Johns Hopkins University, 1941.

Hamm, Homer Alexander. "The Retentivity of Water by Purified Cotton Cellulose." PhD diss., Johns Hopkins University, 1936.

Hammer, Ralph Curtis. "The Homing Instinct of the Chesapeake Shad, Alosa Sapidissima Wilson, as Revealed by a Study of Their Scales." Master's thesis, University of Maryland, 1942.

Hammond, John Clarke. "Effects of Nutrition on Variability in the Growth of Chickens." PhD diss., University of Maryland, 1941.

Haney, Walter Judson. "The Nucleolar Numbers and Attachments in Lilium." PhD diss., University of Maryland, 1943.

Hanna, William Miles. "Related Science in Vocational Agriculture Based on Statements of Successful Dairy Farmers in Harford County." Master's thesis, University of Maryland, 1933.

Harley, Clayton P. "The Influence of Spur Length, Leaf Area, and Carbodydrate Content on Fruit Bud Formation in the Apple." Master's thesis, University of Maryland, 1924.

Harns, Henry G. "Studies Upon the Cheese Skipper, (Piophila Casei L.)." Master's thesis, University of Maryland, 1936.

Harper, Charles Price. "The Administration of the Civilian Conservation Corps." PhD diss., Johns Hopkins University, 1935.

Harper, Floyd Henry. "Forecasting the Acreage, Yield, and Price of Cotton." PhD diss., University of Maryland, 1928.

Harris, Minnie Behm Kraemer. "A Study of Spirochetes in Chickens with Special Reference to Those of the Intestinal Tract." PhD diss., Johns Hopkins University, 1930.

Harris, Marvin Mayer. "A Study of the Bacteriology of Decomposing Crabs and Crab Meat." PhD diss., Johns Hopkins University, 1930.

Harrison, Perry Kips. "Biology and Control of the Red Spider (Tetranychus Telarius)." Master's thesis, University of Maryland, 1930.

Harrison, Robert W. "Classification of Agricultural Areas in Maryland, with Special Reference to Frederick County." Master's thesis, University of Maryland, 1940.

Hartman, Jack D. "The Relation of Input to Output in Milk Production." Master's thesis, University of Maryland, 1939.

Harver, Frederic Fern. "A Comparison of the Institutionalization of Children of Migrant and Children of Native Parents in Harford County." Master's thesis, University of Maryland, 1935.

Haut, I. C. "A Study of after-Ripening in Certain Fruit Tree Seeds." PhD diss., University of Maryland, 1933.

Hauver, W. E. "An Economic Study of 99 Poultry Farms in Maryland." Master's thesis, University of Maryland, 1935.

Hawkins, Lon Adrian. "The Influence of Calcium, Magnesium and Potassium Nitrates Upon the Toxicity of Certain Heavy Metals toward Fungus Spores." PhD diss., Johns Hopkins University, 1913.

Hearn, Mildred Louise. "A Study to Develop New Uses for Maple Sirup with Reference to Their Cost, Technique of Preparation, and the Palatability of the Finished Products." Master's thesis, University of Maryland, 1939.

Heinze, Peter Herman. "A Physiological and Biochemical Study of the Curing Processes in Sweet Potatoes." PhD diss., University of Maryland, 1940.

Henerey, W. T. "Biology and Control of the Mexican Bean Beetle, Epilachna Corrupta Mulsant." Master's thesis, University of Maryland, 1930.

Hensel, Harold Ernest. "Arterial Injections as a Means of Rapidly Curing Hams." Master's thesis, University of Maryland, 1943.

Henson, Paul R. "A Study of the Effect of Starchy Endosperm on the Distribution of Carbohydrates in the Corn Plant." Master's thesis, University of Maryland, 1930.

Hersey, Leroy Harlan. "Contagious Abortion and Its Effect on the Maryland Agricultural Experiment Station Herd." Master's thesis, University of Maryland, 1931.

Hess, Carl William. "Some Genetic Factors Affecting the Efficiency of Feed Utilization in Chickens." Master's thesis, University of Maryland, 1940.

Hesse, Claron Owens. "Some Physical and Chemical Changes Associated with the Maturation of Grimes and Jonathan Apples, on the Tree and During Storage." PhD diss., University of Maryland, 1938.

Heuberger, John W. "The Cytological Phenomena Associated with the Development of the Sporophyte in Sclerotinia Fructicola." PhD diss., University of Maryland, 1934.

———. "Studies on the Mosaic Diseases of Tomatoes." Master's thesis, University of Maryland, 1931.

Higgins, William Burton. "The University of Maryland and Its Relation as a Training Institution to Maryland Rural Life, 1920-1936." Master's thesis, University of Maryland, 1937.

Hildebrandt, Frank Merrill. "A Physiological Study of the Climatic Conditions of Maryland as Measured by Plant Growth. (a Second Contribution from Data Obtained under the Auspices of the Maryland State Weather Service in 1914.)." PhD diss., Johns Hopkins University, 1917.

Hitchcock, Albert Edwin. "Effect of Sodium Nitrate Treatment Upon the Initiation and Stimulation of Sprout Growth in the Potato Tuber." Master's thesis, University of Maryland, 1925.

Hitz, Chester W. "Maturity and Storage Studies with Apples in Relation to the Subsequent Development of Edible and Keeping Qualities." PhD diss., University of Maryland, 1941.

———. "A Study of Certain Maturity Indices in Relation to the Edible and Keeping Qualities of Apples." Master's thesis, University of Maryland, 1938.

Hoadley, Alfred Damon. "A Study of the Botanical Characteristics, Growth, and Composition of Strains of Kentucky Bluegrass, Poa Pratensis L." Master's thesis, University of Maryland, 1938.

———. "A Study of the Nature of Resistance of Dent Corn to Diplodia Zeae (Schw.) Lev. Stalk Rot." PhD diss., University of Maryland, 1942.

Holmes, Myron G. "The Nutrient Value of Soybeans at Various Stages of Growth. The Correct Harvest Date Forage." Master's thesis, University of Maryland, 1924.

Holt, Rebecca Agnew. "A Study of Bacillus Coli as an Index of the Proper Pasteurization of Milk, with Special Reference to Swenarton's Fermentation Test and Standard." Master's thesis, Johns Hopkins University, 1929.

Holtzclaw, Henry Fuller. "The Lumber Industry and Trade." PhD diss., Johns Hopkins University, 1917.

Hoopes, Joseph Darlington. "The Cost of Milk Production in the Maryland Agricultural Experiment Station Herd." Master's thesis, University of Maryland, 1927.

Houghland, Geoffrey V. C. "The Relative Efficiency of Different Forms of Nitrogen and Potassium in Potato Production on the Eastern Shore of Maryland." PhD diss., University of Maryland, 1928.

Howard, Dowell Jennings. "The Value of a Programme in Vocational Agricultural Education as Evidenced by Some Results Attained in the State of Virginia." Master's thesis, University of Maryland, 1926.

Howe, Charles Harold. "A Study of the Physical and Chemical Changes in the Soil Responsible for Decreased Yields Following the Excessive Applications of Calcium Carbonate." Master's thesis, University of Maryland, 1923.

Huffman, Roy E. "An Economic Study of the Maple Products Industry in Garrett County, with Particular Reference to Land Use." Master's thesis, University of Maryland, 1939.

Hunter, Herman A. "Relation of Ear Infection with Certain Micro-Organisms to Internal Cob Discoloration of Corn." Master's thesis, University of Maryland, 1926.

Hurley, Ray. "The Organization and Business Analysis of 282 Farms in the Piedmont Plateau Region of Maryland." Master's thesis, University of Maryland, 1930.

Hyson, Charles D. "Competition of Other Regions with Maryland in the Production and Marketing of Important Truck Crops." Master's thesis, University of Maryland, 1939.

Ingersoll, Mary M. "An Economic Study of 147 Turkey Flocks in Maryland." Master's thesis, University of Maryland, 1933.

Ives, J. Russell. "The Status and Trend of Agricultural Cooperation in Maryland, 1900-1936." Master's thesis, University of Maryland, 1937.

Ives, Mildred. "Suggested Procedure for Incorporated Home Management in the 4-H Club Programme." Master's thesis, University of Maryland, 1936.

Jacob, Walter C. "A Study of Artificial Pollen Germination in the Lima Bean." Master's thesis, University of Maryland, 1938.

Jeffers, Walter Fulton. "Studies on the Wilt of Peas Caused by Fusarium Orthoceras Var. Pisi." Master's thesis, University of Maryland, 1937.

Jehle, Ruth Amanda. "American Square Dancing in Three Sections of Maryland and the District of Columbia." Master's thesis, University of Maryland, 1943.

Jenkins, Harvey Foss. "The Effect of Light and the Relative Length of Day and Night on Growth and Reprodution of the Onion (Allium Cepa, L)." Master's thesis, University of Maryland, 1923.

Johns, Kathryn. "A Study of Certain Factors Regulating Cantaloupe Growth in Tissue Culture Media." Master's thesis, University of Maryland, 1940.

Jones, Charles Archer. "Lime Movement, Buffer and Absortive Capacity in Some Coastal Plain Soils." Master's thesis, University of Maryland, 1927.

Jones, John Paul. "The Relation of Vitamins to Plant Growth." Master's thesis, University of Maryland, 1921.

Jones, Robert Edwin. "Chromosome Number in the Progeny of Triploid Gladiolus with Special Reference to the Contribution of the Triploid, by Robert E. Jones." PhD diss., University of Maryland, 1942.

Jones, Wilbur A. "Land Grants under the Proprietary Government in Maryland - 1633-1775." Master's thesis, University of Maryland, 1937.

Joy, Barnard D. "Some Factors Influencing Length of 4 - H Club Membership." Master's thesis, University of Maryland, 1934.

Kaminsky, Daniel. "On the Source of the Principle Inhibitory to the Normal Acid Production of Streptococcus Lactis." Master's thesis, University of Maryland, 1940.

Kanagy, Mary Tompkins. "Solubilities of Purified Cotton Mulberry Pulp and Corn Stalk Pulp in 1-50 Per Cent Sodium Hydroxide." Master's thesis, University of Maryland, 1933.

Kelley, Carl Williams. "A Study of the Chemical and Physical Changes Produced in a Soil by the Formation of the Organic Colloidal Complex." PhD diss., University of Maryland, 1942.

Kellogg, Frederic Hartwell. "The Role of the Clay Minerals in Soil Mechanics." PhD diss., Johns Hopkins University, 1934.

Kemp, Margaret Cobey. "A Study of the Oxygen Respiration in Corn and Its Relation to the Freezable Water Content." Master's thesis, University of Maryland, 1941.

Kennedy, Cornelia. "The Study of the Possibility of the Isolation of the Anti-Neuritic Substance of Foodstuffs." PhD diss., Johns Hopkins University, 1919.

Kerr, William Leslie. "Cross and Self-Pollination Studies with the Peach." Master's thesis, University of Maryland, 1927.

Kik, Marinus Cornelis. "The Nutritive Value of Haddock and Herring (Clupea Harengus)." PhD diss., Johns Hopkins University, 1927.

Kimbrough, W. D. "Relation of Respiration to Storage and Transportation of Potatoes." Master's thesis, University of Maryland, 1924.

King, R. "A Bacteriological Study of the Organs and Tissue of Apparently Healthy Guinea Pigs." Master's thesis, University of Maryland, 1926.

Klein, T.S. "The Small High School in Maryland, Its Possibilities and Limitations." Master's thesis, University of Maryland, 1932.

Komaroff, Abner. "The Foreign Trade of the United States in Citrus Fruits." PhD diss., Johns Hopkins University, 1933.

Krabill, Verlin C. "A History of Education of Frederick County, Maryland." Master's thesis, University of Maryland, 1929.

Kramer, Amihud. "Mineral Nutrition of the Blueberry." PhD diss., University of Maryland, 1942.

———. "The Use of Ascorbic Acid, Respiratory Enzymes and Pigment Extracts as Measures of Quality of Frozen, Canned and Fresh Lima Beans." Master's thesis, University of Maryland, 1939.

Krantz, John C. "A Study of the Relative Preservative Values of Glycerin and Sugar Solutions in Certain Official Preparations." Master's thesis, University of Maryland, 1924.

Kraybill, H. F. "A Chemical and Pharmalogical Study of the Oxidation of Oils with Special Reference to Fish Meal." PhD diss., University of Maryland, 1941.

Kuhn, Albin O. "The Effect of Removing Different Proportions of Foliage on Strains of Kentucky Bluegrass (Poa Pratensis L.) and on Perennial Rye Grass (Lolium Perenne L.)." Master's thesis, University of Maryland, 1939.

Kunkel, Anne M. "The Relationship between the Chemical Constitution and the Rapidity of Hemolysis of the Sugar Alcohols and Their Anhydrides." Master's thesis, University of Maryland, 1939.

Lachar, George P. "An Economic Study and Analysis of the Marketing of Fresh Fruits and Vegetables on the Marsh Market in Baltimore City." Master's thesis, University of Maryland, 1936.

Lagassé, Felix Scott "The Effect of Nitrogen Application on the Growth Responses and Composition of Jonathan Apple Trees." PhD diss., University of Maryland, 1933.

———. "The Effect of Shade on Apple Trees." Master's thesis, University of Maryland, 1924.

Lampen, Dorothy. "Economic and Social Aspects of Federal Reclamation." PhD diss., Johns Hopkins University, 1929.

Lawrence, Donald B. "Some Features of the Vegetation of the Columbia River Gorge." PhD diss., Johns Hopkins University, 1936.

Leatherman, Martin. "A Study of the Comparative Rates of Neutrallization of Soil Acidity by Different Liming Materials." Master's thesis, University of Maryland, 1925.

Leckie, J. Norman. "Fresh Samples Versus Composite Samples in Milk Testing." Master's thesis, University of Maryland, 1932.

LeCompte, Stuart Burnette. "An Experimental Comparison of Different Instruments and Procedures for Indicating Soil-Moisture Condition with Reference to Drought Injury in Pot-Grown Plants." PhD diss., Johns Hopkins University, 1939.

Leighty, Raymond Vandermark. "A Comparison of Reagents Used in Destroying the Organic Base-Exchange Complex." Master's thesis, University of Maryland, 1940.

Leise, Joshua Melvin. "Differentiation of Human and Chicken Strains of Escherichia Coli on the Basis of Phenol Tolerance." Master's thesis, University of Maryland, 1943.

Lennartson, Roy W. "An Economic Analysis of Marketing Fruits and Vegetables by Motor Truck from Western Maryland Farms." Master's thesis, University of Maryland, 1936.

Levin, Irvin. "The Estimation of Riboflavin in Dairy Products by the Measurement of Fluorescence Using a Photo-Electric Photometer." Master's thesis, University of Maryland, 1940.

Levine, Melvin L. "Methylene Blue Reduction as an Index of the Bacteriological Quality of Eggs." Master's thesis, University of Maryland, 1940.

Lieberman, Samuel. "Numerical Determination of Leucocytes in Milk." Master's thesis, University of Maryland, 1927.

Light, Vernal Earl. "Photoreceptors in Mya Arenaria L. With Special Reference to Their Distribution, Structure and Function." PhD diss., Johns Hopkins University, 1929.

Lincoln, Leonard B. "Retail Margins on Farm Products and Several Manufactured Products in Washington, D.C." Master's thesis, University of Maryland, 1926.

Lindquist, Harry Gotfred. "Variations in Solids - Not - Fat in Cow's Milk." Master's thesis, University of Maryland, 1924.

Lineburg, Bruce. "Reactions to Light in the Honeybee, with Special Reference to Modifiability in Behaviour." PhD diss., Johns Hopkins University, 1924.

Lippy, Grace Elizabeth. "The Vascular System of the Neck Region in the Rabbit, the Rat, the Muskrat, the Squirrel, and the Guinea Pig. With Special Reference to the Left Precaval Vein." Master's thesis, Johns Hopkins University, 1926.

Liu, Ho. "The Effects of Fertilizers on the Chemical Composition and Physical Properties of Tobacco." PhD diss., University of Maryland, 1925.

Lodman, William J. "Short Term Farm Credit on the Lower Eastern Shore of Maryland." Master's thesis, University of Maryland, 1941.

Long, Joseph C. "The Influence of Rooting Media on the Character of Roots Produced by Cuttings." Master's thesis, University of Maryland, 1932.

Loyd, Charles M. "A Modified Babcock Test Bottle for Cream and Ice Cream." Master's thesis, University of Maryalnd, 1939.

Lumsden, David Victor. "Anatomical and Biochemical Changes in Narcissus Bulbs During Summer Storage at Various Temperatures." PhD diss., University of Maryland, 1935.

Lutz, Jacob M. "Physiological Factors Influencing the Ripening of Kieffer Pears." PhD diss., University of Maryland, 1936.

Lynt, Richard K., Jr. "Bacteriological and Heat Penetration Studies on the Blue Crab." Master's thesis, University of Maryland, 1942.

Mack, Warren Bryan. "The Relation of the Partial Pressure of Oxygen to Respiration and Growth in Germinating Wheat." PhD diss., Johns Hopkins University, 1929.

Mackall, John Nathaniel. "Drainage, the Fundamental Principle of Road Construction." CE thesis, Maryland Agricultural College, 1912.

Madigan, George Francis. "A Chemical Investigation of the Cause of Hardpan Formation in Southern Maryland Soils." PhD diss., University of Maryland, 1938.

———. "The Effect of Hydrogen Ion Concentration Upon the Completeness of Precipitation of the Sesquioxides, Calcium, and Magnesium in a Total Soil Analysis." Master's thesis, University of Maryland, 1933.

Mai, Richard Everett. "A Market Quality Study of the Del-Mar-Va Broiler and Fryer." Master's thesis, University of Maryland, 1941.

Manchey, L. L. "A Phytochemical Study of Gillenia Stipulata." Master's thesis, University of Maryland, 1931.

Marks, Louise. "The Effect of Certain Cooking and Holding Methods on the Nutritive Value of the Protein and Vitamins of Cod and Oyster." Master's thesis, University of Maryland, 1943.

Marshall, Housden Lane. "By-Product Potash Its Efficiency as a Source of Potash in Mixed Fertilizer." Master's thesis, University of Maryland, 1926.

Marshall, Robert. "An Experimental Study of the Water Relations of Seedling Conifers with Special Reference to Wilting." PhD diss., Johns Hopkins University, 1930.

Marth, Paul C. "Studies with Growth Regulating Substances on the Vegetative Bud Inhibition of Rose Plants During Common Storage." PhD diss., University of Maryland, 1942.

———. "A Study of the Root Distribution of Apple Trees." Master's thesis, University of Maryland, 1933.

Martin, John H. "A Kernel Classification of the Wheat Varieties of the United States." Master's thesis, University of Maryland, 1921.

Mason, Albert Freeman. "A Physiological Study of the Effects of Different Nitrogen Carriers on the Nitrogen Nutrition of Orchard Plants." PhD diss., University of Maryland, 1929.

Mather, Robert E. "Unlimited Grain Feeding for Milk Production." Master's thesis, University of Maryland, 1941.

Matthews, Earle Dwight. "A Biochemical Study of Soil Organic Matter as Related to Brown Root Rot of Tobacco." PhD diss., University of Maryland, 1935.

———. "The Relation of Sclerotium Bataticola Taub. to Brown Root Rot." Master's thesis, University of Maryland, 1932.

Matthews, William A. "The Economic Value of Certain Cultural Conditions on the Yield and Quality of Raw and Manufactured Tobacco." Master's thesis, University of Maryland, 1930.

Mattingly, J. Philip. "Effect of Heating, under Various Conditions, and of Sprouting on the Nutritive Value of Soybean Oil Meals and Soybeans." Master's thesis, University of Maryland, 1944.

Matz, J. "Artificial Transmission of Sugarcane Mosaic." PhD diss., Johns Hopkins University, 1932.

McCall, Arthur Gillett. "The Physiological Balance of Nutrient Solutions for Plants in Sand Culture." PhD diss., Johns Hopkins University, 1916.

McCann, Wilbur E. "An Economic Study of 180 Dairy Farms in Maryland." Master's thesis, University of Maryland, 1934.

McCarron, Marcus A. "The Marketing of Fluid Milk in Maryland with Special Reference to the Cooperative Movement." Master's thesis, University of Maryland, 1923.

McClung, Marvin Richard. "The Efficiency of Feed Utilization by Pure Bred and Crossbred Bronze and Black Turkeys." Master's thesis, University of Maryland, 1942.

McConnell, Harold S. "Parasites of the Strawberry Leafroller, Ancylis Comptana Froel. In Maryland." Master's thesis, University of Maryland, 1932.

McDougal, Meredith Vernon. "The Boll Weevil Infestation in the South." PhD diss., Johns Hopkins University, 1931.

McKibbin, Reginald Robert. "The Effect of Sulphur on Soil Reaction and Plant Growth." PhD diss., University of Maryland, 1926.

McLean, Forman Taylor. "A Preliminary Study of Climatic Conditions in Maryland as Related to the Growth of Soy Bean Seedlings." PhD diss., Johns Hopkins University, 1915.

McMurtrey, J. E. "A Comparison of the Deficiency Effects of the Different Essential Elements on the Growth of Tobacco Plants in Solution Cultures." PhD diss., University of Maryland, 1931.

———. "Symptoms of Boron Deficiency on the Tobacco Plant." Master's thesis, University of Maryland, 1928.

McNally, Edmund Houston. "The Physiology of Yolk Formation, Especially the Vitelline Membraneand the Mechanism of Ovulation in the Fowl." PhD diss., University of Maryland, 1940.

Mead, Russell K. "Factors Affecting the Sale of Dairy Products at Roadside Markets." Master's thesis, University of Maryland, 1935.

Mecham, C. Marion. "A Study of Some of the Factors Influencing the Whipping Properties of Cream and the Stability of the Finished Product." Master's thesis, University of Maryland, 1935.

Melchor, Beulah H. "The Land Possessions of Howard University, a Study of the Original Ownership and Extent of the Holdings of Howard University in the District of Columbia." Master's thesis, Howard University, 1945.

Melroy, M.B. "The Legume Inocolum Project in the State of Maryland." Master's thesis, University of Maryland, 1925.

Merz, Timothy. "The Effect of Extended Anaerobic Treatments on the Chromosomes of Vicia Faba." PhD diss., Johns Hopkins University, 1927.

Metzger, Jacob Elry. "Agricultural Progress in a Typical Maryland Community, 1865-1924." Master's thesis, Johns Hopkins University, 1925.

Meyers, George Edwin. "Tobacco Exports of Virginia and Maryland, 1698-1715." PhM diss., University of Wisconsin, 1930.

Miller, Charley Baker. "A Survey of the Potato Industry in Garrett County, Maryland with a View of Improving Vocational Teaching." Master's thesis, Pennsylvania State College, 1936.

Miller, Earl E. "Characteristics of Farm Property Assessments in Maryland after the Last Reassessment." Master's thesis, University of Maryland, 1941.

Miller, Erston V. "Biochemical Study of Maturity in the Irish Potato." Master's thesis, University of Maryland, 1921.

Miller, James Zenus. "Ground Versus Unground Soybean Hay for Dairy Cows." Master's thesis, University of Maryland, 1928.

Mitchell, Broadus. "The Rise of Cotton Mills in the South." PhD diss., Johns Hopkins University, 1918.

Mitchell, George Sinclair. "Textile Unionism in the South." PhD diss., Johns Hopkins University, 1926.

Mook, Paul V. "Certain Ecological Factors Affecting the Oyster (Ostera Elongata Solander) in the Cheaspeake Bay, with Special Reference to the Oyster Larvae." Master's thesis, University of Maryland, 1926.

Moore, Oscar Keeling. "Morphological and Physiological Response of Secually Immature Chickens to Injections of Gonadotropic Hormones." Master's thesis, University of Maryland, 1940.

Moore, William Hempstone. "The Effect of Fertilizer on Yield and Quality of Tomatoes." Master's thesis, University of Maryland, 1928.

Moran, John Austin. "The Increased Bacterial Content of Milk Due to Manurial Pollution." Master's thesis, University of Maryland, 1923.

Muma, Martin H. "An Annotated List of the Spiders of Maryland." PhD diss., University of Maryland, 1945.

Mumford, J. Wesley, Jr. "Behaviour and Inheritance in the Fulcaster Wheat Family." Master's thesis, University of Maryland, 1925.

Munsey, Virdell Everard. "An Investigation of the Application of the Neutral Wedge Photometer to the Measurement of Carotenoid Pigments in Flour and Macaroni Products." PhD diss., University of Maryland, 1937.

Myers, Alfred T. "Seasonal Changes in Total and Soluble Oxalates in Leaf Blade and Petiole or Rhubarb (Rheum Hybridium)." Master's thesis, University of Maryland, 1943.

Nance, Nellie Ward. "Revegetation of a Denuded Area of Widewater, Maryland, Below the Great Falls of the Potomac after the Flood of March 19, 1936." Master's thesis, George Washington University, 1938.

Newell, Harry James. "Range of Boron Concentrations in Nutrient Solution for Tomato Plants." Master's thesis, University of Maryland, 1929.

Nichols, Norris Newman. "Quality Control Work, Baltimore Milk Shed." Master's thesis, University of Maryland, 1925.

Nickels, Frank F. "A Study of the Lime Requirement of Chester Soils and Factors Associated with It." Master's thesis, University of Maryland, 1932.

Nickerson, Mark. "An Experimental Analysis of Barred Pattern Formation in Feathers." PhD diss., Johns Hopkins University, 1943.

Norem, Winfred Luther. "Mineral Nutrition of Ageratum in Sand Cultures with Auto-Irrigation." PhD diss., Johns Hopkins University, 1936.

Norris, George Wesley. "The Institutionalization of the Youth of the Third District of Anne Arundel County, Maryland." Master's thesis, University of Maryland, 1934.

Nowotarski, John Stephen. "The Influence of the Mineral Content of the Diet on the Vitamin D Requirement of the Chick." Master's thesis, University of Maryland, 1942.

Nystrom, Paul E. "The Organization and Business Analysis of 267 Farms on the Eastern Shore of Maryland, 1931." Master's thesis, University of Maryland, 1931.

Obando, Nestor. "Land Utilization in Colombia." Master's thesis, University of Maryland, 1944.

Obrist, Jeanette. "Trees of Maryland and District of Columbia." Master's thesis, Catholic University of America, 1932.

Olsen, Marlow William. "Maturation, Fertilization, and Early Cleavage in the Egg of the Domestic Fowl." PhD diss., University of Maryland, 1941.

Olson, Rodney Andreen. "The Relation of Protoplasmic Streaming in the Avena Coleoptile to Respiration and Auxin Transport." PhD diss., University of Maryland, 1939.

———. "Root Formation and the Polarity of Root Development." Master's thesis, University of Maryland, 1937.

Outhause, James B. "The Relative Value of Distillers' Rye Dried Grains and Corn Grain and of Distillers Rye Dried Grains and Clover-Light Timothy Mixed Hay as Determined by Wintering Experiments with Pregnant Ewes." Master's thesis, University of Maryland, 1942.

Pardo, José Heeren. "Nutritional Value of the Ammonium Ion for Higher Green Plants, Especially Agricultural Forms." PhD diss., Johns Hopkins University, 1932.

Parker, Hugh Klemme. "A Study of the Vapor Pressure of Aqueous Solutions of Cane Sugar at Twenty Degrees." PhD diss., Johns Hopkins University, 1921.

Parker, Marion Wesley. "A Comparison of Physical and Colloidal Properties of Translucent and Opague Sweet Corn." Master's thesis, University of Maryland, 1930.

Parks, John J. "The Effect of Solution Change on Carbohydrate Metabolism of the Tomato." Master's thesis, University of Maryland, 1932.

Parks, Lloyd Elwin. "The Chemistry of Menhaden Oil." PhD diss., University of Maryland, 1943.

Paulhus, Norman G. "Factors Affecting the Candling and Broken-out Appearance of Fresh-Shelled Eggs." Master's thesis, University of Maryland, 1940.

Paulus, Max George. "Radiometric Measurements of the Ionization Constants of Methyl Orange and Phenolphthalein." PhD diss., Johns Hopkins University, 1915.

Peltier, Paul Xavier. "The Potato Tuber Moth in Maryland." Master's thesis, University of Maryland, 1928.

Perlmutter, Frank. "The Influence of Various Factors on Blueberry Germination and Seedling Growth." Master's thesis, University of Maryland, 1939.

Pessin, Louis Jeromes. "A Biological Study of Polypodium Polypodioides (Resurrection Fern) as an Air Plant in Mississippi." PhD diss., Johns Hopkins University, 1923.

Pettit, Alfred B. "Hibernation Studies of the Codling Moth in Maryland Orchards." Master's thesis, University of Maryland, 1938.

Phillips, Harold Clair. "The Immediate Effect of Tobacco Smoke on the Learning of Albino Rats." PhD diss., Johns Hopkins University, 1936.

Phillips, Watson D. "The History of Public Education in Cecil County, Maryland, before 1876." Master's thesis, University of Maryland, 1939.

Pierce, C. W. "An Economic Study of 99 Maryland Poultry Farms." Master's thesis, University of Maryland, 1933.

Pierce, Elwood Clifton. "Biochemistry of Salt Absorption and Accumulation by Plants with Special Reference to the Role of Ether Soluble Organic Acids." PhD diss., University of Maryland, 1942.

Poelma, L. J. "The Intravenous Tuberculin Test as a Means of Diagnosing Bovine Tuberculosis." Master's thesis, University of Maryland, 1928.

Poffenberger, Paul R. "An Economic Study of the Broiler Industry in Maryland." Master's thesis, University of Maryland, 1937.

Pollock, George Findlay. "Hay and Forage Available in the United States for Dairy Feeding." Master's thesis, University of Maryland, 1924.

Pope, Merritt Nichol. "Catalese Activity in Relation to the Growth Curve of Barley." PhD diss., University of Maryland, 1929.

Posey, Walter B. "Marketing of Maryland Tobacco." Master's thesis, University of Maryland, 1940.

Potter, Helen. "Federal Protection for the Consumer: An Economic Analysis." PhD diss., Johns Hopkins University, 1942.

Potts, Samuel F. "The Biology and Control of the Pea Aphis (Illinois Pisi Kalt)." Master's thesis, University of Maryland, 1924.

Powell, Burwell B. "The Marketing of Livestock in Maryland with Particular Reference to the Baltimore Market." Master's thesis, University of Maryland, 1928.

Pressley, Helen Wade. "Drinks and Drinking in Colonial Maryland." Master's thesis, University of Maryland, 1943.

Provenza, D. Vincent. "Feeding Injuries Produced on Plants by Certain Coccidae." Master's thesis, University of Maryland, 1941.

Pryor, James Chambers. "On the Presence of Spore-Bearing Aerobic and Anaerobic Bacteria in Washington Market Milk." Master's thesis, Johns Hopkins University, 1913.

Purdy, Daisy I. "A Study of the Bacteriological Changes Produced During the Aging of Cured Hams." PhD diss., University of Maryland, 1931.

Pyle, Milton Allender. "An Investigation of the Original Deed Description of the University of Maryland Property at College Park with Reference to Some of the Oldest Land Marks." PhD diss., University of Maryland, 1937.

Raby, Alfred Benjamin. "Some Basic Facts Pertinent to the Establishment, Organization and Operation of Production Point Egg Auctions in Maryland." Master's thesis, University of Maryland, 1937.

Randall, Gussie. "A Comparative Study of Problems, Situations and Interests of Rural Unmarried Youth, 16-25 Years of Age, in Two Maryland Areas." Master's thesis, University of Maryland, 1939.

Rauchschwalbe, Otto E. "A Statistical Appraisal of Sampling Technique Used in the Disign of Experiments with Wheat Varieties Grown in Field Plots." Master's thesis, University of Maryland, 1940.

Redmond, Paul John, Catholic University of America and Herbarium. "A Flora of Worcester County, Maryland." PhD diss., Catholic University of America, 1932.

Reed, Clyde Franklin. "A Study of the Relations of Seed Germination to Time, Temperature and Submergence." Master's thesis, Johns Hopkins University, 1940.

Reid, Flora Waldman. "A Study to Determine the Uses of Rokusun Soybeans Packed in Brine, and Processed for Varying Lengths of Time, and the Use of Soybeans Packed in Syrup, and Processed for Varying Lengths of Time." Master's thesis, University of Maryland, 1938.

Remsberg, Harold Albert. "A Study of the Reaction, Total Osmotic Concentration and the Nitrate Content of Residual Solutions in Sand Cultures as Related to Yields of Wheat." Master's thesis, University of Maryland, 1925.

Reneger, Cecil Alfred. "Relation of Biological Environment in the Soil to Tobacco Brown Root Rot." Master's thesis, University of Maryland, 1931.

Reynard, George Bergin. "Polycotyledony in the Genus Lycopersicon." PhD diss., University of Maryland, 1943.

Reynolds, Orr E. "The Water, Glycogen, and Fat Metabolism of the Developing Guinea Pig." Master's thesis, University of Maryland, 1943.

Reynolds, R. Selena. "Comparison of the Pure Food Laws of Maryland Concerning Milk, Ice Cream, Cheese, Meat, Butter, and Eggs with Similar Laws of Ohio, Wisconsin, Kentucky, Pennsylvania, Connecticut, California, Illinois, Minnesota, Michigan and Washington." Master's thesis, University of Maryland, 1934.

Riedel, Erna Marta. "History of Home Economics in the State of Maryland." Master's thesis, University of Maryland, 1936.

Robertson, Roy L. "Observations on the Growth Stages in the Common Blue Crab, Callinectes Sapidusrathbun, with Special Reference to Post-Larval Development." Master's thesis, University of Maryland, 1938.

Robey, Carrie Elaine. "An Occupational Survey of the Graduates of Rural Montgomery County, Maryland in Relation to the Curricula Pursued in High School." Master's thesis, University of Maryland, 1941.

Robinson, Harold B. "The Influence of Environmental Factors Upon the Resistance of Some Non-Sporulating Bacteria Commonly Surviving Pasteurization of Milk." Master's thesis, University of Maryland, 1941.

Rosen, Harry. "The Pharmacology of Pyrethrum Flowers." PhD diss., University of Maryland, 1936.

Ross, Hugh. "An Economic Survey of the Black Raspberry Industry of Washington County." Master's thesis, University of Maryland, 1928.

Rothgeb, Russell Grove. "A Statistical Study of Several Morphological Characters Associated with Reproduction and Yield in Maize." PhD diss., University of Maryland, 1928.

Rubin, Max. "The Vitamin A (Carotene) Requirements for Egg Production and Hatchability." PhD diss., University of Maryland, 1942.

———. "Vitamin A Deficiency in Commercial Poultry Flocks." Master's thesis, University of Maryland, 1940.

Ruble, Ralph W. "The Role of Organic Matter in the Total Exchange Capacity of Maryland Soils." Master's thesis, University of Maryland, 1935.

Russell, H. L. "Bacteria in Their Relation to Vegetable Tissue: A Dissertation Presented to the Board of University Studies of the Johns Hopkins University for the Degree of Doctor of Philosophy." PhD diss., Johns Hopkins University, 1892.

Sanders, P. D. "The Plum Curculio in Maryland (Conotrachelus Nunuphar Herbst)." Master's thesis, University of Maryland, 1924.

Sando, Charles Earl. "The Process of Ripening in the Tomato, Considered Especially from the Commerical Standpoint." PhD diss., University of Maryland, 1920.

Sargent, Eloyse. "The Cake and Biscuit Making Gualities of Flour from Certain Varieties of Maryland Wheat." Master's thesis, University of Maryland, 1933.

Scheuch, John D. "The Effect of Different Acid Radicals of Potash Salts on the Chemical Reactions in Differents Depth of Soil." Master's thesis, University of Maryland, 1923.

Schmeisser, Harry Christian. "Leukaemia of the Fowl: Spontaneous and Experimental." PhD diss., Johns Hopkins University, 1914.

Schmidt, E. H. "Quality of Maryland Wheat." Master's thesis, University of Maryland, 1929.

Schmiedeler, Edgar. "The Industrial Revolution and the Home, a Comparative Study of Family Life in Country, Town, and City." PhD diss., Catholic University of America, 1927.

Schnebly, Lewis A. "Fruit and Vegetable Purchasing by Chain Stores in Baltimore." Master's thesis, University of Maryland, 1938.

Schneidereith, Helene M. "Safe-Guarding the Health of the Child in Baltimore and Rural Maryland." Master's thesis, Johns Hopkins University, 1921.

Schopmeyer, C. H. "Analysis of the Management of a Farm Business." Master's thesis, University of Maryland, 1924.

Schopmeyer, Clifford S. "Relative Imbibtion Rates in Varieties of Zea Mays." Master's thesis, University of Maryland, 1935.

Schrader, Albert Lee. "The Concord Grape— Pruning and Chemical Studies in Relation to the Fruiting Habits of the Vine." PhD diss., University of Maryland, 1925.

Schueler, John Edmund, Jr. "Effects of Fertilizer on the Yield Quality, and Composition of Tomatoes." Master's thesis, University of Maryland, 1931.

Schutz, J. Logan. "The Indices of Value and Production of Maryland Agriculture for the Period, 1918-1938." Master's thesis, University of Maryland, 1940.

Scott, L. E. "Boron Nutrition of the Grape." PhD diss., University of Maryland, 1944.

Seabold, Charles Wightman. "Related Science and Vocational Agriculture." Master's thesis, University of Maryland, 1932.

Secrest, John P. "Control of the Corn Ear Worm (Heliothis Armigera) 1939." Master's thesis, University of Maryland, 1941.

Seufferle, Charles H. "The Production and Marketing of Maryland Sweet Potatoes." Master's thesis, University of Maryland, 1940.

Shear, Cornelius B. "Nutritional Studies on the Stalk-Rot of Corn Caused by Diplodia Zeae (Schw.) Lev." Master's thesis, University of Maryland, 1938.

Shearer, Kathleen McCollum. "A Comparative Study of Rural Families Relief and Non-Relief." Master's thesis, University of Maryland, 1940.

Shephard, H. H. "The Biology and Control of the Corn Earworm." Master's thesis, University of Maryland, 1927.

Shillinger, J.E. "The Effect of Certain Parasiticides on the Host Animals." Master's thesis, University of Maryland, 1923.

Shirk, Harold George. "A Study of the Change of Temperature Effect on Respiration in Potatoes as Measured by the Warburg Microrespirometer Technique." Master's thesis, University of Maryland, 1936.

———. "A Study of the Oxygen Respiration in Corn and Wheat Kernels as Measured by the Warburg Manometer Technique." PhD diss., University of Maryland, 1938.

Shive, John Wesley. "A Study of Physiological Balance in Nutrient Media Resulting in a Simplified Culture Solution for Plants." PhD diss., Johns Hopkins University, 1915.

Shoemaker, Henry Reese. "A Standard Department of Vocational Agriculture in Maryland." Master's thesis, University of Maryland, 1926.

Showacre, Jane L. "The Effect of Growth Substances on the Inhibition of the Cotyledonary Buds of Phaseolus Vulgaris." Master's thesis, University of Maryland, 1944.

Shutak, Vladimir G. "Factors Associated with Skin-Cracking of York Imperial Apples." PhD diss., University of Maryland, 1942.

Siegler, Eugene Alfred. "Subject: The Origin of Fleshy Roots Produced on Apple Grafts by the Hairy Root Organism, Phytomonas Rhizogenes." PhD diss., University of Maryland, 1934.

Simmonds, Nina. "The Biological Analysis of Vegetarian Diets." Master's thesis, Johns Hopkins University, 1922.

Simonds, Florence Tucker. "A Cytological Study of the Tomato Leaf Spot and Its Casual Organism, Septoria Lycopersici (Speg.) Sacc." PhD diss., University of Maryland, 1934.

Skilling, F. C. "A Study by the Electrometric Method of the Hydrogen Ion Concentration of Liquid Cultures of the Human, Avian and Bovine Types of the Tubercle Bacilli." Master's thesis, University of Maryland, 1925.

Slacum, Emerson Phillips. "The Maryland Youth Survey and Its Implications for a Guidance Programme in the Rural High Schools of Maryland." Master's thesis, Duke University, 1940.

Slade, Hutton D. "Comparison of Media and Incubation Temperatures for Bacteriological Analysis of Milk." Master's thesis, University of Maryland, 1936.

Slagle, Edgar Apple. "On the Theory of Indicators and Reactions of Phthaleins and Their Salts." PhD diss., Johns Hopkins University, 1909.

Slavin, Elise Estella. "A Survey of the Christ Child Fresh Air Farm at Rockville, Maryland." Master's thesis, Catholic University of America, 1928.

Sloan, Joseph W. "Comparative Vitamin D Studies with Menhaden Oil." Master's thesis, University of Maryland, 1936.

Slocum, Glenn Gerald. "Studies on Food-Poisoning Staphylococci." PhD diss., University of Maryland, 1939.

Smith, Augustine Vivian Pollitt. "The Ecological Relations and Plant Successions in Four Drained Millponds of the Eastern Shore of Maryland." PhD diss., Catholic University of America, 1938.

Smith, Carl B. "An Economic Study of 128 Dairy Farms on the Upper Eastern Shore of Maryland." Master's thesis, University of Maryland, 1938.

Smith, Charles Linton. "A Comparative Study of the Respiratory Responses in Vegetables after Periods of Cold Storage." PhD diss., University of Maryland, 1929.

———. "Studies on Respiration in Parsnips." Master's thesis, University of Maryland, 1927.

Smith, Cornelia Marschall Mrs. "The Development of Dionaea Muscipula. I. The Flower and Seed." PhD diss., Johns Hopkins University, 1928.

Smith, Max Atlee. "Related Science in Vocational Agriculture Based on Statements of Successful Dairy Farmers in Frederick County." Master's thesis, University of Maryland, 1933.

Smith, Paul W. "Tariff Legislation and the Farmer." Master's thesis, University of Maryland, 1930.

Smith, Wilson L. "A Study of Biological Strains of Alternaria Solani (E. & M.) Jones and Grout and Their Pathogenic Effect on Tomatoes." Master's thesis, University of Maryland, 1942.

Specht, Alston Wesley. "The Availability of Boron in Some Maryland Soils." Master's thesis, University of Maryland, 1941.

Speck, Marvin L. "The Bacteriology and Sanitation of Single Service Ice Cream Containers." Master's thesis, University of Maryland, 1937.

Speer, Carl. "Sanitary Engineering Aspects of Shellfish Pollution." Master's Thesis, Johns Hopkins University, 1926.

Spies, Joseph R. "Pt. I.A Study of the Toxicity to Goldfish of a Group of Plants Reputed to Contain Toxic Constituents. Pt. Ii. An Investigation of Cracca (Tephrosia) Cinerea." Master's thesis, University of Maryland, 1931.

Sproat, Ben B. "Certain Effects of Leaf Removal Upon the Strawberry and the Peach." Master's thesis, University of Maryland, 1933.

Stanton, Thomas Ray. "The Inheritance of Plant Height in Oats." Master's thesis, University of Maryland, 1921.

Stanton, William Alexander. "A Study of Pyrethrosin, an Insecticidally Inert Constituent of Pyrethrum Flowers." PhD diss., University of Maryland, 1941.

Stargel, Viola B. "An Analysis of the Factors Affecting Milk Consumption in the Washington, D.C. Area." Master's thesis, University of Maryland, 1942.

Stark, Francis C. "Factors Influencing Growth of Pods and the Development of the Inner Mesocarp in Pods of the Snap Bean (Phaseolus Vulgaris, L.)." Master's thesis, University of Maryland, 1941.

Steinbauer, Clarence E. "Studies on the Rest Period of Tubers of the Jerusalem Artichoke. (Helianthus Tuberosus, L.)." PhD diss., University of Maryland, 1936.

Stephenson, Elizabeth. "Nutritive Value of Canned Soybeans at Different Stages of Development and by Different Methods of Preparation." Master's thesis, University of Maryland, 1944.

Stephenson, R. B. "Effect of Growth Regulating Substances on Development of Seedlings and Excised Parts in Culture." Master's thesis, University of Maryland, 1940.

Stewart, J. Raymond. "An Economic Study of Beef Cattle Farms in Lancaster County, Pennsylvania." Master's thesis, University of Maryland, 1939.

Stier, Howard Livingston. "A Physiological Study of Growth and Fruiting of the Tomato (Lycopersicum Esculentum)—with Reference to the Effect of Certain Climatic and Edaphic Conditions." PhD diss., University of Maryland, 1939.

———. "A Study of Germination and Dormancy in Seeds of the Potato (Solanum Tuberosum)." Master's thesis, University of Maryland, 1937.

Stimpson, Edwin Greenwood. "A Nutritive Study of Vigna Sinensis (Back-Eyed Pea Variety)." PhD diss., University of Maryland, 1937.

Stinnett, Lucile Lactoure. "Construction of Units in Consumer Education for Home Economics Pupils of Mt. Ranier High School." Master's thesis, University of Maryland, 1937.

Stoops, Charles Stratton. "The Corn Wheat Yield Ratio in Maryland." Master's thesis, University of Maryland, 1929.

Stoudt, Harry Nathaniel. "Floral Morphology and Anatomy of Capparidaceae." PhD diss., Johns Hopkins University, 1939.

Strasburger, Minna Elaine. "A Study to Determine the Palability and Cost of Several Cooking Fats." Master's thesis, University of Maryland, 1935.

Stuart, Dorothy Rhett. "The Development of Flower, Seed and Fruit in Sassafras Variifolium." Master's thesis, Johns Hopkins University, 1929.

Stuart, William M. "The Inheritance of Certain Morphological Characters in a Barley Cross." Master's thesis, University of Maryland, 1929.

Swingle, Charles Fletcher. "A Physiological Study of Rooting and Callusing." PhD diss., Johns Hopkins University, 1927.

Syme, William Anderson. "Some Constituents of the Poison Ivy Plant (Rhus Toxicodendron)." PhD diss., Johns Hopkins University, 1906.

Tang, Peisong. "An Experimental Study of the Germination of Wheat Seed under Water, as Related to Temperature and Aeration." PhD diss., Johns Hopkins University, 1930.

Taylor, T. T. "The Relative Efficiencies of Fertilizers on Bent Grasses." Master's thesis, University of Maryland, 1930.

Teeter, Viola C. "Standards of Living of Maryland Farm Families." Master's thesis, University of Maryland, 1938.

Temple, Martha Ross. "The Cake-Making Qualities of Flour Milled by Various Processes from Wheat Grown in Different Sections of Maryland." Master's thesis, University of Maryland, 1932.

Tennent, David Hilt. "A Study of the Life History of Bucephalus Haimeanus, a Parasite of the Oyster." PhD diss., Johns Hopkins University, 1904.

Tepper, A. E. "Studies on Gizzard Ulceration in Chickens." PhD diss., University of Maryland, 1941.

Thompson, Arthur H. "Factors in Dehydration and Subsequent Storage of Apples and Lima Beans Associated with Retention of Ascorbic Acid." PhD diss., University of Maryland, 1945.

Thompson, Ross C. "The Genetic Relations of Some Color Factors in Lettuce (Lactuca Sativa and L. Scariola)." PhD diss., University of Maryland, 1936.

Thompson, Ruth Lee. "Carotene Changes in Seven Varieties of Fresh, Dehydrated, and Reconstituted Sweet Potatoes." Master's thesis, University of Maryland, 1943.

Thornton, Norwood Charles. "The Value of a by-Product Potash Residue for Crop Growth, and Its Effect on the Constituents of Mixed Fertilizers." Master's thesis, University of Maryland, 1928.

Tiller, Richard Edward. "Indications of Compensatory Growth in the Striped Bass (Roccus Saxatilis Walbaum) as Revealed by a Study of the Scales." Master's thesis, University of Maryland, 1942.

Tillson, Albert Holmes. "The Floral Anatomy of the Aurantioideae." PhD diss., University of Maryland, 1938.

———. "The Vascular Anatomy of the Floral Organs of the Valencia Orange, Citrus Sinensis, Osbeck." Master's thesis, University of Maryland, 1935.

Toole, Marguerite Goss. "The Formation of Diploid Plants from Haploid Peppers." Master's thesis, University of Maryland, 1944.

Tottingham, William Edward. "A Preliminary Study of the Influence of Chlorides Upon the Growth of Certain Cultivated Plants." PhD diss., Johns Hopkins University, 1917.

Trelease, Sam Farlow. "The Relation of Salt Proportions and Concentration to the Growth of Young Wheat Plants in Nutrient Solutions Containing a Chloride." PhD diss., Johns Hopkins University, 1917.

Troy, Virgil S. "Determination of the Keeping Quality of Milk by Means of the Sodium Hydroxide Titration Test and Alcohol Precipitation Test and Calorimetric Hydrogen-Ion Concentration Test." Master's thesis, University of Maryland, 1924.

Truitt, Reginald V. "Biological Contributions to the Development of the Oyster Industry in Maryland." PhD diss., American University, 1929.

———. "Investigation of the Oyster Industry of the Chesapeake Bay." Master's thesis, University of Maryland, 1922.

Ulm, Amanda A. "The Multiple Seedlings of Carrot." Master's thesis, University of Maryland, 1945.

Upshall, William Harold. "The Propagation of Apples by Means of Root Cuttings." PhD diss., University of Maryland, 1931.

Urquhart, Norman R. "An Economic Study of 128 Dairy Farms on the Eastern Shore of Maryland." Master's thesis, University of Maryland, 1937.

Van Horn, Clifton W. "Seasonal Development of Vegetative Organs of Strawberry Plants under Field Conditions." Master's thesis, University of Maryland, 1940.

Van Schaack, Eva Blanche. "Studies on May-Apple Rust." PhD diss., Johns Hopkins University, 1937.

Veerhoff, Otto Louis. "Relation of Time and Maintained Temperature to Germination in Flax." PhD diss., Johns Hopkins University, 1937.

Veihmeyer, Frank J. "Some Factors Affecting the Irrigation Requirements of Deciduous Orchards." PhD diss., Johns Hopkins University, 1926.

Verner, Leif. "A Physiological Study of Cracking in Stayman Apples." PhD diss., Johns Hopkins University, 1934.

Vierheller, Albert Frank. "Investigations in the Rooting of Apple Cuttings." Master's thesis, University of Maryland, 1923.

Vincent, Lionel L. "The Organization and Business Analysis of 279 Farms in the Piedmont Plateau Region of Maryland." Master's thesis, University of Maryland, 1932.

Wadkins, Ross F. "Apple Root Rot Disease." Master's thesis, University of Maryland, 1926.

Walker, Earnest Artman. "The Development of Storage Scab of Apple." PhD diss., University of Maryland, 1939.

———. "The Effectiveness of Alcoholic Mercuric Chloride Treatment for Tobacco Seed." Master's thesis, University of Maryland, 1927.

Walker, Wm. Paul. "A Study in Orchard Nutrition." Master's thesis, University of Maryland, 1924.

Wallace, David H. "Length Frequencies of the Striped Bass (Roccus Lineatus) in the Chesapeake Bay, During the Season, 1936-37." Master's thesis, University of Maryland, 1937.

Walls, Edgar Perkings. "The Vascular Anatomy of the Floral Parts of Some Solanaceous Plants." PhD diss., University of Maryland, 1935.

Walzl, Edward McColgan. "Action of Electrolytes on the Beat of the Heart of the Oyster, Ostrea Virginica." PhD diss., Johns Hopkins University, 1935.

———. "Action of Ions on the Beat of the Heart of the Oyster, (Ostrea Virginica)." Master's thesis, Johns Hopkins University, 1934.

Wanlass, William Lawrence. "The United States Department of Agriculture: A Study in Administration." PhD diss., Johns Hopkins University, 1919.

Warman, Henry J. "Population and Land Use of the Manor Counties of Maryland." PhD diss., Clark University, 1945.

Weaver, Elaine Knowles. "A Study of Attitudes of Thirty-Three Families Assisted by the Rural Resettlement Program: Maryland, 1937." Master's thesis, Cornell University, 1937.

Webster, Carolyn I. "The Effect of Ascorbic Acid on the Development of Primary Lesions of Tobacco Mosaic in Nicotiana." Master's thesis, University of Maryland, 1941.

Weiland, Glenn Statler. "A Study of the Factors Influencing the Yield of Ascaridole in Chenopodium Ambrosiodes L. Var. Anthelminticum." PhD diss., University of Maryland, 1933.

Weimer, Winifred R. "Adsorption and Its Effect on the Freezing Point of Hydrogels." Master's thesis, University of Maryland, 1925.

Weinberger, John Howard. "The Effect of Various Potash Fertilizers on the Firmness and Keeping Quality of Apples, Peaches and Strawberries." PhD diss., University of Maryland, 1931.

Weitzell, Everett C. "Probable Economy and Increased Efficiency in Local Governments of Maryland through Redistricting the State." Master's thesis, University of Maryland, 1935.

Wellington, J. W. "Effect of the Shoot Removal Method of Training on the Subsequent Growth of Apple and Peach Trees." Master's thesis, University of Maryland, 1931.

Wenzel, Marie E. "Microorganisms on Nut Meats Used in Ice Cream Mixes." Master's thesis, University of Maryland, 1938.

Whaley, M.S. "The Botany and Culture of Bent Grasses." Master's thesis, University of Maryland, 1927.

Wheaton, Evan. "The Effect of Refrigeration Upon the Bacterial Count of Milk." Master's thesis, University of Maryland, 1927.

Wheelan, Frank N. "An Economic Study of 184 Dairy Farms in Maryland." Master's thesis, University of Maryland, 1933.

White, Philip R. "Studies on the Banana. An Investigation of the Floral Morphology and Cytology of Certain Varieties of the Genus Musa L." PhD diss., Johns Hopkins University, 1928.

Whitehouse, William Edwin. "A Nutritional Study of the Strawberry." PhD diss., University of Maryland, 1928.

Whiteman, Thomas Moore. "The Effect of Various Storage Temperatures on the Carbohydrate Composition of Cooked as Compared with Raw Potatoes." Master's thesis, University of Maryland, 1938.

Wiedemer, Arthur Paul. "Efficiency Tests on a Small Electric Holder Type Pasteurizer." Master's thesis, University of Maryland, 1941.

Wiley, R. C. "Adsorption of Plant Food by Colloidal Silica and Humus." Master's thesis, University of Maryland, 1922.

Williams, Charles Simpson. "Breeding for Candling Quality and Shell Strength." Master's thesis, University of Maryland, 1941.

Note: Contents concern poultry science.

Williams, Gertrude A. C. "A History of Education in Allegany County, Maryland, 1798-1900." Master's thesis, University of Maryland, 1937.

Williams, Ralph Charles. "A Method for Evaluating the Results of Rapid Soil Tests for Maryland Soils." Master's thesis, University of Maryland, 1936.

Wilson, Nathaniel John, Jr. "A Comparison of the Reductase Test, the Catalase Test, and the Plate Count on Market Milk." Master's thesis, University of Maryland, 1926.

Winant, Howard Barr. "A Study of the Potassium-Thiocyanate Method for Determining Soil Activity." Master's thesis, University of Maryland, 1924.

Winbury, Martin M. "Studies on the Golgi Apparatus in the Pancreatic Islets of the Guinea Pig." Master's thesis, University of Maryland, 1942.

Winter, Charles Ernest. "A Comparative Study of the Methods for Detection of Coliform Bacteria in Milk." Master's thesis, University of Maryland, 1945.

Winterberg, Samuel H. "Influence of Cultivation with an Electric Plow on Soil and Crop Response." Master's thesis, University of Maryland, 1930.

Wolf, James McNair. "Soil-Moisture Condition at the Onset of Permanent Wilting and a General Empirical Expression for the Relation of the Water-Supplying Power of a Soil and Its Current Moisture Content." Master's thesis, Johns Hopkins University, 1940.

Woods, Mark W. "Studies in Ring-Spot." Master's thesis, University of Maryland, 1933.

Yedinak, Alec. "The Effect of Salt Water Flooding on Soil Fertility." Master's thesis, University of Maryland, 1934.

Yoder, Roy C. "The Effect of Size of Seed Upon Maturity and Yield of Canning Peas." Master's thesis, University of Maryland, 1928.

9

Dairy Plant Maintenance: Training, Servicing, Equipment and Operations Management

TOWARDS BETTER TRAINING OF DAIRY INDUSTRY ENGINEERS

A well-trained, well-managed team of engineers is vital to the running of a milk plant. If training or management is lacking then the equipment will break down or not work effectively. With short life products such as milk, shortage, stoppages or ineffective machinery can mean disaster. In an ideal world the engineers would be required to be a well-organized, methodical, highly-condensed team devoted to indepth planned maintenance. However, in reality what is in fact often required is a highly reactive force that can respond quickly to the unknown with the minimum of delay. The engineering department needs to contain three teams; engineers, electricians and support team. Each team should be independent in structure but capable of inter-dependence with the other enquiry teams and other working groups within the milk plant. For example, in a creamery receiving 100 000 litres per day, there will be variants depending on the exact produce mix. Assuming that two shifts are worked, the expected manning levels would be six to eight working engineers and electricians per shift. The age and complexity of the milk plant would determine the exact engineer/electrician mix. The structure would probably look as. For a three-shift operation an increase in establishment of four per section per shift would be seen. The extra requirement would cover the rota implications of the extra shift. As with modern equipment there would be a requirement for instrumentation skills. This could be part of the duties of the electrical section, or a separate section altogether, depending on two factors:

(a) the amount of electronic control in the plant, and

(b) whether or not there is a recognition of electronic/instrumentation skills as separate from those of an electrician.

In a highly modern plant separation would probably be an advantage.

Training Needs Analysis (TNA)

If a new plant is to be built or a major expansion is to be carried out then before recruitment or training commences an indepth Training Needs Analysis should be executed. This should also be carried out on existing sites as a review exercise. Training Needs Analysis for an engineering purpose has some major differences from a normal TNA as carried out on a production unit. The first difference is that there are fewer day-to-day tasks on which to base the analysis. Secondly there will exist a number of one-off tasks which must be taken into account regardless of their probability. These will be hard to foresee as information from the manufacturers of equipment will be non-existent. The first step in the TNA is to list the basic skill requirements that will be needed for any particular operation. Once a matrix of basic skills has been formed then the operation is repeated for the higher level skills required. These skills matrices should be carried out for each area, i.e. mechanical, electrical, etc. Once they have been carried out they will then form the basis of a manpower inventory, a skills requirement for an engineer and a training format. The difference between a new development and a redevelopment is that on new projects new recruits must be imported to cover all the skills identified. On redevelopment it may be that by expanding the existing workforce the personnel can be successfully upgraded to the standards required.

After a skills analysis has been carried out an equipment analysis is then necessary. Each piece of major plant is broken down into its known requirements under the headings of skill, theory and plant specific knowledge. From this plant analysis the skill requirements can be calculated in terms of loading, the general theory that will be required for the understanding of processes and equipment, and the amount of indepth specific plant knowledge the engineer must acquire or have at hand. The transfer of this last part must be monitored, as it will show the effectiveness of the commissioning engineer at passing on his knowledge to the work force. This must be tested during the installation and before the commissioning engineer leaves the site. The analysis as explained will require a considerable amount of time, but will in the end save time by ensuring that during recruitment an excellent picture will be formed of what is required. By having the training objectives built in, training plans and scheduling will be easier to control and monitor. This will also provide an excellent vehicle through which to judge when and where the training inputs have been effective. By doing such an analysis it will be possible to foresee problem areas and provide a basis for a planned maintenance schedule.

Recruitment. It is essential to establish a centre core of highly-skilled trouble shooters at an early stage both in the mechanical and electrical areas. The ability to fault-find is not present in all engineers; it is however a skill that can be learnt. This is time-consuming and needs good examples to follow. In demonstrating effective fault-finding skills simple logic is not all that is required. Good fault-finders need to be happy with the theoretical as well as the practical aspects of systems, and demonstrations of this will be required.

Management. The person specification when recruiting Management should include, apart from a good technical background and dairy industry experience, some ostensible indication of high motivation, organizational skills, good communicative ability, a strong training background and a high awareness of production needs and attitudes. It is the human-relations skills rather than the engineering qualifications that should be considered. The Engineering Manager must be consulted when recruiting other engineering management staff. Disharmony within the engineering structure is to be avoided. Also the engineering team must be able to cooperate effectively with other management teams within the unit.

Engineering staff. When recruiting engineers, reference should be made to the skills requirement as defined in the Training Need Analysis. Reliability and the ability to work unsupervised once directed are important factors, as many jobs will require the engineer to work on his own initiative away from the workshop. If recruiting raw beginners such as apprentices, tests for numeracy, spatial ability and mechanical comprehension are very valuable.

Providing training expertise. The training of engineers is time-consuming and requires a large input of skill. This may be available on site. However, time could well be a problem, so training must be carefully planned and use made of all the available resources. These are likely to be:

- Own skilled staff on site
- Local college or skill centre
- Local college operating on site
- Specialist training organizations
- Own specialist trainers.

A great deal will depend on the size of the task, the number to be trained, time constraints and available funding.

Using on-site staff for training. The main constraint is time. It is more than probable that only a limited number of people skilled enough to perform such a task will have been recruited and that they will be fully engaged in their normal tasks. Their instructional ability must also be examined. It is likely that a highly skilled fitter or electrician may not be very good at passing on his own skill and indeed may not want to. Working with a skilled man therefore, although a very valuable experience, must be regarded as only part of the answer. Its success rate can be considerably improved by ensuring that all engineers used as work guides have some form of training in instructional techniques. This will help them understand the learning process. It is essential that supervisory staff have a good understanding of instructional techniques, as they will not only have to instruct, but also test and make judgements on the development of the ability of each individual. If skilled staff are under pressure, or there are a number of personnel who must reach a required standard before working with a skilled man, then there may be justification for an on-site training specialist. Given sufficient numbers this may be the most speedy and effective way of training these personnel to such a standard of skill and knowledge. The advantages over training at a local training centre, i.e. technical college, or bringing in a specialist training company are that it is more flexible in having complete control over input, timing and cost, as well as being geared to the specific plant requirements in skills and detailed knowledge. Local training centres are a good source for providing theory on a general basis and may even have good basic skills programmes at low cost. They need to be carefully examined however to ensure that their practices are not counter to those required on site. It should also be remembered that in many cases their methods are probably not as up-to-date as those used in a new plant. Bringing in a specialist training organization on a short-term basis is both costly and not very effective for basic skills training, but it may be a very good means of quickly uprating a number of skilled staff in new or more advanced methods. An example may well be the upgrading of skilled welders to stainless steel welding.

Skills Training

The best means of ensuring that the basic skills content of any training programme is complete in detail is to use, as a guide, such publications as the Modular Practices published by the Engineering Industry Training Board in the U.K. As an example, their Module J3 Maintenance Factory Services skill specification would include on completion of training that the individual should be capable of the following.

- Applying statutory and safety regulations
- The safe and efficient use of both hand and portable power tools
- The safe and efficient use of test and measuring equipment

- The use of factory recording procedures associated with maintenance and breakdowns
- Working to sketches, engineering drawings
- The cutting, bending and joining of piping and tubing
- Installing, servicing, testing and replacement of factory services pipework and ancillaries
- Applying various methods of securing and fixing components.

This list goes on to include servicing of boilers, the use of lubricants and servicing of air compressors. The module books break down all the necessary skills into their component parts so that any skilled engineer with an instructional background should be able, following the modular pattern, to provide training tasks and theoretical input sufficient to develop a trainee to a good standard of basic skills. Other modules dealing with the basic skills are J1 Mechanical Maintenance and J2 Electrical Maintenance. For a higher skill level, the same system applies using more advanced modules. These are:

- J21 Mechanical Maintenance II
- J22 Electrical Maintenance II
- J4 Electronic Maintenance
- J23 Maintenance Factory Services II
- J28 Instrument Maintenance.

Providing the selection of trainees has been carried out correctly, the modular programmes should provide sufficient information to avoid difficulties and will provide a complete guide to the training of engineering skills.

Dairy product training. It is often not appreciated that an engineer working in a dairy produce environment needs to understand the nature of the products or the processes being applied. That he can repair any individual item of plant is not sufficient. A lack of knowledge in the dairy produce and processing field leads to misunderstanding of the importance of hygiene and clean working practices, usually a major criticism of the engineer by production staff. This can manifest itself in the deposit of foreign matter into produce, e.g. grease streaks in butter, nuts and bolts in cheese blocks and even, more important, it may lead to dangerous practices such as welding in areas where dry milk powder is present. The engineer, who has no knowledge of the processes involved, cannot assist the production staff in trouble shooting process problems and will be at a disadvantage in locating possible defects in modern complex equipment.

Knowledge of the dairy produce and the processing methods will already exist on site and the use of production staff in this type of training is to be encouraged, being both cost-effective and valuable in assisting the development of good relations between production and engineering departments. The areas that need to be covered are milk and its treatments, pasteurization, separation and effects of mishandling, and a brief explanation of test procedure and the microbiological implications would be useful. Buttermaking and cheesemaking processes should be treated in the same way in their basic make-up, outline of the process involved and problem areas. Engineers should also have a good knowledge of packaging, with emphasis of the effects of poor packaging on the product and any legal demands. The production departments may well organize training of this type for their own operators. If so, this would be an excellent opportunity to combine forces.

Health and Safety. In any engineering environment health and safety must be a high priority and, as such, must be included in any training programme. There are special areas within the dairy industry environment which need emphasis. The introduction to health and safety must be during the general induction training and should cover responsibilities and safe working practices. This should be followed up by not only talking about the safety aspects, but also ensuring that trainees follow safe working practices in everything they do. The areas where particular attention should be paid are:

- Electrical supply
- Steam supply
- Acid tanks and supply lines
- Powder areas and equipment.

If there is no permit to work system operating within the milk plant, then there should be some safeguards for engineers working on equipment to prevent the power supply being momentarily turned on. The engineer must be drilled into the habit of checking the electrical and/or steam supply lines before attempting any repairs.

Care in the use of strong acids in storage, or in use for cleaning must be emphasized. The engineer must know the routing of acid, and when and where it is being used. The treatment of acid spillage should be included at an early stage of training as well as part of any training directly connected with acid holding equipment.

Dried milk powder is highly explosive and a considerable live risk. This fact must be known to any engineer who is liable to work within a powder area, especially if the work should include welding in any form.

The use of safe working practices and the wearing of safety equipment cannot be stressed enough. Safety is an area with major training implications.

A systematic approach is as always the key to success. By first carefully analysing both the available resources and plant requirements a clear picture of the final objective may be drawn, and only then can a training plan be formulated and effectively implemented. The early training of supervisory and skilled staff in instructional techniques and the integration of background theory to complement the practical aspects of the training programme are vital in ensuring success.

A major failing of many training programmes is the lack of maintaining momentum. The day-to-day pressure of work takes over all scheduling and effectively destroys any semblance of order. To avoid the problem of work pressure the training programme must be carefully integrated with the development plans and expected day-to-day operation. It is safer to err on the side of caution when scheduling training sections and the use of the TNA sheets will be valuable in maintaining control of progress of the programme. The effectiveness of the milk plant is dependent upon the engineer, and his effectiveness is dependent upon his training.

TOWARDS ADOPTING THE METHODS OF PREVENTIVE MAINTENCE

Principles

The determination of plant efficiency at all essential points of the process is the first step in creating a preventive maintenance system which is fundamental to achieving optimal performance in any milk plant. The precise meaning of the term "preventive maintenance" depends on the concept of the organization of the plant, of its capacity and processing and manufacturing programme as well as on the availability of, and accessibility to specialized services of machinery manufacturers or their agents, who are usually equipped with an ample supply of spare parts. According to Newcomer 1 preventive maintenance in developing countries is a procedure involving:

(a) Planning and scheduling

(b) Proper installation

(c) Periodic inspection

(d) Planned lubrication

(e) Adjustment of machines and instruments

(f) Replacement of worn and damaged parts

(g) Recording and reporting observations, adjustments, repairs, and replacements.

(h) Periodically reviewing records on inspection, lubrication, repairs and performance of equipment

(i) Keeping an adequate supply of spare parts

(j) Determining maintenance costs

(k) Cleaning and painting equipment and buildings

(l) Inspection and maintenance of all emergency, personnel and plant protective equipment

(m) Maintaining full serviceability of all utilities

The obvious benefits of preventive maintenance include:

- Less production interruptions
- Fewer large-scale repairs
- Less raw material and product spoilage
- Increased life expectancy of equipment
- Less standby equipment needed
- Identification of items with high maintenance costs leading to investigation and correction of causes, such as misapplications, operator abuse or obsolescence
- Better spare parts control, greater work safety and lower manufacturing costs.

These benefits are the objectives of preventive maintenance. Although the objectives and principles of preventive maintenance programmes (PMP) are the same for any processing industry and for any plant within an industry, the actual programme of preventive maintenance procedures is an adaption of general rules to the needs of a given plant. Since there are no identical plants, there cannot be identical preventive maintenance programmes. However, the establishment of such an individual programme for any milk plant could be facilitated by giving the dairy engineer and his staff a general guide on the list of issues of which good PMP is composed, on the meaning of these issues and on the methods by which general rules can be transformed into practice. In this chapter an attempt is made to provide such a general guide. Before discussing PMP procedures in more detail, it should be emphasized that the dairy engineer's group usually forms a separate unit within the organizational structure of the milk plant. This unit may take the form of an engineering department in large-scale plants, or the form of a section in smaller enterprises, but in all well-organized milk plants it is a relatively autonomous unit. Its head has the status of a senior executive officer, similar to the position of the heads of the processing group, quality control group, or sales and procurement managers. To perform its duty the dairy engineer's group needs a programme of work, staff, and resources. The preventive maintenance programme is the most essential part of the programme of work of the dairy engineer's group. It is divided into several parts. The first comprises collecting and recording all basic information on machines and installations: the Equipment Records of the plant. This may be classified as the preparatory part of the PMP. The second comprises identification of inspection objectives, frequency and location, and is known as the Inspection Schedule. This includes lubrication schedules, and routine spare parts replacement programmes and may be classified as the plan of operations in which also the recording and reporting systems are defined. The third part of the PMP procedure is the action which starts with the analysis of the records and is followed by decisions on what must be done, by whom, when and by what means.

It also includes decisions on who inspects and accepts the completion of the action ordered. The last component of the action part of the PMP is the maintenance cost estimate. The application of PMP depends on well-equipped workshop facilities and trained staff. The variety of skills needed for performing all the duties specified above implies that training in the dairy engineer's group is an essential requisite of success.

Equipment Records

All plant performance studies require that the rated performance parameters are known, that appropriate instruments are installed to measure those actually achieved and that both are recorded and compared. As a result of this comparison all malfunctions can be detected, the reasons for their occurrence examined and steps taken in order to re-establish proper functioning of the plant. The essential elements to introducing performance studies can be listed as follows:

- All manufacturers' literature such as manuals, drawings and spare parts specifications should be collected, filed according to a selected system and kept under strict control of the engineering department of the plant.
- Flow diagrams of the whole plant and its sections should be prepared with performance check points clearly indicated.
- A recording and reporting system of performance inspections and studies should be introduced and strictly observed so as to make sure that action can be taken immediately after a deviation from the expected performance has occurred.

Manufacturers' literature contains essential technical data for the dairy engineer's use, but it is impracticable to use this in daily routine work. First of all it comprises bulky volumes, from which only parts are of interest for maintenance procedures. Besides it will not allow for recording all changes, major repairs, utilization of spare parts and similar entries essential in the dairy engineer's work. Therefore a method of recording basic information for each piece of equipment of the entire plant has to be adopted. Such a method, or a system, should be simple and, prior to introduction, should be carefully checked on its suitability to the needs of maintenance procedures. Usually card files are used for equipment records. Printed on a suitable format they allow for convenient entries of data and retrieval for future reference. Codes applied for easy reference are another essential part of a recording system. There is no use in a time-consuming preparation of a card file of equipment recording without simultaneously introducing a simple method of identifying the card quickly in the file whenever needed. Cards with perforations on the card edges, termed 'needle cards', are probably the most efficient way of solving this problem. They are widely used in storage registration systems and can be also recommended for equipment recording. An example of a machine record card is given in Fig. 1. As can be seen the record comprises three distinctly different parts. The first gives the identity of the machine and its origin and the second contains information relevant to maintenance procedures and operations. In both parts the information is derived from the manufacturer's literature except for the inventory number which is arbitrarily assigned by the plant. The third part, which is simply the space on the reverse, contains entries made by the plant engineer, and concerns major changes and repairs which the machine has undergone during its life. It is the "health record book" of the machine.

Inspection Schedules

Milk processing or milk products manufacture requires machines and installations by the use of which the raw material is transformed into a product. Each and every piece of equipment of which the plant is assembled must perform its duty. Achieving the exact performance objectives of the plant as a whole requires maximizing the efficiency of all equipment.

The term "efficiency" is used in this context as a comparison between the designed and the actual performance. Should, for instance, the set holding temperature in a pasteurization plant be 72°C (±1°C), the efficiency of the process is determined by checking whether, in the course of pasteurization of the product, any deviation from the wanted temperature levels occurs. In other words determining the efficiency requires collecting data concerning the actual performances and comparing them with those expected. Any equipment is subjected to wear and tear and its efficiency and performance decrease with time. Keeping the plant performance at the required level is the responsibility of the dairy engineer and his staff. To fulfil this responsibility he needs to know what the actual performance is. By detecting a drop in performance he can take corrective action. But his first duty is to prevent any drop in efficiency by taking care of the equipment in a rational way or in other words by servicing the machine. This servicing procedure will be structured according to the needs of the plant but it will always be based on:

- inspection of performance based on log book entries
- scheduled inspection combined with scheduled servicing at selected points considered crucial for plant efficiency
- scheduled inspection of all other sections of the plant aiming at detection of faults not detectable by other routine procedures.

The practice of plant performance inspection and servicing of machines should be carried out with sufficient simplicity and thoroughness to yield the best results. The creation of a routine system in this respect is the first step in establishing good habits of proper care of all items of equipment. It is essential to determine what should be inspected, how often and how to leave out of the procedure any collection of information which does not serve the main purpose. On the other hand the system must be beared toward prompt discovery of each and every deviation from normal in the plant or in the machine operation and also toward immediate action aiming at corrections at the detected sources of the deviation.

A. Inspection Based on Log Book Entries

Log books contain the record of the equipment performance and usually also instructions to and from the operator. They are basically designed to keep records of the daily processing runs and their recording elements are often replaced - at least partly by automatic recording charts. The log books are usually kept by the operators of various sections of the plant. Some or even most of them, may not belong to the engineering group but to the processing group. This should not prevent the engineering staff from inspecting the entries and from drawing conclusions from the figures and remarks recorded. In some cases it may be an advantage to include in the log book recorded observations which are particularly useful for engineering work even when the recorded are basically meant for the processing manager's inspection. Such cooperation between the departments of the plant is essential in many areas. Inspection based on log book entries permits discovery of performance faults occuring during processing and manufacturing operations of the plant. In other words such inspection is based on actual daily performance records. It is a good practice to attach to the log book a simplified sectional flow diagram of the process concerned with all check spots clearly shown and marked in the same order as they are on the recording pages. As can be seen in the example of the performance record there is an instance of an unusually high milk outlet temperature. Simultaneously a high outlet temperature of the ice water was recorded. Inlet temperatures of milk and ice water were at the normal level. This deviation from normal will be noticed by the inspecting engineer (if not called earlier by the operator) and will make him analyse the trouble and identify the cause. The indication is that the flow of ice water was reduced either within the heat exchanger or in the ice water line. An immediate check on the ice water installation performance record should enable the engineer to locate the trouble either in the heat exchanger (choking, leakage) or in the ice water supply (pump defect, high pressure drop in the pipe, leakage). Corrective actions should follow immediately. It is obvious that any instrument used to measure the actual value of a parameter of a process needs to be frequently checked so as to ensure that the readings illustrate the true value of the recorded item. Inspection of log books provides the engineer with invaluable information on the actual performance of the plant, but it cannot indicate whether the equipment is in sufficiently good condition for high performance in the immediate future. Such conclusions require another type of inspection based on different considerations.

B. Scheduled Inspection Combined with Scheduled Servicing

Inspection-cum-servicing schedules are the most crucial task for the engineering team responsible for preventive maintenance. There are three sources of know-how on which these schedules are based and they may be listed as follows:

- Manufacturers' manuals
- Equipment records
- Plant experience.

There is one special category of equipment to which special inspection, servicing and maintenance criteria are applicable and to which strict adherence to special regulations is imposed by law. These include fire protection and safety equipment (fire extinguishers, vacuum and pressure safety valves, emergency relief equipment, emergency alarm devices, etc.). In many countries this type of equipment is periodically inspected by the representatives of official institutions and their recommendations need to be followed strictly and precisely for legal reasons. The manufacturers' manual is an invaluable guide on how to install, how to service, what, and how often to inspect, what and when to replace. The equipment record contains the most essential data deriving from the manufacturers' manual, but it also contains a record of the machine's history during its use in the plant. The records show the age of the machine, major repairs it has undergone and subsequent notifications. By looking at the record the engineer is able to analyse the machine's condition, its ability to continue performing safely and to introduce changes in inspection and servicing schedules concerning any particular piece of equipment. The experience gathered in the plant by the operators and the engineering staff can be an additional but very valuable source of information on which the inspection and servicing schedules can be designed. Even the most common machines in a milk plant are subjected in each plant to different working conditions. These conditions vary for many reasons arising from differences in climate to the difference in operators' ability to handle machines. All this cannot be foreseen by the manufacturer and his manuals will refer to normal operating conditions only. Unavoidable deviations from what the manufacturer considered as normal conditions are known only to the staff of the plant and their experience should be carefully gathered and considered when drawing up a maintenance schedule.

C. Scheduled Inspection of Non-Standard Items

The third group of inspection schedules concerns parts of the plant assets which are not covered by the foregoing two groups and is also based mainly on experience. These schedules cover inspection of civil structures, ducts and pipelines between distant section or buildings, underground structures and mains, water storage tanks, wells, and many other parts of the plant not described in manufacturers' manuals. In this group the inspection of the sanitation of the whole plant is also included which contains anything from painting to insect and rodent control. Neatness and orderliness are important, both inside and outside the plant.

D. Inspection Schedules Recording

All inspection schedules must contain information concerning what to inspect, when and how often. This information needs to be recorded either on overall charts on which equipment is listed in a selected order and the necessary functions of the inspections and servicing, including servicing instructions, are specified, or on individual cards for each section of a machine. An inspection schedules card file is widely considered as advantageous, prticularly in larger scale plants. This contains three parts. The first part gives the name and the identification data of the machine, but its basic component is the check list of items to be inspected. The inspection items are grouped according to the inspection frequency into five sections, A to E. The list should ensure uniform and complete inspection regardless of who does the job. The second part of the inspection card is the confirmation of routine work carried out: the third, on the reverse, is a record of non-routine observations and works. An essential feature of the maintenance inspection schedule cards is the system which ensures that they are pulled out from the file at scheduled periods and that they are distributed to the persons instructed to do the inspection. As already mentioned, coding on perforated edges of the cards is one of the simplest and most convenient to use. The example in Fig. 4 shows and inspection schedule card with perforations on the top edge only. The right hand top corner of each card is cut off to make sure that when placing cards in the file box they will all face the same side of the box. In the example there are 35 holes on the top edge of the card. Perforations for coding purposes could continue along all four edges of the card. The perforations in the example are used for coding the inspection dates and frequencies, the type of the scheduled inspection and the group of machines inspected. The legend of the coding applied is as follows:

Days of the Week

1 Monday

2 Tuesday

3 Wednesday

4 Thursday

5 Friday

6 Saturday

7 Sunday

Decades of the Month

8 First

9 Second

10 Third

Months of the Year

11 January

12 February

13 March

14 April

15 May

16 Juno

17 July

18 August

20 October

19 September

21 November

22 December

Type of Inspection

23 General performance

24 Mechanical

25 Electrical

26 Lubrication

27 Civil and auxiliary

28 Fire protection and work safety

Plant Department

29 Processing

30 Steam raising

31 Refrigeration

32 Water supply and distribution

33 Compressed air and air conditioning

34 Buildings and other civil engineering structures

35 Miscellaneous

By cutting out the space between the perforation and the edge, the card receives a code. Pushing a thin rod or needle through a selected perforation in a pack of cards and lifting the cards on the rod will result in separating all cards coded on this particular perforation (cards not lifted) from those not coded (lifted on the rod). The code specification chosen in the example concerns a maintenance schedule card system in which several (up to six) cards are used for one inspection subject. The check list on each card contains only works referring to one out of the possible six types of inspection works marked on holes 23 to 28. The philospohy of the coding system may be best illustrated by analysing the example in Fig. 4. The cut-outs on Nos 31 and 24 indicate that the card concerns the refrigeration plant or a component part and refers to mechanical maintenance inspection. Cut-outs Nos 11 to 22 indicate that there are items requiring monthly inspection. The cut-out No 9 shows that the monthly inspection for this machine is scheduled for the second decade of each month. The cut-out in No, 3 position shows that weekly inspections are performed on wednesdays. The codding system presented above is simple and helps in the organization of the work, provided it is systematically applied. Every day of the week, cards marked with that day's code should be sorted out from the total by lifting on the rod all those not carrying that code. The next step is putting the rod through holess 23 to 28 - one by one - and distributing the work according to the indications on the cards to mechanics, electricians, lubricators, etc. The cards - if kept in transparent plastic envelopes - can be handed out for easy reference to the inspecting crews, but all entries except initialling should be done by the officer to whom the workmen report after completing the job, otherwise the cards would not last long. It should be noticed that the selection or the day of the week on which the weekly inspection is done as well as the selection of the decade for monthly inspections are the result of work planning. The preventive maintenance staff inspect the equipment and do the routine servicing by going from one group of the machines to another throughout the day and the week, spending as much time on a job as required but basically changing their place of work and returning to it after a specified time. Only operators spend all their time on performaing the same job every day. Cards for monthly inspections are selected on the first, eleventh and twenty-first day of the month concerned. After selecting all cards relating to a given month, those sorted out concern the current decade. The work is then distributed in the same way as for weekly inspections. In the recording part of the card - in the group code register - symbols A, B, C, and D are recorded depending on the type of work done, as illustrated. It is the duty of the officer-in-charge to see that the C and D works are done according to the schedule by analusing the recordings in the past periods. The same applies to the annual inspection by checking the date of the last one recorded at the top of the recording table. On cards of machines not requiring weekly inspections none of the first seven perforations will have the space between the edge of the card cut out. Similarly, when there are no monthly inspections, the perforations 11 to 22 will be coded only for three monthly or semi-annual inspections. One of the important advantages of the card-file system is the fact that the order in which the cards are put back into the file box is immaterial since the uncoded cards will be pulled out by the rod irrespective of the location of the card in the file. Special attention should be given to electric prime movers ane other electrical appliances attached to the machine when considering the preventive maintenance recording system. It often happens that electirc motors are moved from one piece of equipment to another. If the motor data is kept on the equipment card, this could lead in such cases to misinformation. Therefore it is advisable to keep motor data on a separate card attached to the main equipment card, removable and exchangeable whenever required.

E. Lubrication Schedules

These are an integral part of preventive maintenance schedules. Most machines have some elements requiring lubrication, such as gears, bearings, cylinders, chains, etc. Lubrication reduces wear in the lubricated elements. Correct lubrication practices can reduce the overall machine maintenance costs by as much as 20 percent although the costs of lubricants may represent as much as 10 percent of the maintenance costs of a plant. The term lubrication includes lubrication oiling and greasing and the indication on costs svings given above covers all three. Lubrication is such an essential part of the dairy engineer's activities that in larger scale plants one staff member is employed only to take care of this activity, selecting lubricants, teaching mechanics and operators and supervising their work in this respect. Selection of proper lubricants is a very important and difficult task. It is usually done by first recording the recommendations given in manufactirers' manuals for each machine and its cimponents. Next comes the analysis of the recommended list with a view to reducing the variety to a minimum. At this stage and expert opinion from an oil supplying company might be of help and is strongly recommended. Lubrication and selection of proper lubricants are highly specialized fields. Seldom has the dairy plant engineer the opportunity to acquire the necessary background in this field. Calling in a qualified lubrication expert should not be restricted to requests for help in selecting the list of lubricants. He should also be involved in training lubrication personnel in the fundamentals of the work, methods of lubrication, utilization of lubrication tols, methods of marking lubrication points, methods of purification of lubricants and in applying criteria for re-using purigied lubricants. In many instances there are good reasons for establishing a separate card file for lubrication schedules. Designing a lubrication programme requires deciding who is going to do the actual lubrication work. In some plants this work is done by the mcahine operators, in others by specially trained lubrication staff. In the first case proper supervision has to be introduced, particularly in order ot train people in keeping the oil clean and in detecting reasons for quick deterioration of oil quality in gearboxes or other lubricated devices. Finally training is needed in application of properly selected lubricants and in proper marking and recording of lubrication jobs performed. In plants where a lubricator is employed and lubrication schedule cards are used, the routine steps of the lubricator as as follows:

1. Collecting the schedule cards for the day
2. Examining the lubrication programme for the day
3. Collecting necessary lubrication tools and lubricants
4. Lubricating and tagging the machines after completing the job
5. Recording the job on card - adding notes on essential observations
6. Returning the cards to the engineer in charge or to the clerk.

Spare Parts Programme

The spare parts programme is a particularly intricate task for a dairy engineer in a plant in a developing country. In most cases the plants were designed abroad and almost all equipment imported. The supplying companies seldom had sufficient information, not only on the local technical resources form which some parts of the plant could be procured, but even on the legal standards of the recipient countru. In many of the developing countries more than one milk plant had been imported and very often each of them purchased from a different supplier, from a different country, sometiems even from a different continent. As a result there is often practically no standardization of equipment in milk plants in developing countries, except where the complete plant has been installed on a turn key basis. Moreover, even such simple items as bolts, screws and nuts used on a machine imported from the USA differ from the ones used by companies from the European continent, which in turn differ from those purchased in the UK. This variety applies to also to stainless steel pipes and fittings used in milk ducts, where at least five or six standards are in use. In some developing countries they are all represented in one milk plant. With the absence of standardization, with difficulties in acquiring import licences for spare parts and in communication with the suppliers and absence of local specialized servicing agents, the continuous operation of a milk plant and its good performance will depend heavily on the spare parts stock available in the plant for immediate use. It is the dairy engineer's duty to decide on what spares and supplies to keep, on how much should be kept, on how to store and how to register. The engineering stores should contain consumable items such as gaskets, standard lubricants and paints and three basic groups of engineering accessories:

- equipment spare parts
- complete components
- pipes and fittings, bolts, nuts and washers, bars and plates of different metals, electric components and other general types of engineering accessories.

The total number of items stored often amounts to thousands, of which a negligible part or sometimes even the majority may be imported. The selection of spare parts kept in the store is based on the manufacturers' recommendations and on the plants own experience of local operating conditions. In many plants there are several machines of the same type, capacity and make. This must be considered when deciding on the spare parts store, since, although the number of items stored will not be reduced, the quantity of each item may. Moreover such situations may make it convenient and economic to keep in the store complete components as spares, such as pumps for bottle and can washers, milk pumps, gears for tank agitators, electric motors, etc. Such components kept in the store could serve as emergency replacements for a number of machines or functions. The quantities to be kept should ensure regular and trouble-free plant performance without overstocking the stores. In the spare parts lists there are items which are on relatively steady demand such as rubber gaskets for milk pipes, plate gaskets for heat exchangers, graphite and rubber sealings for pumps, electric relays and special bulbs, selected bearings, springs and automatic switches, etc. Also some consumable items may be listed in this group, such as automatic recorder charts and inks, special lubrication components and even packaging materials to which packaging machines are particularly sensitive, such as aluminium capping foils for bottling. It is relatively easy to establish a list of items and of required quantities contained in this group. It is much more difficult to decide on items which are used sporadically. The demand for them is erratic and only experience can tell what deserves to be stored in the plant. This could involve parts of machines which theoretically last for the lifetime of the machine, such as shafts carrying spray discs in milk driers, plates in heat exchangers and pressure and temperature indicators. There are instances when they need quick replacement and should be available in the store. On the other hand large spare parts stores absorb a considerable portion of the capital and may adversely affect the economic situation of the plant. Finding a correct balance between these two aspects of the problem may be facilitated by considering whether local supplies could replace costly and time consuming imports, whether there are components which could be locally reconditioned by removing worn-out parts and replacing them with parts which are locally made or even fabricated in the plant's own workshop. It may be of advantage to the milk plant to make or to purchase tools for spare parts fabrication and to use them when required. This may be illustrated by an example. The local rubber parts manufacturing company may have the skills and facilities to produce rubber parts of required quality, but it may not be interested in undertaking such jobs for the milk plant because of high tool manufacturing costs and of low demand. The rubber moulding tools are the key to the solution of the problem. Making them or paying for their manufacture and keeping for the next order may seem to be an expensive venture, but it may be the safest and cheapest solution in the long run. However, at least one original sample of each spare part should always be kept in the store under special care so that a critical comparison between the original and locally made items is possible and corrections of the latter feasible. Certain spare parts will always have to be imported due to lack of know-how, to the need for special materials, or to high manufacturing costs. A careful survey of the stock position of the relevant items, of their condition on the machines and of their anticipated life should be made with sufficient frequency so as to ensure that ample time is given for the procurement of spare parts before the last pieces are used. It is a well known fact that procurement of spare parts takes often more time than the procurement of a new machine. Sometimes it may take a year or even more to replenish the stocks. Lowest quantity limits on particularly crucial items kept in the store need to be defined in order to avoid the risk of running out of stock before new supplies arrive. The registration system of spare parts is an essential component of a good preventive maintenance programme. Needle cards may also be of use here.

Maintenance Action

The preparatory part of the PMP as well as the plan of action presented in the foregoing leads to the correct performance of the physical maintenance of the whole plant and each of its components. As it has been shown, the inspection schedules - including lubrication - are the guidelines for routine servicing of the machines and in this respect the plan of operation includes an essential part of maintenance: routine servicing during scheduled inspections. However, any routine activity implies action during normal situations in which no deviation from normal has been spotted. But daily review of the inspection records will certainly also reveal that in some of the machines the performance does not meet expectations. In such instances remedial action needs to be taken before the performance decreases further since there is likelihood of reduced output, serious damage to the machine or the situation may develop to a stage at which it becomes hazardous to personnel. The examination of the situation will lead to findings upon which the decision will be taken on further action. Such examinations must be done by members of the engineering staff who are best qualified in the field in which the trouble occurred or by the person actually on duty. The degree of seriousness and the urgency for a solution will influence the way in which the problem is analysed and decision taken. The most dramatic conclusion would be taking the machine out of service because of the hazards in its further operation. This seldom happens and when it does it is most often limited to particularly sensitive machines such as milk centrifuges and refrigeration compressors, of which standby items are usually included in the plant design from the very beginning of the plant's operation. In the majority of instances the decision-making engineer must consider whether:

- the work can be done during the operation of the machine
- the work can be done after normal operation hours without disrupting the normal processing schedules
- the work requires time, skills and resources available in the plant or it is necessary to call in other specialized companies or people.

The decision is followed by work orders. Should the work be carried out by the plant's own crews, the maintenance craftsman or foreman receives a work order which gives details of the work to be done. Usually it will refer to a particular maintenance schedule and will be given together with requisition documents to the stores for issuing materials and spare parts required for carrying out the job. Only in very urgent instances should the craftsmen work to verbal instructions, although discussions on how best to perform the job are of great value to the quality of the work and to the overall working atmosphere in the engineering group. After the completion of the job a Job Report should be submitted in which confirmation of completion is given, and the time and materials used are recorded and recommendations for any further action are specified. The Job Reports are one of the main sources for preparing accounts of the actual costs of machinery maintenance. They may be prepared either within the engineering department of the plant, by the accounts department or in cooperation between both, depending on the organizational system of the factory management. Independent of the organization structure of the plant, it is the dairy engineer's duty to keep his own records on maintenance costs, to review them frequently and to draw conclusions on ways and means by which all necessary expenses on plant maintenance are made. In spite of all the efforts in taking care of the machines, equipment and other assets of the plants, the time will come when overhauling becomes unavoidable. Damage caused by faulty operations, maladjustments and misuse of machines can be minimized under proper management, but deterioration caused by wear and tear of moving parts, by corrosion or by entry of foreign materials between moving parts cannot. Overhauling machines at planned regular intervals is a means to avoid sudden breakdowns. The intervals between planned overhauling of each and every part of the plant's equipment can usually be found in the technical documentation of the plant, including manufacturers' manuals. Here again the plant's own experience may be of even greater value in establishing overhauling schedules of machines. Such schedules are based on the anticipation of breakdowns after a given number of hours of the machine's operation. Most often the planned intervals become shorter when the machine advances in age and in wear and tear. However, a planned schedule cannot be the only indication with respect to the need of major repairs of a machine. During the preventive maintenance inspections the readings of the instruments together with other observations can also indicate the danger of immediate breakdowns. Well trained staff will make it clear in their inspection reporting, but it is for the supervisory staff to spot such dangers in their routine analysis of the reports and to take appropriate actions. It is a good habit to spend 30 minutes each day and to have the written reports of the machinery inspection staff supported by verbal descriptions of the observations. In the long run it will always result in real time saving and also in a better diagnosis of the actual condition of all parts of the entire plant. The time available in a milk plant for overhauling works is usually limited and in well-managed plants the engineering group aims at the reliance on its own resources of which the maintenance workshop is the most essential. The reasons for such an approach become drastically evident when assistance is sought from outside facilities in case of emergency.

Maintenance Workshops and Equipment

Before planning the accommodation and equipment needed for maintenance the external resources available such as spare parts stocklists within the country and the existence of specialist firms which can undertake engineering fabrication, building maintenance or vehicle servicing, should be surveyed and evaluated. If the milk plant is situated in an existing urban community it is likely that builders, motor vehicle agents and probably refrigeration engineers will already be established as these services are needed by so many other enterprises. However comprehensive these outside resources appear to be, it is still essential for the milk plant to include its own maintenance section to operate a preventive maintenance scheme and to cope with emergencies. Although most of the work of inspection and maintenance of the equipment, both service and process, will be carried out in-situ, a base workshop and record office are necessary even in small plants. In large plants this may extend to several workshops each devoted to a particular branch of engineering service work. For example if the plant has to maintain its own vehicles because outside facilities are inadequate, a special workshop suited to this particular requirement may be needed. In small milk plants a single general-purpose workshop should be adequate and this might be sited adjacent to the process area with direct access outside, possibly associated with the refrigeration compressor room, the records office and the spare parts stores. In large plants where more than one workshop may be appropriate a separate building may be needed, possibly in conjunction with the boilerhouse or garage. The workshop should be supplied with the main services including compressed air and should be equipped with benches and a full complement of hand tools appropriate to the various engineering trades, together with special tools required for dismantling specific items of equipment. The latter is especially important in vehicle maintenance. Where repair work or the manufacture of special items of equipment is contemplated, electric and gas welding equipment must be included and, perhaps, basic machine tools, such as lathe, pillar drill and metal working machines.

TOWARDS PROPER MAINTENCE OF PROCESS EQUIPMENT

Pumps and Piping

(i) Centrifugal Pumps

This type of pump requires little maintenance, which is usually confined to disassembly and cleaning, replacement of seals, seal plate and joints. The product seal will obviously in time show signs of wear. The life of a seal depends so much upon its duty and to the product being pumped that it is difficult to generalize. It is however, possible to say that a product seal being used with hot sticky liquids may only give good service for three months. The examination period therefore should be adjusted to suit the conditions, e.g. six monthly intervals for light duties or six weeks for more arduous duties. Premature seal failure or leakage can be caused by numerous means but the most common are:

(a) Foreign matter between seal running faces.

(b) Cracked carbon seal ring, usually caused by damaging during installation or thermal shock.

(c) Exessive wear on seals due to abrasive in-product.

(d) Inorrect installation and setting of product seals.

(e) If water-coiled seal arrangements fitted, then starvation of water to the seal.

The impellor should be examined for blockage in the tangential holes. This problem is usually indicated by the loss of pumping pressure or flow rate. The motors mainly associated with a centrifugal pump require little maintenance but the following points should be noted. For specific requirements, the manufacturers' handbook should be consulted.

(a)Ball and roller bearings should have their grease renewed every 12 months, to ensure that a film of grease is in the rolling elements and that the housing is almost filled with grease in close contact with the bearing faces.

(b)Bearings should be checked each month for temperature and smooth running.

(c)Motor drive shaft for end float should be checked. Evidence of end float would normally indicate bearing wear.

(d)Monthly inspection of cooling fans, ventilation holes and ducts is recommended, cleaning as necessary.

(e)Terminal box connections should be periodically checked for tightness and insulation of leads for hardening, cracking or chafing. Overheating, from whatever cause, is a serious fault. High temperatures can cause deterioration of the motor insulation. The temperature of the motor should be tested by using an open bulb-type thermometer located as close as possible to the motor windings.

(ii) Positive Displacement Pumps

A typical pump comprises a motor, which could be electric, air or hydraulic drive, to which is coupled a gear case assembly which has two rotor shafts. These rotor shafts protrude through the pump body and located on each will be the rotors. The complete assembly is then clamped together via front cap and cap nuts. It is particularly important in handling this type of pump that methods are used which will not allow rotors or housings to become nicked, dented or scratched. The use of rubber mats on which to lay the pump components during disassembling is often practised. The maintenance required on this type of pump is greater than that for centrifugal pumps in that the gear case has many moving parts, bearings and oil seal. The manufacturers's recommendations for specific maintenance requirements for the gearbox must be consulted but a guideline is as follows:

- Lubricants will be required in the gear case and on the bearings; the gear case should be regularly checked for correct levels. It will also be required, usually daily, to drain condensate from the gear case. The shafts should be checked for looseness and adjusted on the bearings as necessary. Oil seals will require replacement as will various 'O' rings located in the gearboxes. Within the pump body, gaskets, 'O' rings and the product seals will require replacing but, as with centrifugal pumps, the nature of the product being pumped will determine their safe working lives.
- The rotors, which are usually rubber coated, will require replacing periodically. Excessive rotor wear could be due to one or more of the following: if the discharge pressure is too high, the pump speed may be excessive or an incorrect rotor material used. Abrasive products can also be a major factor. The incorrect use of the pump, e.g. running the pump dry or pumping foreign objects should be considered. Other factors are abnormalities in the actual pump, such as drive shafts out of line, shafts worn, loose bearings and worn hubs. All of these can cause rapid rotor wear.
- The product seal, if wearing excessively, could be due to a too high discharge pressure, or product temperature. If the pump is mishandled, such as by the use of sharp instruments when installing. or damaged mating parts, problems with seal life will occur. A worn seal housing, loose pump shafts or shaft alignment with the pump body are all mechanical abnormalities within the pump, which will affect seal life. Excessive wear in the gear case may be the result of incorrect oil being used, the failure to drain condensate or failure to change oil. It could also be the result of failure to replace product seals allowing product/cleaning solutions to enter the gear case. Mechanical damage within the pump such as drive alignment, bearing adjustment, or using worn rotors on new shafts, will all contribute to wear in the gear case.
- Finally, poor pumping performance is usually the result of a low pump speed, a 'starved' pump suction condition, or air leaks in the suction line. Product seal condition is obviously also a major factor.

(iii) Valves and Cocks

Various types of valves and valve configurations can be used for product routing. In the main they are pneumatically operated with an exhaust route, internal spring operation being then used to reverse the movement. Valves should be checked as part of a planned maintenance scheme. Factors to be considered for the production of a scheme are the frequency of operation, nature of product, temperature of product, etc. Obviously the product seals should be changed if there are signs of wear or cracking. In conjunction with this, the valve seats should be examined. If they are metal seats they should be refaced if worn; if rubber seats then replacement is necessary. Valves will have an air cylinder of some nature which incorporates the valve return spring, air piston, valve spindle, 'O' rings, etc. As the cylinder may have to to hold internal pressures of around 350 to 700 KPa obviously all components within must be in good order. Valves may be operated via a control box, which houses solenoid valves that control the flow of air to the valve. The sealing of these control boxes is important to stop the ingress of moisture; therefore the seals must be regularly inspected. In conjunction with this the air supplying the valve should be dry and oil free to eliminate possible problems with electrical equipment and valve internals. Feedback on the valve position is a common feature and is usually achieved via micro-limit switches that are activated from the valve shaft. These switches should be inspected and adjusted at regular intervals. For specific assembly/dismantling instructions, reference should be made to manufacturers' instructions. Valve cocks are manually operated valves that can either have a straight through or three-way port configuration. A typical cock will comprise a full bore plug, top and bottom sealing rings, top retaining cap and a handle for operation. Maintenance requirements will include the inspection of the sealing rings together with the valve plug and valve body. When dismantling a cock, great care must be taken not to allow any nicks, scratches or dents to occur in the mating faces of plug and body; these could cause leakage. If any damage is evident then plug and body should be "lapped" to eliminate leakage. When reassembling the cock, a suitable edible vegetable grease should be smeared on plug and body mating faces to ease operation. It should never be necessary to hammer the cock handle to turn the valve.

(iv) Pipework

Interconnecting product piping between various items of plant generally fall into the expanded or welded fabrication categories. An increasing preference is being shown for the welding of pipes and fittings utilizing orbital welding or inert gas purge to reduce the number of pipe unions employed and so minimize potential crevice areas which may prove difficult to clean. Where skilled welders are not available, the expanded type pipe union may be employed utilizing semi-skilled labour. When using the expanding method, it is essential that the correct gauge of tube is used to suit the expanding tool. The tube must be held square with the clamping block and the correct load setting used as appropriate to size of fitting. Use of an incorrect setting for the torque may result in failure; on a low setting, in fittings becoming loose under operational pressure and vibration; on a high setting, in a stretching of the fittings resulting in difficulty in attaining a leakproof union. Plant layour is an important factor with respect to pipework. For ease of support and for an overall pleasing appearance it is usually desirable to run pipework parallel with or at right angles to the enclosing walls. Adequate falls must be arranged to pump suctions and drain points. As a general guide the amount of fall required is 1:120. 'Dead legs' or pockets in pipe layout should be eliminated wherever possible. If unavoidable, they should be positioned so that they are self-draining or in line with the flow. Pipework should be periodically inspected for damage or corrosion/erosion.

Heat Exchangers

A. Plate

By virtue of its passive nature in operation, the plate heat exchanger tends to receive a low priority in terms of maintenance programmes with the result that engineering attendance becomes one to cover emergency or breakdown situations. This is obviously not good practice and logically denies the aims of good maintenance in taking preventive action at an early date to avoid the onset of serious problems. This tendency arises in particular in the food processing industries, where the heat exchanger forms part of a larger plant, the operational control of which is the responsibility of others ranging through production, quality assurance and cleaning and disinfecting. Whilst each of these management areas are required to commit themselves to specific actions to overcome their particular problems, the maintenance engineer must be sure, by means of programmed examination, that actions taken by others in their own disciplines are not detrimental to the well-being of the heat exchanger. The planned maintenance job specification for individual heat exchangers will vary in line with the variety of applications to which they operate. It is, however, recommended that such maintenance specifications be based upon a framework embodying the following ideas, with additional matters being incorporated to deal with the manufacturers' more specific recommendations.

(i) Frame

Where the frame is clad in stanless steel, apply a thin coating of vegetable oil to protect the finish. At the same time this will ease the removal of inadvertent splash markings from product material or cleaning solutions. The plate carrying strip or slots of the top carrying bar should be kept greased with an appropriate agent to minimize drag during plate movement and so reduce risk of damage to the plate hanging eyes. The plate pack tightening device should also be protected against environmental conditions with special attention being given to exposed threads of tie-bars which may be accidentally damaged.

(ii) Plates and Gaskets

Examine plate pack to determine whether any signs of damage are present of either a mechanical or corrosive nature. Mechanical damage may result from:

(a) Damage to the plate hanging eyes resulting in misalignment and gasket damage.

(b) Overtightening of the plate pack to overcome leakage problems due to high pressures or poor gasket condition.

(c) High pressure loads or repeated pressure reversals of great magnitude may emanate from high pressure pumps or centrifuges. Such operational conditions may also lead to the movement and subsequent damage of the plate gasket.

Chemical damage may result from:

(a) The use of aggressive acid solutions during in-place cleaning activities, or from more specialized actions to remove hard scale on product or service sides of plates. Gasket damage may also be present.

(b) The misuse of hypochlorite sterilizing solutions. In this connection the strength of the solution, contact time and temperature are interactive and require close control of their use.

(c) Where calcium chloride brine coolants are used, care must be taken to ensure that the correct pH value is maintained and that any scale build-up on plates is removed frequently. Inadequate draining out of brine solutions and flushing out/neutralizing of residuals will lead to the onset of corrosion during application of high temperatures required for the cleaning and disinfecting activities.

(d) The continued use of high temperatures in excess of those specified for the particular gasket in use will hasten deterioration due to thermal ageing and subsequent loss of electricity.

(iii) Thermal Performance

When problems of loss of thermal performance arise, the maintenance engineer should always consider the institution of heat balance/heat transfer calculations to enable him to identify the source of the problem.

Areas having a detrimental effect on performance are:

(a) Product debris on plates.

(b) Formation of tenacious scale.

(c) Product throughput not under control at specified rate.

(d) Incoming product temperature not to specification.

(e) Process product temperature too high.

(f) Heating/cooling services not to specification in throughput or temperature.

(g) Inspect thermometry.

(h) Arrangement of platage changed - possibly by error.

(iv) Corrosion

Stainless steel plates should undergo detailed examination at frequent intervals to determine whether the improper use of chemical sterilizing solutions of the hypochlorite range or calcium chloride brine cooling liquids has created conditions leading to corrosion attack.

Pitting corrosion may be seen at random areas of the plate as dark individual spots which are not removed during the cleaning process.

Crevice corrosion, if present, will occur at any plate to plate contact point throughout the surface of the plate in the flow path.

Where calcium chloride brine cooling solutions are in use the problem of scale formation is always present, if allowed to remain on the plate. Such scale may harbour residual brine leading to pitting and crevice corrosion.

(v) Regasketing of Plates

The replacement of gaskets for heat exchanger plates is a relatively simple and straightforward operation providing due note is taken of the manufacturers' instructions relative to the type of gasket elastomer being used and the adhesive appropriate for the particular gasket material. Whatever the adhesive system recommended, it is essential that care is exercised in the removal of the worn or damaged gaskets, not only to avoid damage to the gasket retention groove, but to ensure that all old gasket debris is removed easily, prior to cleaning and degreasing of the groove area in preparation for the application of the approved adhesive. Most adhesives produce a satisfactory gasket-to-plate bond strength either by chemical reaction or solvent evaporation provided sufficient plant down time is available. To counter this time factor the application of heat to the completed plate pack may be recommended to ensure that a satisfactory adhesive bond is attained to suit plant operating temperatures and pressures.

B. Swept Surface Heat Exchangers

A swept surface heat exchanger consists of a rotating dasher, with affixed scraper blades totally enclosed with a cylinder. The continual sweep of the blades cleans the cylinder surface of product. Heat transfer is through the cylinder wall. Inlet and outlet product ports are situated at each end of the product cylinders. Individual drives for the dashers may be hydraulic or electric. For specific maintenance requirements of the drive assembly, reference to manufacturers' manuals is recommended. The heat exchanger will normally have a product rotary seal; this seal must be inspected at regular intervals for damage or excessive wear. The dasher should be removed from the cylinder so that examination of the dasher blades together with the upper bearing and seals can be examined. Extreme care must be taken when removing the dasher to eliminate the possibility of damaging cylinder bore or dasher blades. After inspection of all parts, the upper bearing will require lubrication with a sanitary lubricant. The motor/dasher shaft should be checked for alignment at regular intervals. This procedure is explained in detail in the manufacturer's maintenance handbook.

C. Tubular Heat Exchangers

As the name implies the tubular heat exchanger consists of a coil of tubes which is totally enclosed within a larger tube. The product flows through the coil of tubes with the heating cooling medium flowing over the tubes within the larger tube. On most tubular heat exchangers very little maintenance can be carried out as they are of completely welded construction. If the heat exchanger has removable end caps, then the inspection of the smaller coil of these can and should be carried out at regular intervals. Gaskets on the end caps should be replaced if worn or twisted.

Filters

A typical continuous cloth type milk filter will consist of two cylindrical filtration chambers joined to a common inlet via a three-way changeover cock. Both chambers could be mounted on a length of horizontal outlet pipe, which would have non-return valves fitted in the junction points. The non-return valves are to prevent filtered milk feeding back through the chamber not in service. Milk would be directed to either of the two filtration chambers by the three-way cock. Continuous flow through the filter is achieved by alternating the flow between the two chambers. Before starting up, a filter cloth should be fitted to each filtration chamber. Generally, a filter cloth should be removed and washed when the back pressure on the filter increases by 0.14 bar, and replaced with a clean cloth in the chamber. Maintenance of the filter would include stripping down the complete filter and inspection of all joint rings, changing if twisted, nicked or flattened. Attention must be paid to the non-return valves, checking the valve/valve seat for damage, spring tensions and correct alignment of valves. The valves should then be tested to ensure they only allow flow through one way. The three-way cock should be stripped and inspected as described in paragraph (iii), "Valves and cocks", and any inner screens or wire mesh components examined for damage. Milk filters will have drain and vent plugs fitted to enable continuous running of filters during production. These plugs should also be inspected to make sure that their respective seats are clean and seals are in good order.

Milk Storage and Process Tanks

All milk plants have the need for tanks for storing either raw or finished milk. These tanks can come in a variety of shapes and sizes. They may be of the square or rectangular type, particularly in the small sizes, or they could be of the vertical or horizontal cylindrical type. Tanks vary from hundred litres up to about 150 000 litres. Most milk storage tanks have an inner and outer skin with up to 5.08 cm of insulation inbetween. The insulating material is usually cork. The inner skin would be of stainless steel although some tanks are glass lined. Milk storage tanks are designed to withstand a certain amount of pressure or vacuum and can be equipped to be filled by vacuum or emptied by pressure. These types of tanks should have a safety device which will prevent excessive pressure or vacuum. If this type of device is fitted, then regular inspection should be carried out to ensure its effectiveness. Tanks may also be fitted with an air vent which must be of sufficient size to prevent any build-up of excessive pressure or vacuum.

It is important therefore not only from a sanitary point of view, but also from a safety aspect, that the air vent should not become restricted or blocked. Regular inspection of this vent is therefore required. If the tank is fitted with a thermomenter this should be checked against a known standard thermometer for accuracy. The tank will be fitted with a manhole which opens inwardly to minimize danger of leakage. The manhole will have a gasket for sealing which should be checked and replaced if damaged or badly worn. Most milk storage tanks will be fitted with an observation glass. The glass is usually of about 150 mm diameter and situated near the top of the tank. Gaskets sealing this glass should be inspected and replaced if damaged. Finally, the tank should be inspected internally and externally for any defects (cleaning effectiveness, corrosion, etc.). It should be stressed that when internal inspection is taking place all necessary safety precautions should be taken and that rubber-soled boots/shoes worn to prevent damage to surface of tank. Transport tanks will be fitted with top manholes which open outwards.

Agitators and Mixers

High quality standards of design and manufacture are maintained to ensure that this type of equipment gives trouble-free service. Nevertheless, recommended precautions and procedures must be observed in the handling, installation and servicing of all rotating equipment and accessories. Careful application of the manufacturer's recommendations will ensure an optimum performance. Errors or carelessness during installation can often lead to problems or even failure at a later date. At installation, it is therefore necessary that the following points receive consideration:

(a) The manufacturer's instruction and maintenance manual must be made available to all personnel involved with the installation, operation and maintenance of the mixing equipment.

(b) On receipt, all equipment should be carefully checked to ensure that no parts are missing and that no damage has occurred during transport from the factory to the user.

(c) Any discrepancy or damage must be immediately reported to both the carrier and the supplier.

(d) The installation of most standard mixer drives and in-tank mixing equipment can normally be carried out without the presence of the manufacturer's representative. However, it is advised that only experienced tradesmen should be used to install agitators or any rotating equipment.

(e) Care must be taken to ensure that the mounting location of a mixer must be rigid and strong enough to support the mixer drive. This applies equally to direct flange, fabricated stool or structure and concrete flooring. If the unit is to be mounted on a surface which is not horizontal, the manufacturer should be consulted to ensure that the mixer will receive adequate lubrication when mounted in the desired position.

(f) At the start-up of a newly installed agitator, it is sensible to proceed with caution. Regardless of how well an installation is carried out, it is possible for errors or omissions to occur.

(g) Before running the unit, electrical requirements should be checked thoroughly. The voltage range on the motor nameplate should correspond to the supply voltage. The wiring of the unit should be carefully confirmed in accordance with the diagram provided. The direction of rotation must be checked and, if necessary, altered by interchanging the supply leads.

(h) The gearbox should be checked to ensure that it is filled with lubricant to the recommended level. The correct grade of oil as specified in the manual must be used.

(i) On units with mechanical seals, it is necessary to ensure the seal is adjusted in accordance with manufacturer's instructions and that the seal lubrication system is correctly selected and pressurized normally to 1 bar above the vessel pressure before rotation of the unit takes place.

(j) It is of great benefit if the mixer unit can be run under light load conditions for a short period. This permits the gears to be run in. (Instruction manuals cover this point.) The unit can then be operated at normal load conditions.

(k) During service, regular oil changes are essential to ensure efficient performance of the agitator. The following factors should be used to determine the frequency of these oil changes:

(i) oil temperature - unit operating under load

(ii) type of oil - plain or containing additive

(iii) the environment - humidity, dust, etc.

(iv) operating conditions - shock, loading, etc.

Finally, it should be noted that great care has been taken in the design and manufacture of all agitation equipment. If this care is continued by the user through regular checks and recommended maintenance, many years of trouble-free service will be the reward.

Homogenizers

The homogenizer is essentially a reciprocating pump capable of producing very high pressures of up to more than 200 atmospheres. It has a multi-piston cylinder block to minimize cyclical variation of flow and produce the high pressure to force the product between the mating faces of the preloaded homogenizing valve. The homogenizing valve is supplied either as single or two-stage according to the product to be processed or the degree of homogenization control required. The base of the machine carrying the drive end consists of an electric motor connected to the drive shaft by pulley and belts. Further reduction to the eccentric shaft can be by pulleys or gear ratio, which in turn provides the reciprocating motion to the pistons through the connecting rods. The maintenance of a homogenizer should be carried out on a regular basis. A daily inspection should be made of the oil level visible through the sightglass at the rear of the machine. After draining off any condensate from the oil sump through the pet cock, oil should be added if required, before starting up the homogenizer. The oil pressure should be checked to ensure it is above the recommended minimum when the machine is running. Finally, the water lubrication to the pistons and through the oil cooler should be checked to ensure that it is sufficient. More detailed inspection and maintenance is required after each month in addition to the daily inspection. The drive belts should be checked for tightness and condition. Nuts, bolts and fittings within the drive chamber should be tightened and/or checked to torque specification. Connector rod ends and tightness of the crosshead should be inspected and adjusted if necessary. Oil leakage from the piston adaptor oil seals can occur and packings must be tightened or replaced. The piston cylinder block is disassembled by removing the front caps, the piston and packing assemblies, care being taken in the handling of each item and checking for signs of wear and damage, especially in the piston packings. These should be replaced if there is product leakage from the rear of the cylinder block. After removal of the top caps and the inlet and discharge valves, the valve seats within the cylinder block should be carefully inspected for signs of wear and erosion, which will also be seen on the mating valve. Poor condition of these faces can have a marked effect on the product rate and cause erratic homogenizing pressure. Early maintenance and relapping of the valves can avoid a costly recutting exercise. The cap gaskets must be replaced if they become too thin or extruded to a point of leakage at high pressure. This will also apple to the homogenizing pressure gauge and block. The homogenizer valve unit can either be activated by a spring-loaded handwheel or by hydraulic pressure. The handwheel must be manually set on each start-up of the homogenizer whereas the hydraulic unit can be pre-set and will automatically come up to the correct pressure at the push of a button. In each case they must be removed and the valve push rod inspected for signs of wear and damage together with the valve rod seal. This seal is important in keeping a true and steady alignment of the push rod and if worn will cause erratic valve action.

The handwheel thread and the valve rod spring must be checked and lubricated. It is sufficient to check that there is no hydraulic oil leakage from the push rod chamber and adjacent pipework and that the oil level in the main reservoir is correct. The homogenizer valve assembly block or blocks can now be removed and stripped down to show the valve, valve seat and impact ring. The faces of the valve and valve seat must be kept in good serviceable condition to retain an efficient homogenization of the product. This can be recognized by the amount of wear that initiates at the bore of the valve and radiates out across the face. If the wear is up to two-thirds of the surface area, relapping of the valve with carborundum paste between faces can rectify this but further wear will allow erosion grooves to appear, recognized by straight lines radiating outward, and loss of homogenization will rapidly occur. The impact ring is also an essential part of the homogenizing process and must be carefully examined.

Wear of the impact ring will occur on the internal face of the ring in the form of a groove. The erosion of the groove will be acceptable to a depth of 1 mm, when the ring should be replaced to maintain efficient homogenization. Two external factors to the machine should be checked during this maintenance period. Firstly, that the feed pump pressure is retained as recommended and does not fluctuate and secondly, that the pressure relief valve in the discharge pipeline, adjacent to the homogenizer, is set to relieve at the correct pressure. Finally, after six months or 1 000 hours of service, the crankcase should be drained, cleaned and replaced with fresh oil and the oil filter cartridge renewed. The motor bearings should be lubricated according to the manufacturer's instructions.

Plant Instrumentation

A. Temperature Controllers and Recorders

Historically relied upon pneumatic signalling and drive for bellows and linkages, coupled to a mercury filled system comprising a temperature sensing bulb and bourdon tube. The mercury filled system is sealed for life and no maintenance is possible. Operational failure will be due to damage to the capillary between the sensing bulb and bourdon tube or fatigue failure of the material. In both cases the complete system will require replacement. Lubrication of the metal bearings of the linkage mechanism should be carried out at 6-monthly intervals using a good quality clock oil. At the same time the pneumatic control system should be inspected and any contamination due to dirty, wet or oily air supplies should be removed with paraffin or suitable solvent. In more recent years, the tendency has been to design instrumentation embodying solid state techniques, printed circuit boards, etc. with the result that maintenance work has been reduced to the checking of instrument calibration against known standards or, in the event of trouble shooting, the use of diagnostic instruments and subsequent replacement of the failed module or printed circuit board. Inking systems for the temperature recorder pens require inspection at monthly intervals. Various inking systems are employed by instrument manufacturers. Generally these fall into the categories of direct charging of the pen nib using a small ink dropper, or alternatively a capillary feed arrangement from an ink reservoir housed within the instrument. Having ensured that the system is charged with ink, the pen nib should be cleaned to remove dried ink. Where the design of recorder pen permits some manufacturers to utilize ink cartridges with a fibre tip nib, it is only necessary to replace the cartridge when exhausted.

B. Flow Controllers

For process liquids may take the form of:

(i) A device comprising basically a vertical tube with liquid flowing upwards into which is suspended a free piston attached by means of a spindle to a specifically machined disc of a predetermined diameter. This mechanical device has very close tolerances and it is essential to ensure, when taken down daily for cleaning, that no damage occurs by scoring or bruising to the stainless steel components. Due to the close fit of the free piston to the inner tube, a strainer will be fitted either internally as part of the assembly or in the pipeline to the controller. This strainer must be cleaned daily.

(ii) A positive displacement pump usually of the rotary type design. When this type of pump is utilized for control of throughput of a pasteurizer it is more genearly referred to as a "timing" pump. Construction and maintenance requirements are the same as for similar pumps referred to in Section 1, "Pumps and piping".

C. Flow Meters

May be of the mechanical positive displacement, turbine or magnetic design.

(i) The mechanical meter utilizes the energy present as a result of the velocity of fluids passing through the rotor housing to transmit a driving force via linkages to a mechanical counting device. Such meters are of robust construction and require maintenance in respect of moving parts and any lubrication recommended by the supplier. Air entrainment with the product fluid must be avoided as its presence will not be detected by the rotor.

(ii) A turbine device where a multi-blade rotor is housed within a section of tube in the pipeline. Fluid flowing through the tube causes the rotor to rotate and a magnetic pick-up attached to the body registers pulses as the rotor blade interrupts the magnetic flux. The turbine must also be protected by a strainer at its inlet and periodic examination is required to ensure that no damage has occurred to the special bearings supporting the rotor. Cleaning of the whole assembly should be carried out upon conclusion of each process batch to ensure that coagulation of process material around the rotor and bearings is avoided.

(iii) Magnetic flowmeter comprising a tube of non-magnetic materials housing electrodes which create a magnetic field through which a conductive liquid creates small voltages proportional to the flow.

The magnetic type of flowmeter is basically simple in construction and, apart from the associated electronics, care should be taken to ensure that fouling of the electrodes does not occur.

In the case of items (ii) and (iii), additional field instruments will be employed, such as converters and amplifiers, which pass pulses or signals to a control unit. Apart from calibration activity, it is unlikely that the maintenance engineer will be able to take preventive action.

D. Diversion Valves

On process plant will invariably be pneumatically operated in the forward flow mode against a tensioned spring fitted into the air cylinder to effect a fail-safe valve position in the event of air failure. This feature is used to achieve the diversion or recirculation mode of process plant by means of a pre-set alarm setting in the recorder/controller instrument which, in the presence of a temperature condition at variance in either direction of the desired value, will energize/deenergize a solenoid valve situated in the air line or air supply connection of the particular process valve.

Maintenance requirements for the diversion valves will be as explained earlier in basic instrument action or Section 1, "Pumps and piping" insofar as the type of product valve is concerned.

Evaporators and Spray Driers

A. *Evaporators*

(i) Low Vacuum / High Temperature

Should the situation arise where the vacuum is low and the temperature is too high, attention should be paid to the possibility of air entering the effects or being contained in the product or cooling water where a spray condenser is installed. Air leakage may be located at manholes, pumps, covers and seals where joints are installed.

This problem may also be related to lack of cooling water when stuffing boxes and suction pipes should be examined for air leakage, and the water level in the cooling tower checked to ensure sufficient is available in the sump.

Inadequate supplies of product may also be associated with low vacuum and high temperature, and pump stuffing boxes and suction pipes should be checked and distributor plates examined for blocked holes. It is essential that sufficient supplies of product are available at the balance tank.

(ii) Rising Levels in Separators

Rising levels in the separators may be associated with air leakage at pump seals and joints, low steam pressure, fouling of the evaporator tubes, or increased product flow.

(iii) Solids Percentage Not Constant

Failure to maintain the solids percentage in the concentrate may be due to a fluctuation in the steam pressure, the supply or temperature of the cooling water not being constant, or temperature of feed not to specification. The level in the balance tank should also be checked to ensure it is being maintained.

Variation in solids percentage may also be associated with increased steam supply in which case the steam controller should be checked, or with a decrease in the vacuum pump efficiency in which case the temperature or supply of the operation water should be investigated. If the cooling water inlet temperature is too high, and a cooling tower is installed, a check should be made of its proper operation and the water flow increased. If the cooling water outlet temperature is too high, the cooling tower should be checked as above. Where a surface condenser is used a check should be made for fouling of the tubes, too high a condensate level and air pockets. The condensate and vacuum pumps should also be examined. In the case of a spray condenser the strainer may be clogged and should be cleaned.

(iv) Closing of Steam Control Valve

If the steam control valve closes, it may be due to the boiling temperature in the first effect rising above the safety limit, a pump stopping, or an air/electrical failure. In all these cases some fouling of the tubes may have been caused and it is therefore advisable to empty the evaporator of product and clean the plant before production.

(v) Hints for Location of Air Leakage

All the manholes, etc., which have been opened since the last run should be checked. If the source of the leak cannot be located by putting the plant under vacuum and the sealing water in the pumps, it may be possible to find the air leak by listening. If not, connections may be checked by means of smoke or soapy water. If the separators are filled and the ducting flooded, air leakage will be indicated by small air bubbles in the separator. Alternatively, with the plant at atmospheric pressure and no sealing water to the pumps, the separator should be filled with water and the connections and seals of pumps checked for leaking water.

B. Spray Driers

(i) Rising outlet temperature may be due to failure of feedstock supply to atomizer owing to pump stoppage, fouling or leakage from delivery line to drier, blockage of distributor or deposit on outlet sensor.

(ii) Falling outlet temperature may be caused by a large increase in the supply of feedstock to the atomizer, but failure of the outlet sensor should also be checked.

(iii) Fluctuating outlet temperature may be due to erratic feedstocks supply to the atomizer due to low level in feed tanks, or failure of the automatic controls of feed pumps.

(iv) Falling inlet temperature may be the result of loss of control of steam pressure, failure of the temperature sensor, blocking of the condensate outlet, rising inlet temperature, increasing steam temperature or fouling of air filters.

(v) Fluctuating atomizer power consumption may be due to inadequate drive belt tension, damaged atomizer bearings, incorrect setting of the atomizer disc or fluctuating feed stock supply.

(vi) Fluctuating discharger power consumption may be due to mechanical failure of the discharge drive transmission, powder deposits in the chamber, or physical contact between the discharger and the chamber.

(vii) Fouling of cyclones is caused by high water content in the powder and a cold air stream around the cyclone or blockage of the outlet rotary valve.

(viii) Abnormal deposits of powder in the chamber are due to low outlet temperature creating high water content in the powder, which can be due to a fluctuating product supply to the atomizer or incorrect adjustment of the air distributor.

(ix) V-Belts. In order for V-belts to have a long life it is important that they are kept so tight that they hardly slide on the V-belt pulleys. New V-belts must be tightened after 2, 8 and 24 hours and again after 3 days. When changing belts the whole set must be replaced at any one time, and V-belts of exactly the same length must be chosen.

(x) Lubricating. Every month the oil level in the gearboxes is controlled, and is changed once a year. The ball bearings of the motors do not have to be greased.

(xi) General. The capacity of evaporation is dependent on and proportional to the weight of the drying air. Therefore, the air cleaning filter should be changed when the resistance through it has increased to 200 mm W.C., as an increased resistance will have a reducing effect on the air quantity, which means a reduced capacity. Moistening the drying air and the transport air with steam or hot water should be avoided. Cold draughts on the cyclones should also be avoided as cooling these will cause powder deposits on the cold walls. The specific gravity of the product should be kept as uniform as possible.

(xii) Fire in chamber. Theoretically, fire may occur in cases where moist powder deposits are found in the chamber and these deposits are exposed to abnormally high temperatures.

For safety reasons the normal procedure to follow in case of fire in the chamber is as follows:

1. The steam valve is closed.
2. Change from product to water is made.
3. The cleaning switch is changed for the atomizer.
4. The exhaust fan is stopped.
5. The controller is switched to "Man".
6. The feed pump velocity is regulated to maximum.

In-place Cleaning Equipment

The main items of any in-place cleaning plant, namely, pumps, valves, piping and fittings, have been dealt with in an earlier section of this chapter. In addition to those items, numerous electrical/electronic devices may be employed for control functions, ranging from micro-computers, programmed card readers, flow indicators of either solid state or electro-mechanical design, together with conductivity meters and line probes. The maintenance details of all such equipment will be highly variable and, whilst the essence of good maintenance is to ensure correct performance by calibration, the individual supplier's maintenance notes must be consulted.

(i) Stages in Cleaning

The stages in cleaning comprise essentially:

(a) Removal of product residues by flushing with water.

(b) Circulation of an alkaline detergent.

(c) Flushing out alkaline detergent with water.

(d) Circulating an acid descaling agent.

(e) Flushing out acid with water.

(f) Either hot water sterilization or chemical disinfection, followed by flushing out of the disinfectant with cold clean water.

It is appreciated that in some installations one or more stages may be omitted and that many of the larger installations will have the cleaning/disinfecting schedule under automatic/computer control.

(ii) Detergent Cleaning

Brine sections - Heat exchangers which incorporate brine cooling sections necessitate special conditions. Before undertaking the cleaning programme it is essential that the brine section should be filled with dilute (0.25 percent) caustic soda or alkaline detergent to inhibit the corrosive action of the chloride ions at the temperature employed for cleaning.

General precautions -

(a) Detergents should always be carefully measured or dosed into the balance tank to give the desired strength of cleaning solution. If equipment to monitor the conductivity (and hence the strength) of the cleaning solution is not fitted, it is recommended that the solution strengths are checked by titrimetry.

(b) Alkalis (and acids) will attack tinned metals, gunmetal, copper, bronze and solder. If the plant is not constructed entirely in stainless steel, further advice should be sought.

(c) Rinsing must be carried out at sufficiently high flow rates to give turbulent flow in the circuit. High temperatures are not necessary for rinsing. With hard water, final rinsing should be carried out at under 38°C (100°F) and even with soft water there is no need to exceed this. Before final rinsing, detergent solutions should be cooled to this temperature or lower.

(d) Water should not be used for cleaning if it contains more than 100 ppm of chloride salts.

(iii) Acid Descaling

Under no circumstances should sulphuric or hydrochloric acid be used for this operation. The recommended acids are phosphoric acid or nitric acid. Sulphamic acid may be used. Neither nitric nor phosphoric acid is corrosive to austenitic stainless steel under the conditions recommended for cleaning. However, nitric acid can cause deterioration of some types of rubber gaskets and therefore it is strongly recommended that under no circumstances should the concentration exceed 1 percent nor the temperature exceed 85°C (185°F). No such problems exist with phosphoric acid and concentrations of up to 5 percent at temperatures up to 85°C can be used without risk of damage to the gaskets.

(iv) Disinfecting Agents

Whilst hot water is always the preferred disinfecting agent, chemical disinfection is commonly practised and among the chemicals used are the following:

Sodium hypochlorite
Chlorinated sodium phosphate
Organic chlorine containing compounds (e.g. dichloro dimethyl hydantoin)
Quaternary ammonium compounds
Peracetic acid
Iodophors, etc.

It should be borne in mind by maintenance engineers that these brief notes serve only as an introduction to the chemistry of plant cleaning and for more detailed information a study should be made of appropriate publications.

TOWARDS LONG TERM MAINTENCE OF SERVICE EQUIPMENT

Piping Systems

In any milk plant, milk needs to be transported from one part of the plant to another, and this transportation is done in piping systems. Fluids comprise gases, vapours and liquids. Gases and vapours may be transported under increased or reduced pressure. Liquids are transported almost exclusively under pressure higher than that prevailing in the ambient, i.e. atmospheric. Depending on the pressure of the medium transported in the pipe the occurence of leaks will demonstrate itself either by the fluid coming out the pipe (fluids under increased pressure) or by the ambient air being sucked into the pipe (fluids under reduced pressure). In the majority of instances in the dairy industry the pressure inside pipe systems is higher than that in the ambient although there are exceptions such as milk evaporators (vapour ducts) and spray driers (air in the driers, in the cyclones and in the interconnecting components). Milk and liquid products are also transported within the milk plant in piping systems under increased pressure. However, the milk pipes need to meet special requirements not applicable to other fluid pipes in the milk plant. Pressures under which fluids are transported in piping systems in a milk plant are classified by industrial standards as low and medium. To the highest belong steam pressure which in milk powder factories may exceed 2.0 MPa (atmospheric pressure equals about 0.1 MPa) and compressed ammonia in refrigeration plants reaching about 1.2 MPa in single-stage compressors.

Compressed air pressure for pneumatic devices may reach about 0.8 MPa, but air used in all air conditioning systems may be transported in air ducts either under slightly increased, or slightly reduced pressure as compared to the ambient. The pressure in water pipes usually does not exceed 0.3 to 0.5 MPa. Depending on the type of fluid transported in the pipe and on its pressure, the design engineer selects the suitable type of piping and fittings. Pipelines may be flexible (plastic, rubber) or rigid (usually metallic). Various types of fitting are used to interconnect pipes or to connect pipes to machines. Obviously only flexible pipes can be connected to moving components. Worm drive clips and snap-on connector unions are in common use to secure flexible hose to metal pieces and adaptors. Screwed joints, flage joints and welded joints are most commonly used on mild steel piping, but in the case where galvanized pipes (zinc coated) are used - usually for water lines no welding should be applied since it removes the zinc coat and exposes the fluid and the steel to corrosive interactions. Soldered joints are used when lengths of thinwalled copper piping have to be connected and this is very often met in internal connections of automatic controls. Contrary to a frequent attitude, piping systems require the meticulous attention of the engineering staff of the milk plant. This attention should be given not only to the valve , traps, reducers, filters and other more complicated parts of a piping system, but also to the pipelines themselves. They need their share of scheduled inspection, care and maintenance. The trouble most often met in piping systems is the deterioration of the flow of the fluid. Preventive maintenance should be aimed at preventing this from happening. This deterioration is caused by scale accumulation, by condensation inside the system or by leaks. The term "scaling" means depositing solids inside the pipe by which the free way for the passing fluid may be reduced and sometimes completely blocked. This occurs particularly often in water pipes transporting untreated water at elevated temperatures: if the water hardness caused by bicarbonates is high, the scale accumulation is rapid and the deposits are hard like stone. Condensation can occur in vapour (mainly steam) lines and gas (mainly air) piping systems. It always happens when the temperature of the fluid inside the pipe drops below a given value. Leaks are caused by corrosion or by distress cracks, which in turn may be the result of liquid hammer, faulty drainage or faulty pipe installation, particularly in anchoring, supporting and in designing the expansion couplings. Securing high performance of a piping system begins with the proper layout in which all piping elements are well selected and properly assembled into a system. First of all allowance must be made for the fact that all dimensions of the pipeline change with a change of the temperature, increasing when heated and decreasing when cooled. As a result of this expansion the whole system moves and all hangers and supports have to be designed in such a way that they either move together with the pipe (roll or slide) or that they can swing without exposing any stress either on the pipe or on the part of the supporting anchoring structure. Expansion pipe bends are not often used in a milk plant since the "spring effect" of an expansion bend is usually achieved through the frequent change of the direction of the line. However, expansion joints are often in use on steam lines. In vertical runs that the expansion of the pipe may cause the even distribution of the weight of the cold pipe on all rigid hangers will change in such a way that the entire load will be shifted to the bottom hanger. Should, during preventive maintenance inspection, distress cracks on walls and footings near the pipe anchors be detected, the fault diagnosis will indicate improper layout with regard to thermal expansion.

Liquid hammer occurs when a moving column of liquid is rapidly stopped. It exposes the whole piping system to mechanical stresses which may cause serious damage and result in leaks. Slow moving valves do not cause hammering, but pneumatically operated spring loaded valves may. Air chambers are one of the common methods by which the hammer shocks are relieved. Condensate from the steam lines needs to be removed in order to avoid water hammer, but keeping traps in operating condition may not be sufficient when pipes are sagging and creating pockets where condensate can collect. The whole system should be adequately pitched. Condensate will tend to collect above closed valves in vertical lines and in the back of globe valves in horizontal lines. Draining the condensate from all such spots prior to admitting steam to cold pipes is one of the ways of preventing water hammer in steam pipes. Condensation in air pipes leads to rapid corrosion, but first of all it is dangerious because of the introduction of water in a part of equipment in which dry air only is wanted. For instance, air distribution valves for automatics are usually activated by solenoids; when air is discharged from the valve outlet together with water, short circuits in the solenoid wiring systems are unavoidable. Air drying is a must in most milk plants. It is done directly after compression and usually achieved by condensation of the water through cooling the air in a pipe heat exchanger. Prevention from scaling depends much on whether the general technical standard of operations in the entire plant meets the requirements. A water treatment plant, filters and oil separators are the common components of the equipment, permitting fluids to flow smoothly in the pipes without scale formation. They all need proper care and maintenance. The most difficult task in maintaining piping systems at expected levels of performance is the prevention of corrosion. Corrosion is unavoidable, but it can be slowed down by proper operation procedures and proper maintenance. Well-maintained mild steel piping systems may last 12 to 15 years or even longer, but instances of a complete breakdown in 3 to 4 years after commissioning are know. Corrosion can be caused by many factors but they are all of electro-chemical nature. Corrosion is generally caused by atmospheric oxygen dissolved in aqueous liquids, or by dissolved salts such as brines and detergent solution and finally some gases like carbon dioxide and sulphur hydroxide or diozide dissolved in water. The internal corrosion in the pipes can be reduced by strictly observing all rules in handling the fluids prior to their entering the piping system. Such rules must be introduced after a thorough analysis of the fluids, by drawing conclusions from the analytical results and by applying appropriate measures. Most typical in this respect is water treatment, not only for the steam raising plant, but also for general use. If aggressive solutions are transported in the pipes, corrosion can be slowed down by adding to the liquid certain selected chemicals such as sodium silicates. Scale formation and corrosion can be often reduced, particularly in water pipes, by adding up to 20 ppm sodium hexametaphosphate to the water. Basically corrosion is slower in liquids with pH values exceeding 7.0, or in other words in alkaline solutions, as compared to the acidic ones. External corrosion on pipes is always increased on sweating pipes if the pipe is colder than the ambient dew point at which the air moisture condenses on the cold surface. Air and other gases dissolve in the condensate, and cause corrosion. A practical solution to the problem is to prevent the pipe from sweating. Watertight covering applied directly to the pipe (asphaltic coats, thermal insultation, spiral wrapping of strong fabrics) is the simplest remedy. The layout of the piping and the selection of all components should depend on the parameters of the transported fluids and on the wanted capacities. The fluid velocity in a pipe is limited and depends on the type of the fluid: for water usually 2 m/s, for steam up to 50 m/s at pressures reaching 2.0 MPa but only 35 m/s for reduced pressures ranging between 0.2 to 0.6 MPa. The liquid velocity in milk pipes does not usually exceed 1 m/s, except during cleaning. There are several technical rules guiding the installation, operation, care and maintenance of valves, and other components of a piping system. They are a part of the art of proper installation but, particularly concerning care and maintenance, they depend on the design of the component for which a good manufacturer always issues appropriate manuals. Globe-type valves cause higher pressure drops than gate valves, but they are useful for throttling service, and are easier to maintain tight and easier to repair. It is advisable to install whenever possible globe-type valves in such a way that the pressure is above the disc - this prevents vibrations. Automatic drainage needs to be provided in all places in which condensate could accumulate, particularly above valves in vertical lines, and above the discs on stop-check valves. All accessible points of friction of the piping system components, especially valves, should be frequently lubricated (steam threads, yoke sleeves, etc). Any leak discovered in the pipe system should be stopped without delay: stuffing boxes tightened, misalignments corrected, packings replaced, supports repaired or changed if they do not support their share of the load. Any repair on an insulated pipe may adversely affect the insulation. The insulation must be repaired immediately not only to prevent excessive heat exchange with the ambient, but also to prevent rapid outer corrosion.

Centrifugal Pumps, Fans, Air Compressors

Fluids move in a pipeline when there is a pressure difference between the ends of the line. This difference, if not caused by thermal convection, is usually created by pumps when the moving medium is a liquid and by compressors and fans when the moving medium is a gas. In many respects compressors and fans can be considered as gas pumps and there are several similarities in their mechanical design, installation and care. Therefore they could be considered as one group of machines supplying pipelines with fluids. In a milk plant, apart from milk, the most common liquids transported through pipelines are water and water solutions and air is the most common gas. Service pumps in a milk plant are nowadays almost exclusively of the centrifugal type and so are ventilating and exhaust fans used for building ventilation, for supplying air to burners in steam boilers, for air supply in spray driers, can washers, etc. Air compressors used in milk plants are commonly of the small and medium size and are built as single or multi-cylinder piston pumps. Air tanks are standard parts of the machine. In smaller machines the motor and the compressor are usually mounted on the air tank.

Centrifugal pumps. There are several types of centrifugal pumps used in a milk plant of which the simplest are those with motor and pump rotor built on a single shaft. Their care and maintenance is similar to milk pumps which are dealt with in another section below. In all other types of centrifugal pumps, the motor and the pump are joined by a coupling, flexible or rigid. The majority of centrifugal pumps are supplied by the manufacturers with bed-plates which need either to be grouted on a pump foundation or to be fixed on other types of rigid structures. Misalignment is one of the most common sources of pump troubles. Proper alignment should be secured during pump installation and checked frequently throughout the pump's life. It should be noticed that neither a flexible coupling nor delivery of the pump on the manufacturer's bed-plate can be considered as guarantees for proper alignment. Therefore the pump must be checked for alignment - angular and paralleled - and necessary corrections must be done several times during installation: on the bed-plate prior to fixing this to its foundation, after fixing but prior to securing piping and finally after the pipes have been attached to the pump. Absence of air, particularly in suction piping, is the next requisite for satisfactory pump performance. The capacity of a centrifugal pump is reduced by almost half when only 4 percent of free air is present in the pumped liquid. It leads in extreme cases to cavitation and to damage of the pump. The suction side of the pump is the most sensitive to entrained air and proper attention must be given to the following rules guiding the correct installation and operation of a pump:

- The size of the section pipe should never be smaller than that of the discharge pipe; usually it is one size larger.
- As far as feasible valves, filters, excessive fittings, horizontal bends, ells and tees should be avoided on the suction side of the pump but if their installation is necessary, they should not be mounted close to the suction inlet of the pump.
- Air can enter into the pump when liquid is falling directly above the surface close to the pump suction connection: any return line should be brought under the surface of the liquid and as far from the suction connection of the pump as feasible.
- Vapours in hot liquid pipes shorten considerably the life of rotating elements of a pump, particularly in multi-stage centrifugal pumps used commonly in boiler feed systems; vapours can also be created within the pumping system due to many reasons - most often due to operating a pump continuously near the lowest capacity without appropriate installation and operation of by-pass orifices. The by-pass should never be connected straight to the suction pipe.

Proper care of glands is most important during the operation of centrifugal pumps. The main function of a gland on centrifugal pumps is to hold in place a soft packing preventing the liquid from leaking or air from entering the pump casing. Glands may serve also as a device for removing heat from the shaft - they are then called quench glands. They have a hollow portion adjacent to the shaft, through which cooling liquids pass and lower the temperature of the shaft, thus protecting bearings mounted on the shaft. Quench glands must fit the shaft close to the outboard ends. The clearance between shaft and gland has to be kept at the correct tolerance, otherwise the pumped liquid will leak along the shaft. Glands may be treated as replaceable spare parts as a whole, but the necessary clearance may also be re-established by replaceable bushings or auxiliary packings. A normal packed-type stuffing box requires lubrication and cooling. Usually 40 to 60 drops per minute of the cooling/lubricating liquid (which most often is water) flowing out of the stuffing box are considered sufficient for the safe operation of the pump. The maintenance procedures for pumps must be guided - as they are with any other machine - by manufacturers' manuals, and experience, but a few indications may serve as a useful illustration of the work involved:

- the temperature of the moving parts and the general condition of the pump must be inspected weekly or every fortnight; lubricants should be changed every three months;
- alignment must be checked every six months, which is also the most common frequency for replacing packing;
- a thorough inspection on general wear of all major components must be done once a year and the performance of the pump should be checked after the post maintenance reassembly.

Ventilating and Exhaust Fans

Most of the fans used in a milk plant are of much larger dimensions than the pumps and in addition their housings, ducts and blades are built out of relatively thin metal sheets. Because of this they are much more exposed to damage during shipment and installation. They require thorough inspection upon delivery, careful storage prior to installation and extra care in handling, particularly when lifted with a hoist. The installation of larger size fans means reassembling in the milk plant since they are usually disassembled by the manufacturer prior to shipment. The assembly has to be done with proper care being given to match the markings of the manufacturer. Exceptional importance should be attached to the correct direction of rotation. Many of the principles of a correct installation, care and maintenance of centrifugal pumps apply also to fans, particularly concerning alignment, foundations, lubrication of bearings, etc. However, there is a number of installation problems typical for fans only - one of them concerns vibration. Vibration-isolating bases are used in fan installations to reduce the transmission of sound and vibration from a fan to other areas of the plant. The isolating elements consist of resilient material and steel springs which deflect under the weight of the vibrating element and absorb structure-borne sounds and vibration. The isolating elements should support the rigid base on which the motor and the fan are mounted. They should be adequate in number and properly spaced. Flexible connections should be provided between air ducts and the fan. Large fans are usually belt driven and the efficiency of the drive depends on friction adjustable by the V-belt tension. Correct tension is attained when the belts just do not slip when operating at full speed and load. The initial run of a large-scale fan must start under reduced load: outlet dampers and variable inlet vanes should be partially - not completely - closed. Vibration can be measured by instruments indicating displacement: well-running fans will show displacement ranging from 0.05 mm for fans with low (600 rpm) rotation speed to 0.02 mm for those with high (1900 rpm) speed. The maintenance inspection procedures and inspection frequencies for fans are similar to those for centrifugal pumps: in addition isolation of the vibration bases and inspection of V-belts need to be included in the preventive maintenance programmes. Fans also need more devices to protect personnel from contact with rotating elements: protective screens, belt guards and coupling guards.

Air Compressors

In many instances air compressors are assembled by the manufacturer as a unit together with the air tank. Since the highest permitted pressure in the tank is limited to a value declared by the manufacturer and in most countries it is also subject to inspection by legal authorities, the tank is provided with a safety valve and the compressor's operations are automatically controlled by an on-off regulator. The compressor is driven by an electric motor through a V-belt drive. Air filters are installed on the suction side of the compressor. A pressure unloader prevents the machine from starting with air pressure in the compressor's head. The crank case contains oil necessary for lubricating the main drive of the pistons. In many plants the compressed air should contain a very low level of moisture which requires an air cooling and water condensing device mounted between the compressor and the air tank. Moisture-free air is very difficult to obtain and for most purposes the reduction of the compressed air humidity serves the purpose. Even with air cooling installed, some water may condense in the air tank and therefore a drain valve at the bottom of the tank is necessary - preferably automatically operated. Oil separators are mounted on the compressed air pipe, but oil-free air used for milk agitation in large tanks is obtainable only from special type compressors in which the cylinder lubrication is achieved by application of self-lubricating rings. The inspection and maintenance of air compressors include:-

- weekly cleaning of air filters
- weekly check of lubricant level in crank case
- weekly check of the performance of the on-off regulator
- weekly check of condensate level and drainage

- weekly check of the pressure unloader
- monthly cleaning of oil separators
- monthly check and adjustment of V-belt tension
- monthly check of the performance of the compressor by measuring time needed to load the tank from the lowest to the highest pressure.

Power Transmission

The electric motor is practically the exclusive prime-mover for powering all machines in a milk plant. The transmission of the power from the electric motor to the main shaft of the machine is performed by means of a drive. The most commonly used types of drives are gears, belts and chains. The acting part of the driven machine is sometimes mounted directly on the shaft of the electric motor, as in some of the most common types of milk pumps and in some designs of spraying discs in spray driers. Power can also be transmitted from a motor to a main shaft directly by means of couplings. Gears are most commonly used for driving dairy process equipment, belts - almost exclusively V-belts - are widely applied on auxiliary machinery like ammonia compressors, air compressors and fans, whereas chain drives are met in bottle and crate washers and in crate and package transportation equipment.

Gears

Those in most common use are plain spur, helical spur, bevel and worm and worm wheel gears. Plain spur gears transmit power between parallel shafts only. They are becoming less used in milk plant equipment, since helical gears are able to perform similar duties, are quieter in operation and have a higher transmitting capacity. Bevel gears transmit power between shafts meeting at an angle. They are called mitre gears when the angle is 90°. Worm and worm wheel gears provide a high reduction in speed; they operate very quietly and are commonly used as dairy machinery drives. They usually run in an oil bath enclosed in a gear box. Normally the drive can be transmitted only from the worm to the worm wheel. As in any rotating part of a machine, correct alignment is the most crucial requisite of proper performance of a gear drive. It is of paramount importance that the gears mesh correctly. Excessive wear will occur very soon when the shafts of the gears are misaligned, or when the gears are too tight or too loose in mesh. In worm and worm wheel gears there is a heavy load on the worm shaft and adequate thrust bearings are incorporated in the design. They must be checked regularly for wear as should all gearbox bearings. As with all other critical mechanical parts ball and roller bearings should be handled with greatest care to avoid mechanical abuse and corrosion damage. They should be constantly protected from all forms of dirt or foreign matter that might dent or wear the highly polished surfaces of the balls, rollers and races. Dirt causes 90 percent of early bearing failure. The services required on gears - including lubrication - are always described in manufacturers' manuals but local operating conditions may indicate additional sources of trouble and spots to which particular care needs to be given. Experience is a valuable source of information on how to maintain a gear at its optimal performance level.

V-belt Drives

These occupy an important place in the service section of a milk plant. The driving power of the motor can be transmitted by the belts when drive wheels (pulleys) are located on the shafts. V-belt pulleys are held in position by keys or by tapered locking bushes. Before installing belts the tension adjustment should be slackened completely, the driving surfaces of the pulley should be cleaned and the alignment of the pulley checked. Prior to installing a set of V-belts, the belts should be checked that they are a matched set. Standard V-belts are marked with a letter indicating their width across the tip and a number indicating their length. Widths are marked with letters A to E, lengths are stamped with a number which indicates the precisely measured length. A nominal pitch length will be marked, for example, (50), other figures (49, 51) will indicate the deviation from the nominal length. A matched set of V-belts is a set with the same stamped markings. Mismatched V-belts put on a drive will have a short life since the shorter belt will carry all the load whereas the longer ones will remain idle. Belts stretch on the drive, never shorten. Replacing V-belts one at a time (when wear is detected on one only) is a risky operation since the new one - being shorter - is likely to carry most of the load.

For this reason it is always better to install a complete new matched set. The used belts can often be put on less demanding services in machines requiring a smaller number of V-belts. Belts must be tensioned correctly to transfer the drive and prevent unnecessary wear. As they stretch in use, their tension must be regularly checked and adjusted. Testing the V-belt installation on tension is done when it is stopped or when running. The following are suggestions on how to do a simplified testing during routine maintenance inspection. When stopped, a correctly tensioned V-belt should, if pressed firmly with the thumb near the mid-point (half way from centre to centre), depress 3/4 of its own thickness for each one metre centre to centre distance. With belts running at full speed the sag on the slack side of the drive should be checked. Correctly tensioned belts will show a sag equal to the depression shown during stop testing, i.e. about 3/4 of its thickness for each one metre centre to centre distance. It should be noted that the adjustment of the tension of new belts needs checking and rechecking several times during the first 48 hours of operation since new V-belts stretch slightly and settle into the pulley grooves before reaching their working lengths.

Chain Drives

These are subject to wear even when properly selected for installation, properly installed and adequately lubricated. They wear and stretch unevenly during use and checking chain wear means inspecting the complete length and all sprockets. Worn chains or sprockets will cause the chain to jump and possibly come off the sprockets which may mean damage to the machine. Proper alignment of sprockets minimizes chain wear but does not affect chain stretching which stresses the need for checking chain tension frequently and adjusting regularly. Chains need to be adjusted at the tightness tension as they stretch unevenly. The correct selection of the chain drive is given in manufacturers' manuals. Worn sprockets will also adversely affect the performance of a chain driven even with new chains. They should be replaced to ensure proper chain fit on the sprockets. In some cases the life of a worn sprocket may be extended by reversing it on the shaft to bring a new set of working tooth surfaces into use. Properly lubricated chains will not show discolouring at the joints and the connecting link pins will be brightly polished with a very high lustre. Frequent clearing of the chains and sprockets can greatly contribute to extending the life of a chain drive.

Refrigeration

Any refrigeration system forms two circuits. In the primary circuit the refrigerant in the form of low-pressure vapour is compressed in a compressor and the compressed gas is liquified by cooling with air or water in a condenser, from where it passes to the liquid receiver. From the receiver the liquid passes under high pressure to a regulator or expansion valve where the pressure is reduced at its entry into the evaporator coils. In the coils the liquid evaporates by taking heat from the secondary circuit medium in which the coils are immersed. The refrigerant vapours return to the compressor. The secondary circuit can be air in a cold store and brine or water in a container. In most modern milk plants there are usually two secondary circuits in which two sets of evaporator coils are installed: one for evaporation in the air of the cold stores and the second for evaporation in water used for cooling milk or liquid milk products in plate heat exchangers or in jacketed vats. In ice-cream plants, refrigerant evaporation takes place directly in a processing machine (freezer) but then the circuit is usually fed by a separate refrigeration system. There are numerous types of refrigerants, those in most common use are ammonia (R 717), R 12, R 22 and R 502 - the last three known under the trade names "Freon" or"Arcton". They are supplied in cylinders marked with identification colours: R 717 in black cylinders with red and yellow bands, R 12 in grey cylinders with white markings, R 22 in grey cylinders with green markings and R 502 in grey cylinders with orchid markings. Of the four refrigerants listed above ammonia (R 717) is extremely toxic and must be handled under strict controlled conditions. The remaining three are not toxic but they are gases heavier than air and any leak will tend to accumulate at ground level. In poorly ventilated rooms it may cause health hazards such as unconsciousness and suffocation. Before cylinders are returned for recharging they should be completely emptied by venting in an open space. Refrigeration has become in the last few decades a very specialized subject in the engineering field. In many milk plants refrigeration specialists are employed to look after this essential part of a dairy enterprise. Their main responsibility is to maintain the performance of the refrigeration plant at its optimum level and to make it meet the respective requirements. In many milk plants, even in those not manufacturing ice cream, the installed electric power of the refrigeration plant represents one third of the total electric power installed and often even more. The electric power consumption of the refrigeration equipment may come up to half of the total plant's consumption. Therefore the operation of the refrigeration equipment may greatly affect the total energy costs of the plant. A continuous performance checking of the refrigeration system is a requisite of the plant's good overall functioning. The characteristics of the functioning of a refrigeration plant make it necessary to adjust pressures, flows, temperatures, etc. more frequently than in any other part of the milk plant. False readings caused by wear of the gauges and indicators lead to false adjustments. It is of paramount importance to make sure that all of them are checked for accuracy and kept in perfect mechanical condition. Thermostatic expansion valves require checking every month. If they open too wide, through wear or dirt, flooding may cause a frost back to the compressor. Gas leak to the bulbs may cause reduction of the flow to the evaporator and subsequent reduction of the overall capacity of the evaporator. Cartridges in float controls need changing every three months. The float control valve is responsible for the refrigerant level in the evaporator. When closed too much, it reduces the capacity of the evaporator. If it is open too much the compressor may receive liquid on the suction side and get damaged. Back -pressure regulating valves need checking every month. They are responsible for keeping the temperature of the gas within limits in the evaporator. If the temperature falls too low, frost accumulation on the blower units in the cold stores may occur leading to clogging of the units. Solenoid valves are used to stop the flow of either gas or liquid at different points in the system. Their functioning needs occasional checking. The trouble with the solenoid valves occurring most often is the burning of the coil: it should be warm when energized. Scale traps with screens ahead of all controls are important. They should be cleaned every six months. Only stainless steel screens should be allowed in ammonia circuits. Thermostats, which automatically control the system, require checking on proper functioning by comparing the set and the true temperatures. Faulty thermostats can seldom be repaired and they usually have to be replaced with new ones. Relief valves need to be checked frequently on re-seating in order to avoid excessive loss of refrigerant. Items requiring the major attention of maintenance staff in a refrigeration plant may be listed as follows:

1. Purging non-condensable gases, including air. The presence of these gases is checked by comparing the temperature of the vapour with its corresponding pressure, preferably near the expansion valve. Excessive pressure indicates the need for purging. The installation should be shut down during the check.
2. Draining oil traps on the refrigerant discharge side as frequently as indicated by experience.
3. Purging oil out of evaporators and receivers. Oil traps are not always sufficiently effective to prevent oil from the compressor travelling through the condenser pipes, the receiver and the evaporator coils. Oil reaching the evaporator may solidify and obstruct the flow of the refrigerant; in addition even the thinnest oil film on the heat exchange surface reduces the energy transfer.
4. Evaporative condensers are the most common in use in modern refrigeration systems in milk plants. They are usually located outdoors and there is a tendency for dirt accumulation and for growth of algae and fungi on the surfaces and in water basins. The unit needs cleaning with the frequency depending on local conditions under which the condenser operates. Fungicides and algicides are often added to the water basins which also help to keep the outer surfaces of the cooling pipes free of slime deposits.
5. Evaporator coils in cold stores need to be kept free of frost. Defrosting may be achieved by several methods of which passing hot compressed gas from the compressor's discharge side is the most common. Depending on the defrosting arrangements the frost accumulation needs frequent checking in order to keep heat exchange on the evaporator coils as effective as possible. Evaporator coils in brine basins seldom accumulate frost or ice, but in the ice-bank system the equipment is designed to accumulate ice on the evaporator coils. The heat transfer is reduced when ice grows on the pipes but up to a given thickness this reduction is allowed for in the design. The ice-water requirements of the plant vary in the course of the day and are usually nil during the night hours. This makes the thickness of the ice layer on the pipes grow. The plant should be shut down when the thickness of the ice layer reaches the highest calculated value. In many plants automatic switch-off devices are used; their efficiency needs frequent checking in order to operate the refrigeration plant at the lowest possible cost.
6. Compressor stuffing boxes need repacking at regular intervals. The condition of the compressor requires checking at least once a year. Refrigeration compressors are usually supplied with very comprehensive manuals, maintenance instructions and spare parts lists. They should be carefully

studied and all instructions followed precisely. Refrigeration equipment - particularly that operating with ammonia as refrigerant - is long-lasting if well maintained, but any negligence in care and maintenance may lead to drastic accidents and to health hazards to the staff.

The check list on maintenance inspection schedule cards is particularly long for refrigeration equipment. This is also a long list of items for recording in the refrigeration log books. Their contents and mode of preparation are very often suggested by the manufacturers in their manuals. Additional assistance from a reputed refrigeration engineer may be of great help in establishing a preventive maintenance system for the refrigeration plant and will pay for its cost in the long run.

Steam Raising Equipment

There is a great diversity of steam boilers utilized in milk plants. They differ in size from about 100 kg/h steam capacity to tens of metric tons per hour. They differ in the level of pressure of the produced steam, they use different fuels (coal, gas, oil, sometimes electricity) and they may be fully hand operated or fully automatic. The thermal efficiency of the boiler plant may be as low as 40 percent in small coal-fired plants and as high as above 90 percent in large-scale oil or gas-fired boilers. Depending on the type of processing equipment installed and on the layout of the milk plant, the condensate recovery may be nil but it may also exceed 80 percent of the total steam output. Because of this diversity, it is not possible to draw up a universal standard maintenance programme for a steam raising installation. The operation of steam boilers is in most countries subject to official control of legal bodies which also set steam boiler regulations and standards. The diversity of boiler plants is taken into account in these standards and they are considered in the regulations accordingly. They are the first source of guidelines concerning care and maintenance of the plant. The second is the manufacturer's manual. The third could be the standard handbooks' advice on fuel burning, heat transfer in boiler plants, water treatment methods and requirements, feed water handling, etc. Experience of the plant in boiler operations is of course the last but not least important source of indications concerning plant care and maintenance. Only few indications of general nature can be given as a list of first steps to be taken when establishing a preventive maintenance system in the steam boiler section. Periodical checking of the thermal efficiency of the plant will give the most crucial information on the overall performance of the boiler. The thermal efficiency may be considered as the ratio of energy obtained in steam to the energy supplied in the fuel and feed water. Estimating the efficiency requires data on the quantity of produced steam and on its parameters, the parameters of the feed water and the quantity and the caloric value of the fuel used. In large-scale installations instruments for recording most of the values needed for making the heat balance and for calculating efficiency are included. It is particularly easy to calculate the efficiency in boilers utilizing liquid fuels or gas since their calorific values once determined remain relatively constant for the given type of fuel. The situation is more complicated with coal which is difficult to weigh and for which the calorific value may change considerably from one supply to another. Besides, coal is normally stored in the open and is subjected to rain and snow which obviously change its moisture content and subsequently the calorific value. Milk plants seldom possess laboratory facilities for fuel analysis. Such analyses need to be done by specialized laboratories. The problems become even more difficult in small-scale plants in which the team quantity cannot be measured directly and indirect calculations may not give sufficiently accurate results. A common practice is to hire the services of a specialized engineering company or of an individual specialist in this field. The efficiency check should be repeated every year. A useful indication - but indication only - on the overall performance of the boiler plant can be obtained by checking the composition and the temperature of the chimney gases. This can be done by the milk plant staff by means of relatively simple gas analysers and suitable thermometers. Excessive quantities of oxygen or carbon dioxide will indicate incorrect air supply; high temperatures may indicate scale accumulation on the heating surfaces. Scale accumulation is one of the most commonly met reasons for poor boiler performance. It may reduce considerably the steam output and in extreme instances lead to damage of the boiler and be a health hazard to personnel. Scaling on the inner boiler surfaces is caused by improper quality of water which contains calcium, magnesium, iron and other salts.

The quantity of these salts is expressed as water hardness. Hard water deposits scale very quickly reducing the heat transfer from the burning gases to boiling water. Feed water needs treatment prior to entering the boiler and the degree by which the original hardness must be reduced depends on the type of the boiler, although the ideal water for any boiler should be completely free of compounds causing scale deposits. There are many water treatment plants for boiler feed water. The most common in use are continuous ion-exchange water softeners. They are cheap in price and reliable in operations, but the resins need periodical regeneration and general care as instructed in the manuals. Gases dissolved in water such as oxygen and carbon dioxide should also be removed from feed water: suitable equipment is usually included as a component of the feed water treatment plants. It is a very good practice for the laboratory of the milk plant to analyse the feed water quality every day and inform the plant engineer on results. The analyses are very simple and can be carried out in a few minutes. They give essential information on boiler performance evaluation and for maintenance operations. Feed water temperature should be as high as feasible in working installations. The amount of further heating with steam depends on the quantity of condensate collected from the processing sections of the plant and eventually from central heating and air conditioning installations. A very considerable energy saving can be made by collecting the maximum of available condensate: unfortunately this possibility is often neglected. All steam boilers require periodical internal inspection. Daily blowdown and periodic descaling are essential. The relevant instructions given in the manufacturers' manual must be followed strictly. Steam distribution is a system which begins in the boilerhouse and continues through the entire plant. Saturated steam is most suitable for heating processes. Superheated steam is generally used for energizing steam prime movers such as turbines in electric power stations. However, saturated steam becomes superheated in the process of pressure reduction. For economic reasons the pressure in the boiler should be kept at the highest permissible level which in most plants not manufacturing milk powder is about 0.5 MPa to 1.0 Mpa. On the other hand the pressure permitted in most of the dairy machines does not exceed 0.2 MPa to 0.3 MPa, sometimes even lower. Steam pressure reducing valves are therefore installed at several points of the plant to ensure that steam reaches the machine at required pressure levels. The laws of thermo-dynamics must be followed when planning a pressure reduction and steam distribution system. One of the essential principles is to reduce the pressure step-wise when the ratio of high pressure to low pressure required after the reducing valve exceeds the value 1:83. In order to avoid supplying processing equipment with superheated steam (which might happen when the reduction valve is relatively close to the equipment receiving steam) the final length of the team supplying pipe should be left uninsulated. This system allows for cooling the superheated steam and saturating it prior to entering the equipment.

MAINTENCE OPERATIONS

The preventive maintenance programme includes many sections, an essential one of which is the maintenance work done directly on the machine. It needs to be performed by skilled craftsmen and in this chapter an attempt is made to provide the engineering staff of milk plants with a guide to maintenance comprising:

1. Basic rules concerning safety at maintenance work
2. Essential information on hygiene requirements relating to milk plants
3. Diagnosis of faults
4. Examples of routine maintenance procedures

Safety at Work

Regulations concerning health and safety at work may differ from country to country in detail, but some relevant provisions are embodied in all of them. All craftsmen must adhere strictly to the regulations concerning their particular trade, together with the Company rules complying with the country regulations actually in force. It is the duty of every employee while at work to take reasonable care for the health and safety of himself and of other persons, who may be affected by his acts or omissions at work. No person shall intentionally or recklessly interfere with or misuse anything provided in the interests of health, safety or welfare in pursuance of any of the relevant statutory provisions. Every craftsman must be trained to be safety conscious and should know the correct use of protective clothing.

Overalls should be buttoned up, sleeves rolled up above the elbows or cuffs buttoned up, and a protective cap should be worn, as illustrated.

Gloves may not be worn at all times, but a suitable barrier cream should be used to protect the skin. Cotton gloves with reinforced non-slip gripping surfaces are worn when handling oil or grease covered components or materials. Heat-resistant gloves should be used to give protection against burns. Leather gloves give protection against sharp corners when handling bulky or heavy equipment. Rubber gloves are worn when using cleaning fluids and to protect the hands from skin damage due to air blast when cleaning components by means of compressed air.

Footwear worn should be reinforced safety shoes or boots with reinforced toecaps, especially at work demanding the lifting of relatively heavy components. Special care should be taken where the conditions underfoot are hazardous such as excessive water on the floor.

Goggles must be worn when a chisel, a sharpening tool or a grinder are used, and when cleaning with compressed air is performed.

Hazard warning signs, obligations and instructions should be widely displayed throughout the buildings on walls, vehicles, containers, etc. to bring these to the attention of personnel. The series of signs prepared by the British Standards Institute is shown in Figs. 13 A and B. These show the black triangle with a black-on-yellow pictogram indicating warning, the red circle with crossbar and a black-on-white pictogram denoting a prohibition and the blue circle with the pictogram imposed in white denoting obligation or instruction. A green square with the pictogram in white is for information.

Every potentially dangerous part of any machine whether power driven or not must be securely protected. Before work is started on a machine it is essential to ensure that it cannot be accidentally set in motion. The power supply system must be so disconnected that some special action is required for its reconnection. If the machine has individual drive, either the fuses should be removed with precautions to prevent accidental replacement, or the switch should be locked in the 'off' position.

In some cases mechanical isolation might be necessary such as removal of belt from drive. A notice should be displayed at the machine to be worked upon to warn persons against any attempt to set the machine in motion. The power should be connected and the warning notices removed only by the maintenance craftsman after he has completed his work and made the necessary checks. In exceptional circumstances it may be necessary to observe closely the machinery in motion with the guards removed. Such work may only be carried out by qualified persons, and only when permission to do so has been obtained from the supervisory officer on duty. Before a guard or fence is removed from machinery, warning notices should be displayed on all sides where there is access to the unguarded mechanism. Guards and fences should be replaced before the warning notices are withdrawn. This 'permit-to-work' system ensures the safety both of the craftsman and of the equipment on which he works. Certain tasks require an extra strict observance of safe working practices, e.g., when adjacent equipment may still be in use. Such tasks must not be carried out by a craftsman unless he has been issued with written authority to do so in the form of a signed 'permit-to-work'. It is essential that the correct procedures are followed and are accepted as the normal method of working. The practices of a permit-to-work system vary according to local needs, but the contents of such a permit usually include the description of the work involved, the time limits between which the equipment or the area must be made safe, the title of the issuing authority, the officer-in-charge and the necessary precautions required to make the equipment safe to work. The permit must be signed by the officer-in-charge.

General safety rules should be displayed in places where they can be easily and frequently seen by the staff concerned. It is more effective to display them as a list of what should and what should not be done in connection with particular types of jobs. An example of this, referring to safety of electrical works, could be displayed as:

DO		DO NOT	
	- report all electrical faults		- use defective cables, plugs, etc.
	- keep loose cables off the floor		- connect power tools to lamp sockets
	- keep all electrical equipment dry and clean		- attempt to repair electrical electrical equipment
	- use compressed air carefully		- direct compressed air at yourself or others

Hygienic Working Practices

Hygienic working practices include precautions concerning personal hygiene, practices during maintenance operations and finally the CIP (clearing-in-place) system involving cleaning and sterilization of equipment after completing and/or before starting processing and manufacturing operations.

Personal hygiene includes medical clearance for working in food processing plants, which needs to be renewed at pre-determined intervals. Clean protective clothing must be worn, 'no smoking' notices observed and all tools and equipment must be carried in closed tool boxes or bags.

Hygiene precautions during maintenance operations must concentrate on preventing the food from begin contaminated by extraneous material. It is therefore essential that all fastening devices are secured firmly, and that only non-corrosive fixing parts such as washers or split pins are used when in direct contact with the product or passing above the product. Further precautions to be strictly adhered to are:

- Oil grease, solvents and compounds used on food machinery must be those recommended for such use;
- All electrical equipment must be correctly waterproofed;
- Glass instruments such as thermometers must be properly encased, so that they remain in position if broken;
- There must be no leakages of services, especially refrigerants and cleaning solutions;
- Recycled water systems must be frequently trested;
- Cleaning nozzles in washing devices must not be blocked or excessively worn;
- Cold storage units must be adequately ventilated when not in use.

Once a maintenance task has been completed, all debris, tools and equipment must be removed and the plant left ready for cleaning and sterilizing operations.

Cleaning in place (CIP) consists of circulating water and various chemical solutions through assembled process equipment. In pipelines and other enclosed machines the solutions fill the equipment and each wash pushes out the previous solution. Control valves are used to control the critical operations of solution changes. It is of the utmost importance that the control valves are kept in faultless operating condition. Faults in control valve operation can become the source of serious damage to the products by contamination with detergents.

Fault Diagnosis

Skill in diagnosis involves the identification of the faults which arise in a machine or system, and the quick and accurate location of the cause of the faults. Not only should all the signs and symptoms that can indicate a fault and its cause be recognized, but also a suitable method of collecting this information should be adopted. Whenever possible tests should be carried out to determine whether a series of components is working correctly before examining individual components. To reduce diagnosis time to a minimum the 'half split' approach to fault diagnosis should be used. For example, if there is no output from component E, by testing component C the area in which the fault could lie is halved. If C is functioning correctly, the next component to be tested would be D.

Many machine manuals contain a section which includes fault finding or trouble shooting charts covering a number of known common faults. The information contained in these charts is based on previous experience. Predicted or anticipated faults may also be included, based on the knowledge and experience of the design engineer. The charts normally contain the fault, its cause and remedy. They are a useful aid to fault diagnosis, particularly in relation to product or process faults, and they should always be used.

Examples of Routine Maintenance Procedures

Manuals for craftsmen engaged in any given industry usually contain a selection of servicing and maintenance instructions related to the equipment most often installed in the factories concerned. They are used in conjunction with handbooks and general instruction manuals on mechanical maintenance, whose contents cover the job knowledge required by a skilled fitter. In many developing countries such manuals may not be available in the mother-tongue of the craftsman. Moreover, those available may not be easily understood, or may not apply to the needs of a milk plant. In this section two examples are given of the type of information with which the maintenance crew should be provided during on-the-job training. Both are taken from the Instruction Manual for dairy industry fitters, by permission of the Dairy Trade Federation of the United Kingdom.. These examples deal with securing and fixing devices, and bearings, and demonstrate the preparation of training aids and the conveyance of the instructions to the trainees. The efficiency of the training and subsequently the efficiency of the performance of the maintenance work is greatly improved when each trainee is provided with a carefully prepared manual. Such manual preparation may not be an easy task, but it is continuous work and after a few years a manual of valuable information can be made available if proper attention is given to the problem by the engineering department of the plant.

Using the Half-Split Approach to Locate the Cause

Whenever possible, tests should be carried out to determine whether a series of components is working correctly before examining components individually. To reduce diagnosis is time to a minimum the half-split approach to fault diagnosis should be used.

For example, if there is no output from component E, by testing component C the area in which the fault could lie is halved. If C is functioning correctly the next component to be tested would be D.

Other considerations which should be taken into account when using this approach include accessibility of test points and the odds on a fault occurring in a particular component, i.e., if the odds on the fault lying in A are three times as great as for the other components the first component to be tested should be B.

Checking Functions/Components

When checking functions or components inputs and outputs should be considered.

For example: Input from motor to clutch correct. No output from clutch when engaged. Fault lies in clutch.

Note: The casue may be tight or worn bearing in the gearbox. Adjusting the clutch could temporarily remove fault but it would eventually recur. Cause must be determined and removed.

Securing and fixing devices. These include basically screws, nuts, washers and shims.

Screws are identified by the head type, length and the type of thread. The types of screw heads are shown in. The definition of the screw length is also shown. The basic types of threads include metric, British Standard and self tapping. Except for self tapping types, screws are screwed into tapped holes or used with nuts. Blind holes should be checked to see that there is no swarf in the bottom of the hole and that the screw is not too long. The screw should be 1.5 mm shorter than the depth of the thread plus the thickness of the part being secured. The correct type of screwdriver must be used for fitting the screws illustrates the details of screws and their fitments.

Nuts are manufactured as standard and special purpose nuts. Standard hexagon nuts are used either singly with or without a locking washer, or as full nuts or as lock nuts to lock another nut into a fixed position on a thread. A lock nut is placed under the main nut and tightened. The main nut is run down on top of the lock nut and is tightened with one spanner, whilst a second spanner prevents the lock nut from turning as shown in Fig. 18. Special purpose nuts include castellated nuts for use with split pins, self-locking nuts with nylon inserts and spring steel self-locking fasteners. When using castellated nuts the bolt with nut must be assembled, a clearance hole drilled through the bolt shank in line with the slots, a split pin pushed through the hole until the head meets the nut and the ends of the split pin bent tightly around the nut as shown in. Split pins should be used only once.

Washers are of two main types, illustrated. Plain washers are inserted under screw heads or nuts to provide a suitable bearing surface while tightening and to protect the component from damage due to rotation of the nut or screw head. Lock washers are used to prevent loosening due to vibrations, movement or temperature variation. Common types include:

- single coil spring washers normally made of square section spring steel, cut, twisted and chisel-edged;
- thackerey double oil spring washers made of flat section steel or phosphor bronze;
- snakeproof washers punched from spring steel;
- crinkle washers made from berylium copper and used when a lighter spring pressure is required;
- tab washers used when a more positive locking action is required. (The tab is bent up against the nut face, thus preventing it from turning.)

Liquid sealants are used to provide a permanent lock to a fixing point.

Shims, as illustrated, are used as packing between two machine parts. Selecting shims of the appropriate thickness secures the accurate distance between the two parts. The shim must be flat. Holes can be punched in shims using a close fitting plain punch and two plates bolted and dowelled together.

Bearings are illustrated and include plain bearings, split shell bearings and ball and roller bearings. All types are designed to reduce friction between moving and stationary machine parts. Bearings belong to the most delicate parts of machinery and are easily ruined if their surfaces are damaged. Bearings with damaged surfaces should never be used.

Plain bearings are built as shells or bushes in which shafts run. The material of the shell is always softer than the material of the shafts.

When fitting a bush a clean shouldered mandrel should be used, coated with light oil. The bush fitted on to the correct size of mandrel should be aligned with the housing. After checking that it is square with the hole it is gently pressed home with the lever. The internal dimensions must be checked after completing the fitting, as shown.

Seciring amd Fixing Devices: Screws

Screws are used for securing components and assemblies and are identified by:

1. Head type. Flush fitting screws are used when there is little clearance between assemblies or where protruding heads are not desirable.

   The semi-flush type is used mainly for panel assembly or where a pleasing appearance is required.

   The other full head types are used for general assembly work. As these heads are non-registering, some adjustment can be made to the relative positions of the workpieces by using the full clearance.

2. Screw length. Parts of the screw which protrude above the surface being fixed are not included when defining the length of the screw.

3. Thread type. Various types of thread are available such as:

   (a) Metric
   (b) BA (British Association)
   (c) UNF (Unified National Fine)
   (d) UNC (Unified National Coarse)
   (e) Self tapping.

Screws will normally be screwed into tapped holes or used with nuts.

If the hole is blind, ensure (a) that there is no swarf in the bottom of the hole, or (b) the screw is not too long. The screw should be 0.06 inches (1.5 mm) shorter than the depth of thread plus the thickness of the part being secured.

Ensure that the blade of the screwdriver is a good fit in slotted head screws or damage to the slot will result.

Ensure that the correct type of screwdriver is used when fitting hexagon socket/ cruciform slotted screws.

Nuts

Standard hexagon nuts are manufactured in two thicknesses:

(a) Full nut used singly with or without a locking washer.

(b) Lock nut - used to lock another nut into a fixed position on a thread.

Special-purpose nuts include:

(a) Castellated nuts for use with split pins.

(b) Self locking nuts with nylon inserts.

(c) Spring steel self locking fasteners.

Use of Lock Nuts

A lock nut is placed *UNDER* the main nut and tightened.

The main nut is run down on top of the lock nut and is tightened with one spanner while a second spanner prevents the lock nut from turning.

Note: The practice of placing the lock nut on top of the full nut weakens the assembly as the smaller number of threads in the lock nut take the strain.

Use of Castellated Nuts

(a) Assemble bolt with castellated or slotted nut.

(b) Drill clearance hole through bolt shank in line with slots.

(c) Push split pin through hole until head meets nut.

(d) Trim split pin 'ends' to length.

(e) Bend ends tightly around nut.

Note: Do not use split pins more than once.

When used in conjunction with proprietary lock nuts, the screw should protrude by at least one thread.

Washers

Plain washers are inserted under screw heads or nuts to provide a suitable bearing surface while tightening and to protect the component from damage due to rotation of the nut or screw head.

Lock Washers

Locking washers are used to prevent loosening due to vibration, movement or temperature variation. Common types include:

SINGLE COIL spring washers normally made of square section spring steel, cut, twisted and chisel-edged.

THACKERAY DOUBLE COIL spring washers made of flat section steel or phosphor bronze.

SHAKEPROOF WASHERS punched from spring steel. The outer or inner edge is slit radially in a number of places and the segments in between the slits are twisted so that several sharp edges bear against nut and work.

CRINKLE WASHERS made from beryllium copper are used where a lighter spring pressure is required.

TAB WASHERS, used in special applications where a more positive locking action is required. The tab is bent up against the nut face, thus preventing it from turning.

Liquid Sealants

These may be used to provide a permanent lock to a fixing point.

The sealant should be applied to the point where the thread emerges from the nut, using a small brush or the nozzle of the sealant container. Ensure that shim is flat. Holes can be punched in shim, using a close fitting plain punch and two plates bolted and dowelled together.

(a) Select a punch of the correct size.

(b) Place shim between plates and align position of marked hole with hole in plate.

(c) Place punch in hole and strike with hammer.

(d) Remove punch, release plates and remove shim.

Shims Used as Packing

Bearings reduce friction between moving and stationary machine parts. They are designed for minimal wear, and by being replaceable, save wastage of expensive machine parts.

Handling Bearings

The working surfaces of bearings are either honed or very soft; if these surfaces are damaged, the bearing is ruined, therefore:

1. Handle bearings carefully to prevent damage.
2. Keep bearings wrapped until fitted, to keep out dirt.
3. Protect bearings against corrosion during storage, e.g. steel bearings must be oiled.

Installing Bearings

Before any bearing is fitted:

1. Clean the journal or housing thoroughly and the seatings of the locating devices.
2. Inspect the surfaces for damage, do not fit bearings to damaged surfaces.

Plain Bearings

These are shells or bushes in which shafts run. They are made of different materials from the shaft.

Fitting a Bush Using an Arbor Press

Note: A shouldered mandrel must be used when a bush is fitted, to prevent deformation. Use the correct size of mandrel.

(a) Clean the housing, the mandrel and the base of the arbor press thoroughly.

(b) Coat the housing and the mandrel with light oil.

(c) Index the base of the arbor press round to give the most suitable hole size.

(d) Fit the bush on to the mandrel.

(e) Align the bush with the housing and lower the ram on to the mandrel so that the ram holds the bush in position, above the housing.

(f) Check that the BUSH is square to the HOLE.

(g) Gently press the bush home with the lever.

(h) After fitting, check the internal dimensions.

Note: Some bushes are secured by locking compounds and only a light push fit is required. Indexing arbor press.

Types of Bearings

Split shell bearings fit directly on to the shaft. A split shell bearing is fitted after removing the old shells and after cleaning the journal and the shell seatings. When putting the new shells into their housing, care should be taken for proper alignment of the oil holes. Prior to assembling, all parts must be coated with appropriate lubricants. The shaft should be running freely and the bearing should remain cool under running conditions.

Ball and roller bearings have a stationary and a revolving race between which rolling balls or rollers prevent sliding friction. The stationary race is usually a light push fit whereas the revolving race is a press fit; therefore it is fitted to the machine first.

When fitting a bearing, force is applied only to the race in contact with the housing or shaft. The race must be square to the shaft. Preferably arbor presses should be used for fitting bearings. Figs. 22A/22B/22C show the procedures used, illustrating the pressing of a bearing on to the shaft. The square machined tube should touch only the inner race.

Shows the pressing of a shaft into a bearing supported by the inner race only. The force must be applied to the outer race only when the bearing is pressed into the housing. Bearings should only be tapped into place when they cannot be pressed into position as shown in.

In some instances locating devices are used to position bearings on shafts or in housings. They have to be cleaned, coated with light oil prior to use, and must only mate with the face that has to be located, holding the bearing evenly around its circumference.

Removal of bearings requires careful handling so that seatings and bearings for re-use are removed without being damaged. During removal force must be exerted only on the race being extracted, and pullers should never be clamped to the working surfaces of the bearings. Whenever possible arbor presses should be used for bearing removal, as when fitting bearings into position. If arbor presses cannot be used properly, selected pullers can be substituted for them, as shown. Bearing removal screws are used in some housing equipped with tapped and countersunk holes behind the bearing. The holes are fitted with short screws to keep the threads clean. When removing the bearing these screws should be removed from the holes, and the same number of longer screws fitted and tightened equally by small amounts to remove the bearing slowly from its housing.

Handling and maintenance of bearings, particularly ball and roller bearings, require the greatest possible care, as do all critical mechanical parts, mainly to avoid mechanical abuse and corrosion. As far as possible bearings should be protected against moisture. When moisture is present the selection of proper lubricants can help to run the bearings successfully under these more severe conditions. Discoloration and metal smearing is an indication of a bearing operating with inadequate lubrication. Temperature is usually considered a fair indication of the condition of the bearing, although grease lubricated bearings can operate at temperatures as high as 80° to 90°C. The temperature at which the bearing can work depends mainly on the type of lubricant used. In most cases in milk processing machines the temperature of the bearings does not exceed 50°C under normal operating conditions. This can be felt by placing the land lightly against the housing or the shell of the bearing. If it feels hot, the bearing should be inspected more closely. Wear of properly selected and properly lubricated bearings occurs only when foreign matter mixes with the lubricant and distorts the alignment of the races. Dirt causes most of the early bearing failures.

Ball and Roller Bearings

These bearings have balls or rollers which prevent sliding friction by rolling between the races. As bearings are used between stationary and revolving machine parts, such bearings have a stationary and a revolving race.

Normally, the revolving race is a press fit, so it is fitted to the machine part first, and the stationary race is a light push fit, so it is assembled with the machine part.

Bearings are precision-made of hard, brittle materials. Therefore, when force is applied to fit them:

1. Coat the journal or housing with clean, light oil.
2. Only apply force to the race in contact with the housing or shaft.
3. Keep the bearing square to the housing or shaft.

Fitting a Bearing, Using an Arbor Press

1. Whenever possible, use an arbor press, because it keeps the bearing square to the shaft or housing. Decide which is the best way to set the job up on the press:

   (a) Pressing a bearing onto a shaft. Use a tube to distribute the force evenly to keep the bearing square to the shaft. The tube must only touch the inner race and its ends must be machined square.

   (b) Pressing a shaft into a bearing. The bearing must be supported by the inner race only.

   (c) Pressing a bearing into a housing. A tube or flat block must be used so that the force is applied to the outer race only.

2. Operating the press.

   (a) Set up the part on the base of the press.

   (b) Align the part to be pressed into place and lower the ram to just hold it.

   (c) Check that the parts are square to each other.

   (d) Gently press the bearing home.

   (e) Check the bearing is positioned correctly.

Tapping Bearings into Place

Bearings should only be tapped into place when they cannot be pressed into position. Decide which is the best method.

1. Using the tube and striking block. This is the best method, since the bearing can more easily be kept square to its seating.
2. Using a drift. Tap evenly around the race being fitted. Take care to keep bearing square to seating.

The method is useful when the seating is in an awkward situation. Take care to prevent foreign matter from entering the bearing.

(a) Align the bearing with its seating and start to fit it by hand. A soft mallet may be used if necessary.

(b) Tap the bearing home gently, stopping frequently to check that it is square.

Thrust Races

Thrust bearings are ball or roller bearings specially designed to take end loads only.

Using Bearing Pullers

When using bearing pullers take care to keep the bearing square to the shaft.

(a) Select the most suitable puller:

(i) Impact pullers are only suitable for small bearings; the weight must be held when it is tapped against the stop.

(ii) Screw pullers are suitable for most purposes: take care to keep the puller square when turning the screw.

(iii) Hydraulic pullers are to be used when a large force is required.

(b) Arrange the pullers inside or outside the bearing as required.

(c) Fit the puller and tighten it to hold it in position.

(d) Check that the puller is square and bearing only on the correct surfaces.

(e) Pull the bearing slowly. Stop frequently to check that the bearing puller is square.

Bearing Pullers

When using bearing pullers take care to keep the bearing square to the shaft.

Select the most suitable puller:

1. Impact pullers are only suitable for small bearings; the weight must be held when it is tapped against the stop.
2. Screw pullers are suitable for most purposes: take care to keep the puller square when turning the screw.
3. Hydraulic pullers are to be used when a large force is required.
4. Arrange the pullers inside or outside the bearing as required.

5. Fit the puller and tighten it to hold it in position.
6. Check that the puller is square and bearing only on the correct surfaces.
7. Pull the bearing slowly. Stop frequently to check that the bearing puller is square.

Puller Plates

When a bearing mounted on a shaft is to be re-used, a puller plate should be used so that the force is exerted against the inner race.

Bearing Removal Screws

Some housings have tapped and countersunk holes behind the bearing. These holes are normally fitted with short screws to prevent the threads becoming blocked with dirt.

To remove the bearing:

(a) Remove the screws from the holes.
(b) Select the same number of screws that will fit the holes and are long enough to remove the bearing.
(c) Fit the longer screws and tighten them equally by small amounts to remove the bearing slowly from its housing. Take care to keep the bearing square to its housing.

10

Modern Day Dairy Technology and Milk Supply in India: An Educational Approach

MODERN DAY DAIRY AND MILK SUPPLY: AN OVERVIEW

The development of modern organised dairying and milk supply has been given attention since the inception of the First Five Year Plan in 1951-52. Dairy projects are being set up so as to provide an assured and remunerative market to the producers of milk in the rural areas and to make available depending quality to milk and milk products at reasonable prices to the consumers in the urban areas. Starting with a modest effort in the First Plan, the programmes have been substantially expanded in the subsequent Plans. Presently, there are 123 dairy plants in the country comprising 69 liquid milk plants, 10 milk product factories and 44 pilot milk scheme/rural dairy centres. Cooperatives are playing a significant role in dairy development in the country. Of the 123 plants, 34 are in the cooperative sector. In many States, cooperative constitute a major source of supply of milk to the dairy plants. The pioneering work done by the Kaira District Cooperative Milk Producers' Union, Anand, is being emulated widely. In addition, 26 dairy projects are under different stages of implementation. The average daily through put of milk from all the plants during 1971-72 was 2.4 million litres as compared to 0.7 million litres in 1960-61. In order to make the country self-sufficient in respect of milk products as such as, better, cheese, milk powder, condensed milk, infant milk food and malted milk food, the private sector was also encouraged to set up factories for milk products in potential milk producing areas. The country is now almost self-sufficient in respect of all these milk products except milk powder which is imported to overcome the seasonal shortage in milk production and to maintain the level of milk distribution to the public in the major cities and towns. Increase of milk production will be achieved by provision of such technical inputs as ready mixed concentrates and green fodder, artificial insemination, veterinary services and medicines. The milk cooperative movement is a new phenomenon in India which has in it the seeds of a social revolution. Not only will it inculcate the spirit of cooperative endeavour in the village community; but the cooperatives could become effective instruments for dissemination of information on modern farm techniques, dairy development, family planning, literacy drive and above all, healthy environment. A national milk grid is proposed to be set up in India on the lines of water and power grid. This will be based in areas like Haryana where more milk is produced and will help sending milk to scarcity areas instead of sending milch cattle there. Three methods of distribution of milk were now under consideration. They are the existing system of bottled milk, distribution by bulk vending units and distribution through tetrapack single service disposal packages. Distribution by bulk vending units was being organised particularly for the sake of the poor. Under this system milk will be available to the consumers seven paise cheaper per litre. An assured supply of milk through the development of the dairy industry and animal husbandry on a planned scientific basis is not only an urgent necessity to enable millions of the coming generation to develop a healthy body but is also imperative for stablising the income of rural folk. This is what Operation Flood stands for. If a break-through in agriculture, ushering in the green revolution, was achieved by the use of high-yielding varieties of seeds evolved by Indian and international scientists, a white revolution of milk production, marketing and distribution is sought to be achieved smoothly by buttressing local milk supply with considerable quantities of imported milk powder and butteroil supplied by the World Food Programme, simultaneously with a determined and organised effort at improving the yields of the cattle and buffaloes. In the field of dairy equipment manufacturing, the country has become almost self-sufficient. All equipment, except certain items of higher capacity and a few specialised and sophisticated items, required for the establishment of dairy plants, are now being manufactured in the country. This has enabled the country to save considerable foreign exchange.

Analysing Objectives and Outlays

As some of the existing dairy projects are operating at a loss, one of the principal tasks in the Fourth Plan is to review the work of the projects and to take corrective measures. This will include changes in the milk pricing policy introduction of modern management practices. The desirability of changing the public sector projects from departmental to corporate management will have to be pursued. It will also be necessary to establish a direct link between the small producers and the public sector milk plants through cooperative organizations. Dairy projects will need to be encouraged to take up extension work under their own auspices. At present, a number of dairy projects balance their operations by using imported milk powder. A phased programme is intended to be drawn up to increase production in the milk shed areas and gradually eliminate dependence on imported milk powder. The organised sector of dairy industry will be extended to smaller towns with emphasis on milk production in the rural areas. Measures will be taken to ensure that dairy projects are economically viable and, as far as possible, organised in the cooperative sector. The financing of dairy development will be based on three principal sources, namely Plan outlays, institutional finance and counterpart funds generated by the sale of commodity gifts under the World Food Programme. As far as institutional sources are concerned, the Agricultural Refinance Corporation has already entered the field and financed one dairy project. Further suitable schemes will have to be formulated for financing by ARC. As regard the commodity gifts from the World Food Programme, details are given later. Briefly, it is expected that funds of the order of nearly Rs. 95.40 crores will be generated and will form part of the Plan outlays. On this basis, the total outlay under the Plan will be of the order of Rs 138.97 crores. In the Fourth Plan, the first priority will be to complete the dairy schemes numbering 33 which split over from the earlier period. Organised dairy industry will be extended by taking up 24 new schemes in towns with a population of about 50,000. Four milk product factories are proposed to be established. In addition, 64 rural dairy centres will be organised in areas with a population of less than 50,000 with a view of providing chilling and marketing facilities is isolated pockets of milk production.

Evaluating "Operation Flood" and White Revolution in Modern Day Context

With the cooperation of the World Food Programme (WFP) the Department of Agriculture formulated a project for stimulating milk marketing and dairy development in India. Under this project, known as 'Operation Flood' and launched in 1970-71 the WFP has agreed to supply, free of cost, during the five year period from 1970-71 to 1974-75, 126,000 tonnes of skimmed milk powder and 42,000 tonnes of butter oil, worth Rs 41.90 crores at international price. After recombination of the skimmed milk powder and butter oil into liquid milk at the public sector dairies at Bombay, Calcutta, Delhi and Madras, the milk is being sold and the sale proceeds from the quantity estimated at Rs 95.40 crores will be used for increasing milk processing facilities of the public sector dairies from 1.00 million litres at present to 2.75 million litres per day at the end of five-year project period. The generated funds will also be used for increasing milk production and procurement in the Union Territory of Delhi and the ten neighbouring states. This will be achieved by the provision of technical inputs which will include production of ready mixed concentrates and green fodder, artificial insemination, veterinary services and medicines, calf rearing assistance, development of improved milch animals and organization of rural procurement of milk. The project will also provide for the resettlement of the city kept cattle and buffaloes in the adjacent rural areas. For the implementation of this project, Government has set up an Indian Dairy Corporation with headquarters at

Baroda. Up to the end of March 1972, the World Food Programme has supplied 16,500 tonnes of skimmed milk powder and 3,700 tonnes of butter oil. Funds to the extent of Rs 8.58 crores have been generated and advances aggregating to Rs 4.06 crores have been released to various States for expansion of milk plants, establishment of cattle feed mixing plants and for new dairies.

In order to meet the requirements of the first phase of the expansion programme of the public sector milk plants, estimated to cost Rs. 1.91 crores, the Indian Dairy Corporation has placed indents amounting to Rs. 1.80 crores, on UNICEF for the import of dairy processing equipment and components. The balance of the amount is being made available from free foreign resources. The requirements of plant and machinery for the second phase estimated to cost Rs. 25.26 crores have been discussed with the indigenous manufacturers and the Director General of Technical Development. Operation Flood, perhaps the most spectacular and certainly the largest development project the WFP has undertaken in any part of the world, has other enduring objectives in view. For the first time, farmers have understood their inherent strength through milk cooperatives. They depend no more on the middle man to market their milk and milk products. In specific areas where in project is in operation, they sell their surplus milk twice daily through milk cooperatives in their respective villages and get paid for it on the same day or the next day. With an assured daily income, they are able to feed themselves and their cattle in most cases much better than they did in the past.

Recent studies have shown that farmers have doubled their income in the areas where milk cooperatives are functioning effectively. Milk production will be increased through this and other programmes from 22.5 is 50 million tonnes to ensure that per capita per day availability is increased from the present 108 grammes to 280 grammes necessary for a balanced diet. With the pressure on land, all people living in the villages cannot depend entirely on crops for sustenance. Some could easily be diverted to dairy development. It is estimated that two million people will find jobs as a result of Operation Flood, while ten million people will have additional nutrition. Since dairy development is a labour-intensive industry, it has provided a ready answer for tackling the problem of growing rural unemployment. Moreover, the drift of population to urban areas from rural areas, which is causing considerable concern to the authorities, has chances of being arrested. In view of our ability to grow more food in much less area than in the past through high-yielding varieties of seeds, we are in a position to spare land for cattle and fodder development on an organised and scientific basis. Large scale cross-breeding of cows is expected to increase cow milk production although implementation of the programme already initiated will take some three or four years. The programme of rearing high-yielding cattle involves growing of fodder and producing other animal feed in areas close to milk sheds. About 18,75,000 milch animals will be involved in the project. The Punjab Government has decided to set up a milk plant in every district. The output is to be doubled in the next ten years. In addition to the existing units a dairy complex is being established at Ludhiana with UNICEF assistance at a cost of Rs 2 crores with an installed capacity of 75,000 litres a day. Another milk complex is planned for Bhatinda with an investment of Rs 1.5 crores. To absorb and canalise milk produced in the rural areas, composite milk plants are proposed to be set up at Jullundur, Gurdaspur and Patiala. The Amritsar factory is to be expanded to handle 2 lakh litres of milk per day. The total cost of all these schemes is estimated to be of the order of Rs 44 crores. Haryana's plans are no less ambitious. Its Government has provided for the commissioning of a chain of milk plants within the next three years. The State's milch animal population is nearly 75 per cent that of a premier dairy country like New Zealand. India's first wholly indigenous sweetened condensed milk plant became operative a Bhiwani in October last. A composite milk plant was commissioned at Jind in December 1970. The third is nearing completion at Ambala. Besides Punjab and Haryana, several other States have taken steps to develop the dairy industry. The Rajasthan Government has a Rs. 6.5 crore dairy development plan. It envisages the setting up of milk banks at Jodhpur and Bikaner, each handling 100,000 litres of milk every day. Bharatpur and Alwar will have milk feeder centres, each handing 60,000 litres per day. West Bengal's Rs 2.4 crores scheme for milk production has already been put into operation and envisages a gross output of 13.10 lakh litres per day by the end of the Fifth Plan. Kerala proposes to turn the State into a large milk-shed by the end of this decade, the aim being to provide each citizens with at least 10 ounces of milk daily.

The Jawaharlal Agricultural University at Jabalpur is engaged in experiments to initiate a milk revolution in Madhya Pradesh. Its experts claim that artificial mating of even the normal village cow with an exotic breed can substantially raise milk yields of the progeny, especially if backed up with nutritious fodder.

THE NATIONAL DAIRY RESEARCH INSTITUTE: CASE STUDY

The best for a break-through in the dairy industry is the region comprising Punjab, Haryana, Western U.P. and Rajasthan. Before partition. India's finest cattle breeding tracts were in Sind, Montgomery and Lyallpur which now from part of Pakistan. Fortunately for us, sub continent's premier and first dairy research station, then known as the Imperial Institute of Animal Husbandry and Dairying, had been located at Bangalore in 1923. After independence it was shifted to Karnal and renamed as the National Dairy Research Institute (NDRI). Bangalore became the Southern Regional Station of the Institute. The Western and Eastern Regional Stations are located at Aarey (Bombay) and Kalyani (near Calcutta). The existence of the NDRI at Karnal has given a big boost to dairy development in the Punjab-Haryana region. The two States are varying with each other in raising their output of milk and its products. At present Punjab-produces 50 lakh litres a day while Haryana's daily output is 35 lakh litres. The Karnal institute has been helping dairy farmers, particularly in Punjab and Haryana, in a big way. During the 50 years of its existence, it has concentrated on research leading to the development of high yielding dairy animals and fodder cropping programme to provide nutrients necessary for milk production in the most efficient manner and at minimum cost. The NDRI's most outstanding product is a new strain of dairy cattle, Karan Swiss. It incorporates exotic inheritance from Brown Swiss and Zebu inheritance from the famous Sahiwal and Red Sindhi. A nucleus herd of 140 Karan Swiss cows along with 187 female followers and bulls has been developed for providing germ plasm. The Karan Swiss female calves weigh about 27 kgt a birth. They are 25 to 30 per cent heavier than their contemporary pure bred Zebu. The average age at first calving for Karan Swiss is 30 months as against 40 months for Zebu. On an average Karan Swiss produces 60 per cent more milk than the Sahiwal.

Understanding Milk Chemistry

One of the successful fodder cropping programmes evolved at the institute's experimental farm consists of a mixture of jowar, bajra, sudan, teosnite and cowpea in the kharif season and a mixture of bersem, mustard and oats in the rabi season, yielding a gross of 200 tonnes of green fodder a year per hectare sufficient to support 10 cows. Another programme comprising four crops a year has a yield of 150 tonnes per hectare. Special attention has been paid to reducing the cost of milk production. While in the Sahiwal herd averaging 2,000 litres of milk per lactation it is Rs. 1.10 per litre, for the Karan Swiss herd averaging 3,000 litres of milk per lactation it is Re 0.72. The work of the institute has clearly established that for dairying to become an industry, the cost of production of milk has to be lowered to give adequate margin to the producer. The NDRI has worked on buffalo milk chemistry and technology to provide the base for developing processes for manufacturing milk products from buffalo milk as satisfactorily as from cow milk. Sweetened condensed milk and evaporated milk of satisfactory quality cannot be produced from buffalo milk in view of its different chemical composition and behaviour when conventional methods of manufacturing are used.

Re-inventing Long Term Dairy Technology

A problem of considerable importance in handling milk is its rapid transport from the place of production to the dairy. The work at the institute suggested the feasibility of concentrating milk for long distance transport and this innovation has been adopted by the Delhi Milk Scheme in getting its supplies from Gujarat. With a view to finding new outlets for surplus milk, researches have demonstrated the feasibility of keeping Dahi over long intervals at room temperatures, preparation of cheese with various combinations of condiments and fruits, "complete" coffee, ready-made tea in powder form, sterilised cream, preserving butter milk and preparing refreshing drinks from milk and sour milk. The research programme at the NDRI is being further strengthened by providing a larger herd of buffaloes so that all aspects of the nutritional physiology and breeding behaviour of the buffalo are studied. So far, cross-breeding between the different breeds of buffaloes has not been attempted on scientific lines. A new bio-manufacturing sector will be set up to prepare on a semi-commercial scale some of the materials which have shown promising results in the laboratories. One such material is animal and microbial 'rennet'. Animal rennet is prepared in the Western countries by slaughtering young calves. The NDRI has shown that rennet can be prepared from living calves by putting a fistula in the young animal which could be removed at a later date without any ill-effects on the subsequent growth of the animal. At present, there is a shortage if rennet even in European countries and the finding has good commercial potentials. Process for the manufacture of microbial rennet has also been standardised and rennet from this source could also be prepared on a commercial scale. Recently, a machine has been devised for the continuous manufacture of Khoa from milk which dispenses with considerable amount of manual labour required now. The packaging of milk products is a subject which has hardly been in India. A dynamic marketing programme is closely related to packaging. Emphasis will be given to this aspect which will also include gamma-rays for sterilising the material.

Milk Bottle

Milk bottles are reusable glass bottles used mainly for doorstep delivery of fresh milk by milkmen. Customers are expected to rinse the empty bottles and leave on the doorstep for collection. The standard size of a bottle varies with location; common sizes are 1 litre, 1 pint or 1 quart, although cream may be delivered in smaller bottles. They are a popular collector's item.

For centuries, milk has been carried and stored in glass containers. Before milk bottles, milkmen filled the customers' jugs. For many collectors, milk bottles carry a nostalgic quality of a bygone age. The most prized milk bottles are embossed or pyroglazed (painted) with names of dairies on them, which were used for home delivery of milk so that the milk bottles could find their way back to the dairy for reuse. The color, picture, dairy, and condition all contribute to the value of the milk bottle.

It is not clear when the first milk bottles came into use. However, the New York Dairy Company is credited with having the first factory that produced milk bottles, and the first patents for a milk container is held by the Lester Milk Jar on January 29, 1878 - US patent number 199837, filed on September 22, 1877). There are many other similar milk containers from around this period, including the Mackworh Pure Jersey Cream crockery type jar, the Manorfield Stock Farm, the Manor, the Pa glass wide mouth jar, and the Tuthill's Dairy Unionville, NY.

Lewis P. Whiteman holds the first patent for a glass milk bottle with a small glass lid and a tin clip (US patent number 225,900, filed on January 31, 1880). The next earliest patent is for a milk bottle with a dome type tin cap and was granted September 24, 1884 to Whitemen's brother, Abram V. Whiteman (US patent number 305,554, filed on January 31, 1880). This bottle has been found with cream line marks and is very valuable. The Whiteman brothers produced milk bottles based on these specifications at the Warren Glass Works Company in Cumberland, Maryland and sold them through their New York sales office.

The Original Thatcher is one of the most desirable milk bottles for collectors. The patent for the glass dome lid is dated April 27, 1886. There are several variations of this early milk bottle and many reproductions. During this time period, many types of bottles were being used to hold and distribute milk. These include a pop bottle type with a wire clamp, used by the Chicago Sterilized Milk Company, Sweet Clover, and others. Fruit jars were also used, but only the Cohansey Glass Manufacturing plant made them with dairy names embossed on them.

The commonsense milk bottle with the first cap seat was developed as an economical means for sealing a reusable milk bottle by the Thatcher Manufacturing Company around 1900. Most bottles produced after this time have a cap seat.

Milk bottles before the 1930s were a round shaped bottle. In the 1940s, a square squat bottle become the more popular style. Milk bottles since the 1930s have used pyroglaze or ACL (Applied Color Label) to identify the bottles. Before the 1930s, names were embossed on milk bottles using a slug plate. The name was impressed on the slug plate, then the plate was inserted into the mold used to make the bottle - the result was the embossed name on the bottle. By the 1960s, glass bottles had been replaced with paper cartons.

Chronology

- 1880 - British milk bottles were first produced by the Express Dairy Company. They were delivered by horse-drawn carts and delivered four times a day. The first bottles used a porcelain stopper top held on by wire.
- 1894 - Anthony Hailwood developed a pasteurization process for milk which allowed it to be sterilized and be safely stored for longer periods. Milk could now be delivered once a day.
- 1920 - Advertisements began to appear on milk bottles. A sand-blasting technique was used to etch them on the glass.

- mid 1950s - Cardboard tops were deemed unhygienic and banned in some locations.
- early 1990s - The advertising largely disappeared with the introduction of infrared bottle scanners designed to check cleanliness.

Present Day

In some locations, silver, red, blue or yellow aluminium tops on today's bottles indicate the fat content. Unpasteurised is green-topped. Other dairies use other color designations. Bottles may also be marked or stamped with the name of the dairy.

Modern dairies may also use refillable plastic bottles, as well as plastic bottle tops.

Developing Future Approach

Complete success in the field of dairying is possible only when milk, which is an essential item of food, is made available in adequate quantities to the different sections of the population including children, expectant and nursing mothers and invalids. To reach this target, which is an increase of about 100 per cent over the present production, a comprehensive integrated programme is required to improve the productive efficiency in cattle and buffaloes through improved breeding, feeding and management practices. The recent launching of six all-India coordinated research projects in the breeding of cattle and buffaloes, and on the technology of milk and milk products by the Indian Council of Agricultural Research is the beginning of a new era in dairy research. Milk trade is at present generally disorganised and only about three per cent of the milk total production is in the hands of the organised sector of the industry. The effectiveness of the process of diverting the milk from unorganised sector will depend not only on increased milk production, but also on the effectiveness of the link between dairy plants and the milk producers in their milk shed areas. This link can be established more efficiently by setting up small dairies in towns with a population ranging from 25,000 upwards. The problem of starting small dairies is closely linked up with the procurement of surplus milk from villages which are cut off from the cities due to inadequate transportation facilities and milk collection agencies and it is expected that this will be solved through a project known as Operation Flood Programme launched by the Department of Agriculture through the funds made available from World Food Programme. Apart from the development of dairy industry, a great deal of fundamental work in dairy research is needed to improve the quality of milk products like dahi, khoa, cheese, chhenna, ghee, butter, ice cream, milk powder, etc. It is of upmost importance that there should be coordination between research institutions and the results of research should percolate into actual operations through the channels of extension and management. The speedy implementation of the dairy projects in the coming decade will, it is hoped, substantially contribute to the development and progress of dairy industry and will help India occupy a leading position in the Dairy Map of the world.

Revisiting the Problem of Stray Cattle in India

The magnitude of the problem of stray cattle in India is anybody's guess. According to one estimate their number in the country is 461 thousand only. *(The Dying Cow* by Bajrang Bahadur Singh Bhadri, pp. 2-3, published by Deptt, of Animal Husbandry, Govt. of Himachal Pradesh.) This estimate, however, seems to grossly err on the conservative side. On the basis of an average of five stray cattle per inhabited village in the country the number of stray cattle found in rural India alone would come to (5,64,718 inhabited villages x 5) about 28,24,000. Assuming that all male and female cattle over three years and not in use for breeding or work, all female dry cattle and the cattle not calved even once, and all female youngstock, in urban India are likely to be stray cattle, their total would come to 44,19,000 heads. That means the total number of stray cattle for the entire country is likely to be in the neighbourhood of 72,43,000 heads. On the other hand the percentage of stray cattle to the total cattle in rural India is as low as 1.3, here the assumption being that there are no an average only five stray cattle per inhabited village. Taking the entire country (urban + rural) the percentage of stray cattle to the total number of cattle is about 2.4 assuming that (i) all male cattle over three years and not in use for breeding or work, (ii) all female cattle over three years excluding those in milk and those used for work, and (iii) all female youngstock, so astray or are likely to do so. The percentage of stray cattle to the total cattle would go down a bit if we exclude from the former all male and female cattle not used for work or breeding. The incidence of stray cattle varies widely not only between urban and rural India, but also between different states of the Indian Union. According to Bajrang Bahadur Singh, the incidence of stray cattle is the highest in Uttar Pradesh, closely followed by Madhya Pradesh and Punjab, and the lowest in Maharashtra, Jammu & Kashmir and Delhi. But, as stated at the very out-set of this chapter, all estimates of stray cattle, whether in terms of their magnitude or incidence, are anybody's guess. There exists no authentic study on the subject. What is however, fairly obvious is that the problem of stray cattle is primarily an urban problem. In rural India the incidence of stray cattle is, however, likely to be large during the years when, and in areas where, rains fail, droughts set in, crops are poor or fail, giving rise to a serious scarcity of *bhusa* and other roughages, and also bringing about in its wake an upward spiral of their prices. As a result being unable to feed their cattle, the more hard-up among the farmers let them stray. This also happens when there are widespread floods or a wave of some epidemic hitting cattle as well as those tending them.

(a) Raison D'etre of Stray Cattle

In rural India a person does not normally buy food for his cattle daily. If he is a farmer, he puts aside straw, *bhusa,* etc., for his cattle's requirement of a year after he has harvested his crops. He, in fact, seldom incurs any out-of-pocket expenditure on roughages for his cattle. He certainly has not to do it every day as does the owner of a cattle in a town or a city. Besides, in rural India there are common/ *panchayat* grazing grounds where all the villagers send their cattle for the day on a nominal monthly charges. This arrangement is conspicuous by its absence in urban India. Towns and cities do not have common grazing ground over by municipality or Corporation where its cattle can graze and be left unattended every day either free of cost or at nominal charge to their owners. Besides, as stated above, their owners with urban avocations, have to buy *bhusa, kutti,* etc. every day for these cattle. This out-of-pocket expenditure irks. It irks all the more when a cattle yields no milk. There is, thus, a strong temptation for their owners to let them fend for themselves, if possible. This temptation becomes irresistable when their cattle is dry. It is this economic squeeze, abetted by the absence of common grazing grounds for cattle, that largely explains the *raison d'etre* for the presence of stray cattle in urban India.

(b) Economic Aspects of Stray Cattle

The cost of upkeep of stray cattle can be looked at from the point of view of the individual, *i.e.* the owner, as also from the point of view of the society, *i.e.* the country. From an individual's point of view, stray cattle fend for themselves. Their owners let them loose at day-break and tie them to their pegs at the end of the day. They feed them with practically nothing so long as these cattle are dry or not calved. Their cost of upkeep to them is, thus, practically nil or negligible. From the point of view of society, however, it is not so. Stray cattle, as other cattle, fill their belly. They eat mostly wild grass, rubbish and such feeds as are given to them out of charity or which they can poach upon. The monetised cost of this varied feed is a cost on the GNP. Stray cattle, thus, present the paradox of their cost of upkeep being nil or negligible to their owners but real, as of other cattle, to the country, *i.e.* on the GNP.

The per capita cost of upkeep of stray cattle on the GNP is, however, not of the same magnitude as of the non-stray cattle or cattle regularly maintained by a farmer or owner. It is so because stray cattle often eat up what would have in all probability otherwise gone waste or unutilised. Also stray cattle is seldom fed with concentrates. It is thus observed that the per capita cost of "upkeep" of stray cattle is in all probability likely to be much lower than the per capita cost of upkeep of cattle regularly kept and maintained. In short the per capita cost of upkeep of stray cattle even to the country is not likely to be of any significant magnitude as compared to the per capita cost of upkeep of regular cattle. While the per capita cost of upkeep of stray cattle to the Nation is much less as compared to the per capita cost of upkeep of regularly maintained cattle, the former give the Nation almost all those benefits which the latter do. Thus, dung, meat, bones, hides, hoofs and horns of stray cattle are of as much use as are of the regularly maintained cattle. In some cases they also give as much milk as the other do. In short stray cattle are almost as useful to the Nation as are the regularly maintained cattle, and besides at a lower per capita cost, the presumption being that they cattle are normally healthy, productive, etc. The view is held that if stray cattle does not get adequate feed for itself through foraging it is likely with every lactation its milk yield would go down and its capacity to breed and beget efficient draught bullock would suffer. ("Dairying for development of the cow," L.C. Sikka, *Building From Below—Essay on India's Cattle Economy,* I.C.A.R. Publication, pp. 20.) Also, it is often presumed that stray cattle might not be picking up adequate for its needs. Both these views, however, seem to be based apparently on commonsense only for there is not enough statistical evidence in support of these. If anything the available evidence points to the contrary. Even Sikka has observed that green grass is generally picked up by stray cattle in its most succulent state, when it has high nutrition value." *(Ibid., p.* 25.) With an abundant feeding of livestock, researchers have noticed, more of females of the species are born and *vice versa.* ("Boy or Girl?" Igar Goldman, *The Hindustan Times,* Sunday, April 28, 1968, p. 8.) If what Goldman points out is correct, and if the presumption that stray cattle must not be picking up adequate for its needs, is valid, the chances are that stray cattle would give birth to more of males of the species and the well-tended kept cattle to more of the females of the species. The whole subject of stray cattle and the issues relating to it thus remain to be explored and examined. Much work needs to be done on the subject.

COURSE AND INSTITUTES IN INDIA ON DAIRY TECHNOLOGY: AN EYE-OPERATION

Dairy Technology deals with all methods of handling milk from production and consumption and includes processing, packaging, storage, transport and physical distribution. Based on the sciences of biochemistry, bacteriology, and nutrition, Dairy Technology employs the principles of engineering. Its objectives are to prevent spoilage, improve quality, increase shelf-life, and make milk palatable and safe for human consumption.

The foundation of Anand Milk Union Ltd, more well known by its acronym AMUL, in 1946 led to the development of the dairy industry and gave momentum to education in dairying. Before that, there was not even a single college offering exclusive graduate degree programme in dairying. Traditionally it was the part of veterinary and animal husbandry courses. The first dairy science college was established at the National Dairy Research Institute in Karnal (Haryana) which now offers diploma, undergraduate and postgraduate courses in dairy technology.

Educational Opportunities

At present, there are eleven dairy science colleges which offer first degree level courses leading to BTech (Dairy Technology) of BSc (Dairy Technology). Of these, National Dairy Research Institute (Karnal and Bangalore) and the Sheth MC College of Dairy Science (Anand) have been identified as centres of excellence. The Karnal and Anand colleges have established modern commercial dairies with the financial support of National Dairy Development Board (NDDB). An automated dairy plant having a capacity of one lakh litres named Vidya Dairy became operational at the Dairy Science College, Anand in 1994. All operations in this dairy are carried out by students under the supervision of dairy staff and teachers.

The Bachelor's degree course is of four-year duration. The eligibility requirement is a pass in 10+2 with physics, chemistry and mathematics. In most of the States the admission is made on the basis of entrance test common to other agricultural and animal science courses. There are also diploma level courses in the subject. Three well-known diploma courses leading to Indian Dairy Diploma (IDD), are offered by the Dairy Science Institute (Aaray, Mumbai), Allahabad Agricultural Institute, and the State Institute of Dairying (Haringhata, West Bengal), NDRI (Bangalore) has a National Dairy Diploma (NDD) course of two-year duration. Many general universities now offer dairy science as vocational subject at the BSc level. Dairying is also available as a vocational subject for the 10+2 level education.

Master's degree courses in dairy science and dairy technology are offered by 17 institutions which include several agricultural colleges, colleges of veterinary science and animal husbandry. The nomenclature of the awards varies viz., MSc (Dairy Science), MVSc (in Animal Husbandry and Dairying). Admission to postgraduate programmes in such disciplines as Dairy Chemistry, Dairy Microbiology. Quality Control is also offered by Dairy Science Colleges at Karnal and Anand which are also open to general science graduates. In the Indian Institute of Technology (Kharagpur), Dairy Engineering is offered as part of BTech (Hons) in Agricultural and Food Engineering and MTech in Dairy and Food Engineering besides PhD in Dairy Engineering and Dairy Technology. Six universities have introduced PhD courses. Annexure 1 gives a comprehensive list of institutions offering courses at different levels.

Career Opportunities

The establishment in 1965 of the National Dairy Development Board (NDDB) and the promotion of the "Operation Flood" scheme by it, gave a great fillip to the dairy industry in India. The spectacular growth of the dairy industry in the last two decades had created various demands for indigenous production of dairy equipment, increased quality standards and production of varieties of milk products. This has resulted in the progressive development of dairy equipment manufacturing industry and technical consultancy organizations and have also given impetus to research and teaching. Such notable expansion of the dairy processing and related industries has opened up vast career opportunities for dairy technologists and engineers.

There are now more than 400 dairy plants in the country making various types of milk products. They need good qualified and well trained personnel to run the plants efficiently. The growth of the dairy processing industry has also encouraged the indigenization of dairy equipment. Most requirements of the dairy industry are now met indigenously. Presently, there are over 170 dairy equipment manufacturers. Many of them undertake turnkey jobs. The industry has two distinct kinds of jobs: (1) equipment design and fabrication and plant design; and (2) project execution. These jobs can be performed by a dairy technologist with an aptitude in engineering.

The dairy technologists can also undertake consultancy work. A successful consultant, however, needs several years of working experience in dairy firms to understand the nitty gritty of the work. Besides opportunities for teaching and research, dairy technologists can start their own enterprises such as small-scale milk plants, creamery, ice-cream units. Dairy India (A – 25 Priydarshini Vihar, Delhi 110092) is an excellent source of information about dairy education and industry. It is frequently updated.

List of Institutions

1. Archarya N.G. Ranga Agricultural University, College of Veterinary Science (Dairy Technology Programme), Tirupati - 517 502
2. Allahabad Agricultural Institute, Allahabad - 211 007
3. Andhra Pradesh Agricultural University, Rajendra Nagar, Hyderabad - 500 030
4. Dairy Science Institute, Asrey Milk Colony, (Government of Maharashtra), Mumbai - 400 064
5. Dr. Panjabrao Deshmuk Krishi Vidyapeeth, College of Dairy Technology, Warud (Pasad), PostMoha - 445 204
6. Jawaharlal Nehru Krishi Vishwavidyalaya, College of Agricultural Engineering, Adhartal, Jabalpur - 4
7. Kerala Agricultural University, College of Veterinary and Animal Sciences, Mannuthi, Thrissur

8. National Dairy Research Institute, Indian Council of Agricultural Research, Karnal - 132 001
9. National Dairy Research Institute, Southern Regional Station, Bangalore
10. Rajasthan Agricultural University, College of Dairy Science, Udaipur - 313 001
11. Sanjay Gandhi Institute of Dairy Technology, Lohianagar, Patna - 800 020
12. Sheth M.C. College of Dairy Science, Anand Campus, Anand - 388 110
13. University of Agricultural Sciences, Bangalore Dairy Science College, Hebbal, Bangalore - 560 024
14. West Bengal University of Animal & Fishery Sciences, Faculty of Dairy Technology, Mohanpur Campus, PO Krishi Vishwavidyalaya - 741 252.

CAREER CENTER IN DAIRY TECHNOLOGY & MANAGEMENT: AN INDIAN PERSPECTIVE

For most of us, doodhwala is the first person to come across every morning. But did we ever wonder about the various roles he performs? He is not only breeder of the cattle, but also milcher of the cattle, distributor, and even promoter - all combined into one person.

But today with the advancement of higher education, the professional skills are being taught at different level of studies. The dairy management is one such diversified field consisting of many tasks and involving many principles. Also with the introduction of modern technology, the diary sector has now been given the status of a full-fledge industry.

Application of the principles of management sciences to the field of Dairy industry is the main feature of modern dairy management. India is the second largest milk producer in the world and the white revolution has made milk as one of the pillars of the Indian economy. It plays a dynamic role in India's agro-based economy. Whereas earlier there were few cooperatives, now with the active role played by the state, there are cooperatives of milk producers in every state. These cooperatives not only protect the small milk producers but also manage to bring a change in the production and storage and transportation to selling of milk.

In the past few decades Government of India has taken great care of dairy production and as a result management in dairy farming has been modeled on professional line. Dairy industry is now not only caring about small dairy farmers but about big business. Right from the procurement of the cattle to its breeding, great care is taken of. The processing of milk, its packing, its transportation to various markets, its constant supply etc. requires a professional approach.

The focus of dairy management course is on providing basic input to students about production, planning and management of dairy farms, entrepreneurship development in milk preservation, entrepreneurship development in dairy processing and management of dairy farm, co-operative and industry.

Eligibility

Candidates who wish to opt for various courses in dairy technology should have passed 10+2 with Science with a minimum required percent of aggregate marks. The selection for such courses is done on the basis of an entrance examination conducted at the National or state level. After successful graduation in dairy technology courses, one can pursue a master's programme in dairy engineering or any other dairy technology discipline. For postgraduate courses like M.Tech in dairy Technology, generally universities ask for B.Sc or B.Tech degree qualification.

Moreover, Candidates who have completed B.V.Sc. and Biotechnology courses are exempted from graduation courses in dairy technology. Such candidates can directly opt for master's degree programmes.

Note: Dairy technology and management is taught both at undergraduate and postgraduate level. The study gives emphasis both on technological and management aspects of dairy science.

Remuneration

Though, initially pay packet in this field of work is not as lucrative as other management professions, but there is tremendous satisfaction associated with this field. One can expect a starting salary somewhere between Rs 6,000-10,000 per month.

Institutes/Universities

Diploma in Dairy Husbandry and Dairy Technology course of two years duration is offered by Allahabad Agricultural Institute.

Contact Information

Allahabad Agricultural Institute Deemed University

Allahabad 211007, U.P

Phone: +91 532 2684281,

Fax: +91 5322684394,

E-mail: registrar@aaidu.org

Website: http://www.indiaeducation.net/rd.asp?url=http://www.aaidu.org

National Dairy Diploma (Dairy Tech.) of 2 years duration is offered at Southern Regional Station, National Dairy Research Institute, Bangalore.

Contact Information
NDRI Deemed University
Karnal-132 001
Phone (Office): 0184-2252800,2259002
Fax: 0184-2250042
Website: http://www.indiaeducation.net/rd.asp?url=http://www.karnal.nic.in/res_ndri.asp

Archarya N.G. Ranga Agricultural University

Contact Information
Acharya N G Ranga Agricultural University
Rajendranagar,
Hyderabad-500 030, Andhra Pradesh.
EPBAX: 040-24015011 to 040-24015017
E-mail: angrau@ap.nic.in
Website: http://www.indiaeducation.net/rd.asp?url=http://www.angrau.net

College of Dairy Technology, Nagpur

(Maharashtra Animal and Fishery Sciences University),
Contact Information
Warud(Pusad), Distt. Yaovatmal 445 204
Seminary Hills, Nagpur- 440 006,
E-mail: dtc@mafsu.in

Jawaharlal Nehru Krishi Vishwavidyalaya

Contact Information
College of Agricultural Engineering,
Adhartal, Jabalpur – 4
Phone: 0761- 2681235 (O), 0761- 2361230, 9893287025
Fax: 0761- 2681235
E-mail: dswjnkvv@rediffmail.com

Kerala Agricultural University

Contact Information
Kerala Agricultural University
Mannuthy - 680 651
Thrissur, Kerala, India
Phone: 0487 2370051
Fax: 91 0487 2370150 E-mail: pro@kau.in
Website: http://www.indiaeducation.net/rd.asp?url=http://www.kau.in

Rajasthan Agricultural University,
Bikaner,Rajasthan ,India
Phone: 0151-2250025
Fax: 0151-2250025
Email: reg@raubikaner.org
Website: http://www.indiaeducation.net/rd.asp?url=http://www.raubikaner.org

University of Agricultural Sciences,
Krishinagar, Dharwad
Postal Pin: 580005
Karnataka
Phone: +91-836-2447422
Fax: +91-836-2745276
Website: http://www.indiaeducation.net/rd.asp?url=http://www.uasd.edu/

Bibliography

Abbott, F. V., Franklin, K. B., and Connell, B. (1986). The stress of a novel environment reduces formalin pain: possible role of serotonin. *European Journal of Pharmacology* 126, 141-144.

Abbott, F. V., Franklin, K. B., and Westbrook, R. F. (1995). The formalin test: scoring properties of the first and second phases of the pain response in rats. *Pain* 60, 91-102.

Abbott, F. V., Ocvirk, R., Najafee, R., and Franklin, K. B. (1999). Improving the efficiency of the formalin test. *Pain* 83, 561-569.

Afzal, S., R. Ahmad, T. Hussain, and G. Jilani. 1994. Developments for EM-technology to replace chemical fertilizers in Pakistan. In Third Conference on Effective Micro-organisms (EM). Kyusei Nature Farming Center. Saraburi, Thailand. pp. 45-66.

Amer, P., and G. Fox. 1995. Imputing input characteristic values from optimal commercial breed or variety choice decisions: comment. *American Journal of Agricultural Economics* 77 (4): 1054-1058.

Anand, K.J.S. & McGrath, P.J. (Eds) (1993) *Pain in neonates.* Publ: Elsevier Science, Amsterdam.

Anderson, S. 2003. Animal genetic resources and sustainable livelihoods. *Ecological Economics* 45 (3): 331-339.

Angniman, P.A. 1996. *Privatization of veterinary services within the context of structural adjustment in Mali, Cameroon and Chad.* Rome, FAO.

Ashley, S.D., Holden, S.J. & Bazeley, P.B.S. 1996. *The changing role of veterinary services: a report of a survey of chief veterinary officers' opinions.* Somerset, UK, Livestock in Development.

Association of American Feed Control Officials, Feline Nutrition Expert Subcommittee revised their Cat Food Nutrient Profiles, 1995. CARE SOPs: 542, 544, 605, 307, 538 and 547.

Ayalew, W., J. M. King, E. Bruns, and B. Rischkowsky. (2003). Economic evaluation of smallholder subsistence livestock production: lessons from an Ethiopian goat development programme. *Ecological Economics* 45 (3): 473-485.

Babjee, A.M. 1996. *Privatization with reference to the veterinary services in Malaysia*. Paper presented at the FAO/GTZ Workshop on Systematic Improvement of the Efficiency of Public and Private Livestock Services in Asia, 22-26 April 1996, Bangkok, Thailand.

Back, W. & Clayton, H. (2000) *Equine Locomotion*. Publ: W.B. Sanders, Harcourt Health Sciences, London.

Baker, J.L. (1995). *Profile of veterinary services in New Zealand*. Rome, FAO.

Barbano, D (2007). bST Fact Sheet. Monsanto. Retrieved on 2008-01-16.

Bateson, P. (1991). Assessment of pain in animals. *Animal Behaviour*. 42, 827-839.

Baur, J.A.; Pearson, K.J.; Price, N.L.; Jamieson, H.A.; Lerin, C.; Kalra, A.; Prabhu, V.V.; Allard, J.S.; Lopez-lluch, G.; Lewis, K.; Others, (2006). "Resveratrol improves health and survival of mice on a high-calorie diet". *Nature* 444: 337-342. doi:10.1038/nature05354.

Berriatua, E., French, N. P., Broster, C. E., Morgan, K. L., and Wall, R. (2001). Effect of infestation with Psoroptes ovis on the nocturnal rubbing and lying behaviour of housed sheep. *Applied Animal Behaviour Science* 71, 43-55.

Braum's Milk - We Believe in Natural. (press release). Braum's (2006). Retrieved on 2008-01-29.

Breeds of Livestock - Sheep: (Ovis aries). *Oklahoma State University*. Retrieved on 2006-08-02.

Brent RL (2004). "Utilization of animal studies to determine the effects and human risks of environmental toxicants (drugs, chemicals, and physical agents)". *Pediatrics* 113 (4 Suppl): 984–95.

Brown SL, Brett SM, Gough M, Rodricks JV, Tardiff RG, Turnbull D (1988). "Review of interspecies risk comparisons". *Regul. Toxicol. Pharmacol.* 8 (2): 191–206.

Campbell, B. M., D. Doré, M. Luckert, B. Mukamuri, and J. Gambiza. 2000. Economic comparisons of livestock production in communal grazing lands in Zimbabwe. *Ecological Economics* 33 (3): 413-438.

Chan JM, Stampfer MJ, Giovannucci E, *et al* (1998). "Plasma insulin-like growth factor-I and prostate cancer risk: a prospective study". *Science* 279 (5350): 563–6. doi:10.1126/science.279.5350.563. PMID 9438850.

Cheneau, Y. 1984. Towards new structures for the development of animal husbandry in Africa south of the Sahara. *Rev. Sci. Tech. Off. Int. Epiz.*, 3(3): 621-627.

Cheneau, Y. 1985. The organization of veterinary services in Africa. *Rev. Sci. Tech. Off. Int. Epiz.*, 5(1): 107-154.

Cicia, G., E. D'Ercole, and D. Marino. 2003. Costs and benefits of preserving farm animal genetic resources from extinction: CVM and bio-economic model for valuing a conservation programme for the Italian Pentro horse. *Ecological Economics* 45 (3): 445-459.

CIP-UPWARD. 2003. *Conservation and sustainable use of agrobiodiversity: a source book.* Los Baños, Laguna, Philippines: CIP-UPWARD in collaboration with GTZ, IDRC, IPGRI and SEARICE.

Clark, C. W. 1973. Profit maximization and the extinction of animal species. *The Journal of Political Economy* 81 (4): 950-961.

Clutton Brock Juliet, *Horse power: a history of the horse and donkey in human societies*, National history Museum publications, London 1992

Clutton Brock Juliet, *The walking larder. Patterns of domestication, pastoralism and predation*, Unwin Hyman, London 1988

Coderre, T. J., Fundytus, M. E., McKenna, J. E., Dalal, S., and Melzack, R. (1993). The formalin test: a validation of the weighted-scores method of behavioural pain rating. *Pain* 54, 43-50.

Collier RJ, Miller MA, McLaughlin CL, Johnson HD, Baile CA (2008). "Effects of recombinant bovine somatotropin (rbST) and season on plasma and milk insulin-like growth factors I (IGF-I) and II (IGF-II) in lactating dairy cows". *Domest. Anim. Endocrinol.*. doi:10.1016/j.domaniend.2008.01.003. PMID 18325721.

Cook, C. J. (2002). Rapid noninvasive measurement of hormones in transdermal exudate and saliva. *Physiology & Behaviour* 75, 169-181.

Corke, M. J. and Broom, D. M. (1999). The behaviour of sheep with sheep scab, Psoroptes ovis infestation. *Veterinary Parasitology.* 83, 291-300.

Crooker, BA; et al. (1994). Dairy Research and Bovine Somatotropin. University of Minnesota. Retrieved on 2008-01-16.

CTA (Technical Centre for Agriculture and Rural Cooperation). 1985b. *Proceedings of the Seminar on Primary Animal Health Care in Africa.* 25-28 September 1985, Balantyre, Malawi.

Dalal, Rooshin et al. *Replacement Alternatives in Education: Animal*-Free'" *Teaching* abstract from Fifth World Congress on Alternatives and | }Animal Use in the Life Sciences, Berlin, August 2005.

DARIGOLD - Choose Local, Choose Fresh (2006). Retrieved on 2008-04-21.

Darling, Jennifer Dorland (ed.) (2002). *Better Homes and Gardens New Cook Book*, 12th ed., Des Moines, IA: Meredith Corp..

de Haan, C. & Bekure, S. 1991. *Animal health services in sub-Saharan Africa: initial experiences with alternative approaches.* World Bank Technical Papers No. 134. Washington, DC, World Bank.

Dimitri, C; Greene, C. Recent Growth Patterns in the U.S. Organic Foods Market (pdf). Economic Research Service. Retrieved on 2008-01-29.

Dohoo IR, DesCôteaux L, Leslie K, *et al* (2003). "A meta-analysis review of the effects of recombinant bovine somatotropin. 2. Effects on animal health, reproductive performance, and culling". *Can. J. Vet. Res.* 67 (4): 252-64. PMID 14620861.

Dohoo, I.; Leslie, K.; Descôteaux, L.; Shewfelt, W. (2003). "A meta-analysis review of the effects of recombinant bovine somatotropin". *Can J Vet Res* 67 (4): 241-251. Retrieved on 2008-01-16.

Dolan, S. K. and Nolan, A. M. (2000). Behavioural evidence supporting a differential role for spinal group I and II metabotropic glutamate receptors in inflammatory hyperalgesia in sheep. *Neuropharmacology* 39, 1132-1138.

Drucker, A. G. 2004. *The economics of farm animal genetic resource conservation and sustainable use: why is it important and what have we learned?* Commission on Genetic Resources for Food and Agriculture Background Study Paper 21.

Drucker, A. G. 2006. An application of the use of safe minimum standards in the conservation of livestock biodiversity. *Environment and Development Economics* 11 (1): 77-94.

Drucker, A. G., and R. Scarpa. 2003. Valuing animal genetic resources. *Ecological Economics* 45 (3).

Drucker, A. G., and S. Anderson. 2004. Economic analysis of animal genetic resources and the use of rural appraisal methods: lessons from Southeast Mexico. *International Journal of Agricultural Sustainability* 2 (2): 77-97.

Drucker, A. G., E. Bergeron, U. Lemke, L. T. Thuy, and A. Valle Zrate. 2006. Identification and quantification of subsidies relevant to the production of local and imported pig breeds in Vietnam. *Tropical Animal Health and Production* 38 (4): 305-322.

Drucker, A. G., V. Gómez, and S. Anderson. 2001. The economic valuation of farm animal genetic resources: a survey of available methods. *Ecological Economics* 36 (1): 1-18.

Drucker, A. G., V. Gómez, N. Ferraes-Ehuan, O. Rubio, and S. Anderson. 1999. Comparative economic analysis of criollo, crossbreed and imported pigs in backyard production of Yucatan, Mexico. FMVZ-UADY Mimeo.

Dubuisson, D. and Dennis, S. G. (1977). The formalin test: a quantitative study of the analgesic effects of morphine, meperidine, and brain stem stimulation in rats and cats. *Pain* 4, 161-174.

Earley, B and Crowe, M. A. (2002). Effects of ketoprofen alone or in combination with local anesthetic during the castration of bull calves on plasma cortisol, immunological, and inflammatory responses. *Journal of Animal Science* 80, 1044-1052.

Eckersall, P. D. (2000). Recent advances and future prospects for the use of acute phase proteins as markers of disease in animals. *Revue de Medecine Veterinaire.* 151, 577-584.

Eckersall, P. D., Young, F. J., McComb, C., Hogarth, CJ, Safi, S, Weber, A, McDonald, T, Nolan, A. M., and Fitzpatrick, J. L. (2001). Acute phase proteins in serum and milk from dairy cows with clinical mastitis. *Veterinary Record* 148, 35-41.

Erika N. Ringdahl, "Treatment of Recurrent Vulvovaginal Candidiasis", *American Family Physician* 61:11 (June 1, 2000)

Eshraghi, HR, Zeitlin, IJ, Fitzpatrick, J. L., Ternent, H, and Logue, D. N. (1999). The release of bradykinin in bovine mastitis. *Life Sciences* 64, 1675-1687.

Executive Summary - Report of the Royal College of Physicians and Surgeons of Canada Expert Panel on Human safety of rbST

Falconi, C. A., S. W. Omamo, G. d'Ieteren, and F. Iraqi. 2001. An ex ante economic and policy analysis of research on genetic resistance to livestock disease: trypanosomosis in Africa. *Agricultural Economics* 25 (2/3): 153-163.

FAO. 1991. *Guidelines for strengthening animal health services in developing countries.* Rome.

FAO. 1994. *Expert Consultation on Application of effective herd health and production programmes to increase livestock productivity in developing countries.* 30 March-1 April 1993, Rome.

FAO. 1996. *World livestock production systems - current status, issues and trends.* FAO Animal Production and Health Paper No. 127. Rome.

FAO/RLAC. 1992. *Report of the round table on privatization of the veterinary services in Latin America and the Caribbean.* 12-14 November 1991, Santiago, Chile.

FAO/RNEA. 1994. *Study on privatization of veterinary services in the Near East region.* Report of a meeting, 28-30 June 1994, Cairo, Egypt.

Fassi-Fehri, B.E.C. & Bakkoury, M. 1995. *Profile of the veterinary services in Morocco.* Rome, FAO.

Faulkner, P. M. and Weary, D. M. (2000). Reducing pain after dehorning in dairy calves. *Journal of Dairy Science* 83, 2037-2041.

Firth, A. M. and Haldane, S. L. (1999). Development of a scale to evaluate postoperative pain in dogs. *JAVMA* 214, 651-659.

Fisher, A. D., Crowe, M. A., Nuallain, E. M. O., Monaghan, M. L., Prendiville, D. J., OKiely, P., and Enright, W. J. (1997). Effects of suppressing cortisol following castration of bull calves on adrenocorticotropic hormone, in vitro interferon- gamma production, leukocytes, acute-phase proteins, growth, and feed intake. *Journal of Animal Science* 75, 1899-1908.

Fisher, A. D., Knight, T. W., Cosgrove, G. P., Death, A. F., Anderson, C. B., Duganzich, D. M., and Matthews, L. R. (2001). Effects of surgical or banding castration on stress responses and behaviour of bulls. *Australian Veterinary Journal.* 79, 279-284.

Fitzpatrick, J. L., Nolan, A. M., Scott, E. M., Harkins, LS., and Barrett, D. C. (2002). Observers perception of pain in cattle. *Cattle Practice.* 10, 209-212.

Flecknell, P. & Waterman-Pearson, A.(2000) *Pain Management in Animals*. Publ: W.B.Sanders, Harcourt Health Sciences, London.

Fleming G., Guzzoni M., *Storia cronologica delle epizoozie dal 1409 av. Cristo sino al 1800*, in Gazzetta medico-veterinaria, I-II, Milano 1871-72

Fotsis T, Murphy C, Gannon F (1990). "Nucleotide sequence of the bovine insulin-like growth factor 1 (IGF-1) and its IGF-1A precursor". *Nucleic Acids Res.* 18 (3): 676. doi:10.1093/nar/18.3.676. PMID 2308858.

Gandini, G. 1997. What economic value for local livestock breeds? Proceedings of the 48th Annual of the European Association of Animal Production. The Netherlands: Wageningen Academic Publishers.

Gatongi, P.M., Scott, M.E., Ranjan, S., Gathuma, J.M., Munyna, W.K., Cheruiyot, H. & Prichard, R.K. 1993. Effects of three nematode anthelmintic treatment regimes on flock performance of sheep and goats under extensive management in semi-arid Kenya, *Vet. Parasitology,* 68: 323-336.

Graf, B. and Senn, M. (1999). Behavioural and physiological responses of calves to dehorning by heat cauterization with or without local anaesthesia. *Applied Animal Behaviour Science* 62, 153-171.

Graham, M.J., Kent, J.E. and Molony, V. (1997) Effects of four analgesic treatments on the behavioural and cortisol re-sponses of 3-week old lambs to tail docking. *British Veterinary Journal* 153, 87-97.

Grant, C. (2002) The safety and efficacy of intramuscular Xylazine for pain relief in sheep and lambs. Faculty of Health Sciences, Adelaide University, South Australia. MSc Thesis.

Grant, C. and Upton, R. N. (2001). The anti-nociceptive efficacy of low dose intramuscular xylazine in lambs. *Research in Veterinary Science* 70, 47-50.

Groen, A. F. 1988. Derivation of economic values in cattle breeding: a model at the farm level. *Agricultural Systems* 27: 195-213.

Gruys, E., Obwolo, MJ, and Toussaint, MJM (1994). Diagnostic significance of the major acute phase proteins in veterinary clinical chemistry: a review. *Veterinary Bulletin* 64, 1009-1018.

Guide for the Care and Use of Laboratory Animals. National Research Council; National Academy Press, Washington, DC, 1996.

Hall S, Clutton Brock Juliet, *Two hundred years of British farm livestock*, Natural History Museum Publications, London 1988

Halliday, Tim R.; Kraig Adler (eds.) (1986). *Reptiles & Amphibians*. Torstar Books, p. 101. ISBN 0-920269-81-8.

Hamath, A. S., M. D. Faminow, G. V. Johnson, and G. Crow. 1997. Estimating the values of cattle characteristics using an ordered probit model. *American Journal of Agricultural Economics* 79: 463-476.

Hansen, M (2003-02-11). Dr. Michael Hanson on rBGH & Monsanto's Recent Intimidation Tactics. Organic Consumers Association. Retrieved on 2008-01-16.

Harada, D.Y., M.J.A. Jorge, A.B. Sanches, and H. Tokeshi. 1995. Interaction between microorganisms, soil physical structure, and plant diseases. Mimeographed sheet. Universidade do São Paulo, Peracicaba, Brazil.

Harkins, LS., Fitzpatrick, J. L., Nolan, A. M., Barrett, D. C., and Scott, E. M. Perception of pain in Sheep. *Sheep Veterinary Society* 57, 17-20. 2002.

Hay, M., Vulin, A., Genin, S., Sales, P. and Prunier, A (2003). Assessment of pain induced by castration in piglets: behavioural and physiological responses over the 5 subsequent days. *Applied Animal Behaviour Science* 82, 201-218.

Hellebrekers, L.J. (2000) *Animal Pain. A practical-oriented approach to an effective pain control in animals.* Publ: van der Wees Uitgevery, Utrecht, The Netherlands.

Higa, T. 1996. An Earth Saving Revolution. (English translation.) Sunmark Publishers, Inc. Tokyo, Japan.

Holden, S., Ashley, S. & Bazeley, P. 1996. *Improving the delivery of animal health services in developing countries: a literature review.* Somerset, UK, Livestock in Development.

Holden, S., Ashley, S. & Bazeley, P. 1996. *Improving the delivery of animal health services in developing countries: a literature review.* Somerset, UK, Livestock in Development.

Holton, L. L., Scott, E. M., Nolan, A., Reid, J., Welsh, E., and Flaherty, D. (1998). Comparison of three methods used for assessment of pain in dogs. *Journal of the American Veterinary Medical Association.* 212, 61-67.

Holton, L., Reid, J., Scott, E. M., Pawson, P., and Nolan, A. (2001). Development of a behaviour-based scale to measure acute pain in dogs. *Veterinary Record* 148, 525-531.

Hosie, B. D., Carruthers, J., and Sheppard, B. W. (1996). Bloodless castration of lambs: results of a questionnaire. *British Veterinary Journal.* 152, 47-55.

House, Dawn. "Wal-Mart milk hormone-free, but labels are mum", The Salt Lake Tribune, 2008-03-24. Retrieved on 2008-04-04.

Howard, J.R. 1969. Costs of feedlot disease and prevention programmes. *J. Am. Vet. Med. Assoc.,* 154(10): 1166-1167.

IASP (1991) *Core curriculum for professional education in pain.* Ed. H.L. Fields. Publ: IASP Publications, Seattle, USA.

Ilemobade, A.A. 1996. *Privatization of veterinary services within the context of structural adjustment in Zimbabwe, Namibia and Ghana.* Rome, FAO.

International Association for the Study of Pain (IASP) (1979) Pain terms: a list with definitions and notes on usage. *Pain* 6, 249-252.

Jabbar, M. A., and M. L. Diedhiou. 2003. Does breed matter to cattle farmers and buyers? evidence from West Africa. *Ecological Economics* 45 (3): 461-472.

Jackson, R E., Molony V., and Kent, J. E. (1999) Behavioural effects of chronic neuropathic pain from the tail in lambs. p144. *Proceedings of the 10th World Congress on Pain.* Vienna, Austria, August 1999.

Janick Jules, Noller Carl H., Rhykerd Charles L., *The Cycles of Plant and Animal Nutrition*, in Food and Agriculture, Scientific American Books, San Francisco 1976

Jemal, A. & Hugh-Jones, M.E. 1995. Association of tsetse control with health and productivity of cattle in the Didessa Valley, western Ethiopia. *Prev. Vet. Med.,* 22: 29-40.

Jerram L. Brown, "Helping and Communal Breeding" in *Birds* 70-75 (1987)

Kania, B. F., Zaremba-Rutkowska, M., and Romanowicz, K. (1999). Experimental intestinal stress induced by duodenal distention in sheep. *Journal of Animal & Feed Sciences.* 8, 233-245.

Karugia, J. T., O. A. Mwai, R. Kaitho, A. Drucker, C. B. A. Wollny, and J. E. O. Rege. 2001. *Economic analysis of crossbreeding programmes in sub-Saharan Africa: a conceptual framework and Kenyan case study*. FEEM Working Paper 106.2001.

Kent, J. E. (1992) Acute phase proteins: their use in veterinary diagnosis. Guest Editorial for British Veterinary J. 148 279-281.

Kent, J. E. and Goodall, J. (1991) Assessment of an immunotur-bidimetric method for measuring equine serum haptoglobin concen-tration. Equine veterinary J. 23 59-66.

Kent, J. E., Jackson, R. E, Molony, V., and Hosie, B. D. (2000). Effects of acute pain reduction methods on the chronic inflammatory lesions and behaviour of lambs castrated and tail docked with rubber rings at less than two days of age. *The Veterinary Journal* 160, 33-41. (Abstract)

Kent, J. E., Meikle, L., Molony V., and McKendrick, I. J. (2001b) Qualitative versus quantitative assessment of an acute pain in lambs. Sheep Veterinary Society 25, 65-66.

Kent, J. E., Molony V., and Graham, M. J. (2001a). The effect of different bloodless castrators and different tail docking methods on the response of lambs to the combined Burdizzo rubber ring method of castration. *The Veterinary Journal* 162, 250-254.

Kent, J. E., Molony, V., and Graham, M. J. (1998). Comparison of methods for the reduction of acute pain produced by rubber ring castration or tail docking of week-old lambs. *The Veterinary Journal*. 155, 39-51 (

Kent, J. E., Molony, V., Jackson, R. E., and Hosie, B. D. (1999). Chronic inflammatory responses of lambs to rubber ring castration: are there any effects of age or size of lamb at treatment. *Occasional Publication - British Society of Animal Science.* 23, 160-162.

Kent, J. E., Thrusfield, M. V., Molony, V., Hosie, B. D., and Sheppard, B. W. (2004). A randomised, controlled field trial of two new techniques for castration and tail docking of lambs less than two days of age. *Veterinary Record* 154, 193-200..

Kent, J.E. (1977) The effect of road transportation on young cattle. University of Liverpool. MSc Thesis.

Kent, J.E., Molony, V. and Robertson, I.S. (1993) Changes in plasma cortisol concentration in lambs of three ages after three methods of castration and tail docking. *Research in Veterinary Sciences* 55 246-251.

Kent, J.E., Molony, V. and Robertson, I.S. (1995) Comparison of Burdizzo and rubber ring methods of castration and tail dock-ing of lambs. *Veterinary Record* 136 192-196.

Korach, Kenneth S. *Reproductive and Developmental Toxicology*, CRC Press, 1998, p. 55

Kosgey, I. S., J. A. M. Van Arendonk, and R. L. Baker. 2004. Economic values for traits in breeding objectives for sheep in the tropics: impact of tangible and intangible benefits. *Livestock Production Science* 88: 143-160.

Krajcovic, V., E. Uhliarová, M. Michalec, and M. Zimková. 2003. Evaluation of some environmentally-friendly private livestock farms in Central Slovakia. In *Livestock farming systems in Central and Eastern Europe*, ed. A. Giban, and S. Mihina. Wageningen, The Netherlands: Wageningen Academic Publishers.

Kroger to complete transition to certified rBST-free milk by early 2008 (press release). Kroger (2007). Retrieved on 2008-01-29.

Kurian, V. 2001. Value-based crop-livestock production systems for the future in the semi-arid tropics. In *Future of agriculture in the semi-arid tropics*, ed. V. Kurien, M. C. S. Bantilan, P. Parthasarathy Rao, and R. Padmaja. Proceedings of an International Symposium on Future of Agriculture in Semi-Arid Tropics, November 14, 2000, in ICRISAT, Patancheru, India. Patancheru, India: International Crops Research Institute for the Semi-Arid Tropics.

Ladd, G., and C. Gibson. 1978. Microeconomics of technical change: what's a better animal worth? *American Journal of Agricultural Economics* 60 (2): 236-240.

Leonard, D.K. 1987. The supply of veterinary services: Kenyan lessons. *Agric. Admin. & Extension,* 26: 219-236.

Leonard, D.K. 1987. The supply of veterinary services: Kenyan lessons. *Agric. Admin. & Extension,* 26: 219-236.

Leonard, D.K. 1993. Structural reform of the veterinary profession in Africa and the new institutional economics. *Dev. & Change,* 24: 227-267.

Leonard, D.K. 1993. Structural reform of the veterinary profession in Africa and the new institutional economics. *Dev. & Change,* 24: 227-267.

Lester, S. J., Mellor, D. J., and Ward, R. N. (1991a). Effects of repeated handling on the cortisol responses of young lambs castrated and tailed surgically. *New Zealand Veterinary Journal.* 39, 147-149.

Lester, S. J., Mellor, D. J., Holmes, R. J., Ward, R. N., and Stafford, K. J. (1996). Behavioural and cortisol responses of lambs to castration and tailing using different methods. *New Zealand Veterinary Journal.* 44, 45-54.

Lester, S. J., Mellor, D. J., Ward, R. N., and Holmes, R. J. (1991b). Cortisol responses of young lambs to castration and tailing using different methods. *New Zealand Veterinary Journal.* 39, 134-138.

Ley, S. J., Livingston, A., and Waterman, A. E. (1991). Effects of chronic lameness on the concentrations of cortisol, prolactin and vasopressin in the plasma of sheep. *Veterinary Record.* 129, 45-47.

Ley, S. J., Waterman, A. E., Livingston, A., and Parkinson, T. J. (1994). Effect of chronic pain associated with lameness on plasma cortisol concentrations in sheep: a field study. *Research in Veterinary Science.* 57, 332-335.

Liles, J. H. and Flecknell, P. A. (1993a). The effects of surgical stimulus on the rat and the influence of analgesic treatment. *British Veterinary Journal* 149, 515-525.

Liles, J. H. and Flecknell, P. A. (1993b). The influence of buprenorphine or bupivacaine on the post-operative effects of laparotomy and bile-duct ligation in rats. *Laboratory Animals* 27, 374-380.

Lopez, M.F. 1995. Evaluación Técnica y Proyecto Mejoras Tratamiento Aguas Residuales Compañia Mundimar–Guapiles. EASA Consultores, S.A., San José, Costa Rica.

Malcolm Rowland (February 2006). Microdosing and the 3Rs. National Center for the Replacement, Refinement, and Reduction of Animals in Research (NC3Rs). Retrieved on 2007-09-22.

Malloy, D. "State reverses on dairy labeling, allows hormone claims", Pittsburgh Post-Gazette, 2008-01-18. Retrieved on 2008-01-29.

Manson, F. J. and Leaver, J. D. (1988). The influence of concentrate amount on locomotion and clinical lameness in dairy cattle. *Animal Production* 47, 185-190.

McCorkle, C.M. & Mathias, E. 1996. Paraveterinary healthcare programmes: a global overview. *In* K.-H. Zessin, ed. *Livestock production and diseases in the tropics: livestock production and human welfare.* Proceedings of the VII International Conference of Institutions of Tropical Veterinary Medicine, Vol. II, p. 544-549. 25-29 September 1994, Berlin. Feldafing, Deusche Stifung fur Internationale Entwicklung.

McInerney, J.P., Howe, K.S. & Schepers, J.A. 1992. A framework for the economical analysis of disease in farm livestock. *Prev. Vet. Med.,* 13: 137-154.

Mehraban, A.B., Barker, T. & Ward, D.E. 1996. *Community-based animal health and production in Afghanistan - a delivery system.* Paper presented at the Asian-Australian Animal Production Meeting, 13-18 October 1996, Makuhari, Japan.

Mellor, D. J. and Murray, L. (1989a). Changes in the cortisol responses of lambs to tail docking, castration and ACTH injection during the first seven days after birth. *Research in Veterinary Science.* 46, 392-395.

Mellor, D. J., Stafford, K. J., Todd, K. S., Jr., Lowe, TE, Gregory, N. G, Bruce, R. A., and Ward, R. N. (2002). A comparison of catecholamine and cortisol responses of young lambs and calves to painful husbandry procedures. *Australian Veterinary Journal* 80, 228-233.

Melton, B. E., W. A. Colette, and R. L. Willham. 1994. Imputing input characteristic values from optimal commercial breed or variety choice decisions. *American Journal of Agricultural Economics* 76 (3): 478-491.

Melzack, R. (1975). The Mcgill pain questionaire:major properties and scoring methods. *Pain* 1, 277-299.

Mendelsohn, R. 2003. The challenge of conserving indigenous domesticated animals. *Ecological Economics* 45 (3): 501-510.

Mitchell, G., C. Smith, M. Makower, and P. J. W. N. Bird. 1982. An economic appraisal of pig improvement in Great Britain 1. Genetic and production aspects. *Animal Science (Animal Production)* 35 (2): 215-224.

Mitchell, G., P. J. W. N. Bird, M. Makower, and C. Smith. 1983. An economic appraisal of pig improvement in Great Britain. *Oxford Development Studies (formerly Oxford Agrarian Studies)* (12): 48-62.

Molony, V. and Kent, J.E. (1997) Assessment of acute pain in farm animals using behavioural and physiological measurements. Journal of Animal Science 75 266-272. Written paper of presentation given by V. Molony at 87th Annual meeting of the American Society of Animal Science, Florida, USA, July 1995.

Molony, V., Kent, J. E., and McKendrick, I. J. (2002). Validation of a method for the assessment of an acute pain in lambs. *Applied Animal Behaviour Science* 76, 215-238.

Molony, V., Kent, J. E., Fleetwood-Walker, S. M., Munro, F., and Parker, R. M. C. (1993b) Effects of Xylazine and L659874 on behaviour of lambs after tail docking. p80. Proceedings of the 7th IASP World Congress on pain, Paris.

Molony, V., Kent, J.E. and Robertson, I.S. (1993a) Behavioural responses of lambs of three ages in the first three hours after three methods of castration and tail docking. *Research in Veterinary Science* 55 236-245.

Molony, V., Kent, J.E. and Robertson, I.S. (1995) Assessment of acute and chronic pain after different methods of castration of calves. *Applied Animal Behavioural Sciences* 46 33-48.

Molony, V., Kent, J.E., Hosie, B. and Graham, M.J. (1997) Reduction in pain suffered by lambs at castration. *British Veterinary Journal* 153 205-213.

Mostl, E., Maggs, J. L., Schrotter, G., Besenfelder, U., and Palme, R. (2002). Measurement of cortisol metabolites in faeces of ruminants. *Veterinary Research Communications.* 26, 127-139.

Mukhebi, A.W., Kariuki, D.P., Mussukuya, E., Mullins, G., Nugumi, P.N., Thorpe, W. & Perry, B.D. 1995. Assessing the economic impact of immunization against East Coast fever: a case study in Coast Province. Kenya. *Vet. Rec.*, 137: 17-22.

Mullan, W.M.A. (2002). Science and technology of modified atmosphere packaging. [On-line] UK: Available: file:///D:/dairy-w/Dairy-27.htm. Accessed: 16 May 2008.

Mwacharo, J. M., and A. G. Drucker. 2005. Production objectives and management strategies of livestock keepers in South-East Kenya: implications for a breeding programme. *Tropical Animal Health and Production* 37 (8): 635-652.

Nakamura, Yasunori; Naoyuki Yamamoto, Kumi Sakai, and Toshiaki Takano (1995). "Antihypertensive Effect of Sour Milk and Peptides Isolated from It That are Inhibitors to Angiotensin I-Converting Enzyme" (PDF). *Journal of Dairy Science* 78 (6): 1253-7. PMID 7673515. Retrieved on 2007-06-30.

Noonan, G. J., Rand, J. S., Priest, J., Ainscow, J., and Blackshaw, J. K. (1994). Behavioural observations of piglets undergoing tail docking, teeth clipping and ear notching. *Applied Animal Behaviour Science.* 39, 203-213.

North, R (2007-01-10). Safeway & Chipotle Chains Dropping Milk & Dairy Derived from Monsanto's Bovine Growth Hormone. Oregon Physicians for Social Responsibility. Retrieved on 2008-01-29.

O. Adolfsson *et al.*, "Yogurt and gut function", *American Journal of Clinical Nutrition* 80:2:245-256 (2004).

O'Callaghan, K (2002). Lameness and associated pain in cattle - challenging traditional perceptions. *In Practice* April, 212-219.

OIE (International Office *of* Epizootics). 1995. *Recommendation of the OIE: appropriate privatization of veterinary practice in Africa.* Rabat, Morocco.

Ott, S.L., Hillberg Seistzinger, A. & Hueston, W.D. 1995. Measuring the national economic benefits of reducing livestock mortality. *Prev. Vet. Med.,* 24: 203-211.

Otto, K. A. and Short, C. E (1998). Pharmaceutical control of pain in large animals. *Applied Animal Behaviour Science* 59, 157-169.

Ouma, E. A., A. O. Gideon, and S. J. Staal. 2004. The socio-economic dimensions of smallholder livestock management in Kenya and its effects on competitivenessof crop-livestock systems. Paper contributed to the NARO conference "Integrated agricultural research for development-achievements, lessons learnt and best practice," September 1-4, 2004, Kampala, Uganda.

Ouma, E., A. Abdulai, and A. G. Drucker. 2007. Measuring heterogeneous preferences for cattle traits amongst cattle keeping households in East Africa. *American Journal of Agricultural Economics* , forthcoming.

Owens, M. E. (1984). Pain in Infancy: Conceptual and methodological issues. *Pain* 20, 213-230.

Peagram, R.G., James, A.D., Bamhare, C., Dolan, T.T., Hove, T., Kan, G.K. & Latif, A.A. 1996. *Effects of immunization against Theileria parva on beef cattle productivity and economics of control options.* Trop. Anim. Health Prod., 28: 99-11.

Peagram, R.G., James, A.D., Bamhare, C., Dolan, T.T., Hove, T., Kan, G.K. & Latif, A.A. 1996. *Effects of immunization against Theileria parva on beef cattle productivity and economics of control options.* Trop. Anim. Health Prod., 28: 99-11.

Peers, A., Mellor, D. J., Wintour, E. M., and Dodic, M (2002). Blood pressure, heart rate, hormonal and other acute responses to rubber ring castration and tail docking of lambs. *New Zealand Veterinary Journal* 50, 56-62.

Pepys, M.B., Baltz, M.L., Tennent, G.A., Kent, J., Ousey, J and Rossdale, P.D. (1989) Serum amyloid A protein (SAA) in horses: objective measurement of the acute phase response. Equine Veterinary J. 21 106-109.

Petrie, N. J., Mellor, D. J., Stafford, K. J., Bruce, R. A., and Ward, R. N. (1996). Cortisol responses of calves to two methods of disbudding used with or without local anaesthetic. *New Zealand Veterinary Journal* 44, 9-14.

Phillips, J.M., and S.R. Phillips. 1995. The EM Application Manual for APNAN Countries. EM Technologies, Inc., Tucson, Arizona, USA

Pollard, J. C., Littlejohn, R. P., Johnstone, P, Laas, FJ, Corson, ID, and Suttie, J. M. (1992). Behavioural and heart rate responses to velvet antler removal in red deer. *New Zealand Veterinary Journal* 40, 56-61.

Pritchett, LC, Ulibarri, C, Roberts, MC, Schneider, RK, and Sellon, DC (2003). Identification of potential physiological and behavioural indicators of postoperative pain in horses after exploratory celiotomy for colic. *Applied Animal Behaviour Science* 80, 31-43.

Raloff, J (2003-11-01). "Hormones in Your Milk". *Science News* 164 (18). Retrieved on 2008-01-29.

Rege, J. E. O. 1999. Economic valuation of animal genetic resources. Proceedings of an FAO/ILRI Workshop held at FAO Headquarters, March 15-17, 1999, in Rome, Italy. Nairobi, Kenya: International Livestock Research Institute.

Rege, J. E. O., and J. P. Gibson. 2003. Animal genetic resources and economic development: issues in relation to economic valuation. *Ecological Economics* 45 (3): 319-330.

Ripudaman S. Beniwal, *et al.*, "A Randomized Trial of Yogurt for Prevention of Antibiotic-Associated Diarrhea", *Digestive Diseases and Sciences* 48:10:2077-2082 (October, 2003) doi:10.1023/A:1026155328638

Robertson, I.S., Kent, J.E. and Molony, V. (1994) Effect of different methods of castration on behaviour and plasma cortisol in calves of three ages. *Research in Veterinary Sciences* 56, 8-17.

Roosen, J., A. Fadlaoui, and M. Bertaglia. 2002. Economic evaluation and biodiversity conservation of animal genetic resources. University of Kiel, Germany.

Roosen, J., A. Fadlaoui, and M. Bertaglia. 2005. Economic evaluation for conservation of farm animal genetic resources. *Journal of Animal Breeding and Genetics* 122 (2005): 217-228.

Ross, M.R. and Dyson, S.J. (2003) Diagnosis and Management of Lameness in Horses. Publ. Sanders, Elsevier Science, USA.

Roughan, J. V. and Flecknell, P. A. (2001). Behavioural effects of laparotomy and analgesic effects of ketoprofen and carprofen in rats. *Pain* 90, 65-74.

Royal College of Physicians and Surgeons of Canada (1999-01-01). Report of the Royal College of Physicians and Surgeons of Canada Expert Panel on Human safety of rbST. Health Canada. Retrieved on 2008-01-16.

Ruane, J. 1999. A critical review of the value of genetic distance studies in conservation of animal genetic resources. *Journal of Animal Breeding and Genetics* 116 (5): 317-323.

Ruane, J. 1999. Selecting breeds for conservation. In *Genebanks and the conservation of farm animal genetic resources*, ed. J. K. Oldenbroek. The Netherlands: IDO-DL Press.

Russell, W.M.S. and Burch, R.L.. The Removal of Inhumanity: The Three R's. Retrieved on 2007-05-24.

S.P. Otto and D.B. Goldstein. "Recombination and the Evolution of Diploidy". Genetics. Vol 131 (1992): 745-751.

Saltini Antonio, *Storia delle scienze agrarie*, 4 vols, Bologna 1984-89, ISBN 88-206-2412-5, ISBN 88-206-2413-3, ISBN 88-206-2414-1, ISBN 88-206-2414-X

Sanford, J.; Ewbank, R.; Molony, V.; Tavernor, W.D.; Uvarov, O. (1986) *Guidelines for the recognition and assessment of pain in animals. Veterinary Record* 118, 334-338.

Savage, Thomas F. (September 12, 2005). A Guide to the Recognition of Parthenogenesis in Incubated Turkey Eggs. *Oregon State University*. Retrieved on 2006-10-11.

Scarpa, R., A. G. Drucker, S. Anderson, N. Ferraes-Ehuan, V. Gómez, C. R. Risopatrón, and O. Rubio-Leonel. 2003. Valuing genetic resources in peasant economies: the case of 'hairless' creole pigs in Yucatan. *Ecological Economics* 45 (3): 427-443.

Scarpa, R., E. S. K. Ruto, P. Kristjanson, M. Radeny, A. G. Drucker, and J. E. O. Rege. 2003. Valuing indigenous cattle breeds in Kenya: an empirical comparison of stated and revealed preference value estimates. *Ecological Economics* 45 (3): 409-426.

Schatz, S and Palme, R. (2001). Measurement of faecal cortisol metabolites in cats and dogs: A non-invasive method for evaluating adrenocortical function. *Veterinary Research Communications.* 25, 271-287.

Schillhorn van Veen, T.W. & de Haan, C. 1995. Trends in the organization and financing of livestock and animal health services. Prev. *Vet. Med.,* 25: 225-240.

Schillhorn van Veen, T.W. & de Haan, C. 1995. Trends in the organization and financing of livestock and animal health services. Prev. *Vet. Med.,* 25: 225-240.

Schmidt, R.F. & Thews, G. (Eds) (1989) *Human Physiology*. Publ: Springer-Verlag, Berlin, New York.

Schreuder, B.E.C., Moll, H.A.J., Noorman, N., Halimi, C., Kroese, A.H. & Wassink, G. 1995. A benefit-cost analysis of veterinary intervention in Afghanistan based on a livestock mortality study. *Prev. Vet. Med.,* 26: 303-314.

Schreuder, B.E.C., Moll, H.A.J., Noorman, N., Halimi, C., Kroese, A.H. & Wassink, G. 1995. A benefit-cost analysis of veterinary intervention in Afghanistan based on a livestock mortality study. *Prev. Vet. Med.,* 26: 303-314.

Scott, P. R., Dun, K., Penny, C. D., Strachan, W. D., and Keeling, N. (1996). Field assessment of lamb behaviour after xylazine hydrochloride epidural injection for castration using rubber rings. *Agri-Practice.* 17, 19-22.

Secchi A, Deligianni V (2006). "Ocular toxicology: the Draize eye test". *Curr Opin Allergy Clin Immunol* 6 (5): 367–72.

Select Committee on Animals in Scientific Procedures Report, House of Lords, Chapter 3: The purpose and nature of animal experiments.

Sidahmed, A.E. 1995. *Livestock development and rangelands management cluster: privatization of veterinary services in sub-Saharan Africa.* Rome, International Fund for Agricultural Development (IFAD).

Signorello, G., and G. Pappalardo. 2003. Domestic animal biodiversity conservation: a case study of rural development plans in the European Union. *Ecological Economics* 45 (3): 487-499.

Simianer, H. 2000. Valuation of indigenous farm animal populations and breeds in comparison with imported exotic breeds with a focus on sub-Saharan Africa.

Simianer, H., S. B. Marti, J. Gibson, O. Hanotte, and J. E. O. Rege. 2003. An approach to the optimal allocation of conservation funds to minimize loss of genetic diversity between livestock breeds. *Ecological Economics* 45 (3): 377-392.

Smith LL (2001). "Key challenges for toxicologists in the 21st century". *Trends Pharmacol. Sci.* 22 (6): 281–5.

Smith, C. 1984. *Estimated costs of genetic conservation of farm animals*. FAO Animal Production and Health Paper 44/1. Rome: Food and Agriculture Organization of the United Nations.

Smith, C. 1984. Genetic aspects of conservation in farm livestock. *Livestock Production Science* 11: 37-48.

Smith, C. 1985. Scope for selecting many breeding stocks of possible economic value in the future. *Animal Science (Animal Production)* 41: 403-412.

Smith, C., A. Gibson, and G. Mitchell. 1982. An economic appraisal of pig improvement in Great Britain 2. Factors affecting estimated benefits. *Animal Science (Animal Production)* 35 (2): 225-230.

Smith, J (2004). Whistleblowers, Threats, and Bribes: A Short History of Genetically Engineered Bovine Growth Hormone. Council for Responsible Genetics. Retrieved on 2008-01-29.

Sprecher, D. J., Hostetler, D. E., and Kaneene, J. B. (1997). A lameness scoring system that uses posture and gait to predict dairy cattle reproductive performance. *Theriogenology.* 47, 1179-1187.

Stafford, K. J., Mellor, D. J., Todd, S. E., Gregory, N. G., Bruce, R. A., and Ward, R. N. (2002). Effects of local anaesthesia or local anaesthesia and nonsteroidal anti-inflammatory drug on cortisol responses of calves to castration by five different methods. *Research in Veterinary Science* 73, 61-70.

Stafford, K. J., Mellor, D. J., Todd, S. E., Ward, R. N. and McMeekan, C.M. (2003) The effect of different combinations of lignocaine, ketoprofen, xylazine and tolazoline on the acute cortisol response to dehorning in calves. *New Zealand Veterinary Journal* 51, 219-226. (Abstract)

Statement and Q&A-Starbucks Completes its Conversion – All U.S. Company-Operated Stores Use Dairy Sourced Without the Use of rBGH. Starbucks Corporation. Retrieved on 2008-04-04.

Steinman G (2006). "Mechanisms of twinning: VII. Effect of diet and heredity on the human twinning rate". *J Reprod Med* 51 (5): 405–10. PMID 16779988.

Stephen T. Emlen & Sandra L. Vehrencamp, "Cooperative Breeding Strategies among Birds", in *Perspectives in Ornithology*, 93 (Alan H. Brush & George A. Clark, Jr. eds., 1983)

Sukhov, SV; Kalamkarova LI, Il'chenko LA, Zhangabylov AK (1986). "Microfloral changes in the small and large intestines of chronic enteritis patients on diet therapy including sour milk products" (in Russian). *Voprosy pitaniia* Jul-Aug (4): 14-7. PMID 3765530. Retrieved on 2007-06-30.

Sutherland, M. A., Mellor, D. J., Stafford, K. J., Gregory, N. G., Bruce, R. A., and Ward, R. N. (2002a). Modification of cortisol responses to dehorning in calves using 5-hour local anaesthetic regimen plus phenylbutazone, ketoprofen or adrenocorticotropic hormone injected prior to dehorning. *Research in Veterinary Science* 73, 115-123.

Sutherland, M. A., Mellor, D. J., Stafford, K. J., Gregory, N. G., Bruce, R. A., and Ward, R. N. (2002b). Effect of local anaesthetic combined with wound cauterization on the cortisol response to dehorning in calves. *Australian Veterinary Journal* 80, 165-167.

Swallow, B.M., Mulatu, W. & Leak, S.G.A. 1995. Potential demand for a mixed public-private animal health input: evaluation of a pour-on insecticide for controlling tsetse-transmitted trypanosomiasis in Ethiopia. *Prev. Vet. Med.,* 24: 265-275.

Swallow, B.M., Mulatu, W. & Leak, S.G.A. 1995. Potential demand for a mixed public-private animal health input: evaluation of a pour-on insecticide for controlling tsetse-transmitted trypanosomiasis in Ethiopia. *Prev. Vet. Med.,* 24: 265-275.

Sylvester, S. P., Stafford, K. J., Mellor, D. J., Bruce, R. A., and Ward, R. N. (1998). Acute cortisol responses of calves to four methods of dehorning by amputation. *Australian Veterinary Journal.* 76, 123-126.

Tano, K., M. Kamuanga, M. D. Faminow, and B. Swallow. 2003. Using conjoint analysis to estimate farmers' preferences for cattle traits in West Africa. *Ecological Economics* 45 (3): 393-407.

Tarbourton, IS, Bray, AR, and Wilson, JA (2002). Incidence and perceptions of cryptorchid lambs in 2000. *Proceedings of the New Zealand Society of Animal Production* 62, 334-336.

Thornton, P. D. and Waterman-Pearson, A. E. (1997). Castration in young lambs produces changes in mechanical nociceptive threshold responses and behaviour as assessed by a dynamic and interactive visual analogue scale. *Journal of Veterinary Anaesthesia.* 21, 11.

Thornton, P. D. and Waterman-Pearson, A. E. (2002). Behavioural responses to castration in lambs. *Animal Welfare* 11, 203-212.

Till, J.E. 1995. Building credibility in public studies. *Am. Sci.*, 83(Sept-Oct): 468-473.

Tisdell, C. 2003. Socioeconomic causes of loss of animal genetic diversity: analysis and assessment. *Ecological Economics* 45 (3): 365-376.

Tobler, M. & Schlupp,I. (2005) Parasites in sexual and asexual mollies (Poecilia, Poeciliidae, Teleostei): a case for the Red Queen? Biol. Lett. 1 (2): 166-168.

Toledo, I. 1996. Tratamiento de las aguas residuales del proceso de fabricación de puré de banano en la planta Mundimar, S.A. (Guapiles, Limon) Utilizando Microorganismos Eficientes (EM). EARTH College, Limon, Costa Rica.

Top Cat Breeds for 2004

Umali, D. & Schwartz, L. 1994. *Public and private agricultural extension: beyond traditional frontiers.* World Bank Discussion Papers No. 236, Washington, DC, World Bank.

Umali, D., Feder, G. & de Haan, C. 1992. *The balance between public and private sector activities in the delivery of livestock services.* World Bank Discussion Papers No. 63. Washington, DC, World Bank.

US patent 1356340, *"Method of Drying Buttermilk"*, granted 1920-10-19, assigned to Collis Products Company

US patent 3978243, *"Process for preparing gelled sour milk"*, granted 1976-08-31, assigned to Kobenhavns Pektinfabrik

Vaccaro, L. P. d. 1973. Some aspects of the performance of European pure bred and crossbred dairy cattle in the tropics: Part I: reproductive efficiency in females. *Animal breeding Abstracts* 41: 571-589.

Vaccaro, L. P. d. 1974. Some aspects of the performance of European pure bred and crossbred dairy cattle in the tropics: Part II: mortality and culling rates. *Animal breeding Abstracts* 42: 93-103.

Wall, P.D. & Melzack, R. (Eds) (1994) *Textbook of Pain.* Publ: Churchill Livingstone, Edinburgh.

Wallace MS; et al. (1990) Gastric ulceration in the dog secondary to the use of non-steroid anti-inflammatory drugs. Journal of American Animal Hospital Association 26, 467-472.

Wamukoya, J.P.O., Gathuma, J.W. & Mutiga, E.R. 1995. *Spontaneous private veterinary practices evolved in Kenya since* 1988. Rome, FAO.

Watkins JB (1989). "Exposure of rats to inhalational anesthetics alters the hepatobiliary clearance of cholephilic xenobiotics". *J. Pharmacol. Exp. Ther.* 250 (2): 421–7. .

Weary, D. M., Braithwaite, L. A., and Fraser, D. (1998). Vocal response to pain in piglets. *Applied Animal Behaviour Science* 56, 161-172.

Wells, S. J., Trent, A. M., Marsh, W. E., and Robinson, R. A. (1993). Prevalence and severity of lameness in lactating dairy cows in a sample of Minnesota and Winsconsin herds. *Journal American Veterinary Medicine Association* 202, 78-82.

Welsh, E. M. and Nolan, A. M. (1995). Effect of flunixin meglumine on the thresholds to mechanical stimulation in healthy and lame sheep. *Research in Veterinary Science.* 58, 61-66.

Welsh, E. M., Gettinby, G., and Nolan, A. M. (1993). Comparison of a visual analogue scale and a numerical rating scale for assessment of lameness, using sheep as a model. *American Journal of Veterinary Research.* 54, 976-983.

Wemelsfelder, F., Hunter, E. A. Mendl, M. T. Lawrence, A. B (2000)The spontaneous qualitative assessment of behavioural expressions in pigs: first explorations of a novel methodology for integrative animal welfare measurement. *Applied Animal Behaviour Science.* 67, 193-215

Wesselmann, U. and Lai, J (1997) Mechanisms of referred visceral pain: uterine inflammation in the adult virgin rat results in neurogenic plasma extravasation in the skin. *Pain* 73, 309-317

White, R. G., DeShazer, J. A., Tressler, C. J., Borcher, G. M., Davey, S., Waninge, A., Parkhurst, A. M., Milanuk, M. J., and Clemens, E. T. (1995). Vocalization and physiological response of pigs during castration with or without a local anesthetic. *Journal of Animal Science.* 73, 381-386.

Wickenheiser, M. "Oakhurst Sued by Monsanto Over Milk Advertising", Portland Press Herald, 2003-07-08. Retrieved on 2008-01-29.

Wickham, B., and G. Banos. 1998. Impact of international evaluations on dairy cattle breeding programmes. Proceedings of 6th World Congress on Genetics applied to livestock production. *Armidale* 23: 315-322.

Wididana, G.N. 1994. Preliminary experiment of EM technology on waste water treatment. In Third Conference on Effective Microorganisms (EM). Kyusei Nature Farming Center, Saraburi, Thailand. pp. 1-6.

Williams G C. 1975. Sex and Evolution. Princeton (NJ): Princeton University Press.

Woodbury, M. R., Caulkett, N. A., Baumann, D., and Read, M. R. (2001). Comparison of analgesic techniques for antler removal in wapiti. *Canadian Veterinary Journal-Revue Veterinaire Canadienne* 42, 929-935.

Wurzinger, M., D. Ndumu, R. Baumung, A. G. Drucker, A. M. Okeyo, D. K. Semambo, N. Byamungu, and J. Sölkner. 2006. Comparison of production systems and selection criteria of Ankole cattle by breeders in Burundi, Rwanda, Tanzania and Uganda. *Tropical Animal Health and Production* 38 (7-8): 571-581.

Zander, K. 2006. *Modelling the value of the borana cattle in Ethiopia - an approach to justify its conservation*. Bonn, Germany: Center for Development Research.

Zander, K., and J. Mburu. 2004. *Compensating pastoralists for conserving animal genetic resources: the case of borana cattle in Ethiopia.* Bonn, Germany: Center for Development Research.

Zimmer, Carl. "Parasite Rex: Inside the Bizarre World of Nature's Most Dangerous Creatures", New York: Touchstone, 2001.

Zimmermann, M. (1986) Behavioural investigations of pain in animals. In: Assessing pain in farm animals. Eds. I.J.H.Duncan and Molony V. pp16-27. Luxembourg, Commission of the European Communities.

Index

A

B

M

N

O

P

■■■

International Encyclopaedia of DAIRY TECHNOLOGY AND ANIMAL HUSBANDRY

Volume 1

P.K. Saraswat

ANMOL PUBLICATIONS PVT. LTD.
NEW DELHI - 110 002 (INDIA)

ANMOL PUBLICATIONS PVT. LTD.
H.O.: 4374/4B, Ansari Road, Daryaganj,
New Delhi-110 002 (India)
Ph.: 23278000, 23261597
B.O.: No. 1015, Ist Main Road, BSK IIIrd Stage
IIIrd Phase, IIIrd Block,
Bangalore - 560 085 (India)
Visit us at: www.anmolpublications.com

International Encyclopaedia of Dairy Technology and Animal Husbandry

First Edition, 2009

ISBN 978-81-261-3793-0 (Set)

PRINTED IN INDIA

Printed at Mehra Offset Press, Delhi.

Contents

Preface

Husbandry implies application of scientific principles to agriculture, especially to animal breeding. Animal husbandry is the looking after and breeding of animals, particularly livestock. Animal husbandry has been around for over 20,000 years. Animal husbandry today is the agricultural practice of breeding and raising livestock. Examples of animal husbandry include beekeeping, dog breeding, farming, horse breeding. Livestock are domesticated animals intentionally reared in an agricultural setting to make produce such as food or fibre, or for their labour. Livestock include pigs, cattle, goats, deer, sheep, yaks and poultry. The type of livestock reared varies worldwide and depends on factors such as climate, consumer demand, native animals, local traditions, and land type. Veterinary Science is vital to the study and protection of animal production practices, herd health and monitoring spread of widespread disease. Veterinary medicine is the application of medical, diagnostic, and therapeutic principles to companion, domestic, exotic, wildlife, and production animals. It requires the acquisition and application of scientific knowledge in multiple disciplines and uses technical skills towards disease prevention in both domestic and wild animals. Human health is protected by veterinary science working closely with many medical professionals by the careful monitoring of livestock health as well as its unique training in epidemiology and emerging zoonotic diseases worldwide. Veterinary medicine is informally as old as the human/animal bond but in recent years has expanded exponentially because of the availability of advanced diagnostic and therapeutic techniques for most species. Animals nowadays often receive advanced medical, dental, and surgical care including insulin injections, root canals, hip replacements, cataract extractions, and pacemakers. Veterinarians assist in ensuring the quality, quantity, and security of food supplies by working to maintain the health of livestock and inspecting the meat itself. Veterinary scientists are very important in chemical, biological, and pharmacological research.

This publication titled "International Encyclopedia of Dairy Technology and Animal Husbandry" deals primarily with animal husbandry, dairy technology and animal breeding. It also provides an introductory overview of animal husbandry, veterinary science and dairying in India, including an analysis of initiatives taken at centre and state levels. Issues and options for India and rest of the world with respect to veterinary science, animal health and livestock improvement are dealt in detail. The plans and projects related to intensive cattle development in India have been reflected briefly, analyzing various schemes and evaluation processes in detail. An introductory overview of global dairy science and technology has been provided for the readers. An in-depth analysis of dairy farming, cattle types and milking has been done with focus on dairy products, biosynthesis, grading and defects. An analysis of dairy microbiology, starter cultures, inhibitors and antimicrobial proteins has been undertaken in detail. Other issues which have been discussed at great length include dairy plant maintenance with focus on training, servicing, equipment and operations management. An educational approach has been followed while dealing with modern day dairy technology and milk supply in India. Appendix, glossary of terms, acronyms/abbreviations, bibliography and links have been provided to make this publication a user friendly one.

—Editors

Glossary of Terms

Accuracy: A measure of how good and close a calculated estimate of an animal's genetic value is compared to the unknown true genetic value.

Acoustic demarcation: When an animal makes sounds to signal its presence examples include singing of birds, calling of amphibians, hooting of some primates, howling of wolves and coyotes.

Across Breed Comparison (ABC): An estimate of genetic merit. It indicates the expected difference between the performance of progeny of an individual ram and the average of all rams tested regardless of breed.

Adaptation traits: Complex of traits related to reproduction and survival of the individual in a particular production environment. Adaptation traits contribute to individual fitness and to the evolution of animal genetic resources. By definition, these traits are also important to the ability of the animal genetic resource to be sustained in the production environment.

Additive gene action: The effect where the expression of a trait is controlled by one or more pairs of genes, each of which act in an additive manner.

Additive genes effects: Members of a gene pair that have the ability to be expressed equally are additive genes. The gene pair is expressed as the sum of the individual effects of the genes in the pair.

Adjusted 100 day weight: This weight is the 50–day adjusted weight plus the postweaning 50-day weight gain (postweaning ADG x 50).

Adjusted 50-day weight: This weight is calculated from an actual weight taken between 35 and 65 days of age. All lamb weights are then adjusted to the equivalent of a 50–day old ram lamb raised as a single from a mature ewe (4–5 years of age).

Aestivation: The condition of dormancy or torpidity.

Agrobiodiversity or agricultural biological diversity: That component of biodiversity that contributes to food and agriculture production. The term agrobiodiversity encompasses within-species, species and ecosystem diversity.

Alleles: The variant forms of each gene.

Allergen: A substance, usually a protein, that can cause an allergy or allergic reaction in the body.

Allergy: A reaction by the body's immune system after exposure to a particular substance, often a protein.

Alpha animal: Individual or animal that takes a lead role and occupies the dominant position in the group.

Animal genetic resources databank: A databank that contains inventories of farm animal genetic resources and their immediate wild relatives, including any information that helps to characterize these resources.

Animal genome (gene) bank: A planned and managed repository containing animal genetic resources. Repositories include the environment in which the genetic resource has developed, or is now normally found (in situ) or facilities elsewhere (ex situ – in vivo or in vitro). For in vitro, ex situ genome bank facilities, germplasm is stored in the form of one or more of the following: semen, ova, embryos and tissue samples.

Animal model: This is a system for genetic evaluations that estimates breeding values of bulls and cows at the same time. The system uses production data on all known relatives in calculating a genetic evaluation.

Anthropomorphic: Attributing to animals the emotions of humans.

Appeasement behaviour: Actions or body language that may send a danger, threat or challenge signal to another animal. Zookeepers must strive to avoid such behaviour.

Appetitive behaviour: An introductory phase of a specific behaviour pattern that preceeds instinctive behaviour. e.g. roaming before taking chase of a fleeing animal.

Artiodactyls: Terrestrial herbivorous mammals characterized by having an even number of toes and by having the main limb axes pass between the third and fourth toes examples include: hippos (4 toes), deer, sheep, antelope, goats (2 toes).

Average Daily Gain (ADG): ADG is the amount of body weight change of an animal per day. For example, between the 50 and 100–day weighings: (100 Day Weight – 50 Day Weight) ÷ 50 Days or ADG = WT change/day

AZA: American Zoological Association.

Bacillus thuringiensis (Bt): A soil bacterium that produces toxins that are deadly to some pests. The ability to produce Bt toxins has been engineered into some crops. See Bt crops.

Back crossing: A form of crossbreeding in which an offspring is mated to either of its parents. The mating of a hybrid to a purebred of a parent breed or line. The mating of an individual (purebred or hybrid) to any other individual (purebred or hybrid) with which it has one or more ancestral breeds or lines in common. A mating system where the crossbred female progeny are mated to a male of either parental breed.

Backcross: Backcross is the mating of a two–breed animal back to one of the two parental breeds.

Backliner: An externally applied medicine, applied along the backline of a freshly-shorn sheep to control lice or other parasites. In the British Isles called "pour-on".

Bale: A wool pack containing a specified weight of wool as regulated by industry authorities.

Behaviour: What an animal does, how it cares for itself and how it reacts to other organisms and the environment in which it lives.

Bellwether: Originally an experienced wether given a bell to lead a flock; now mainly used figuratively for a person acting as a lead and guide.

Best linear unbiased prediction: A method of genetic prediction that is particularly appropriate when performance data come from genetically diverse contemporary groups.

Beta animal: Subordinate or second animal in the social group.

Biodiversity OR biological diversity: The variety of life in all its forms, levels and combinations, encompassing genetic diversity, species diversity and ecosystem diversity.

Biological rank: When a dominance exists between different species, which compete for food, water or space.

Biopharming: The production of pharmaceuticals such as edible vaccines and antibodies in plants or domestic animals.

Boiling Point Elevation: One of the colligative properties. The boiling point of a solution is increased over that of water by the presence of solutes, and the extent of the increase is a function of both concentration and molecular weight.

***Bos indicus*:** The breeds of cattle referred to as tropical or humped breeds, e.g. Zebu breeds including Brahman and Sahiwal.

***Bos taurus*:** The temperate, British and European breeds of cattle, e.g. Hereford, Angus, Murray Grey and Charolais.

Bottle lamb or cade lamb: An orphan lamb reared on a bottle. Also "poddy lamb" or "pet lamb".

Break: A marked thinning of the fleece, producing distinct weakness in one part of the staple.

Breed at risk: Any breed that may become extinct if the factors causing its decline in numbers are not eliminated or mitigated. Breeds may be in danger of becoming extinct for a variety of reasons. Risk of extinction may result from, inter alia, low population size; direct and indirect impacts of policy at the farm, country or international levels; lack of proper breed organization; or lack of adaptation to market demands. Breeds are categorized as to their risk status on the basis of, inter alia, the actual numbers of male and/or female breeding individuals and the percentage of pure-bred females. FAO has established categories of risk status; critical, endangered, critical-maintained, endangered-maintained, and not at risk.

Breed complementarity: An improvement in the overall performance of crossbred offsprings resulting from crossing breeds of different but complementary biological types.

Breed not at risk: A breed where the total number of breeding females and males is greater than 1 000 and 20 respectively; or the population size approaches 1 000 and the percentage of pure-bred females is close to 100 percent, and the overall population size is increasing.

Breed: Either a sub specific group of domestic livestock with definable and identifiable external characteristics that enable it to be separated by visual appraisal from other similarly defined groups within the same species or a group for which geographical and/or cultural separation from phenotypic ally similar groups has led to acceptance of its separate identity.

Breeding objective: A weighted combination of traits defining aggregate breeding value for use in an economic selection index. A general goal for a breeding programme - a notion of what constitutes the best animal.

Breeding value: The value of an individual as a (genetic) parent. The part of an individual's genotypic value that is due to independent and therefore transmittable gene effects.

Breedplan. A performance recording system, operated by the National Beef Recording Scheme, which uses an animal's own performance plus the performance of all known relatives in the herd and all genetically correlated (or related) traits to estimate breeding values.

Broken-mouth or broken-mouthed: An old sheep which has lost some of its incisor teeth.

Bt crops: Crops that are genetically engineered to carry a gene from the soil bacterium Bacillus thuringiensis (Bt). The bacterium produces proteins that are toxic to some pests but non-toxic to humans and other mammals. Crops containing the Bt gene are able to produce this toxin, thereby providing protection for the plant. Bt corn and Bt cotton are examples of commercially available Bt crops.

Butt: An underweight bale of greasy wool in a standard wool pack.

Canids: Belonging to the Canidae family of carnivorous mammals including dogs and their allies.

Carrier: A heterozygous animal with one recessive and one dominant gene for a particular gene pair.

Central test: A central testing location where animals from different flocks are gathered to evaluate differences in performance traits under uniform management and environmental conditions.

CFA or cast for age: Sheep culled because of their age. Also see “cull ewe”.

Characterization of animal genetic resources: All activities associated with the description of AnGR aimed at better knowledge of these resources and their state. Characterization by a country of its AnGR will incorporate development of necessary descriptors for use, identification of the country's sovereign AnGR; baseline and advanced surveying of these populations including their enumeration and visual description, their comparative genetic description in one or more production environments, their valuation, and ongoing monitoring of those AnGR at risk.

Chromosome: A DNA molecule. Genes are located on the chromosomes. Sheep have 27 pairs of chromosomes. The self-replicating genetic structure of cells, containing genes, which determines inheritance of traits. Chemically, each chromosome is composed of proteins and a long molecule of DNA.

Circadian rhythm: A rhythmic process within an organism occurring independently of external synchronizing signals.

CITES: Convention on Intentional Trade in Endangered Species places restrictions on the importing or exporting of species of plants or animals that are rare or endangered.

Classical and operant conditioning: Involves rewarding an animal for a particular action while at the same time providing a separate and distinct stimulus. Pavlov's working giving dog food, which resulted in salivation while at the same time ringing a bell was repeated. Then he found just ringing the bell would elicit salivation. This operant conditioning is used to train dolphins.

Clip: All the wool from a flock (in Australian Wool Classing).

Clone: A genetic replica of an organism created without sexual reproduction.

Cloner: A genetic replica of another organism obtained through a non-sexual (no fertilization) reproduction process. Cloning by nucleus transfer involves the transfer of a donor nucleus from (cultured) cells of embryonic, fetal or adult origin into the recipient cytoplasm of an enucleated oocyte or zygote, and the subsequent development of embryos and animals. These clones usually have different mitochondria: genomes.

Closed flock: An existing flock that does not introduce any outside breeding stock.

Collateral relatives: Relatives that are not progeny or parents (descendants or ascendants). For example, brothers.

Colligative properties: Properties which depend on the number of molecules in solution, a function of concentration and molecular weight, rather than just on the total percent concentration. Such properties include boiling point elevation, freezing point depression, and osmotic concentration.

Comeback: The progeny of a mating of a Merino with a British longwool sheep.

Complementarity: The use of two or more breeds in roles that are consistent with the strengths and weaknesses of the breeds, e.g. cows of moderate size and optimum milk mated to sires noted for high growth rate.

Composite (synthetic) breed: A hybrid with at least two and typically more breeds in its background. Composites are expected to be bred to their own kind, retaining a level of hybrid vigour normally associated with traditional crossbreeding systems

Composite: A breed resulting from the matings of two or more existing breeds and animals are selected from within the progeny to continue the breed e.g. Belmont Red, Droughtmaster.

Conception rate: The number of ewes that lamb compared to the number of ewes exposed for breeding. This is usually expressed as a percentage.

Congenital: A condition that exists at birth.

Conservation of farm animal genetic resources: Refers to all human activities including strategies, plans, policies and actions undertaken to ensure that the diversity of farm animal genetic resources is being maintained to contribute to food and agricultural production and productivity, now and in the future.

Contemporary group: A genetically similar group of sheep born in a particular time period and raised under the same management and environmental conditions.

Coprophagy: Feeding on dung or excrement.

Correlation coefficient: A measure of the interdependence of two random variables that ranges in value from "1 to +1, indicating perfect negative correlation at "1, absence of correlation at zero, and perfect positive correlation at +1. It determines the degree to which two variable's movements are associated.

Correlation: A measure of how two traits relate to each other. Correlation coefficients are expressed in the range –1.00 to +1.00.

Covariance functions: A function that gives covariance between records taken at times ti and tj as a function of the times. They are infinite dimensional equivalent to covariance matrices.

Crepuscular: Active during twilight hours e.g. Nighthawk.

Critical breed: A breed where the total number of breeding females is less than 100 or the total number of breeding males is less than or equal to five; or the overall population size is close to, but slightly above 100 and decreasing, and the percentage of pure-bred females is below 80 percent.

Critical distance: A component of the flight distance, distance at which the animal feels forced to defend itself.

Critical-maintained breed and endangered-maintained breed: Categories where critical or endangered breeds are being maintained by an active public conservation programme or within a commercial or research facility.

Crossbreeding: Mating of sires of one breed or breed combination to dams of another breed or breed combination

Crossbreeding: Mating system in which two or more straight breeds are combined.

Cross-pollination: Fertilization of a plant with pollen from another plant. Pollen may be transferred by wind, insects, other organisms, or humans.

Crutching: Shearing parts of a sheep (especially the hind end of some woollier breeds such as Merino), to prevent fly-strike. Also see "dagging".

Cryo-preservation: A process where cells or whole tissues are preserved by cooling to low sub-zero temperatures, such as (typically) -80°C or -196°C (the boiling point of liquid nitrogen). At these low temperatures, any biological activity, including the biochemical reactions that would lead to cell death is effectively stopped. Maintenance of the viability of excised tissues or organs by freezing at extremely low temperatures

Cull ewe: A ewe no longer suitable for breeding, and sold for meat.

Dagging: Clipping off dags. Also see "crutching".

Dags: Clumps of dried dung stuck to the wool of a sheep, which may lead to fly-strike. (Hence "rattle your dags!", meaning "hurry up!", especially used in New Zealand.)

Desirable genes: Genes which, when present, improve the animal performance.

Deviation: The difference between the average performance of the group and the performance of an individual in the group.

Diagnostic alleles: These are unique alleles to certain breeds, e.g. allele unique to indicine breeds or taurine breeds. They are used to determine the purity of breeds, the introgression by one breed type into a population and to determine the genetic composition of breeds

Dip: To immerse sheep in a plunge or shower dip to kill external parasites. Backliners are now replacing dips.

Direct Effect: That portion of a trait due to an individual's genotype.

Direct use value (DUV): The financial value of resources used domestically

Displacement activity: An animal be respond to a releaser in an alternative way under certain conditions e.g. stimulus to attach another animal may be redirected to another activity less dangerous.

Distribution or range: Is the geographic distribution of the species.

Diurnal active: During the day.

DNA (deoxyribonucleic acid): The chemical substance from which genes are made. DNA is a long, double-stranded helical molecule made up of nucleotides which are themselves composed of sugars, phosphates, and derivatives of the four bases adenine (A), guanine (G), cytosine (C), and thymine (T). The sequence order of the four bases in the DNA strands determines the genetic information contained. The long chain of molecules in most cells that carries the genetic message and controls al cellular functions in most forms of life.

DNA bank: Storage of DNA, which may or man not be the complete genome, but should always be accompanied by inventory information. (Note: at the present time, animals cannot be reestablished from DNA alone).

Domestic animal diversity (DAD): The spectrum of genetic differences within each breed, and across all breeds within each domestic animal species, together with the species differences; all of which are available for the sustainable intensification of food and agriculture production.

Dominance: A dominant gene affects the appearance of the sheep (phenotype). A dominant gene from only one parent will change the phenotype, even if the corresponding gene from the other parent is different.

Draft ewe: A ewe too old for rough grazing (such as moorland), "drafted" (selected) out of the flock to move to better grazing, usually on another farm. Generally spelt "draft", but in British Isles either as "draft" or "draught".

Drench: An oral veterinary medicine administered by a drenching gun (usually an anthelmintic).

Driving: Walking animals from one place to another.

Dynamic optical demarcation: When an animal uses a special signalling device in a stereotypical movement (the waving of the fiddler crabs claw) to alert other members of its species.

Earmark: A distinctive mark clipped out of the ear (or sometimes a tattoo inside the ear) to denote ownership.

Eartag: Plastic or metal tag clipped to ear, with number or name, or (in European Union) the official flock mark.

Ecological niche: Describes the role of the animal within its community. It considers what the animal does in relation to the food chain, plant and animal associations and energy flow. The niche might be compared to an occupation.

Embryo transfer: A reproductive technology in which embryos from donor females are collected and transferred in fresh or frozen form to recipient females.

Emulsion: Liquid droplets dispersed in another immiscible liquid. The dispersed phase droplet size ranges from 0.1 - 10 μ m. Important oil-in-water food emulsions, ones in which oil or fat is the dispersed phase and water is the continuous phase, include milk, cream, ice cream, salad dressings, cake batters, flavour emulsions, meat emulsions, and cream liquers. Examples of food water-in-oil emulsions are butter or margarine. Emulsions are inherently unstable because free energy is associated with the interface between the two phases. As the interfacial area increases, either through a decrease in particle size or the addition of more dispersed phase material. i.e. higher fat, more energy is needed to keep the emulsion from coalescing. Some molecules act as surface active agents (called surfactants or emulsifiers) and can reduce this energy needed to keep these phases apart.

Endangered breed: A breed where the total number of breeding females is between 100 and 1 000 or the total number of breeding males is less than or equal to 20 and greater than five; or the overall population size is close to, but slightly above 100 and increasing and the percentage of pure-bred females is above 80 percent; or the overall population size is close to, but slightly above 1 000 and decreasing and the percentage of pure-bred females is below 80 percent.

Environment: All conditions that affect an animal's performance that are not genetic.

Environmental variance: The variance due to environmental variation.

Enzyme-linked immunosorbent assay (ELISA): A technique using antibodies for detecting specific proteins. Used to test for the presence of a particular genetically engineered organism.

Escape reaction: Animals that sense a predator may run, flee or stay motionless in order to avoid contact.

Estimated Breeding Value (EBV): An estimate of an animals value as a parent for a particular production trait, eg growth rate and is calculated from the measured performance of the animal and its close relatives for that trait. In other words, a prediction of a breeding value. See breeding value.

Ethology: The study of the behaviour of animals in their normal environment.

Evaluation: Measurement of the characteristics that are important for production and adaptation, either of individual animals or of populations, most commonly in the context of comparative evaluation of the traits of animals or of populations.

Ewe: A female sheep capable of producing lambs. In areas where "gimmer" or similar terms are used for young females, may refer to a female only after her first lamb. In some areas "yow".

Ex situ conservation of farm animal genetic diversity: All conservation of farm animal genetic material in vivo, but out of the environment in which it developed, and in vitro including, inter alia, the cryoconservation of semen, oocytes, embryos, cells or tissues. Note that ex situ conservation and ex situ preservation are considered here to be synonymous.

Expected progeny difference (EPD): The expected difference between the progeny performance of a breed average animal and the performance of an individual's progeny for any given trait(s).

Experience: This is the basis for trial and error learning. An example is the frog that snaps up a wasp and gets its tongue stung. The frog learns to associate the color and pattern of the wasp with pain and will not again try to eat a wasp. When a young insectivorous bird leaves its nest, it picks at pebbles, leaves or sticks until it finds and eats an insect. Once the bird has eaten the insect, a reward, it greatly reduces the frequency with which it picks at pebbles, sticks or other non-rewarding items.

Extinct breed: A breed where it is no longer possible to recreate the breed population. Extinction is absolute when there are no breeding males (semen), breeding females (oocytes), nor embryos remaining.

F1: Progeny resulting from the mating of a purebred ram and a purebred ewe of different breeds.

Farm animal genetic resources (AnGR): Those animal species that are used, or may be used, for the production of food and agriculture, and the populations within each of them. These populations within each species can be classified as wild and feral populations, landraces and primary populations, standardized breeds, selected lines, and any conserved genetic material.

Field trial: A test of a new technique or variety, including biotech-derived varieties, done outside the laboratory but with specific requirements on location, plot size, methodology, etc.

Fitness traits: Traits related to reproduction, survival/mortality of cattle.

Fixed action patterns: These are complex instinctive behaviours. An example might be the series of behavioural activities engaged in by birds that are not sexually dimorphic that allows mate selection. It may include bowing, mutual vocalization, mutual preening and posturing. Sometimes a series of dozens of behavioural acts, each stimulating a response, follow one another.

Fleece: The wool covering of a sheep.

Flight distance: Is recognized in many prey species which, while they may defend a territory against conspecifics, flee when a perceived potential predator gets within this distance. Flight distance can be modified by habituation.

Flock: A group of sheep (or goats). All the sheep on a property (in Australian Wool Classing); also all the sheep in a region or country. Sometimes called *herd* or *mob*.

Flushing (eggs/embryo): Removing unfertilised or fertilised egg from an animal; often as part of an embryo transfer procedure. In other words, providing especially nutritious feed in the few weeks before mating to improve fertility, or in the period before birth to increase lamb birth-weight.

Foam: A gas dispersed in a liquid where the gas bubbles are the discrete phase. There are many food foams including whipped creams, ice cream, carbonated soft drinks, mousses, meringues, and the head of a beer. A foam is likewise unstable and needs a stabilizing agent to form the gas bubble membrane.

Fold (or sheepfold): A pen in which a flock is kept overnight to keep the sheep safe from predators, or to allow the collection of dung for manure.

Folding: Confining sheep (or other livestock) onto a restricted area for feeding, such as a temporarily fenced part of a root crop field, especially when done repeatedly onto a sequence of areas.

Foot rot: Infectious pododermatitis, a painful hoof disease commonly found in sheep (also goats and cattle), especially when pastured on damp ground.

Freezing point depression: Of a solution is a colligative property associated with the number of dissolved molecules. The lower the molecular weight, the greater the ability of a molecule to depress the freezing point for any given concentration. For example, in ice cream manufacturing, monosaccharides such as fructose or glucose produce a much softer ice cream than disaccharides such as sucrose, if the concentration of both is the same.

Gene (DNA) sequencing: Determining the exact sequence of nucleotide bases in a strand of DNA to better understand the behaviour of a gene.

Gene expression: The result of the activity of a gene or genes which influence the biochemistry and physiology of an organism and may change its outward appearance.

Gene flow: The movement of genes from one individual or population to another genetically compatible individual or population.

Gene mapping: Determining the relative physical locations of genes on a chromosome. Useful for plant and animal breeding.

Gene: The basic genetic unit of heredity that occurs in pairs in an individual, but is passed to progeny from one parent as a single unit. In other words, the fundamental physical and functional unit of heredity. A gene is typically a specific segment of a chromosome and encodes a specific functional product (such as a protein or RNA molecule). The unit of heredity transmitted from generation to generation during sexual or asexual reproduction. More generally, the term "gene" may be used in relation to the transmission and inheritance of particular identifiable traits.

Generation interval: Average age of parents when the animals that will replace them in the flock are born. This represents the average turnover rate of a flock.

Genetic correlation: Relationships between traits that arise because some of the same genes affect both traits.

Genetic distances: This is a measure of gene differences between populations (and hence genetic relationships among them) measured by some numerical quantity and usually refer to the gene differences as measured by a function of gene frequencies

Genetic distancing: The collection of the data on phenotypic traits, marker allele frequencies or DNA sequences for two or more populations, and estimation of the genetic distances between each pair of populations. From these distances, the best representation of the relationships among all the populations may be obtained.

Genetic engineering: Manipulation of an organism's genes by introducing, eliminating or rearranging specific genes using the methods of modern molecular biology, particularly those techniques referred to as recombinant DNA techniques.

Genetic gain: Genetic gain is the amount of increase in performance that is achieved through genetic selection after one generation of selection

Genetic improvement programme: A systematic plan to enhance the genetic of animal performance

Genetic markers: A gene or DNA sequence having a known location on a chromosome and associated with a particular gene or trait. A gene phenotypically associated with a particular, easily identified trait and used to identify an individual or cell carrying that gene. Genetic markers associated with certain diseases can be detected in the blood and used to determine whether an individual is at risk for developing a disease.

Genetic merit: The genetic worth of an animal for a particular trait can be expressed as an EPD, EBV or ABC.

Genetic modification: The production of heritable improvements in plants or animals for specific uses, via either genetic engineering or other more traditional methods. Some countries other than the United States use this term to refer specifically to genetic engineering.

Genetic progress: An increase in the average genetic merit of a population for a particular trait as a result of selective breeding.

Genetic variance: Phenotypic variance resulting from the presence of different genotypes in the population. Within a population, the measure of how much of the variance of a particular phenotype is due to genotype variation. In the context of the key equation for genetic change, variability of breeding values within a population for a trait under selection

Genetic: A measure of the genetic similarity between any pair of populations. Such distance may be based on phenotypic traits, allele frequencies or DNA sequences. For example, genetic distance between two populations having the same allele frequencies at a particular locus, and based solely on that locus, is zero. The distance for one locus is maximum when the two populations are fixed for different alleles. When allele frequencies are estimated for many loci, the genetic distance is obtained by averaging over these loci.

Genetically engineered organism (GEO): An organism produced through genetic engineering.

Genetically modified organism (GMO): An organism produced through genetic modification.

Genetics: The study of the patterns of inheritance of specific traits.

Genome: All the genetic material in all the chromosomes of a particular organism. The complete set of genes and non-coding sequences present in each cell of an organism, or the genes in a complete haploid set of chromosomes of a particular organism.

Genomic library: A collection of biomolecules made from DNA fragments of a genome that represent the genetic information of an organism that can be propagated and then systematically screened for particular properties. The DNA may be derived from the genomic DNA of an organism or from DNA copies made from messenger RNA molecules. A computer-based collection of genetic information from these biomolecules can be a "virtual genomic library."

Genomics: The mapping and sequencing of genetic material in the DNA of a particular organism as well as the use of that information to better understand what genes do, how they are controlled, how they work together, and what their physical locations are on the chromosome.

Genotype by environment interaction: A dependent relationship between genotypes and environments in which the difference in performance between two or more genotypes changes from one environment to another. In other words, changes in the ranking of performance of genotypes in different environments. For example, one genotype may perform the best in one environment and only average in another environment.

Genotype: The genetic identity of an individual. Genotype often is evident by outward characteristics, but may also be reflected in more subtle biochemical ways not visually evident.

Gimmer: A young female sheep, usually before her first lamb (especially used in the north of England). Also "theave".

Guard llama: A llama (usually a castrated male) kept with sheep as a guard. The llama will defend the flock from predators such as foxes and dogs.

Habitat: Habitat is the area within the range that contains the environmental factors and conditions needed to support the species.

Habituation: This is the suppression of response to a repeated harmless stimulus. An example might be animals learning to ignore a train or sound of a horn near the enclosure. The stimulus results in no reward or punishment.

Half-sibs: Individuals who have the same sire or dam (i.e. half brothers and half sisters).

Hefting (or heafing): The instinct in some breeds of keeping to a certain *heft* (or *heaf* – a small local area) throughout their lives. Allows different farmers in an extensive landscape such as moorland to graze different areas without the need for fences, the ewes remaining scattered evenly over a wide area. Lambs usually learn their heft from their mothers.

Herbicide-tolerant crops: Crops that have been developed to survive application(s) of particular herbicides by the incorporation of certain gene(s) either through genetic engineering or traditional breeding methods. The genes allow the herbicides to be applied to the crop to provide effective weed control without damaging the crop itself.

Hereditary: A condition controlled or influenced to some degree by gene action. This is in contrast to characters which are entirely controlled by environmental variables.

Heredity: The transmission of characteristics from parents to their offspring through genes.

Heritability: Is the proportion of a measured or observed trait that is transmitted to the offspring by genes. Therefore, the higher the heritability, the more likely the individual's actual performance will be passed to offspring and response due to selection for that trait will be faster. That fraction of the total variation for any trait in a population which is due to additive genetic effects. The efficiency of transmission of a trait from parent to offspring.

Heterosis (Hybrid Vigour): The increase in performance observed for a specific crossbred above the average performance of the parental lines. An increase in the performance of crossbreds over that of purebreds, most noticeably in traits like fertility and survival ability

Heterozygous: Each gene of a specific gene pair is different in an individual (i.e. Aa).

High-input production environment: A production environment where all rate-limiting inputs to animal production can be managed to ensure high levels of survival, reproduction and output. Output and production risks are constrained primarily by managerial decisions.

Hogget or hogg: A young sheep of either sex from about 9 to 18 months of age (until it cuts two teeth). Also the meat of a hogget.

Home range: Is the living area normally occupied and inhabited by an animal.

Homozygous: Each gene of a specific gene pair is the same in an individual (i.e. AA).

Husbandry: Application of scientific principles to agriculture, especially to animal breeding.

Hybrid: The offspring of any cross between two organisms of different genotypes.

Ideal model: A model that is close to a true model based on an understanding of the problem.

Identity preservation: The segregation of one crop type from another at every stage from production and processing to distribution. This process is usually performed through audits and site visits and provides independent third-party verification of the segregation.

Imprinting: This is a genetically programmed form of learning in which a newly hatched or new-born learns to identify with its species within a finite time by following or being exposed to stimuli it learns to identify. The sensitive period for imprinting baby mallard ducks to follow a man instead of other ducks is 13 to 16 hours after hatching. This is the time when the female typically leaves the nest and , by following, the ducklings learn they are ducks.

In lamb: Pregnant.

In situ conservation of farm animal genetic diversity: All measures to maintain live animal breeding populations, including those involved in active breeding programmes in the agro-ecosystem where they either developed or are now normally found, together with husbandry activities that are undertaken to ensure the continued contribution of these resources to sustainable food and agricultural production, now and in the future.

***In situ* conservation:** On-site conservation. It is the process of protecting an species in its natural habitat, either by protecting or cleaning up the habitat itself, or by defending the species from predator

Inbreeding: A system of mating in which mates are more closely related than average individuals of the population to which they belong. Production of progeny from closely related parents. Inbreeding increases the number of homozygous gene pairs and decreases the number of heterozygous gene pairs. Inbreeding increases prepotency and the expression of undesirable recessive genes.

Indirect use values (IUV): These are values of functions of livestock which do not have monetary quantification for consideration in a commercial market.

Individual heterosis: The advantage of the crossbred individual relative to the average of the purebred individuals. The positive difference in performance of the individual compared to the average of its parents often in respect to the turnoff animal. Usually referred to as 'nicking' by stud breeders.

Innate: Any behaviour that is performed without prior learning.

Insecticide resistance: The development or selection of heritable traits (genes) in an insect population that allow individuals expressing the trait to survive in the presence of levels of an insecticide (biological or chemical control agent) that would otherwise debilitate or kill this species of insect. The presence of such resistant insects makes the insecticide less useful for managing pest populations.

Insect-resistance management: A strategy for delaying the development of pesticide resistance by maintaining a portion of the pest population in a refuge that is free from contact with the insecticide. For Bt crops this allows the insects feeding on the Bt toxin to mate with insects not exposed to the toxin produced in the plants.

Insect-resistant crops: Plants with the ability to withstand, deter or repel insects and thereby prevent them from feeding on the plant. The traits (genes) determining resistance may be selected by plant breeders through cross-pollination with other varieties of this crop or through the introduction of novel genes such as Bt genes through genetic engineering.

Insight or reasoning: This is the most sophisticated form of learning where the solution to an entirely different problem is used to solve a new challenge. Chimpanzees using straws to fish termites from burrows be an example.

Instinct: Any behaviour that is performed without prior learning and that is inherited. Suckling by new-born animals is an instinctive behaviour.

Intellectual property rights: The legal protection for inventions, including new technologies or new organisms (such as new plant varieties). The owner of these rights can control their use and earn the rewards for their use. This encourages further innovation and creativity for the benefit of us all. Intellectual property rights protection includes various types of patents, trademarks, and copyrights.

Introgression: A breeding strategy for transferring specific favourable alleles from a donor population to a recipient population. This would, for example, be of great interest for genes responsible for disease resistance genes, which could be introgressed into a susceptible but otherwise economically superior breed.

ISIS: International Species Inventory System - a computer based information system for animals in captivity.

Ked, or sheep ked: *Melophagus ovinus*, a species of louse-fly, a nearly flightless biting fly infesting sheep.

Kinesis: Induction of movement in an animal where the movement lacks directional orientation and depends on the stimulus intensity producing a change of rate of random movements.

Lagomorphs: Order of animals including hares, rabbits and pikas, differentiated from rodents by two pairs of upper incisors covered by enamel.

Lamb (pronounced /læm/): A young sheep in its first year. In many eastern countries there is a looser use of the term which may include hoggets. Also the meat of younger sheep.

Lamb marking: The work of earmarking, docking and castration of lambs.

Lamb's fry: Lamb's liver (as food).

Lambing jug: A small pen to confine ewes and newly born lambs.

Lambing percentage: The number of lambs successfully reared in a flock compared with the number of ewes – effectively a measure of the success of lambing and the number of multiple births. May vary from around 100% in a hardy mountain flock (where a ewe may not be able to rear more than one lamb safely), to 150% or more in a well-fed lowland flock (whose ewes can more easily support twins or even triplets).

Lambing: The process of giving birth in sheep. Also the work of tending lambing ewes (shepherds are said to "lamb" their flocks).

Lek: Males defend and area or lek among several other male's territories during mating.

Line breeding: A form of inbreeding in which an effort is made to maintain high relationship in subsequent generations with a favoured ancestor.

Linear functions: function which satisfies and for all and in the domain, and all scalars .

Linebreeding: A form of inbreeding which increases the average relationship of the individuals in a flock to an outstanding ancestor or line of ancestors.

Linecross: Progeny produced by crossing two or more inbred lines.

Livestock guarding dog or livestock guardian dog: A dog bred and trained to guard sheep from predators such as bears, wolves, people or other dogs. Usually a large type of dog, often white and woolly, apparently to allow them to blend in with the sheep. Sometimes given a spiked collar to prevent attack by wolves or dogs. Does not usually herd the sheep. Sometimes called a *sheepdog* – but also see separate entry for this.

Livestock revolution: is a convenient expression that summarizes a complex series of interrelated processes and outcomes in production, consumption, and economic growth arising comes from the participation of developing countries in large-scale transformations that had previously occurred mostly in the temperate zones of developed countries

longitudinal data: Data resulting from the observation of a population on a number of variables over time. Whenever observations are made more than once, the data is considered to be longitudinal

Low-input production environment: A production environment where one or more rate-limiting inputs impose continuous or variable severe pressure on livestock, resulting in low survival, reproductive rate or output. Output and production risks are exposed to major influences which may go beyond human management capacity

Lug mark: Local term in Cumbria for "earmark".

Management of farm animal genetic resources: The sum total of technical, policy, and logistical operations involved in understanding (characterization), using and developing (utilization), maintaining (conservation), accessing, and sharing the benefits of animal genetic resources.

Market-oriented dairy technology: Dairy technologies geared towards markets e.g. fermentation, pasteurization e.t.c.

Marking knife: A knife with a clamp and hook made for lamb marking.

Mass selection: A form of artificial selection in which only individuals with phenotypic values greater or less than some threshold level are used for breeding.

Maternal breed: The breed of the dam, which is often selected for its reproductive and adaptive traits.

Maternal effect: The effect that a dam's maternal ability may have on a trait. For example, the effect of milk yield on weaning weight.

Maternal heterosis: Positive differences in the performance of the crossbred female which are used to give further performance benefits in an additive way to the progeny. The advantage of the crossbred mother over the average of purebred mothers.

Maternal sires: Sires that are used in a crossbreeding programme to sire replacement females for the flock. These sires have strong maternal traits to produce the next generation of productive ewes.

Mean: The average value for a trait of a group.

Microsatellites: Short, noncoding DNA sequence (a tandemly repetitive DNA sequence) that is repeated many times within the genome of an organism.

Many repeats tend to be concentrated at the same locus Repetitive stretches of short sequences of DNA used as genetic markers to track inheritance in families.

Mob: A group or cohort of sheep of the same breed that have run together under similar environmental conditions since the previous shearing (in Australian Wool Classing).

Molecular biology: The study of the structure and function of proteins and nucleic acids in biological systems.

Molecular genetics: The branch of genetics that deals with hereditary transmission and variation on the molecular level. It deals with the expression of genes by studying the DNA sequences of chromosomes.

Monorchid: A male mammal with only one descended testicle, the other being retained internally. Monorchid sheep are less fertile than full rams, but have leaner meat than wethers.

Mule: A type of cross-bred sheep, both hardy and suitable for meat (especially in northern England). Usually bred from a Bluefaced Leicester ram on hardy mountain ewes such as Swaledales. May be qualified according to the female parent: for example a Welsh mule is from a Blue-faced Leicester ram and a Welsh Mountain ewe.

Mulesing: A practice in Australia of cutting off wrinkles from the crutch area of Merinos, to prevent fly-strike. Controversial, and illegal in other parts of the world. Named after a Mr Mules.

Multiple Trait Selection: Selection for more than one trait at the same time.

Muster: To round up all of the sheep (cattle or horses) into one mob.

Mutation: Any heritable change in DNA structure or sequence. The identification and incorporation of useful mutations has been essential for traditional crop breeding.

Mutton: The meat of an older sheep. Also an older female sheep to be used for this purpose. May refer to goat meat in eastern countries. Derived from the Anglo-Norman French word *mouton* ("sheep").

Nocturnal: Active at night e.g. owls.

Nonadditive Gene Effects: Favourable effects produced by specific gene combinations in which the members of the gene pairs are not equally expressed. This is the primary cause of heterosis.

Normal Distribution (Bell Curve): A graph of all possible performance levels for a trait. The number of individuals displaying each performance level usually forms a normal distribution.

Nucleotide: A subunit of DNA or RNA consisting of a nitrogenous base (adenine, guanine, thymine, or cytosine in DNA; adenine, guanine, uracil, or cytosine in RNA), a phosphate molecule, and a sugar molecule (deoxyribose in DNA and ribose in RNA). Many of nucleotides are linked to form a DNA or RNA molecule.

Nucleus herds: These are herds at the top of the tier in the breeding scheme and are the source of superior genetic material to the cooperating herds.

Observed heterozygosity: The percentage of loci heterozygous per individual or the number of individuals heterozygous per locus.

Old-season lamb: A lamb a year old or more.

Olfactory demarcation: An animal signalling its presence via scent which can be in its faeces, urine and products of special scent glands. Generally male scent glands are more elaborate than those of the females.

Open nucleus schemes: A nucleus breeding scheme in which the flow of germ plasm is bidirectional- from the nucleus to cooperating herds or flocks and from cooperating herds or flocks to the nucleus

Operant conditioning: A process of behaviour modification in which the likelihood of a specific behaviour is increased or decreased through positive or negative reinforcement each time the behaviour is exhibited, so that the subject comes to associate the pleasure or displeasure of the reinforcement with the behaviour.

Operational model: a generic of an ideal model usually simplified due to missing information or computational problems. The model allows a certain level of accuracy in predictions and is based on some underlying principles and assumptions about an ideal model.

Option values (OV): These are values derived from the value given to safeguard an asset for the option of using it in future. It is a kind of insurance value against the occurrence of, for example, a new animals disease or drought changes.

Orf: A disease of sheep (and goats) causing sores, especially in the mouth.

Organic agriculture: A concept and practice of agricultural production that focuses on production without the use of synthetic inputs and does not allow the use of transgenic organisms. USDA's National Organic Programme has established a set of national standards for certified organic production which are available online.

Osmotic pressure: A chemical force caused by a concentration gradient. It is a colligative property and the principle behind membrane processing. pH:

pH is a measure of the activity of the hydronium ion (H3O+) which, according to the Debye-Huckel expression, is a function of the concentration of the hydronium ion [H3O+], the effective diameter of the hydrated ion and the ionic strength (μ m) of the solvent. For solutions of low ionic strength (μ m < 0.1) hydronium ion activity is nearly equivalent to [H3O+] which is normally abbreviated to [H+]. Then, for a weak acid (HA) dissociating to H+ and A- with a dissociation constant, Ka and pKa equal to -log10 Ka, the most important relationships are defined the following two equations: Ka = [H+] [A-] / [HA] pH = log 1 / [H+] = pKa + log [A-] / [HA] Reynold's Number: a dimensionless expression used in predicting flow patterns: Stoke's Equation: The velocity at which a sphere will rise or fall in a liquid varies as the square of its diameter: For example, a fat globule with a diameter of 2 microns will rise 4 times faster than a fat globule with a diameter of 1 micron.

Outbreeding: A system of mating in which mates are less related than average individuals of the population being intermated.

Outcrossing: Mating between different populations or individuals of the same species that are not closely related. The term "outcrossing" can be used to describe unintended pollination by an outside source of the same crop during hybrid seed production.

Peck orders: Animals social position within a group, allows one to discern different levels of dominance.

Pedigree: A list of an individual's ancestors.

Percent choice: The percentage of animals grading choice in the American carcase grading system (a fat grade).

Performance Test: Systematic collection of performance information for use in decision making to improve genetic merit, productivity, efficiency and profitability.

Personal or intimate distance: Is the individual distance within which conspecifics are tolerated. In some species such as horses the distance may be 0. Personal distance is often considered a component of territoriality.

Personal space: In group species this is the distance inside of which the intrusion of a fellow member of the species creates stress where that intruder is a mate or immature. In humans this distance is about an arm's length in western societies but is somewhat less in Mediterranean countries.

Pesticide resistance: The development or selection of heritable traits (genes) in a pest population that allow individuals expressing the trait to survive in the presence of levels of a pesticide (biological or chemical control agent) that would otherwise debilitate or kill this pest. The presence of such resistant pests makes the pesticide less useful for managing pest populations.

Pest-resistant crops: Plants with the ability to withstand, deter or repel pests and thereby prevent them from damaging the plants. Plant pests may include insects, nematodes, fungi, viruses, bacteria, weeds, and other.

Phenotype: The observed or measured expression of a trait for an individual (i.e. weaning weight, birth type, etc.). Phenotype is equal to genotype plus environment effects. visible and/or measurable characteristics of an organism (how it appears outwardly).

Phenotypic correlation: A correlation between two traits caused by genetic plus environmental factors influencing both traits.

Plant breeding: The use of cross-pollination, selection, and certain other techniques involving crossing plants to produce varieties with particular desired characteristics (traits) that can be passed on to future plant generations.

Plant pests: Organisms that may directly or indirectly cause disease, spoilage, or damage to plants, plant parts or processed plant materials. Common examples include certain insects, mites, nematodes, fungi, molds, viruses, and bacteria.

Plant-incorporated protectants (PIPs): Pesticidal substances introduced into plants by genetic engineering that are produced and used by the plant to protect it from pests. The protein toxins of Bt are often used as PIPs in the formation of Bt crops.

Pleiotropy: The control by a single gene of several distinct and seemingly unrelated phenotypic effects.

Poddy lamb or pet lamb: An orphan lamb reared on a bottle. Also "bottle lamb" or "cade lamb".

Polymerase chain reaction (PCR): A technique used to create a large number of copies of a target DNA sequence of interest. One use of PCR is in the detection of DNA sequences that indicate the presence of a particular genetically engineered organism.

Pour-on: An externally applied medicine, applied along the backline of a freshly-shorn sheep to control lice or other parasites. Also called "backliner".

Precocious: Early maturing, ahead of development.

Prepotent: The ability of a parent to pass traits to it's progeny so that they resemble the parent more than usual. Inbred sheep are usually more prepotent than outbred sheep because more homozygous gene pairs are present.

Production environment: All input-output relationships, over time, at a particular location. The relationships will include biological, climatic, economic, social, cultural and political factors, which combine to determine the productive potential of a particular livestock enterprise.

Production traits: Characteristics of animals, such as the quantity or quality of the milk, meat, fibre, eggs, draught, etc. they (or their progeny) produce, which contribute directly to the value of the animals for the farmer; and that are identifiable or measurable at the individual level. Production traits of farm animals are generally quantitatively inherited, i.e. they are influenced by many genes whose expression in a particular animal also reflects environmental influences.

Progeny testing: A test used to help predict an individual's breeding values involving multiple matings of that individual and evaluation of its offspring. The evaluation of an individual's genotype using the performance records of it's progeny.

Promoter: A region of DNA that regulates the level of function of other genes.

Protein: A molecule composed of one or more chains of amino acids in a specific order. Proteins are required for the structure, function, and regulation of the body's cells, tissues, and organs, and each protein has a unique function.

Pure breeding: Also straightbreeding. The mating of purebreds of the same breed.

Pyschology: Is the science of mental processes and behaviour in humans.

Qualitative traits: Traits which show a sharp distinction between phenotypes. For example, polled and horned. Usually only one or few gene pairs are involved.

Quantitative trait loci (QTL): A loci that affect a quantitative trait(s)

Quantitative traits: Traits that do not show a sharp distinction between phenotypes. There is a gradual variation from one phenotype to another, for example, birth weight. Usually, environmental effects as well as many gene pairs are involved.

Raddle: Coloured pigment used to mark sheep for various reasons, such as to show ownership, or to show which lambs belong to which ewe. May be strapped to the chest of a ram, to mark the backs of ewes he tups (different rams may be given different colours). Also a verb ("that ewe's been raddled"). Also "ruddy".

Ram: An uncastrated adult male sheep. Also *tup*.

Random mating: A mating system where all ewes have the same chance of being mated to any ram used.

Recessive genes: Recessive genes only affect the phenotype when present in a homozygous condition. Therefore, the recessive gene must be received from both parents before the recessive phenotype will be observed.

Recombinant DNA (rDNA): A molecule of DNA formed by joining different DNA segments using recombinant DNA technology.

Recombinant DNA technology: Procedures used to join together DNA segments in a cell-free system (e.g. in a test tube outside living cells or organisms). Under appropriate conditions, a recombinant DNA molecule can be introduced into a cell and copy itself (replicate), either as an independent entity (autonomously) or as an integral part of a cellular chromosome.

Redirected activity: An animal responding to a releaser e.g. to fight may alter its behaviour to carry out another activity e.g. male deer may scratch a tree with its antlers.

Reflexes: These are responses of part of the body to a stimulus that does not involve the higher brain centers. The blinking of the eye or knee jerk are examples of reflexes.

Regression coefficient: The slope of a line obtained using linear least square fitting.

Releaser: A behaviour pattern or structure, which serves to trigger a behaviour pattern e.g. sounds of young or color patterns.

Ribonucleic Acid (RNA): A chemical substance made up of nucleotides compound of sugars, phosphates, and derivatives of the four bases adenine (A), guanine (G), cytosine (C), and uracil (U). RNAs function in cells as messengers of information from DNA that are translated into protein or as molecules that have certain structural or catalytic functions in the synthesis of proteins. RNA is also the carrier of genetic information for certain viruses. RNAs may be single or double stranded.

Riggwolter: A sheep that has fallen onto its back and is unable to get up (usually because of the weight of its fleece).

Ring: A mob of sheep moving around in a circle.

Ringing: The removal of a circle of wool from around the pizzle of a male sheep.

Rooing: Removing the fleece by hand-plucking. Done once a year in late spring, when the fleece begins to loosen naturally, especially in some breeds.

Rotational Crossbreeding: Systematic crossing of two or more breeds in which the crossbred ewes are mated to rams of the breed contributing the least genes to that ewe's phenotype.

Rotational: A crossbreeding system which systematically uses three or more breeds crossbreeding in rotation.

Ruddy: Local Cumbrian term for "raddle".

Rut: Males of certain species of mammals become sexually active and aggressive during a period prior to mating e.g. antlered an horned hoof species and some elephants.

***Sanga*:** The Sanga breeds appear to have developed thousands of years ago from a mixture of *Bos indicus* and *Bos taurus* breeds.

Scab or sheep scab: A skin disease of sheep caused by attack by the sheep scab mite *Psoroptes ovis*, a psoroptid mite.

Selectable marker: A gene, often encoding resistance to an antibiotic or an herbicide, introduced into a group of cells to allow identification of those cells that contain the gene of interest from the cells that do not. Selectable markers are used in genetic engineering to facilitate identification of cells that have incorporated another desirable trait that is not easy to identify in individual cells.

Selection criteria: Phenotypic values or other pieces of information that form the basis for selection decisions

Selection differential: The difference between the average for a trait in replacement animals and the average of the group from which the replacements were chosen. The expected response from selection equals selection differential times heritability of a trait.

Selection index: A linear combination of phenotypic information and weighting factors used for genetic prediction when performance data comes from generally similar contemporary groups.

Selection intensity: A measure of how particular breeders are in deciding which individuals are selected. The difference between the mean selection criterion of those individuals selected to be parents and the average selection criterion of all potential parents, expressed in standard deviation units

Selection: The process of deciding which animals will be parents of the next generation.

Selective breeding: Making deliberate crosses or matings of organisms so the offspring will have particular desired characteristics derived from one or both of the parents.

Sequence breeding: Breeding systems in which males of two or more breeds are used in sequence on crossbred female population.

Shearing: Cutting off the fleece, normally done in two pieces by skilled shearers. A sheep may be said to have been either *sheared* or *shorn*, depending on dialect.

Sheep: The species, or members of it. The plural is the same as the singular, and it can also be used as a mass noun. Normally used of individuals of any age, but in some areas only for those of breeding age.

Sheepdog or shepherd dog: A dog used to move and control sheep, often very highly trained. Other types of dog may be used just to guard sheep.

Sheepwalk: An area of rough grazing occupied by a particular flock or forming part of a particular farm.

Shepherd: A stockperson or farmer who looks after sheep while they are in the pasture.

Shepherding: The act of shepherding sheep, or sheep husbandry more generally.

Sibs (Full Sibs): Brothers and sisters from the same sire and dam.

Slink: A very young lamb.

Smallholder: This is a producer who typically keep 1-3 animals

Social rank: Is the position an animal holds in a social group of the same species.

Species-specific: Any behaviour that is the same in all members of a species.

Specific action potential: Instinctive behaviour is triggered by internal factors, hormonal or otherwise, that build to a certain level and allow the behaviour to take place. This build up is called the specific action potential. It is responsible for an animal performing one behaviour in preference to other behaviours.

Stag: A ram castrated after about 6 months of age.

Static-optical demarcation: Result arises from the presence or appearance of the animal's body. Visual demarcation of animals territory so other members of the species can recognize it using the whole body or even parts of it. Form and color are often important.

Stereotyped: Any behaviour that an animals repeats in the same way.

Store: A sheep (or other meat animal) in good average condition, but not fat. Usually bought by dealers to fatten for resale.

Straightbred: Animals of a particular breed as opposed to crossbred animals.

Sucker: An unweaned lamb.

Super-ovulation: The administration of a hormone causing a female to develop and release more eggs than normal

Sustainable intensification of national production systems: The manipulation of inputs to, and outputs from, livestock production systems aimed at increasing production and/or productivity and/or changing product quality, while maintaining the long- term integrity of the systems and their surrounding environment, so as to meet the needs of both present and future human generations. Sustainable agricultural intensification respects the needs and aspirations of local and indigenous people, takes into account the roles and values of their locally adapted genetic resources, and consider the need to achieve long-term environmental sustainability within and beyond the agro-ecosystem.

Synthetic breed: A new breed established from the mating of the progeny of two or more existing breeds.

Taxes: This name is given to behaviour that involves movement to or away from a stimulus. An example would be cockroaches fleeing from light.

Teg: A sheep in its second year. Also see "shearling".

Terminal cross: The progeny of a particular crossbred mating that are sold for slaughter and bred for market and growth attributes.

Terminal sires: Sires that are used in a crossbreeding programme where all offspring are marketed. This system allows for maximum heterosis and breed complementarity. Usually an F1 female with strong maternal traits is bred to a terminal sire that imparts carcass quality and growth ability to the offspring to be marketed. But, a supply of F1 females is necessary.

Territory: Within the home range of an animal there is often a defended area which is identified as its territory.

Theave or theaf (plural of either: theaves): A young female sheep, usually before her first lamb (used especially in lowland England). Also "gimmer".

Titratable acidity: A measure of titaratable hydrogen ions. Includes H+ ions free in solution and those associated with acids and proteins.

Total economic value: Total economic value = direct use value + indirect use value + optional value + non-use value

Traditional breeding: Modification of plants and animals through selective breeding. Practices used in traditional plant breeding may include aspects of biotechnology such as tissue culture and mutational breeding.

Trait: Any measurable or observable characteristic of an animal.

Transgene: A gene from one organism inserted into another organism by recombinant DNA techniques.

Transgenic organism: An organism resulting from the insertion of genetic material from another organism using recombinant DNA techniques.

True model: A model that describes the pattern of the data perfectly. This is usually unknown

Tup: An alternative term for "ram".

Tupping: Mating in sheep, or the mating season (autumn, for a spring-lambing flock).

Utilization of farm animal genetic resources: The use and development of animal genetic resources for the production of food and agriculture. The use in production systems of AnGRs that already possess high levels of adaptive fitness to the environments concerned, and the deployment of sound genetic principles, will facilitate sustainable development of the AnGRs and the sustainable intensification of the production systems themselves. The wise use of AnGRs includes a broad mix of ongoing activities that must be well planned and executed for success, and compounded over time, hence with high value. It requires careful definition of breeding objectives, and the planning, establishment and maintenance of effective and efficient animal recording and breeding strategies.

Vacuum activity: Sometimes when the specific action potential is high an animal may act without a stimulus that we are aware of.

Variation: The amount of difference observed or measured for a trait in a group of animals.

Variety: A subdivision of a species for taxonomic classification also referred to as a 'cultivar.' A variety is a group of individual plants that is uniform, stable, and distinct genetically from other groups of individuals in the same species.

Vector: 1. A type of DNA element, such as a plasmid, or the genome of a bacteriophage, or virus, that is self-replicating and that can be used to transfer DNA segments into target cells. 2. An insect or other organism that provides a means of dispersal for a disease or parasite.

Vigour: Increased productivity of an animal as part of healthy growth.

Warner-Bratzler shear: The force required to sever the muscle fibres of a piece of meat and used as an indication of tenderness.

Weaner: A young animal that has been weaned, from its mother, until it is about a year old.

Wether: A castrated male sheep (or goat).

Wigging: The removal of wool from around a sheep's eyes to prevent wool blindness.

Wool pack: A standard-sized woven nylon container manufactured to industry specifications for the transportation of wool.

Wool-blind: When excessive wool growth interferes with the normal sight of a sheep.

Yoke: Two crossed pieces of timber fixed to the neck of a habitually straying sheep in an attempt to prevent it breaking through hedges and fences.

1

Animal Husbandry and Animal Breeding: An Introductory Overview

Animal husbandry, also called animal science, stockbreeding or simple husbandry, is the agricultural practice of breeding and raising livestock. The science of animal husbandry is taught in many universities and colleges around the world. Students of animal science may pursue degrees in veterinary medicine following graduation, or go on to pursue master's degrees or doctorates in disciplines such as nutrition, genetics and breeding, or reproductive physiology. Graduates of these programmes may be found working in the veterinary and human pharmaceutical industries, the livestock and pet supply and feed industries, or in academia. Historically, certain sub-professions within the field of animal husbandry are specifically named according to the animals that are cared for.

Types

A swineherd is a person who cares for hogs and pigs (older English term: swine). A shepherd is a person who cares for sheep. A goatherd cares for goats. A cowherd cares for cattle. In previous years, it was common to have herds which were made up of sheep and goats. In his case, the person tending them was called a shepherd Camels are also cared for in herds. In Tibet yaks are herded. In Latin America, llamas and alpacas are herded. In more modern times, the cowboys or *vaqueros* of North and South America ride horses and participate in cattle drives to watch over cows and bulls raised primarily for food. In Australia many herds are managed by farmers on motorbikes and in helicopters. Today, herd managers often oversee thousands of animals and many staff. Farms and ranches may employ breeders, herd health specialists, feeders, and milkers to help care for the animals. Techniques such as artificial insemination and embryo transfer are frequently used, not only as methods to guarantee that females are bred, but to help improve herd genetics. This may be done by transplanting embryos from stud-quality females, into flock-quality surrogate mothers - freeing up the stud-quality mother to be reimpregnated. This practice vastly increases the number of offspring which may be produced by a small selection of stud-quality parent animals. This in turn improves the ability of the animals to convert feed to meat, milk, or fiber more efficiently and improve the quality of the final product.

Ethical Aspects

There are contrasting views on the ethical aspects of breeding animals in captivity, with one debate being in relation to the merits of allowing animals to live in natural conditions which are reasonably close to those of their wild ancestors, compared to the view that considers natural pressures and stresses upon wild animals from disease, predation, and the like as vindication for captive breeding. Some techniques of animal husbandry such as factory farming, tail docking, the Geier Hitch, and castration, have been condemned by animal welfare activists and groups such as Compassion In World Farming. Genetic engineering is also controversial though it does not necessarily involve suffering. Animal rights activists are often opposed to all forms of animal husbandry. Some domesticated species of animals, such as the vechur cow, are rare breeds and are endangered. They are the subject of conservation efforts.

ANIMALS IN CAPTIVITY

Animals that live under human care are in captivity. Captivity can be used as a generalizing term to describe the keeping of either domesticated animals (livestock and pets) or wild animals. This may include for example farms, private homes and zoos. Keeping animals in human captivity and under human care can thus be distinguished between three primary categories according to the particular motives, objectives and conditions.

Animal Husbandry

Keeping and breeding livestock domesticated for economic reasons in farms, stud farms and similar establishments.

Wild Animal Keeping

Keeping wild, non-domesticated animals in menageries, zoos, aquaria, marine mammal parks or dolphinariums and similar establishments for various reasons:

- prestige (illustration of wealth and power)
- entertainment and amusement
- profit

- science
- education
- conservation biology

Pet Keeping

Keeping pets domesticated for personal reasons mostly at private homes.

Behaviour of Animals in Captivity

Captive animals, especially those which are not domesticated, sometimes develop repetitive, apparently purposeless motor behaviours called *stereotypical behaviours*. These behaviours are thought to be caused by the animals' abnormal environment. Many who keep animals in captivity, especially in zoos and related institutions and in research institutions, attempt to prevent or decrease stereotypical behaviour by introducing novel stimuli, known as environmental enrichment.

History

The domestication of animals is the oldest documented keeping of animals in captivity. The result was habituation of wild animal species to survive in the company of, or by the labor of, human beings. Domesticated species are those whose behaviour, life cycle, or physiology has been altered as a result of their breeding and living conditions under human control for multiple generations. Probably the earliest known domestic animal has been the dog, likely as early as 15000 BC among hunter-gatherers in several locations. Throughout history not only domestic animals as pets and livestock were kept in captivity and under human care, but also wild animals. Some were failed domestication attempts. Also, in past times, primarily the wealthy, aristocrats and kings collected wild animals for various reasons. Contrary to domestication, the ferociousness and natural behaviour of the wild animals were preserved and exhibited. Today's zoos claim other reasons for keeping animals under human care: conservation, education and science.

SHEEP HUSBANDRY

Sheep husbandry is the raising and breeding of domestic sheep, and a subcategory of animal husbandry. Sheep farming is primarily based on raising lambs for meat, or raising sheep for wool. Sheep may also be raised for milk. Some farmers specialize in breeding sheep to sell to other farmers.

Animal Care

Health Care

Sheep, particularly those kept inside, are vaccinated when they are newborn lambs. The lambs receive their first antibodies via their mother's colostrum in the first few hours of life, and then via a vaccination booster every six weeks for next three months and then by booster every six months. Weaning is a critical period in the life of young sheep as it is this time when more problems occur than at any other stage of a sheep's life. Sheep of this age need careful observation as to their general health by noting any weaners that are hollow, have a pale skin or are falling behind the mob etc. Weaners are very susceptible to the deadly Barbers Pole worm (*Haemonchus contortus*), fly strike (Myiasis), scabby mouth, mycotic dermatitis, occasionally pneumonia, fluctuations in feed availability and general ill thrift. Farmers work with animal nutritionists and veterinarians to keep sheep healthy and to manage animal health problems. Lambs may be castrated and have their tails docked for easier shearing, cleanliness and to help protect them from fly strike. Shearers or farmers need to remove wool from the hindquarters, around the anus, so that droppings do not adhere. In the southern hemisphere this is called *dagging* or crutching.

Shelter and Environment

Sheep are kept in mobs in paddocks; in pens or in a barn. In cold climates sheep may need shelter if they are freshly shorn or have baby lambs. Hoggets, especially, are very susceptable to wet, windy weather and will succumb to exposure very quickly. Sheep have to be kept dry for one to two days before shearing so that the fleece is dry enough to be pressed and to protect the health of the shearers.

Water, Food and Air

Sheep need fresh water from troughs or ponds, except that in some countries, such as New Zealand, there is enough moisture in the grass to satisfy them much of the time. Upon being weaned from ewe's milk, they eat hay, grains and grasses. The lambs are weaned due to increasing competition between the lamb and ewe for food. Sheep are active grazers where such feed is available at ground or low levels. They are usually given feed twice a day from troughs or they are allowed to graze in a pasture. Sheep are most comfortable when the temperature is moderate, so fans may be needed for fresh air if sheep are kept in barns during hot weather. In Australia, sheep in pasture are often subjected to 40 °C (104 °F), and higher, daytime temperatures without deleterious effects. In New Zealand sheep are kept on pasture in snow for periods of 3 or 4 days before they have to have supplemental feeding.

Goals of Flock Management

A sheep farmer is concerned with keeping the correct ratio of male to female sheep, selecting traits for breeding, and controlling under-/over-breeding based on the size and genetic diversity of the flock. Other tasks include sheep shearing, crutching and lambing the sheep. Sheep breeders look for such traits in their flocks as high wool quality, consistent muscle development, quick conception rate (for females), multiple births and quick physical development. Another concern of a sheep farmer is the protection of livestock. Sheep have many natural enemies, such as coyotes (North America), foxes (Europe), dingoes (Australia), and dogs. Newborn lambs in pasture are particularly vulnerable, also falling prey to crows, eagles and ravens. In addition, they are susceptible in some areas to flystrike which in itself has led to invention of practices such as mulesing. Sheep may be kept in a fenced-in field or paddock. The farmer must ensure that the fences are maintained in order to prevent the sheep from wandering onto roads or neighbours' property. Alternatively, they may be "heafed" (trained to stay in a certain area without the need for fences). The hardy Herdwick breed is particularly known for its affinity for being heafed.

A shepherd and a sheep dog may be employed for protection of the flock. On large farms, dogs and riders on horseback or motorcycles may muster sheep. Marking of sheep for identification purposes is often done by means of sheep tags - a type of ear tag. In some areas sheep are still identified through the use of notches cut in the ear known as ear marking, using either specially designed tools (ear marking pliers) or other cutting implements.

Lambing

Lambing is term for the management of birth in domestic sheep. In agriculture it often requires assistance from the farmer or shepherd because of breeding, climate or the individual physiology of the ewe. Australian farmers generally arrange for all the ewes in a mob to give birth (the *lambing season*) within a period of a few weeks often in early Autumn. As ewes sometimes fail to bond with newborn lambs, especially after delivering twins or triplets, it is important to minimize disturbances during this period. In order to more closely manage the births, vaccinate lambs, and protect them from predators, shepherds will often have the ewes give birth in "lambing sheds"; essentially a barn (sometimes a temporary structure erected in the pasture) with individual pens for each ewe and her offspring.

Flock Management Styles

Generally speaking, there are four general styles of sheep husbandry to serve the varied aspects of the sheep industry and the needs of a particular shepherd. Commercial sheep operations supplying meat and wool are usually either "range band flocks" or "farm flocks". Range band flocks are those with large numbers of sheep (often 1,000 to 1,500 ewes) cared for by a few full-time shepherds. The pasture-which must be of large acreage to accommodate the greater number of sheep-can either be fenced or open. Range flocks usually require the shepherds to live with the sheep as they move throughout the pasture, as well as the use of sheepdogs and means of transport such as horses or motor vehicles. As range band flocks move within a large area in which it would be difficult to supply a steady source of grain, almost all subsist on pasture alone. This style of sheep raising accounts for most of the sheep operations in the U.S., South America, and Australia. Farm flocks are those that are slightly smaller than range bands, and are kept on a more confined, fenced pasture land. Farm flocks may also be a secondary priority on a larger farm, such as by farmers who raise a surplus of crops to finish market lambs on, or those with untillable land they wish to exploit. However, farm flocks account for many farms focused on sheep as primary income in the U.K. and New Zealand (due to the more limited land available in comparison to other sheep-producing nations). The farm flock is a common style of flock management for those who wish to supplement grain feed for meat animals. An important corollary form of flock management to the aforementioned styles are specialized flocks raising purebred sheep. Many commercial flocks, especially those producing sheep meat, utilize cross-bred animals. Breeders raising purebred flocks provide stud stock to these operations, and often simultaneously work to improve the breed and participate in showing. Excess lambs are often sold to 4-H groups. The last type of sheep keeping is that of the hobbyist. This type of flock is usually very small compared to commercial operations, and may be considered pets. Those hobby flocks which are raised with production in mind may be for subsistence purposes or to provide a very specialized product, such as wool for handspinners. Quite a few people, especially those who emigrated to rural areas from urban or suburban enclaves, begin with hobby flocks or a 4-H lamb before eventually expanding to farm or range flocks

Life Cycle

Ewes are pregnant for just under five months before they lamb, and may have anywhere from one to three lambs per birth. Some ewes can have seven or eight lambs. Twin and single lambs are most common, triplets less common. A ewe may lamb once or twice a year. Lambs are weaned at three months. Sheep are full grown at two years weighing between 60 and 125 kilograms. Sheep can live to eleven or twelve years of age.

Sheep Production Worldwide

According to the Food and Agricultural Organization of the United Nations, the top ten "indigenous sheep meat" producing countries in order of quantity are:

1. Australia
2. New Zealand
3. Iran

4. UK
5. Turkey
6. Syria
7. India
8. Spain
9. Sudan
10. Pakistan

China actually has the greatest number of sheep in terms of number of livestock. While New Zealand rates number 2 on the list of total quantity of "indigenous sheep meat" produced, it has the highest number of sheep per-capita (outside of the Falkland Islands).

CAT HUSBANDRY: PROCEDURES AND PRODUCTIONS

The intent of this procedure is to describe the routine husbandry procedures at Cornell University's Cat Facilities. This procedure is intended for use by all animal care staff. This SOP is approved by the Institutional Animal Care and Use Committee (IACUC) and the Cornell Center for Animal Resources and Education (CARE). Any exemption must be approved by the IACUC prior to its application.

Introduction

This SOP provides general husbandry guidelines for cats housed in research facilities at Cornell.

Materials

(a) Feed, water and bowls
(b) Litter and litter pan
(c) Enrichment equipment
(d) Sanitation equipment

Procedures

(a) Animal Care

(i) Observe all animals to assess their health and well-being daily, including weekends and holidays
(ii) Identify animals by ear tattoos and cage cards.
(iii) Records and documentation

(iv) Feeding routine

(a) Feed animals to meet current National Research Council recommendations for feline nutrition.

(b) Store and deliver feeds in a manner that minimizes their contamination or spoilage.

(c) Offer water *ad lib*.

(d) When cats are housed in cages, ensure that the bowls are returned to the same cage, to prevent cross contamination.

(e) Monitor water regularly to ensure that it is free of contaminants that could potentially expose animals to chemical or infectious agents.

(v) Social and environmental enrichment

(a) Group house cats whenever possible and allow them to roam free within their room.

(b) Include enrichment devices in the environment (e.g., ladders, perches, toys, cardboard boxes, and blankets).

(vi) Cleaning and sanitation of cages

(a) Sanitize the cage/room frequently, as is appropriate to keep the animals clean and dry and to provide a healthy environment. The maximum time between cage sanitizing is two weeks.

(b) Scoop litter daily and remove soiled litter. Perform a total litter change at least once per week. Observe for abnormal feces and call CARE staff if necessary.

(c) Sanitize water bowls, feeders/food bowls and litter pans at least once per week.

(b) Animal Health

(i) Quarantine procedures

(a) Cats from approved vendors (i.e., Liberty) are not quarantined. Cats from unapproved vendors are quarantined for two weeks.

(ii) Euthanasia and disposal of dead animals

(a) Euthanasia is conducted by trained and experienced staff.

(b) Dispose of dead animals by incineration at the College of Veterinary Medicine.

(c) Facilities

(i) Maintain room temperatures between 64 and 84° F and relative humidity between 30% and 70%.

(ii) Lights are controlled by timers with a photoperiod of 12 to 14 hours of light. Check the timer performance regularly.

(iii) Space requirements

(a) Group house cats in a room with a minimum of four square feet of floor space per cat.

(b) When housed in cages, use an appropriate sized cage for the size and number of cats, meeting or exceeding recommendations.

(iv) Pest control

(v) Cleaning and sanitation of equipment and enrichment devices

(a) Clean and sanitize equipment thoroughly at least every two weeks, more frequent if needed.

(b) IMPORTANT: Whenever quaternary ammonias are used, thoroughly rinse and dry all surfaces before cats are allowed to come in contact with them. Cats are very susceptible to chemical burns from this compound.

(c) Enrichment devices that are not sanitizable, must be replaced every two weeks or more frequently as needed.

(vi) Transportation

(a) Outside of Cornell, transport cats in filtered commercial cat transport boxes.

(b) Within Cornell, transport cats either in filtered crates or unfiltered crates wrapped in a material which prevents animal allergens from escaping into the environment.

(vii) Waste management

(a) Deposit non-regulated Medical Waste and soiled litter in dumpsters.

(b) Regulated Medical Waste is incinerated at the Vet College, following EH&S policy.

Safety

(a) Handle cats carefully as they may bite or scratch. Any incident of this nature must be reported as an accident report to EH&S http://prp.ehs.cornell.edu/Acc-Inj/.

(b) Pregnant woman must seek counseling from Gannett before working with cats.

Contingencies

(a) Emergency veterinary care is available at all times including after working hours and on weekends and holidays through CARE.

(b) For facility emergencies, contact the facility manager.

Cat Body Type Genetic Mutations

Cats, like all living organisms, occasionally have mutations that affect their body type. Sometimes, these changes in body type are striking enough that humans select for and perpetuate them. This is not always in the best interests of the cat, as many of these mutations are harmful; some are lethal in their homozygous form.

This page gives a selection of cat body type mutant alleles and the associated mutations with a brief description.

Tail Types

Jb = Japanese bobtail gene (dominant with incomplete penetrance). Cats heterozygous for this gene have abnormal tails, but unlike the Manx cat there are no associated skeletal disorders and the gene is not associated with lethality.

M = Manx gene (dominant). Cats with the homozygous genotype (MM) die before birth, and stillborn kittens show gross abnormalities of the central nervous system. Cats with the heterozygous genotype (Mm) show severely shortened tail length, ranging from taillessness to a partial, stumpy tail. Some Manx cats die before 12 months old and exhibit skeletal and organ defects. People have suggested that the Manx gene, because it was discovered in naturally occurring populations of cats, is a gene conferring some kind of selective advantage to the cats. The trait also occurred and died out in Cornwall (mainland England), but became fixed in the island population where outbreeding was not possible due to isolation.

There are numerous other bobtail types in the cat population, most of which are identical to the Japanese Bobtail or the variably expressed Manx mutation. However, some may be novel mutations that have not been investigated.

There are numerous types of curly-tailed cats whose tails loop over the back or form tight corkscrews. One such mutation has been developed into the American Ringtail but others have been regarded as curiosities and not perpetuated. The gene(s) responsible have not been fully investigated.

Limbs

Mk = Munchkin gene (dominant). Cats heterozygous for this gene (Mkmk) have shortened legs, but are not disabled. They have a ferret-like gait. The homozygous form (MkMk) may be lethal as litter sizes are smaller than average. Although there was initial concern that Munchkin-type cats would have impaired mobility or spinal problems, this was based on comparison with dog breeds and proved to be unfounded due to the cat's more flexible spine. The mutation has occurred naturally in many locations and has also been perpetuated in feral cats without human intervention (Robinson's Genetics for Cat Breeders).

The mutation has proven not to be achondroplasia, but is most likely to be either hypochodroplasia or pseudochondroplasia which affect the long bones of the leg while leaving other bodily proportions, especially the head, unchanged.

Paws

Sh = Split Foot (Syndactyly). A dominant gene that reduces the number of toes resulting in a "lobster-claw" appearance. This is considered an undesirable mutation.

Polydactyl (extra-toed) cats. There are probably many genes, both dominant and recessive, that cause polydactyly in cats. Most cases of polydactyly in cats are perfectly harmless.

Pd = Thumb-cat polydactyly gene. The Pd gene (dominant with incomplete penetrance) causes the benign, pre-axial form of polydactyly where one or more extra toes occur near the dew claw. Often, the dew claw is converted into a thumb. There are occasional problems such as fused claws or claws facing in the wrong direction, but generally, this form of polydactyly is harmless.

On the other hand, the "hamburger-feet" polydactyly gene is associated with gene for radial hypoplasia (RH). The 1995 European Convention for the Protection of Pet Animals considers RH an impairing condition. In a scandal in the late 1990s, an experimental breeder in Texas tried to perpetuate this deformity as the "Twisty Cat" breed. Mild RH can cause the post-axial form of polydactyly - enlarged paws, extra three-jointed toes on the outer, little-toe side of the paws, and no thumb. X-rays can determine the structure of the extra toes and whether the cat has the gene for RH. Cats with the gene for RH should never be bred. Cats with severe RH have unusually short front legs. They move like a ferret and they tend to sit like a squirrel or kangaroo and are colloquially known as squittens. In some RH cats, the forelegs are twisted with the long bones either severely shortened or absent. All polydactyl cats are banned from German cat shows, possibly because of confusion with the impairing form of polydactyly associated with RH.

Polydactyl cats are relatively common in southwest Britain, Norway, Sweden, and the eastern coast of the USA and Canada, and some parts of Asia. Sailors thought they were lucky. There were, and are, many myths surrounding polydactyl cats:

- That they are superior mousers and ratters,
- That they have better balance on ships in stormy weather,
- That their paws are natural snowshoes,
- That their opposable thumbs (in the thumb-cat form of polydactyly) give them a survival advantage.

Ernest Hemingway supposedly collected polydactyl cats, and the reported descendants of his collection may still be found at the Ernest Hemingway House on Key West.

Ear Types

Cu = American Curl gene (dominant). Cats with this gene have ears that start out normal, but gradually curl backwards. So far, no harmful defects have been associated with this gene.

Fd = Scottish Fold gene (dominant with incomplete penetrance). Cats with this gene have ears that curl forward. There are different degrees of folding, and more genes may be involved in the expression of the Fd gene. This gene is associated with bone and cartilage defects such as thickened tail and swollen feet. The homozygous form (FdFd) is probably lethal.

Australian Curl - a curl-eared mutation occurred in a female stray cat in Australia, but was not inherited by her offspring. When the original cat became ill, necessitating spaying, it was impossible to test-mate her sons back to her to identify a possibly recessive curled-ear mutation.

Sumxu - extinct Chinese Lop-eared cat breed reported between 1700 and 1938 around Peking, most descriptions are based on a specimen in a German museum. The mode of inheritance of its pendulous ears is not known.

Some information reproduced with permission from Messybeast.com

Main Type of Cat Breeds

The following is a list of cat breeds recognized by various cat registries.

Cats can also be grouped by type according to appearance or function.

Longhair and Semi-longhair

- Aegean cat
- American Longhair
- Asian Semi-longhair (or Tiffanie)
- Balinese
- Birman
- British Longhair
- Chantilly/Tiffany cat
- Himalayan (USA) or Colourpoint (rest of world).
- Javanese
- Maine Coon
- Nebelung
- Norwegian Forest Cat
- Oriental Longhair
- Persian
- Ragdoll (and Ragamuffin)
- Siberian
- Turkish Van
- Turkish Angora

Shorthair

- Abyssinian
- American Shorthair
- Australian Mist
- Bombay
- Brazilian Shorthair
- British Shorthair
- Burmese
- Burmilla
- California Spangled Cat
- Chartreux
- Colorpoint Shorthair

- Egyptian Mau
- European Shorthair
- Exotic
- Havana Brown
- Ichabod
- Korat
- Ocicat
- Oriental Shorthair
- Russian Blue
- Russian
- Siamese (and Traditional Siamese or Applehead Siamese)
- Singapura
- Snowshoe
- Sokoke
- Somali
- Tonkinese
- Ukrainian Levkoy

Breeds Based on Mutations

- American Bobtail
- American Curl
- American Polydactyl, aka Hemingway Cat or Mitten Cat
- American Wirehair
- Cornish Rex
- Cymric
- Devon Rex
- German Rex
- Japanese Bobtail
- LaPerm
- Manx
- Munchkin
- Ojos Azules

- Peterbald
- Pixie-bob
- Selkirk Rex
- Scottish Fold
- Sphynx

Breeds Derived from Hybridizations between Domestic Cats & Wild Felids

- Bengal : domestic Cat / Asian Leopard Cat (*Prionailurus bengalensis*)
- Bristol : domestic Cat / Margay (*Leopardus wiedii*)
- Chausie aka Stone Cougar : domestic Cat / Jungle Cat aka swamp-lynx (*Felis chaus*)
- Cheetoh : Ocicat / Bengal
- Jungle-Bob : Pixie-bob / Jungle Cat aka swamp-lynx (*F. Chaus*)
- Jungle-Curl : Hemingway Curl aka American Curl / Jungle Cat aka swamp-lynx (*Felis chaus*)
- Machbagral and/or Viverral : domestic Cat / Fishing Cat (*Prionailurus viverrinus*)
- Pantherette : Pixie-bob / Asian Leopard Cat (*Prionailurus bengalensis*)
- Punjabi : (domestic Cat with Indian Desert-Cat aka Asiatic Wildcat (*Felis s. ornata*)
- Safari : (domestic Cat with Geoffroy's Cat (*Leopardus geoffroyii*)
- Savannah and/or Ashera : (domestic Cat with Serval (*Leptailurus serval*)
- Serengeti : Oriental / Bengal
- Toyger : domestic Cat / Bengal
- Ussuri (cat) : domestic Cat / Amur Asian Leopard Cat (*Prionailurus b. euptailura*)
- Caracat (proposed name) : Domestic Cat / Caracal (accidental Hybridization, Moscow Zoo, 1998)
- Oncicat (proposed name) : Domestic Cat / Oncilla (Little Spotted Cat/ Tiger Cat)
- Domestic Cat / Black-footed Cat (*F. nigripes*)
- Domestic Cat / Rusty-spotted Cat (*Prionailurus rubiginosus*) (wild-occurring Hybrids, India)

Breeds Derived from Multiple Hybridizations between Domestic Cats & Felids

- Afro-Chausie (proposed name) : Chausie / African Wildcat (*F. s. lybica*)
- Euro-Chausie : Chausie / European Wildcat (*F. s. silvestris*)
- Scottie-Chausie (proposed name) : Chausie / Scottish Wildcat (*F. s. grampia*)

Attempted or Unconfirmed Hybridization between Domestic Cats & Felids

- Jaguarundi Curl (alleged name) aka Mandalan Jaguar (proposed name) : Domestic Cat / Jaguarundi
- Domestic Cat / Canada Lynx
- Domestic Cat / Bobcat (*Felis rufus*)
- Domestic Cat / Pallas Cat (*Otocolobus manula*)

FISH HUSBANDRY: THE FEAP CODE OF CONDUCT FOR EUROPEAN AQUACULTURE

Any person who owns farmed fish, or has farmed fish under his or her control, and every person engaged in the overseeing of farmed fish shall, according to their responsibilities, ensure that every step is taken to safeguard the health and welfare of such fish.

Water

The water supply should be of sufficient quality and quantity to ensure the well-being of the species being farmed.

Fish Stocks

The intake of live fish stocks into an aquaculture system must be of good health and known origin.

(a) Genetically Modified Organisms

The FEAP does not endorse the use of genetically modified fish in aquaculture since it is concerned about the maintenance of the natural characteristics of the products, in addition to the environmental qualities of biodiversity. However, the results of genetic research may play an important part in the future development of global food production. The FEAP may review its position on this topic if such developments are acceptable to the consumer and do not pose any safety or environmental problems.

Fish Health

The responsibilities concerning the optimization of fish health include:

1. Avoidance of unnecessary stress of the fish - all measures should be taken to ensure that the media and conditions in which the population is held are optimised for the reduction of stress.
2. Regular inspections - the fish should be inspected frequently enough to ensure that significant behavioural and physical changes would be discovered and acted upon immediately.
3. Avoidance of the introduction of diseases - fish brought into an aquaculture system must be of good health and certified origin. Adequate precautionary measures should be taken to avoid inter-farm contamination through direct physical contact.
4. Seeking proper diagnosis if disease presence is suspected.
5. The use and application of therapeutic agents should observe the prescribed dosage and where appropriate, withdrawal times, in order to avoid the accumulation of residues in the flesh.
6. When required, only licensed or approved therapeutic agents should be used.
7. Avoidance of spreading of diseases - farmers have the responsibility to minimise the risk of the spread of diseases beyond their farms into the ecosystem where wild fish and other farms may be affected.
8. Regardless of the reason for mortalities, any dead or dying fish require prompt removal from the growing area, in a way that does not affect the welfare and health of the remaining stock.
9. The disposal of dead fish should be done carefully and effectively, in a way that does not affect the environment negatively.

Food and Feeding

Correct feeding practises reduce wastage, assuring better water quality, good health and farm performance.

1. All fish should receive adequate quantities of feed, using the correct nutritional formulation for the species farmed.
2. Such feeds should be properly composed and manufactured and, where possible, labelled and providing the correct granular or pellet sizes for the size of the fish.

3. Daily rations should be appropriate for the species and the growing conditions available in the site facilities.
4. Feed distribution methods should ensure that all individuals have sufficient access to the feeds supplied.
5. Excessive feeding should be avoided since this can result in feed wastage that may cause water quality deterioration.

Handling and Transportation

1. For the avoidance of unnecessary stress and injury to live fish, the handling of live fish should be kept to a minimum and should be done using the least stressful method.
2. The movement and transport of live fish should be done as quickly as possible and with an adequate oxygen supply.
3. The strictest control procedures should be applied to fish that are transferred between farms and freshwater catchment areas in order to reduce the potential transfer of disease to a minimum.

Predators

Many predators affecting aquaculture are species that are protected by legislation. Whenever possible, predators should be excluded from the areas where live fish are held. Where this is not possible, lethal methods of predator control shall only be used when this action is legally permissible for the predator species in question.

Stocking Density

The stocking density for fish should be adjusted to the specific requirements of the species and include respect for

1. The average live weight of the fish,
2. The population's health and behavioural needs,
3. The population's demands on the growing environment, in particular their behavioural needs, the availability of an adequate oxygen supply and the removal of wastes to avoid the excessive accumulation of substances that may cause stress or toxic effects (e.g. CO_2 and ammonia).

Slaughter

1. All fish should be fasted sufficiently before slaughter so as to induce a completely empty digestive system.

2. Fish should be killed quickly and humanely, referring to national regulations for guidance.

Monitoring and Record Keeping

1. Fish farms should aim to be self-regulating. To achieve this, proper systems of monitoring and recording are required so that problems can be averted before they arise.
2. Written records are essential for the farmers to ensure good husbandry and welfare of the fish.
3. The use of computer-assisted monitoring of stocks and record-keeping is to be encouraged given the benefits of:
   (a) Optimal feed distribution,
   (b) Use of therapeutic agents and their traceability
   (c) HACCP (Hazard Analysis and Critical Control Point) systems.
4. Effective self-regulation can be achieved through the routine monitoring of:
   (a) Water quality (on and off-farm),
   (b) The quality of other inputs and resources used in the production process,
   (c) Off-farm environmental parameters that are of immediate and direct relevance to the production process,
   (d) Environmental standards and objectives that, ideally, are agreed with local authorities,
   (e) Product quality and safety standards.

List of Common Fishes

- African glass catfish
- African lungfish
- aholehole
- airbreathing catfish
- airsac catfish
- Alaska blackfish
- albacore
- alewife

- alfonsino
- algae eater
- alligatorfish
- Amago
- American sole
- Amur pike
- anchovy
- anemonefish
- angelfish
- angel shark
- anglemouth
- angler
- angler catfish
- anglerfish
- Antarctic cod
- Antarctic dragonfish
- Antarctic icefish
- antenna codlet
- arapaima
- archerfish
- Arctic char
- armored catfish
- armored gurnard
- armored searobin
- armorhead
- armorhead catfish
- arowana
- arrowtooth eel
- aruana
- Asian carps
- Asiatic glassfish

- Atka mackerel
- Atlantic cod
- Atlantic eel
- Atlantic herring
- Atlantic salmon
- Atlantic saury
- Atlantic silverside
- Atlantic trout
- Australasian salmon
- Australian grayling
- Australian herring
- Australian lungfish
- Australian prowfish
- Ayu
- Baikal oilfish
- Bala shark
- ballan wrasse
- bamboo shark
- banded killifish
- bandfish
- bango
- bangu
- banjo catfish
- barb
- barbol
- barbeled dragonfish
- barbeled houndshark
- barbelless catfish
- barfish
- barracuda
- barracudina

- barramundi
- barred danio
- barreleye
- basking shark
- bass
- basslet
- batfish
- bat ray
- beachsalmon
- beaked salmon
- beaked sandfish
- beardfish
- beluga sturgeon
- bengal danio
- bent-tooth
- betta
- bichir
- bigeye
- bigeye squaretail
- bighead carp
- bigmouth buffalo
- bigscale
- bigscale fish
- bigscale pomfret
- billfish
- bitterling
- black angelfish
- black bass
- black dragonfish
- blackchin
- blackfish

- blacktip reef shark
- black mackerel
- black pickerel
- black prickleback
- black scalyfin
- black sea bass
- blacksmelt
- black swallower
- black tetra
- black triggerfish
- bleak
- blenny
- blind goby
- blind shark
- blue catfish
- blue danio
- blue-redstripe danio
- blue eye
- bluefin tuna
- bluefish
- bluegill
- blue gourami
- blue shark
- blue triggerfish
- blue whiting
- bluntnose knifefish
- bluntnose minnow
- boafish
- boarfish
- bobtail snipe eel
- bocaccio

- boga
- Bombay duck
- bonefish
- bonito
- bonnetmouth
- bonytail chub
- bonytongue
- bottlenose
- bowfin
- boxfish
- bramble shark
- bream
- bristlemouth
- bristlenose catfish
- broadband dogfish
- brook lamprey
- brook trout
- brotula
- brown trout
- buffalofish
- bullhead
- bullhead shark
- bull shark
- bull trout
- burbot
- buri
- burma danio
- burrowing goby
- butterfish
- butterfly ray
- butterflyfish

- California flyingfish
- California halibut
- California smoothtongue
- canary rockfish
- candiru
- candlefish
- capelin
- cardinalfish
- carp
- carpetshark
- carpsucker
- catalufa
- catfish
- catla
- cat shark
- cavefish
- Celebes rainbowfish
- central mudminnow
- cepalin
- chain pickerel
- channel bass
- channel catfish
- char
- cherry salmon
- chimaera
- chinook salmon
- Cherubfish
- chub
- chubsucker
- chum salmon
- cichlid

- cisco
- climbing catfish
- climbing gourami
- climbing perch
- clingfish
- clownfish
- clown loach
- clown triggerfish
- cobbler
- cobia
- cod
- cod icefish
- codlet
- codling
- coelacanth
- coffinfish
- coho salmon
- coley
- collared carpetshark
- collared dogfish
- Colorado squawfish
- combfish
- combtail gourami
- combtooth blenny
- common carp
- common tunny
- conger eel
- convict blenny
- cookie-cutter shark
- coolie loach
- cornetfish

- cowfish
- cownose ray
- cow shark
- crappie
- creek chub
- crestfish
- crevice kelpfish
- croaker
- crocodile icefish
- crocodile shark
- crucian carp
- cuchia
- cuckoo wrasse
- cusk-eel
- cuskfish
- cutlassfish
- cutthroat eel
- cutthroat trout
- dab
- dace
- daggertooth pike conger
- damselfish
- danio
- darter
- dartfish
- dealfish
- Death Valley pupfish
- deep sea anglerfish
- deep sea bonefish
- deep sea eel
- deep sea smelt

- deepwater cardinalfish
- deepwater flathead
- deepwater stingray
- delta smelt
- demoiselle
- denticle herring
- desert pupfish
- Devario
- devil ray
- discus
- diver: New Zealand sand diver or Long-finned sand diver
- dogfish
- dogfish shark
- dogteeth tetra
- dojo loach
- Dolly Varden trout
- dorab
- dorado
- dory
- dottyback
- dragonet
- dragonfish
- dragon goby
- driftfish
- driftwood catfish
- drum (fish)
- duckbill
- duckbilled barracudina
- duckbill eel
- dusky grouper
- dwarf gourami

- dwarf loach
- eagle ray
- earthworm eel
- eel
- eelblenny
- eel cod
- eel-goby
- eelpout
- eeltail catfish
- elasmobranch
- electric catfish
- electric eel
- electric knifefish
- electric ray
- electric stargazer
- elephantfish
- elephantnose fish
- elver
- emperor
- emperor angelfish
- emperor bream
- escolar
- eucla cod
- eulachon
- European chub
- European eel
- European flounder
- European minnow
- false brotula
- false cat shark
- false moray

- false trevally
- fangtooth
- fathead sculpin
- featherback
- featherfin knifefish
- fierasfer
- filefish
- finback cat shark
- fingerfish
- fire bar danio
- firefish
- flabby whalefish
- flagblenny
- flagfin
- flagfish
- flagtail
- flashlight fish
- flatfish
- flathead
- flathead catfish
- flat loach
- flier
- flounder
- flying characin
- flying gurnard
- flyingfish
- footballfish
- forehead brooder
- four-eyed fish
- french angelfish
- freshwater eel

- freshwater flyingfish
- freshwater hatchetfish
- freshwater herring
- freshwater shark
- frigate mackerel
- frilled shark
- frogfish
- frogmouth catfish
- fusilier fish
- galjoen fish
- Ganges shark
- gar
- garden eel
- garibaldi
- garpike
- ghost flathead
- ghost carp
- ghost knifefish
- ghost pipefish
- ghoul
- giant danio
- giant gourami
- giant sea bass
- giant wels
- gianttail
- gibberfish
- Gila trout
- gizzard shad
- glass catfish
- glassfish
- glass knifefish

- glowlight danio
- goatfish
- goblin shark
- goby
- golden dojo
- golden loach
- golden trout
- goldeye
- goldfish
- goldspotted killifish
- gombessa
- goosefish
- gopher rockfish
- gouramie
- grass carp
- graveldiver
- gray eel-catfish
- grayling
- gray mullet
- gray reef shark
- great white shark
- green swordtail
- greeneye
- greenling
- grenadier
- grideye
- ground shark
- grouper
- grunion
- grunt
- grunter

- grunt sculpin
- gudgeon
- guitarfish
- gulf menhaden
- gulper eel
- gulper
- gunnel
- guppy
- gurnard
- haddock
- hagfish
- hairtail
- hairyfish
- hake
- half-gill
- halfbeak
- halfmoon
- halibut
- halosaur
- hamlet
- hammerhead shark
- Hammerjaw
- handfish
- hardhead catfish
- harelip sucker
- hatchetfish
- hawkfish
- herring
- herring smelt
- hillstream loach
- hog sucker

- hoki
- horn shark
- horsefish
- houndshark
- huchen
- humuhumu-nukunuku-apua'a
- hussar
- icefish
- ide
- ilisha
- inanga
- inconnu
- Indian mullet
- iniom
- jack
- jackfish
- Jack Dempsey
- Japanese eel
- jawfish
- jellynose fish
- jewelfish
- jewel tetra
- jewfish
- john dory
- Kafue pike
- kahawai
- kaluga
- kanyu
- kelp perch
- kelpfish
- killifish

- king of herring
- king-of-the-salmon
- kissing gourami
- knifefish
- knifejaw
- koi
- kokanee
- kokopu
- kuhli loach
- labyrinth fish
- ladyfish
- lagena
- lake trout
- lake whitefish
- lampfish
- lamprey
- lancetfish
- lanternfish
- large-eye bream
- largemouth bass
- largenose fish
- leaffish
- leatherjacket
- lefteye flounder
- lemon shark
- lenok
- leopard danio
- lightfish
- lighthousefish
- limia
- lined sole

- ling
- ling cod
- lionfish
- livebearer
- lizardfish
- loach
- loach catfish
- loach goby
- loach minnow
- longfin
- longfin dragonfish
- longfin escolar
- long-finned char
- long-finned pike
- longjaw mudsucker
- longneck eel
- longnose chimaera
- longnose dace
- longnose lancetfish
- longnose sucker
- longnose whiptail catfish
- long-whiskered catfish
- lookdown catfish
- loosejaw
- Lost River sucker
- louvar
- loweye catfish
- luminous hake
- luderick
- luminous hake
- lumpsucker

- lungfish
- lyretail
- mackerel
- mackerel shark
- madtom
- mahi-mahi
- mahseer
- mail-cheeked fish
- mako shark
- manefish
- man-of-war fish
- Manta Ray
- marblefish
- marine hatchetfish
- marlin
- masu salmon
- medaka
- medusafish
- megamouth shark
- menhaden
- merluccid hake
- Mexican blind cavefish
- Mexican golden trout
- midshipman
- milkfish
- minnow
- Modoc sucker
- mojarra
- mola
- molly
- monkeyface prickleback

- monkfish
- mooneye
- moonfish
- Moorish idol
- mora
- moray eel
- morid cod
- morwong
- Moses sole
- mosquitofish
- mosshead warbonnet
- mouthbrooder
- Mozambique tilapia
- mrigal
- mud catfish (mud cat)
- mudfish
- mudminnow
- mudskipper
- mudsucker
- mullet
- mummichog
- murray cod
- muskellunge
- mustache triggerfish
- mustard eel
- naked-back knifefish
- nase
- needlefish
- neon tetra
- New World rivuline
- New Zealand smelt

- nibbler
- noodlefish
- North American darter
- North American freshwater catfish
- North Pacific daggertooth
- northern anchovy
- northern clingfish
- northern lampfish
- northern pearleye
- northern pike
- northern sea robin
- northern squawfish
- northern Stargazer
- Norwegian Atlantic salmon
- nurseryfish
- nurse shark
- oarfish
- ocean perch
- ocean sunfish
- oceanic flyingfish
- oceanic whitetip shark
- oilfish
- oldwife
- Old World knifefish
- Old World rivuline
- olive flounder
- opah
- opaleye
- orange roughy
- orangespine unicorn fish
- orangestriped triggerfish

- orbicular batfish
- orbicular velvetfish
- Oregon chub
- oreo
- Oriental loach
- Owens pupfish
- Pacific albacore
- Pacific argentine
- Pacific cod
- Pacific hake
- Pacific herring
- Pacific lamprey
- Pacific salmon
- Pacific saury
- Pacific trout
- Pacific viperfish
- paddlefish
- paperbone
- paradise fish
- parasitic catfish
- parrotfish
- peacock flounder
- peamouth
- pearleye
- pearlfish
- pearl danio
- pearl perch
- pejerrey
- peladillo
- pelagic cod
- pelican eel

- pelican gulper
- pencil catfish
- pencilfish
- pencilsmelt
- perch
- Peter's elephantnose fish
- pickerel
- pigfish
- pike characid
- pike conger
- pike eel
- pike
- pikeblenny
- pikehead
- pikeperch
- pilchard
- pilot fish
- pineconefish
- pink salmon
- píntano
- pipefish
- piranha
- pirarucu
- pirate perch
- plaice
- platy
- platyfish
- pleco
- plownose chimaera
- plunderfish
- poacher

- pollyfish
- pollock
- pomfret
- pompano
- pompano dolphinfish
- ponyfish
- poolfish
- popeye catafula
- porbeagle shark
- porcupinefish
- porgy
- Port Jackson shark
- powen
- priapumfish
- prickleback
- pricklefish
- prickly shark
- prowfish
- pufferfish
- pumpkinseed
- pupfish
- pygmy sunfish
- queen danio
- queen parrotfish
- queen triggerfish
- quillback
- quillfish
- rabbitfish
- raccoon butterfly fish
- ragfish
- rainbow trout

- rainbowfish
- rasbora
- ratfish
- rattail
- ray
- razorback sucker
- razorfish
- red snapper
- redfin
- redfish
- redhorse sucker
- redlip blenny
- redmouth whalefish
- redside
- redtooth triggerfish
- red velvetfish
- red whalefish
- reedfish
- reef triggerfish
- regal whiptail catfish
- remora
- requiem shark
- ribbon eel
- ribbon sawtail fish
- ribbonbearer
- ribbonfish
- rice eel
- ricefish
- ridgehead
- riffle dace
- righteye flounder

- Rio Grande perch
- river loach
- river shark
- river stingray
- rivuline
- roach
- roanoke bass
- rock bass
- rock beauty
- rock cod
- rocket danio
- rockfish
- rockling
- rockweed gunnel
- rohu
- ronquil
- roosterfish
- ropefish
- rough pomfret
- rough scad
- rough sculpin
- roughy
- roundhead
- round herring
- round stingray
- round whitefish
- rudd
- rudderfish
- ruffe
- Russian sturgeon
- sabalo

- sabertooth
- saber-toothed blenny
- sabertooth fish
- sablefish
- sailback scorpionfish
- sailbearer
- sailfin silverside
- sailfish
- salamanderfish
- salmon
- salmon shark
- sandbar shark
- sandburrower
- sand dab
- sanddiver
- sand eel
- sandfish
- sand goby
- sand knifefish
- sand lance
- sandperch
- sandroller
- sand stargazer
- sand tiger
- sand tilefish
- sarcastic fringehead
- sardine
- sargassumfish
- sauger
- saury
- sawfish

- saw shark
- sawtooth eel
- scabbard fish
- scaleless black dragonfish
- scaly dragonfish
- scat
- scissor-tail rasbora
- scorpionfish
- sculpin
- scup
- scythe butterfish
- sea bass
- sea catfish
- sea chub
- seadevil
- seadragon
- seahorse
- sea lamprey
- seamoth
- sea raven
- searobin
- sea snail
- sea toad
- Sevan trout
- seatrout
- sergeant major
- shad
- shark
- sharksucker
- sharpnose pufferfish
- sheatfish

- sheepshead
- sheepshead minnow
- shell-ear
- shiner
- shortnose chimaera
- shortnose greeneye
- shortnose sucker
- shovelnose sturgeon
- shrimpfish
- Siamese fighting fish
- sillago
- silver carp
- silver dollar
- silver driftfish
- silver hake
- silverside
- sind danio
- sixgill ray
- sixgill shark
- skate
- skate moss
- skilfish
- skipjack tuna
- skipping goby
- slender barracudina
- slender mola
- slender snipe eel
- sleeper
- sleeper shark
- slickhead
- slimehead

- slimy mackerel
- slimy sculpin
- slipmouth
- smalleye squaretail
- smalltooth sawfish
- smelt
- smelt-whiting
- smooth dogfish
- smoothtongue
- snailfish
- snake eel
- snakehead
- snake mackerel
- snake mudhead
- snapper
- snipe eel
- snipefish
- snoek
- snook
- snubnose eel
- snubnose parasitic eel
- soapfish
- sockeye salmon
- soldierfish
- sole
- South American darter
- South American Lungfish
- southern Dolly Varden
- southern flounder
- southern grayling
- southern hake

- southern sandfish
- southern smelt
- spadefish
- spaghetti eel
- Spanish mackerel
- spearfish
- speckled trout
- spiderfish
- spikefish
- spinefoot
- spiny-back
- spiny basslet
- spiny dogfish
- spiny dwarf catfish
- spiny eel
- spinyfin
- splitfin
- spookfish
- spotted danio
- spotted dogfish
- sprat
- springfish
- squarehead catfish
- squaretail
- squawfish
- squeaker
- squirrelfish
- staghorn sculpin
- stargazer
- starry flounder
- steelhead

- stickleback
- stingfish
- stingray
- stonecat
- stonefish
- stoneroller minnow
- straptail
- stream catfish
- streamer fish
- striped bass
- striped burrfish
- sturgeon
- sucker
- suckermouth armored catfish
- summer flounder
- Sundaland noodlefish
- sunfish
- surf sardine
- surfperch
- surgeonfish
- swallower
- swamp-eel
- swampfish
- sweeper
- swordfish
- swordtail
- tadpole cod
- tadpole fish
- tailor
- taimen
- tang

- tapetail
- tarpon
- tarwhine
- telescopefish
- temperate bass
- temperate ocean-bass
- temperate perch
- tench
- tenpounder
- tenuis
- tetra
- thorny catfish
- thornfish
- thornyhead
- threadfin
- threadfin bream
- threadsail
- threadtail
- three spot gourami
- threespine stickleback
- three-toothed puffer
- thresher shark
- tidewater goby
- tiger barb
- tigerperch
- tiger shark
- tiger shovelnose catfish
- tilapia
- tilefish
- titan triggerfish
- toadfish

- tommy ruff
- tompot blenny
- tonguefish
- tope
- topminnow
- torpedo
- torrent catfish
- torrent fish
- trahira
- treefish
- trevally
- triggerfish
- triplefin blenny
- triplespine
- tripletail
- tripod fish
- trout
- trout cod
- trout-perch
- trumpeter
- trumpetfish
- trunkfish
- tubeblenny
- tube-eye
- tube-snout
- tubeshoulder
- tui chub
- tuna
- turbot
- turkeyfish
- unicornfish

- upside-down catfish
- velvet-belly shark
- velvet catfish
- velvetfish
- vendace
- vimba
- viperfish
- wahoo
- walking catfish
- wallago
- walleye
- walleye pollock
- walu
- warbonnet
- warmouth
- warty angler
- waryfish
- wasp fish
- weasel shark
- weatherfish
- weever
- weeverfish
- wels catfish
- whale catfish
- whalefish
- whale shark
- whiff
- whiptail gulper
- whitebait
- white croaker
- whitefish

- white marlin
- white shark
- whitetip reef shark
- whiting
- wobbegong
- wolf-eel
- wolffish
- wolf-herring
- woody sculpin
- worm eel
- wormfish
- wrasse
- wrymouth
- yellow-and-black triplefin
- yellowbanded perch
- yellow bass
- yellow-edged moray
- yellow-eye mullet
- yellowhead jawfish
- yellowfin croaker
- yellowfin cutthroat trout
- yellowfin grouper
- yellowfin pike
- yellowfin surgeonfish
- yellowfin tuna
- yellow jack
- yellowmargin triggerfish
- yellow moray
- yellow perch
- yellowtail
- yellowtail amberjack

- yellowtail barracuda
- yellowtail clownfish
- yellowtail horse mackerel
- yellowtail kingfish
- yellowtail snapper
- yellow tang
- yellow weaver
- zander
- zebra bullhead shark
- zebra danio
- zebrafish
- zebra lionfish
- zebra loach
- zebra oto
- zebra pleco
- zebra shark
- zebra tilapia
- ziege
- zingel

UNDERSTANDING THE OUTCROSSING PRACTICE

Outcrossing is the practice of introducing unrelated genetic material into a breeding line. It increases genetic diversity, thus reducing the probability of all individuals being subject to disease or reducing genetic abnormalities(only within the first generation). It actually can serve to increase the number of individuals who carry a disease recessively. It is used in line breeding to restore vigor or size and fertility to a breeding line. Outcrossing is now the norm of most purposeful breeding, contrary to what is commonly believed. The outcrossing breeder intends to remove the traits by using "new blood". With dominant traits, one can still see the expression of the traits and can remove those traits whether one outcrosses, line breeds or inbreds. With recessives, outcrossing allows for the recessive traits to migrate across a population. It may actually increase the number of individuals carrying a disease. The outcrossing breeder then may have individuals that have many deleterious genes that are expressed by placing their animals against a similarly outcrossed individual. There is now a gamut of deleterious genes within each individual in many breeds.However one may increase the variance of genes within the gene pool by outcrossing, protecting against extinction by a single stressor from the environment. In cats, there is currently a study running to determine the genetic diversity within the cat breeds. Outcrossing is believed to be the "norm" in the wild. However, it is not logical as migration occurs by necessity. Feral cats, for example are one of the most inbred as individuals remain nearby their original homes, unless environmental stresses drive them to migration. Breeders inbreed within their genetic pool, attempting to maintain desirable traits and to cull those traits that are undesirable. When undesirable traits begin to appear, breedings are selected to determine if a trait is recessive or dominant. Removal is accomplished by breeding two individuals of known genetic status, usually they are related. In nature, where breeding is not managed, outcrossing rates may be estimated by genetic analysis. This allows calculation of the amount of genetic exchange between populations, and thus provides insights into the biogeography and phytogeography of species. Gregor Mendel used outcrossing in his experiments with flowers for his breeding stock. He then used the resulting offspring to chart inheritance patterns, using the crossing of siblings, and backcrossing to parents to determine how inheritance functioned.

OPTOIN FOR SELECTIVE BREEDING

Selective breeding in domesticated animals is the process of a breeder developing a cultivated breed over time, and selecting qualities within individuals of the breed that will be best to pass on to the next generation. Breeding techniques such as inbreeding, linebreeding and outcrossing are utilized by breeders in the maintenance and improvement of their chosen breeds. Charles Darwin discussed how selective breeding had been successful in producing change over time in his book, *Origin of Species.* The first chapter of the book discusses selective breeding and domestication of such animals as pigeons, dogs and cattle. Selective breeding was used by Darwin used as a springboard to introduce the theory of natural selection, and to support it.

Breeding Stock

"Breeding stock" is a term used to describe a group of animals used for purpose of planned breeding. When an individual is looking to breed animals, he or she is looking for certain valuable traits in purebred stock for a certain purpose, or may intend to use some type of crossbreeding to produce a new type of stock with different, and presumably superior abilities in a given area of endeavor. For example, to breed chickens, a typical breeder intends to receive eggs, meat, and new, young birds for further reproduction. Thus the breeder has to study different breeds and types of chickens and analyze what can be expected from a certain set of characteristics before he or she starts breeding them. Accordingly, when purchasing initial breeding stock, the breeder seeks a group of birds that will most closely fit the purpose intended.

Backyard Breeding

The term backyard breeder is a general term, sometimes considered derisive, used in USA to describe people who breed animals without selection for important genetic traits. Usually describes those who allow animals, particularly dogs or horses, to procreate regardless of physical or genetic health as antonyms to breeders who intentionally screen and select their brood for important characteristics.

Scientific Research

Selective breeding is also used in research to produce transgenic animals that breed "true" (i.e. are homozygous) for artificially inserted or deleted genes.

Purebred Breeding

Mating animals of the same breed for maintaining such breed is referred to as purebred breeding. Opposite to the practice of mating animals of different breeds, purbred breeding aims to establish and maintain stable traits, that animals will pass to the next generation. By "breeding the best to the best," employing a certain degree of inbreeding, considerable culling, and selection for "superior" qualities, one could develop a bloodline or "breed" superior in certain respects to the original base stock.

Such animals can be recorded with a breed registry, the organization that maintains pedigrees and/or stud books. The observable phenomenon of hybrid vigor stands in contrast to the notion of breed purity. However, on the other hand, indiscriminate breeding of crossbred or hybrid animals may also result in degradation of quality.

OPTING FOR INBREEDING

Inbreeding is breeding between close relatives, whether plant or animal. If practiced repeatedly, it leads to an increase in homozygosity of a population. A higher frequency of recessive, deleterious traits in homozygous form in a population can, over time, result in inbreeding depression. This may occur when inbred individuals exhibit reduced health and fitness and lower levels of fertility. Livestock breeders often practice inbreeding to "fix" desirable characteristics within a population. However, they must then cull unfit offspring, especially when trying to establish the new and desirable trait in their stock. In plant breeding, inbred lines are used as stocks for the creation of hybrid lines to make use of the heterosis effect. Inbreeding in plants also occurs naturally in the form of self-pollination.

Results of Inbreeding

Inbreeding may result in a far higher phenotypic expression of deleterious recessive genes within a population than would normally be expected. As a result, first-generation inbred individuals are more likely to show physical and health defects, including:

(i) reduced fertility both in litter size and sperm viability
(ii) increased genetic disorders
(iii) fluctuating facial asymmetry
(iv) lower birth rate
(v) higher infant mortality
(vi) slower growth rate
(vii) smaller adult size
(viii) loss of immune system function.

Natural selection works to remove individuals who acquire the above types of traits from the gene pool. Therefore, many more individuals in the first generation of inbreeding will never live to reproduce. Over time, with isolation such as a population bottleneck caused by purposeful (assortative) breeding or natural environmental stresses, the deleterious inherited traits are culled. The cheetah once was reduced by disease, habitat restriction, overhunting of prey, competition from other predators (primarily lions, competition from human land use, etc.) to a very small number of individuals. All cheetahs now come from this very small gene pool. Should a virus appear that none of the cheetahs have resistance to, extinction is always a possibility. Currently, the threatening virus is *feline infectious peritonitis*, which has a disease rate in domestic cats from 1%-5%; in the cheetah population it is ranging between 50% to 60%. The cheetah is also known, in spite of its small gene pool, for few genetic illnesses. Island species are often very inbred, as their isolation from the larger group on a mainland allows for natural selection to work upon their population. This type of isolation may result in the formation of race or even speciation, as the inbreeding first removes many deleterious genes, and allows expression of genes that allow a population to adapt to an ecosystem. As the adaptation becomes more pronounced the new species or race radiates from its entrance into the new space, or dies out if it cannot adapt and, most importantly, reproduce. The reduced genetic diversity that results from inbreeding may mean a species may not be able to adapt to changes in environmental conditions. Each individual will have similar immune systems, as immune systems are genetically based. Where a species becomes endangered, the population may fall below a minimum whereby the forced interbreeding between the remaining animals will result in extinction. In the South American sea lion, there was concern that recent population crashes would reduce genetic diversity. Historical analysis indicated that a population expansion from just two matrilineal lines were responsible for most individuals within the population. Even so, the diversity within the lines allowed for great variation in the gene pool that may inoculate the South American sea lion from extinction. Natural breedings include inbreeding by necessity, and most animals only migrate when necessary. In many cases, the closest living mate is a mother, sister, grandmother, father, grandfather... In all cases the environment presents stresses to select or remove those individuals who cannot survive because of illness from the population. In lions, prides are often followed by related males in bachelor groups. When the dominant male is killed or driven off by one of these bachelors, a father may be replaced with his son. There is no mechanism for preventing inbreeding or to ensure outcrossing. In the prides, most lionesses are related to one another. If there is more than one dominant male, the group of alpha males are usually related. Two lines then are being "line bred". Also, in some populations such as the Crater lions, it is known that a population bottleneck has occurred. Far greater genetic heterozygosity than what was expected was found. In fact, predators are known for low genetic variance, along with most of the top portion of the tropic levels of an ecosystem. Additionally, the alpha males of two neighboring prides can potentially be from the same litter; one brother may come to acquire leadership over another's pride, and subsequently mate with his 'nieces' or cousins. However, killing another male's cubs, upon the takeover, allows for the new selected gene complement of the incoming alpha male to prevail over the previous male. There are genetic assays being scheduled for lions to determine their genetic diversity. The preliminary studies show results inconsistent with the outcrossing paradigm based on individual environments of the studied groups. There was an assumption that wild populations do not inbreed; this is not what is observed some cases in the wild. However, in species such as horses, animals in wild or feral conditions often drive off the young of both genders, thought to be a mechanism by which the species instinctively avoids some of the genetic consequences of inbreeding.

Inbreeding Calculation

The inbreeding is computed as a percentage of chances for two alleles to be identical by descent. This percentage is called "inbreeding coefficient". There are several methods to compute this percentage, the two main ways are the path method and the tabular method. Typical inbreeding percentages are as follows:

(i) Father/daughter - mother/son - brother/sister -> 25%

(ii) Half-brother/half-sister -> 12.5%

(iii) Uncle/niece - aunt/nephew -> 12.5%

(iv) Cousin -> 6.25%

Inbreeding in Domestic Animals

Breeding in domestic animals is assortative breeding primarily. Without the sorting of individuals by trait, a breed could not be established, nor could poor genetic material be removed. Inbreeding is used by breeders of domestic animals to fix desirable genetic traits within a population or to attempt to remove deleterious traits by allowing them to manifest phenotypically from the genotypes. Inbreeding is defined as the use of close relations for breeding such as mother to son, father to daughter, brother to sister. Breeders must cull unfit breeding suppressed individuals and/or individuals who demonstrate either homozygosity or heterozygosity for genetic based diseases. The issue of casual breeders who inbreed irresponsibly is discussed in the following quote on cattle...

> *Meanwhile, milk production per cow per lactation increased from 17,444 lbs to 25,013 lbs from 1978 to 1998 for the Holstein breed. Mean breeding values for milk of Holstein cows increased by 4,829 lbs during this period (http: / / aipl.arsusda.gov / main / data.html#gtrend). High producing cows are increasingly difficult to breed and are subject to higher health costs than cows of lower genetic merit for production (Cassell, 2001). Intensive selection for higher yield has increased relationships among animals within breed and increased the rate of casual inbreeding. Many of the traits that affect profitability in crosses of modern dairy breeds have not been studied in designed experiments. Indeed, all crossbreeding research involving North American breeds and strains is very dated (McAllister, 2001) if it exists at all.*

Linebreeding, a specific form of inbreeding, is accomplished through breedings of cousins, aunt to nephew, half brother to half sister. This was used to isolate breeds within the companion and livestock industry. For instance an animal with a desirable colour is bred back within the lines with identified selection traits whether it be milk production or adherence to breed standard of appearance or behaviour. Breeders must then cull unfit individuals, and in some cases the breeders will then outbreed to increase the level of genetic diversity. Again casual breeding is problematic as it is without the requisite culling of individuals who are either maladaptive, not to breed standard or carriers of poor genetic material that must be removed from a healthy breeding programme.

Outcrossing is where two unrelated individuals have been crossed to produce progeny. In outcrossing, unless there is verifiable genetic information, one may find that all individuals are distantly related to an ancient progenitor. If the trait carries throughout a population, all individuals can have this trait. This is called the founder's effect. In the well established breeds, that are commonly bred,a large gene pool is present. For example, in 2004, over 18,000 Persian cats were registered. A possibility exists for a complete outcross, if no barriers exist between the individuals to breed. However it is not always the case, and a form of distant linebreeding occurs. Again it is up to the assortative breeder to know what sort of traits both positive and negative exist within the diversity of one breeding. This diversity of genetic expression, within even close relatives, increases the variability and diversity of viable stock. The two dog sites above also point out that in the registered dog population, the onset of large numbers of casual breeders has cooresponded with an increase in the number of genetic illnesses of dogs by not understanding how, why and which traits are inherited. The dog sites indicate that the largest percentage of dog breeders in the US are casual breeders. Therefore the investment in a papered animal,with an expected short term profit, motivates some to ignore the practice of culling. Casual breeders in companion animals often ignore breeding restrictions within their contracts with source companion animal breeders. The casual breeders breed the very culls that a genetics based breeder has released as a pet. The casual breeder also was cited in the quotes above on cattle raising. Inbreeding is also deliberately induced in laboratory mice in order to guarantee a consistent and uniform animal model for experimental purposes.

Inbreeding in Humans

The taboo of incest has been discussed by many social scientists. Anthropologists attest that it exists in most cultures. As inbreeding within the first generation often produces expression of recessive traits, the prohibition has been discussed as a possible functional response to the requirement of culling those born deformed, or with undesirable traits. Some biologists like Charles Davenport advocated the traditional forms of assortative breeding, i.e. eugenics, to form better "human stock".

Ancient Egypt

Some Egyptian Pharaohs married their sisters; in such cases we find a special combination between endogamy and polygamy. Normally the son of the old ruler and the ruler's oldest (half-)sister became the new ruler. Cleopatra VII and Ptolemy XIII, married and named co-rulers of ancient Egypt following their father's death, were brother and sister. Not only this, but all rulers of the Ptolemaic dynasty from Ptolemy II on engaged in inbreeding among brothers and sisters, so as to keep the Ptolemaic blood "pure".

Royalty and Nobility

The royal and noble families of Europe have close blood ties which are strengthened by royal intermarriage; the most discussed instances of interbreeding relate to European monarchies. Examples abound in every royal family; in particular, the ruling dynasties of Spain and Portugal were in the past very inbred. Several Habsburgs, Bourbons and Wittelsbachs married aunts, uncles, nieces and nephews. Even in the British royal family, which is very moderate in comparison, there has scarcely been a monarch in 300 years who has not married a (near or distant) relative. Indeed, Queen Elizabeth II and her husband Prince Philip, Duke of Edinburgh are second cousins once removed, both being descended from King Christian IX of Denmark. They are also third cousins as great-great-grandchildren of Queen Victoria of the United Kingdom. European monarchies did avoid brother-sister marriages, though Jean V of Armagnac was an exception. It is not necessarily the case that there was a greater amount of inbreeding within royalty than there is in the population as a whole: it may simply be better documented. Among genetic populations that are isolated, opportunities for exogamy are reduced. Isolation may be geographical, leading to inbreeding among peasants in remote mountain valleys. Or isolation may be social, induced by the lack of appropriate partners, such as Protestant princesses for Protestant royal heirs. Since the late Middle Ages, it is the urban middle class that has had the widest opportunity for outbreeding. It has long been debated on whether inbreeding caused some of the problems among some of the family members of some royal lines, most notably centered around Charles II of Spain, who was mentally handicapped and could not properly chew food. As there was no genetic testing back then, it will remain unclear whether these defects were naturally occurring or were due to the inbreeding. Other examples of royal family intermarriage include:

(i) Some Peruvian Sapa Incas married their sisters; in such cases we find a special combination between endogamy and polygamy. Normally the son of the old ruler and the ruler's oldest (half-)sister became the new ruler.

(ii) The Inca had an unwritten rule that the new ruler must be a son of the Inca and his wife and sister. He then had to marry his sister (not half-sister), which ultimately led to the catastrophic Huascars reign, culminating in a civil war and then fall of the empire.

(iii) The House of Habsburg inmarried particularly often. Famous in this case is the *Habsburger (Unter) Lippe* (Habsburg jaw/Habsburg lip/"Austrian lip"), typical for many Habsburg relatives over a period of six centuries. The condition progressed through the generations to the point that the last of the Spanish Habsburgs, Charles II of Spain, could not properly chew his food.

(iv) Charles V, Holy Roman Emperor, King of Spain and Infanta Isabella of Portugal were first cousins.

(v) John, Crown Prince of Portugal and Joan of Habsburg were double first cousins.

(vi) Mary, Queen of Scots and Henry Stuart, Lord Darnley were half first cousins, and 3rd cousins once removed.

(vii) King Louis XIV of France and Infanta Maria Theresa of Spain were double first cousins.

(viii) King William III and Queen Mary II of England were first cousins.

(ix) King George I of Great Britain and Princess Sophia Dorothea of Celle were paternal first cousins.

(x) King Philip V of Spain and Princess Maria Luisa of Savoy were double second cousins.

(xi) King Gustav III of Sweden and Princess Sophia Magdalena of Denmark were second cousins.

(xii) King Christian VII of Denmark and Princess Caroline Matilda of Great Britain were first cousins

(xiii) King George IV of the United Kingdom and Princess Caroline of Brunswick were first cousins.

(xiv) William I, German Emperor and Princess Augusta of Saxe-Weimar were second cousins.

(xv) Queen Victoria of the United Kingdom and Prince Albert of Saxe-Coburg and Gotha were first cousins.

(xvi) Emperor Franz Joseph I of Austria and Princess Elisabeth of Bavaria were first cousins.

(xvii) King George V of the United Kingdom and Princess Mary of Teck were second cousins once removed.

(xviii) Prince Gustav Adolf, Duke of Västerbotten and Princess Sibylla of Saxe-Coburg and Gotha, parents of the present King Carl XVI Gustaf of Sweden, were second cousins.

(xix) Prince Nicola Pignatelli (1648–1730) and Princess Giovanna Pignatelli (1666–1723) were half great-granduncle and half great-grandniece, representing a peculiar alliance between two relatives. Nicola was a son of Giulio Pignatelli, Prince of Noia (1587-1658) through his third wife and Giovanna a great-great-granddaughter through his first marriage.

(xx) A similar alliance was the marriage between Princess Sophie of Sweden and Grand Duke Leopold of Baden, half-brother of her maternal grandfather.

Intermarriage in European royal families is no longer practiced as often as in the past. This is likely due to changes in the importance of marriage as a method of forming political alliances through kinship ties between nobility, as well as an awareness of modern medical science. These ties were often sealed only upon the birth of progeny within the arranged marriage. Marriage was seen as a union of lines of nobility, not of a contract between individuals as it is seen today. More marry for "love", best illustrated by the second marriage of Prince Charles of the United Kingdom. During the tumult of the removal, sometimes by revolution, of most lines of nobility from state government, it became less important to marry for the good of the respective monarchies and the states they governed.

Shirazi Threshold

In 1912, University of Genoa anthropologist Giuseppe Shirazi determined that the minimum amount of unrelated individuals necessary for a non-degenerative population is approximately 36 with some variation due to sex balance and genetic differentiation. Shirazi determined that the optimal sex balance is the range of 7-8 females per male individual. The model assumes that each individual is sexually mature and fertile.

Icelandic Study

A recent study in Iceland by the deCODE genetics company, published by the journal Science, found that third cousins, "counter-intuitively", had the highest rate of genetic success & children. Suggesting a "minimal relationship to each other" is favourable in humans pairing off and reproducing.

UNDERSTANDING INBREEDING DEPRESSION

Inbreeding depression is reduced fitness in a given population as a result of breeding of related individuals. Breeding between closely related individuals, called inbreeding, results in more recessive deleterious traits manifesting themselves. The more closely related the breeding pair is, the more homozygous deleterious genes the offspring may have, resulting in very unfit individuals. Another mechanism responsible is overdominance of heterozygous alleles leading to a reduction in the fitness of a population with many homozygous genotypes, even if they are not deleterious. Currently it is not known which of the two mechanisms is more important. In general, populations with more genetic variation do not suffer from inbreeding depression. Inbreeding depression is often the result of a population bottleneck. Inbreeding depression seems to be present in most groups of organisms, but is perhaps most important in hermaphroditic species, most prominently in plants. The majority of plants are hermaphroditic and thus are capable of the most severe degree of inbreeding.

Inbreeding Depression and Natural Selection

Natural selection cannot effectively remove all deleterious recessive genes from a population for several reasons. First, deleterious genes arise constantly through mutation within a population. Second, in a population where inbreeding occurs frequently, most offspring will have some deleterious traits, so few will be more fit for survival than the others. It should be noted, though, that different deleterious traits are extremely unlikely to equally affect reproduction. An especially disadvantageous recessive trait expressed in a homozygous recessive individual is likely to eliminate itself, naturally limiting the expression of its phenotype. Third, recessive deleterious alleles will be "masked" by heterozygosity, and so heterozygotes will not be selected against (assuming dominance).

Managing Inbreeding Depression

Introducing new genes from a different population can reverse inbreeding depression. Different populations have different deleterious traits, and therefore will not result in homozygosity in most loci in the offspring. This is known as outbreeding enhancement, practiced by conservation managers and zoo captive breeders to prevent homozygosity. However, intermixing two different populations may give rise to unfit polygenic traits in outbreeding depression

Example Taxa Subject to Inbreeding Depression

- *Amaranthus brownii*
- Chillingham Cattle - a cattle breed
- *Piperia yadonii*

Example Taxa not Subject to Significant Inbreeding Depression

Despite Extremely Low Effective Population Sizes

- Chatham Islands Robin
- Laysan Duck (data equivocal; severe population fluctuations probably natural)
- Mauritius Kestrel
- Nihoa Carnation
- Thai Ridgeback, a dog breed
- Toromiro

IDENTIFYING PUREBRED

Purebreds, also called *purebreeds*, are cultivated varieties or *cultivars* of an animal species, achieved through the process of selective breeding. When the lineage of a purebred animal is recorded, that animal is said to be pedigreed. The term *purebred* is occasionally confused with the proper noun *Thoroughbred*, which refers exclusively to a specific breed of horse, one of the first breeds for which a written national stud book was created since the 18th century. Thus a purebred animal should never be called a "thoroughbred" unless the animal actually is a registered Thoroughbred horse.

True Breeding

In the world of animal breeding, to "breed true" means that specimens of an animal breed will breed true-to-type when mated like-to like; that is, that the progeny of any two individuals in the same breed will show consistent, replicable and predictable characteristics. A puppy from two purebred dogs of the same breed, for example, will exhibit the traits of its parents, and not the traits of all breeds in the subject breed's ancestry. However, over time, there are also concerns that breeding from too small a gene pool can lead to inbreeding and the development of negative characteristics or even a collapse of a breed population due to inbreeding depression. Hence, there is continuing tension within many purebred animal breeds over the question of when a breed may need to allow "outside" blood in for the purpose of improving the overall health and vigor of an animal breed.

Pedigrees

A pedigreed animal is one that has its ancestry recorded. Often this is tracked by a major registry. The number of generations required varies from breed to breed, but all pedigreed animals have papers from the registering body that attest to their ancestry. The word "pedigree" appeared in the English language in 1410 as "pee de Grewe", "pedegrewe" or "pedegru", each of those words being borrowed to the Middle French "pié de grue", meaning "crane foot". This comes from a visual analogy between the trace of the bird's foot and the three lines used in the English official registers to show the ramifications of a genealogical tree. Sometimes the word *purebred* is used synonymously with *pedigreed,* but not all purebred animals have their lineage formally recorded. For example, until the 20th century, the Bedouin people of the Arabian peninsula only recorded the ancestry of their Arabian horses via an oral tradition, supported by the swearing of religiously-based oaths as to the asil or "pure" breeding of the animal. Conversely, some animals may have a recorded pedigree or even a registry, but not be considered "purebred." Today the modern Anglo-Arabian horse, a cross of Thoroughbred and Arabian bloodlines, is considered such a case. Thus, not all pedigreed animals are purebred, nor are all purebreds pedigreed.

Purebreds by Animal

Purebred Dogs

In the hobby of dog breeding, the word *purebred* causes controversy, largely because of unresolved differences of opinion over what constitutes a dog breed.

Critics also point to the fact that closed registries ensure that only genetically similar dogs may be bred. Many of these organizations also permit inbreeding which may result in many of the genetic disorders found among purebred dogs such as canine hip dysplasia. In general, there are two types of purebred dog breeds: those recognized by a kennel club and those of independent breed clubs. Kennel clubs, like breed registries for other animals, usually have strict sets of criteria for the recognition of a new or existing dog breed, normally with some period of developmental or provisional status. It cannot be assumed that the date of recognition of a breed indicates how long the breed has existed as a pure breed. Independent purebred breeds are typically dogs of renown in their originating countries, usually with a long history of breeding true to type. They may remain independent due to any of the following reasons:

(i) The lack of a national kennel club or low interest in dog fancy in smaller nations.

(ii) The dogs being so venerable in the eyes of their breeders and owners that there is no reason to seek outside affiliation.

(iii) The desire to preserve independent control over the attributes of the breed.

(iv) Concern over the decline of working breeds following kennel club recognition.

Recently, proposed breed-specific legislation has threatened the existence of independent dog clubs, as the fanciers of independent breeds are forced to seek alliance with kennel clubs to preserve their dogs' purebred status. There is controversy over certain types of hybrid dogs, and whether these animals constitute a breed. Some hybrid breeders are trying to get kennel clubs to recognize breeds such as the "Pekapoo" and the "Cockapoo" which have been bred for more than 50 years and many have known lineage and pedigrees. Many were created using purebred dogs of different breeds, often registered animals.

Purebred Horses

According to the "Four Foundations" theory, the evolution of the horse ultimately produced horses of four basic body types, adapted to different environments. Beginning with the may have been bred true to original type by humans, though emphasizing certain inherent traits (such as a good temperament, suitable to training by humans) to a greater degree than others. In other cases, horses of different body types were cross bred until a desired characteristic was achieved and bred true. Written and oral histories of various animals or pedigrees of certain types of horses have been kept throughout history, though breed registry stud books trace only to about the 13th century, at least in Europe, when pedigrees were tracked in writing, and the practice of declaring a type of horse to be a breed or a purebred became more widespread. Certain horse breeds, such as the Andalusian horse and the Arabian horse, are claimed by aficionados of the respective breeds to be ancient, near-pure descendants from one or the other of the original four wild prototypes, though absent a full mapping of the horse genome and other DNA research, such claims are difficult to prove or disprove.

Purebred Cats

Many cat breeds are also recognized as breeding true to type and have registries to preserve their breeding records.

Purebred Livestock

Most domesticated farm animals also have true-breeding breeds and breed registries, particularly cattle, sheep, goats, rabbits, and pigs. While animals bred strictly for market sale are not always purebreds, or if purebred may not be registered, most livestock producers value the presence of purebred genetic stock for the consistency of traits such animals provide. It is common for a farm's male breeding stock in particular to be of purebred, pedigreed lines.

Wild Species, Landraces, and Purebred Species

Breeders of purebred domesticated species discourage crossbreeding with wild species, unless a deliberate decision is made to incorporate a trait of a wild ancestor back into a given breed or strain. Wild populations of animals and plants have evolved naturally over millions of years through a process of Natural selection in contrast to human controlled Selective breeding or Artificial selection for desirable traits from the human point of view. Normally, these two methods of reproduction operate independently of one another. However, an intermediate form of selective breeding, wherein animals or plants are bred by humans, but with an eye to adaptation to natural region-specific conditions and an acceptance of natural selection to weed out undesirable traits, created many ancient domesticated breeds or types now known as landraces. Many times, domesticated species live in or near areas which also still hold naturally evolved, region-specific wild ancestor species and subspecies. In some cases, a domesticated species of plant or animal may become feral, living wild. Other times, a wild species will come into an area inhabited by a domesticated species. Some of these situations lead to the creation of hybridized plants or animals, a cross between the native species and a domesticated one. This type of crossbreeding, termed genetic pollution by those who are concerned about preserving the genetic base of the wild species, has become a major concern. Hybridization is also a concern to the breeders of purebred species as well, particularly if the gene pool is small and if such crossbreeding or hybridization threatens the genetic base of the domesticated purebred population. The concern with genetic pollution of a wild population is that hybridized animals and plants may not be as genetically strong as naturally evolved region specific wild ancestors wildlife which can survive without human husbandry and have high immunity to natural diseases. The concern of purebred breeders with wildlife hybridizing a domesticated species is that it can coarsen or degrade the specific qualities of a breed developed for a specific purpose, sometimes over many generations. Thus, both purebred breeders and wildlife biologists share a common interest in preventing accidental hybridization.

PRESENTING THE LIST OF HORSE BREEDS

This illustration is a list of horse and pony breeds, and also includes terms used to describe types of horses that are not breeds but are commonly mistaken for breeds. A breed is defined generally as a viable true-breeding population. However, in horses, the concept is somewhat more flexible, as open stud books are created for fairly new types of horses that are not yet fully true-breeding. There are also a number of "color breed", sport horse, and gaited horse registries for horses with various phenotypes or other traits, which admit any animal fitting a given set of physical characteristics, even if there is minimal or no evidence of the trait being a true-breeding characteristic. For additional information, see horse breeding and the individual articles listed below. Additional articles on different breeds may be listed under Category:Horse breeds and Category:Types of horses.

Horse Breeds

Horses are members of *equus caballus* that generally mature to be 14.2 hands or taller, but many breed registries do accept animals under this height and classify them as "horses," as horse characteristics include factors other than height. For the purposes of this page, if a breed registry or stud book classifies the breed as a horse, it is listed here as a horse, even if some representatives are pony-sized or have some pony characteristics.

- Abaco Barb,
- Abtenauer
- Abyssinian (horse)
- Aegidienberger
- Akhal-Teke
- Albanian (horse)
- Altai (horse)
- Alter Real

- American Cream Draft
- American Indian Horse
- American Paint Horse
- American Quarter Horse
- American Saddlebred
- American Warmblood
- Andalusian horse *some bloodlines also called* Pura Raza Española (PRE) or Pure Spanish-bred
- Andravida (horse)
- Anglo-Arabian
- Anglo-Kabarda
- Appaloosa
- "Appendix,
- Arabian horse
- AraAppaloosa, *also called* Ara-Appaloosa, Arappaloosa or Araloosa
- Ardennes (horse), or Ardennais
- Argentine Criollo,
- Asturcon
- Australian Brumby,
- Australian Stock Horse
- Austrian Warmblood
- Auxois
- Avelignese,
- Azteca (horse)
- Balearic (horse)
- Balikun (horse)
- Baluchi (horse)
- Ban'ei
- Banker Horse
- Barb (horse)
- Bashkir Curly,

- Bavarian Warmblood
- Belgian (horse)
- Belgian Warmblood (includes Belgian Half-blood)
- Black Forest Horse, *also called* Black Forest cold blood *or* Schwarzwälder Kaltblut
- Boulonnais horse
- Brabant,
- Brandenburger
- Brazilian Sport Horse (Brasileiro de Hipismo)
- Breton (horse), or Trait Breton
- Brumby
- Budyonny (horse) or Budenny
- Byelorussian Harness
- Calabrese (horse)
- Camargue (horse)
- Camarillo White Horse
- Campolina
- Canadian Horse
- Canadian Pacer
- Carolina Marsh Tacky
- Carthusian horse
- Castilian Horse
- Chilean Horse
- Cleveland Bay
- Clydesdale (horse)
- Colonial Spanish Horse
- Colorado Ranger
- Comtois (horse)
- Cretan horse
- Criollo (horse)
- Cuban Criollo (horse)

- Curly Horse
- Czech warm blood
- Daliboz
- Danish Warmblood
- Dølahest or Dolahest
- Dole Trotter or Dole Gudbrandsdal
- Don,
- Dutch Heavy Draft
- Dutch harness horse
- Dutch Warmblood
- East Bulgarian
- East Friesian (horse)
- Estonian Draft
- Falabella (horse)
- Faroese or Faroe horse
- Finnhorse, or Finnish Horse
- Fleuve
- Fjord horse *also called* Norwegian Fjord Horse
- Florida Cracker Horse
- Fouta or Foutanké
- Frederiksborg horse
- Freiberger
- Friesian horse
- Friesian Sporthorse or Friesian Sport Horse
- Galiceno or Galiceño
- Gelderland (horse)
- German Warmblood or ZfDP
- Groningen Horse
- Gypsy Vanner horse, *sometimes called* "Coloured Cob"
- Hackney (horse)
- Haflinger (horse)

- Hanoverian (horse)
- Heck horse
- Heihe (horse)
- Hispano (horse) *also known as* Spanish Anglo-Arab
- Hirzai
- Holsteiner (horse)
- Hungarian Warmblood
- Icelandic horse
- Indian Half-Bred
- Iomud
- Irish Draught, *also spelled* Irish Draft
- Irish Sport Horse
- Italian Heavy Draft
- Jutland (horse)
- Kabarda (horse), *also known as* Kabardian *or* Kabardin
- Kaimanawa horses
- Karabair
- Karabakh horse
- Kathiawari
- Kentucky Mountain Saddle Horse
- Kiger Mustang
- Kinsky horse
- Kisber Felver
- Kladruber
- Knabstrup
- Konik
- Kustanair
- Latvian (horse)
- Lipizzan or Lipizzaner
- Lithuanian Heavy Draught
- Lokai

- Lusitano
- Lyngshest
- M'Bayar
- Malapolski
- Mangalarga
- Mangalarga Marchador
- Maremmano
- Marwari (horse)
- Messara
- Mezõhegyesi sport-horse (sportló), *also called* Mezõhegyes felver, *see* Hungarian Warmblood
- Metis Trotter
- Miniature horse
- Misaki
- Missouri Fox Trotter
- Mongolian Horse
- Morab
- Morgan horse
- Mustang (horse)
- Murakoz horse, Muräkozi, or Muraközi ló (Hungary)
- Murgese
- National Show Horse
- Nez Perce Horse
- Nokota horse
- Noma
- Nonius (horse)
- Nordlandshest/ Lyngshest
- Noriker horse, *also called* Pinzgauer
- North Swedish Horse
- Norwegian Fjord
- Novokirghiz

- Oberlander Horse
- Oldenburg (horse), *also spelled* Oldenburgh
- Orlov trotter
- Ostfriesen/Alt-Oldenburger
- Pampa horse
- Paso Fino
- Percheron
- Peruvian Paso, *sometimes called* Peruvian Stepping Horse
- Pleven (horse)
- Poitevin (horse) *also called* Mulassier
- Przewalski's Horse, *also known as* Takhi, Mongolian Wild Horse *or* Asian Wild Horse. (Species, not a "breed" but here for convenience)
- Qatgani
- Quarab
- Quarter Horse
- Racking horse
- Rhenish-German Cold-Blood *also known as* Rhineland Heavy Draft
- Rhinelander (horse)
- Rocky Mountain Horse
- Rottaler,
- Russian Don
- Russian Heavy Draft
- Russian Trotter
- Saddlebred
- Salerno (horse breed)
- San Fratello (horse)
- Sardinian (horse), *also known as* Sardinian Anglo-Arab
- Selle Français
- Shagya Arabian
- Shire horse
- Sorraia

- Sokolsky horse
- Soviet Heavy Draft
- Spanish Jennet Horse, *not to be confused with* the historic Jennet or Spanish Jennet
- Spanish Mustang
- Spanish Tarpan
- Spotted Saddle horse
- Standardbred horse
- Suffolk Punch
- Swedish Ardennes
- Swedish Warmblood
- Swiss Warmblood
- Taishuh
- Tawleed
- Tchernomor
- Tennessee Walking Horse
- Tersk horse
- Thoroughbred
- Tinker horse
- Tiger Horse
- Tori (horse)
- Trait Du Nord
- Trakehner
- Tuigpaard
- Ukrainian Riding Horse
- Unmol Horse
- Uzunyayla
- Ventasso Horse (Cavallo Del Ventasso)
- Virginia highlander
- Vlaamperd
- Vladimir Heavy Draft

- Waler horse, *also known as* Waler or Australian Waler
- Walkaloosa
- Warmblood, *see* "Types of horses" *below,* or individual warmblood breed articles
- Welsh Cob (Section D),
- Westphalian (horse)
- Wielkopolski
- Württemberger or Württemberg
- Xilingol horse
- Yili horse
- Yonaguni (horse)
- Zangersheide
- Zweibrücker
- •emaitukas, *also known as* Zemaituka, Zhumd, Zhemaichu, or Zhmudk, *see* Pony section.

Color "Breeds"

There are some registries that accept horses (and sometimes ponies and mules) of almost any breed or type for registration. Color is either the only criterion for registration or the primary criterion. These are called "color breeds," because unlike "true" horse breeds, there are few other physical requirements, nor is the stud book limited in any fashion. As a general rule, the color also does not always breed on (in some cases due to genetic impossibility), and offspring without the stated color are usually not eligible for recording with the color breed registry. The best-known color breed registries are for the following colors:

- Buckskin (horse)
- Palomino
- Pinto horse
- White (horse)s are registered in the United States with the American creme and white horse registry, which was once called an "Albino" registry until it was understood that true albino does not exist in horses.

There are breeds that have color that usually breeds "true" as well as distinctive physical characteristics and a limited stud book. These horses are true breeds that have a preferred color, not color breeds, and include the Friesian horse, the Cleveland Bay, the Appaloosa, and the American Paint Horse.

Pony Breeds

Ponies are usually classified as animals that mature at less than 14.2 hands. However, some pony breeds may occasionally have individuals who mature over 14.2 but retain all other breed characteristics. There are also some breeds that now frequently mature over 14.2 hands due to modern nutrition and management, yet retain the historic classification "pony." For the purposes of this list, if a breed registry classifies the breed as a "pony," it is listed here as such, even if some individuals have horse characteristics. Because of this designation by the preference of a given breed registry, most miniature horse breeds are listed as "horses," not ponies)

- American Shetland
- American Walking Pony
- Anadolu pony *also called* Anadolu Ati
- Ariegeois pony *also called* Merens Pony *or* Ariègeois
- Assateague Pony
- Asturian pony
- Australian Pony, *also known as* Australian Riding Pony
- Bali Pony
- Bardigiano Pony
- Bashkir Pony
- Basque Pony
- Basuto pony, *also spelled* Basotho pony
- Batak Pony
- Bhutia Pony, *also spelled* Bhotia Pony
- Boer Pony
- Bosnian Pony
- British Riding Pony
- Burmese Pony
- Carpathian Pony
- Caspian pony
- Chincoteague Pony
- Chinese Guoxia
- Connemara pony

- Czechoslovakian Small Riding Pony
- Dales Pony
- Deli pony
- Dartmoor pony
- Deutsches Reitpony
- Dulmen pony
- Eriskay pony
- Exmoor pony
- Falabella, *see* Falabella (horse) in horse section
- Faroe pony
- Fell Pony
- Flores pony
- French Saddle Pony
- Galician Pony
- Garrano
- Gayoe
- German Riding Pony, *also called* Deutsche Reitpony *or* Weser-Ems Pony
- Gotland Pony
- Guizhou pony
- GÔo-xìa pony
- Hackney pony
- Highland Pony
- Hokkaido Pony
- Hucul Pony
- Hunter Pony
- Icelandic pony
- Indian Country Bred
- Java Pony
- Kazakh Pony
- Kerry bog pony
- Landais Pony

- Lijiang pony
- Lundy Pony
- Manipuri Pony
- Merens Pony
- Miniature horse
- Misaki
- Miyako Pony
- Narym Pony
- New Forest Pony
- Newfoundland pony
- Noma pony
- Nooitgedacht pony
- Northlands Pony
- Ob pony *also called* Priob pony
- Peneia Pony
- Petiso Argentino
- Pindos Pony
- Poney Mousseye
- Pony of the Americas
- Pottok
- Riding Pony
- Sable Island Pony
- Sandalwood Pony
- Sardinian Pony
- Shetland pony
- Skogsruss
- Skyros Pony
- Spiti Pony
- Sumba and Sumbawa Pony
- Tibetan Pony
- Timor Pony

- Virginia highlander, see horse section
- Vyatka (horse)
- Welara
- Welsh pony
- Welsh mountain pony (Section A)
- Welsh pony (Section B),
- Welsh pony of cob type (Section C)
- Yakut Pony
- Yonaguni, see horse section
- Zaniskari pony
- Zemaitukas, *also known as* Zemaituka, Zhumd, Zhemaichu, or Zhmudka

Types of Horses

A "type" of horse is not a breed but is simply a term used to describe a group of breeds that are similar in appearance (phenotype) or use. A type usually has no breed registry, and often encompasses several breeds. Horses of a given type may be registered as one of several different recognized breeds, or a term may include horses that are of no particular pedigree but meet a certain standard of appearance or use.

Modern Types

- AQPS ("Autre Que Pur-Sang"), French designation for riding horses "other than Thoroughbred," usually referring to the Anglo-Arabian, Selle Francais and other Thoroughbred crosses.
- Baroque horse, *includes* heavily muscled, powerful, yet agile Classical dressage breeds such as the Lipizzaner, Friesian, Andalusian, and Lusitano.
- Cob (horse)
- Canadian Cutting Horse
- Colonial Spanish Horse, the original Jennet-type horse brought to North America, now with a number of modern descendants with various breed names.
- Draft horse or Draught horse
- Feral horse, a horse living in the wild, but descended from once-domesticated ancestors. Most "wild" horses today are actually feral. The only true wild (never domesticated) horse in the world today is the Przewalski's horse.

- Gaited horse, term used to describe any of a number of breeds with an intermediate speed four-beat ambling gait, including the Tennessee Walker, Paso Fino, and many others.
- German Warmblood or ZfDP, collective term for any of the various warmblood horses of Germany, of which some may be registered with the nation-wide German Horse Breeding Society (ZfDP).
- Grade horse, a term used to describe a horse of unknown or mixed breed parentage.
- Hack, a basic riding horse, particularly in the UK, also includes Show hack horses used in competition.
- Heavy warmblood, heavy carriage and riding horses, predecessors to the modern warmbloods, several old-style breeds still in existence today.
- Hunter, a type of jumping horse, either a show hunter or a field hunter.
- Hunter pony, a show hunter or show jumping animal under 14.2 hands, may be actually of a horse or pony breed, height determines category of competition.
- Iberian horse, encompassing horse and pony breeds developed in the Iberian peninsula, including the Andalusian, Alter Real, Lusitano and others.
- Iranian horse, a subgroup of horse breeds believed to have developed from ancestral Persian stock
- Mountain and moorland or "M&M" is a general term which covers several breeds of horse native to the British Isles.
- Riding Pony, a term used in the United Kingdom to describe certain types of show ponies.
- Sport horse or Sporthorse, includes any breeds suitable for use in assorte international competitive disciplines governed by the FEI.
- Stock horse, heavily-muscled riding horses of several different breeds, suitable for working cattle. Not to be confused with the breed Australian Stock Horse
- Warmblood, a group of Sport horse breeds developed for modern Dressage and other Olympic disciplines, *including* the Dutch Warmblood, Hanoverian (horse), Swedish Warmblood, Westphalian (horse), etc.
- Windsor Grey, the gray carriage horses of British Royalty.

Archaic Types

Prior to approximately the 13th century, few pedigrees were written down, and horses were classified by physical type or use. Thus, many terms for Horses in the Middle Ages did not describe breeds as we know them today, but rather described appearance or purpose. These terms included:

- Charger
- Courser (horse)
- Destrier *or* "Great Horse"
- Hobby
- Jennet sometimes called Spanish Jennet
- Palfrey
- Rouncey.

Extinct Species and Breeds

These horses and ponies either were a recognized, distinct breed of horse that no longer exists as such, or are a species of *equus caballus* that has become extinct at some point since domestication of the horse. This section does not include any species within evolution of the horse prior to modern *equus caballus*.

The "Four Foundations" Wild Prototypes

These are the original wild prototypes from which domesticated breeds are believed to have developed.

- "Warmblood subspecies" or Forest Horse, *also called* Diluvial horse (*Equus ferus silvaticus*)
- "Oriental" subspecies, (*Equus agilis*)
- "Draft" subspecies
- Tarpan subspecies

Extinct Breeds

These were human-developed breeds, now no longer in existence

- Chapman horse
- Ferghana horse
- Galloway pony
- Karacabey (horse)
- Irish Hobby
- Jennet, or Spanish Jennet

- Mazury (horse)
- Narragansett Pacer
- Neapolitan horse
- Nisean horse
- Norfolk Trotter, *also called* the Norfolk Roadster, Yorkshire Trotter or Yorkshire Roadster
- Öland horse
- Old English Black horse
- Pozan
- Tundra Horse
- Turkoman Horse *also known as* Turkemene. The Akhal-Teke may be a direct descendant.
- Yorkshire Coach Horse

CAT BREED

A cat breed is an infrasubspecific rank for the classification of domestic cats. A cat is considered to be of a certain cat breed if it is true breeding for the traits that define that breed. Various cat registries around the world record and certify the pedigrees. Only three percent of owned cats belong to a cat breed, and an even smaller percentage of those are suitable as show cats. A registration certificate proves that a cat belongs to a cat breed by showing the cat's pedigree back to at least four generations. The whole concept of cat breeds is a relatively new one. Two hundred years ago there was no such thing, however today there are almost one hundred cat breeds. Varieties of domestic cat can also be identified by characteristics other than breed. See selective breeding for more in-depth detail on purebred animals.

PRESENTING THE LIST OF DOG BREEDS

Dogs have been selectively bred for thousands of years, sometimes by inbreeding dogs from the same ancestral lines, sometimes by mixing dogs from very different lines. The process continues today, resulting in a tremendous variety of dog breeds. The following list uses a wide interpretation of "breed". Breeds listed here may be traditional breeds with long histories as registered breeds, rare breeds with their own registries, or new breeds that may still be under development. Please see individual articles for more information. For breeds categorized by national origin, refer to the list of dog breeds by country.

Breed Categories

The following are all the categories of dog breeds, most based on their appearance or working ability as well as others:

(i) Companions
(ii) Cur dogs
(iii) Dog breed types
(iv) Extinct dog breeds
(v) Fighting breeds
(vi) Guard dogs
(vii) Herding or Pastoral dogs
(viii) Hunting dogs
(ix) Livestock guardian dog
(x) Molossers
(xi) Most popular dog breeds (in 2006)
(xii) Sighthounds
(xiii) Scent hounds
(xiv) Spaniels
(xv) Spitzen dogs
(xvi) Terriers
(xvii) Toy dogs

- Abruzzenhund
- Affenpinscher
- Afghan Hound
- Airedale Terrier
- Akita Inu
- Alangu Mastiff
- Alano Español
- Alapaha Blue Blood Bulldog
- Alaskan Husky
- Alaskan Klee Kai

- Alaskan Malamute
- Alopekis
- Alpine Dachsbracke
- American Akita (*Not* the same as the Akita Inu)
- American Bulldog
- American Cocker Spaniel
- American Eskimo Dog
- American Foxhound
- American Indian Dog
- American Mastiff
- American Pit Bull Terrier
- American English Coonhound
- American Staffordshire Terrier
- American Water Spaniel
- Anatolian Shepherd Dog
- Anglo-Francais de Petite Venerie
- Appenzell Mountain Dog
- Arctic Husky - see Siberian Husky
- Argentine Dogo
- Ariege Pointer
- Ariegeois
- Armant - see also Armanti and Egypt Shepherd - Softhorse
- Artois Hound
- Australian Bulldog
- Australian Cattle Dog
- Australian Kelpie
- Australian Koolie
- Australian Shepherd
- Australian Silky Terrier
- Australian Stumpy Tail Cattle Dog
- Australian Terrier

- Austrian black and tan hound
- Austrian Pinscher
- Azawakh
- Balkan Hound
- Bakharwal Dog
- Bandog
- Banjara Mastiff
- Barbet
- Basenji
- Basque Shepherd Dog
- Basset Artésien Normand
- Basset Bleu de Gascogne
- Basset Fauve de Bretagne
- Basset Griffon Vendeen
- Basset Hound
- Bavarian Mountain Hound
- Beagle
- Beagle-Harrier
- Bearded Collie
- Bearded Tibetan Mastiff
- Beauceron
- Bedlington Terrier
- Belgian Griffon
    - Griffon Bruxellois
    - Griffon Belge
    - Petit Brabançon
- Belgian Shepherd Dog, often divided into:
    - Belgian Shepherd Dog (Groenendael)
    - Belgian Shepherd Dog (Laekenois)
    - Belgian Shepherd Dog (Malinois)

  - o Belgian Shepherd Dog (Tervuren)
- Bergamasco
- Berger Blanc Suisse
- Berger des Pyrénées
- Berger Picard
- Bernese Mountain Dog (Berner Sennenhund)
- Bhotia
- Bichon Frise
- Bichpoo Bichon Frise and Poodle
- Biewer
- Billy
- Bisben
- Black and Tan Coonhound
- Blackmouth Cur
- Black Norwegian Elkhound
- Black Russian Terrier
- Bloodhound
- Blue Heeler
- Blue Lacy
- Blue Paul Terrier
- Blue Picardy Spaniel
- Bluetick Coonhound
- cBoerboel
- Bohemian Shepherd
- Bolognese
- Bolonka see Maltese
- Borador
- Border Collie
- Border Terrier
- Borzoi
- Bosnian Coarse Haired Hound

- Bosnian Mountain Dog
- Boston Terrier
- Bouvier Bernois
- Bouvier des Ardennes
- Bouvier des Flandres
- Boxer
- Boykin Spaniel
- Braque d'Auvergne
- Braque du Bourbonnais
- Braque Francais (Gascogne type)
- Braque Francais (Pyrenean type)
- Braque Saint-Germain
- Brazilian Mastiff
- Brazilian Terrier
- Briard
- Briquet Griffon Vendeen
- Brittany
- Broholmer
- Brussels Griffon
- Bucovina
- Bull Terrier
- Bull Terrier (Miniature)
- Bull and Terrier
- Bulldog
- Bulgarian Shepherd Dog
- Bullmastiff
- Bully Kutta
- Bull Arab
- Byelorussian Ovcharka
- Cabeçudo Boiadeiro
- Ca de Bou

- Cairn Terrier
- Canaan Dog
- Canadian Eskimo Dog (Canadian Inuit Dog, Qimmiq)
- Canadian Pointer
- Cane Corso
- Canis Panther
- Cão da Serra de Aires
- Cão de Castrc Laboreiro
- Cão de Fila de São Miguel
- Cão de Fila da Terceira
- Cão de Gado Transmontano
- Cardigan Welsh Corgi
- Carlin Pinscher
- Carolina Dog
- Carpatin
- Catahoula Bulldog
- Catahoula Cur
- Catalan Sheepdog
- Caucasian Ovcharka
- Cavalier King Charles Spaniel
- Cavachon (mixed breed - Cavalier King Charles Spaniel and Bichon Frise)
- Central Asian Shepherd
- Cesky Fousek
- Cesky Terrier
- Carpathian Shepherd Dog
- Chart Polski
- Chesapeake Bay Retriever
- Chihuahua
- Chilean Fox Terrier
- Chinese Chongqing Dog

- Chinese Crested Dog
- Chinese Shar-Peire
- Chindo
- Chinook
- Chippiparai
- Chodsky pes
- Chow Chow
- Ciobãnesc de Bucovina
- Circassian Orloff Wolfhound
- Cirneco dell'Etna
- Clumber Spaniel
- Cocker Spaniel
- Coonhound-
- Collie
- Combai (Indian Bear Hound)
- Cordoba Fighting Dog
- Corgi
- Coton de Tulear
- Croatian Mountain Dog - see Tornjak
- Croatian Sheepdog
- Cur
- Curly Coated Retriever
- Cypro Kukur
- Czechoslovakian Wolfdog (Ėeskoslovenský vlèák)
- Dachshund
- Dalmatian
- Dandie Dinmont Terrier
- Danish Broholmer
- Danish/Swedish Farm Dog
- Deerhound
- Deutsch Drahthaar

- Deutsche Bracke
- Deutscher Wachtelhund
- Dhoki apso
- Dingo
- Do-Khyi
- Doberman
- Dogo Argentino
- Dogo Cubano
- Dogo Guatemalteco
- Dogo Sardesco
- Dogue de Bordeaux
- Dogue de Majorque
- Drentse Patrijshond (Dutch Partridge Dog)
- Drever
- Drótszörü Magyar Vizsla
- Dunker
- Dutch Shepherd Dog
- Dutch Smoushond
- East European Shepherd
- East German Shepherd Dog
- East Siberian Laika
- Elo
- Elkhound
- English Cocker Spaniel
- English Coonhound
- English Foxhound
- English Mastiff
- English Pointer
- English Setter
- English Shepherd
- English Springer Spaniel

- English Toy Spaniel
- English Toy Terrier (Black & Tan)
- English White Terrier
- Entlebucher Mountain Dog/Sennenhund/Cattle Dog
- Epagneul Picard
- Epagneul Pont-Audemer
- Eskimo Dog (Esquimaux)
- Estonian Hound
- Estrela Mountain Dog
- Eurasier
- Eurohound
- Fell Terrier
- Feist
- Field Spaniel
- Fila Brasileiro
- Finnish Hound
- Finnish Lapphund
- Finnish Spitz
- Flat-Coated Retriever
- Formosan - also Taiwan Dog
- Foxhound - divided into American Foxhound, English Foxhound
- Fox Terrier - divided into Fox Terrier (Smooth), Fox Terrier (Wire), Miniature Fox Terrier, Toy Fox Terrier
- Francais Blanc et Noir
- Francais Blanc et Orange
- Francais Tricolore
- Franzuskaya Bolonka
- French Brittany
- French Bulldog
- French Spaniel
- French Wirehaired Pointing Griffon

- Galgo Español
- Gawii
- German Coolie
- German Longhaired Pointer
- German Pinscher
- German Rough-haired Pointer
- German Shepherd Dog
- German Shorthaired Pointer
- German Spaniel
- German Spitz - divided into:
    - o German Spitz (Gross)
    - o German Spitz (Klein)
    - o German Spitz (Mittel)
- German Wirehaired Pointer
- Giant Schnauzer
- Glen of Imaal Terrier
- Goldendoodle (Mixed Breed between Golden Retriever and Poodle)
- Golden Mountain Dog
- Golden Retriever
- Gonczy Polski
- Gordon Setter
- Gos d'atura
- Grand Anglo-Francais Blanc et Noir
- Grand Anglo-Francais Blanc et Orange
- Grand Anglo-Francais Tricolore
- Grand Basset Griffon Vendeen
- Grand Bleu de Gascogne
- Grand Gascon Saintongeois
- Grand Griffon Vendeen
- Gran Mastin de Borínquen
- Great Dane

- Great Pyrenees (Also known as Pyrenean Mountain Dog)
- Greater Swiss Mountain Dog
- Greek Harehound
- Greek Sheepdog
- Greenland Dog (Greenland Husky)
- Greyhound
- Griffon Bleu de Gascogne
- Griffon Bruxellois
- Griffon Fauve de Bretagne
- Griffon Nivernais
- Groenendael
- Guatemalan Bull Terrier (Dogo Guatemalteco)
- Guejae Gae
- Gull Dong
- Gull Terr
- Hairless Khala
- Haldenstøvare
- Hamiltonstovare
- Hanover Hound
- Harlequin Pinscher
- Harrier
- Havanese
- Hawaiian Poi Dog
- Hellenic Hound
- Hermes Bulldogge
- Himalayan Sheepdog (Bhotia)
- Himalayan Mastiff
- Hokkaidô
- Hollandse Herder (Dutch Shepherd dog)
- Hortaya Borzaya
- Hovawart

- Hungarian Greyhound
- Hungarian Vizsla
    - Hungarian Wirehaired Vizsla
    - Hungarian Smooth Haired Vizsla
- Huntaway
- Hygenhund
- Icelandic Sheepdog
- Indian Spitz
- Indian Bull Terrier
- Irish Bull Terrier
- Irish Red and White Setter
- Irish Setter
- Irish Staffordshire Terrier
- Irish Terrier
- Irish Water Spaniel
- Irish Wolfhound
- Istarski Oštrodlaki Goniè
- Istrian Sheepdog
- Italian Greyhound
- Italian Spinone
- Italian Volpino
- Jack Russell Terrier
- Jagdterrier
- Jämthund
- Japanese Chin
- Japanese Mastiff,
- Japanese Spitz
- Japanese Terrier
- Jindo
- Jonangi
- Kaikadi

- Kai Ken
- Kangal Dog
- Kangaroo Dog
- Kanni
- Karafuto Ken
- Karelian Bear Dog
- Karelo-Finnish Laika
- Kars Dog
- Keeshond
- Kelpie
- Kelb-tal Fenek
- Kerry Beagle
- Kerry Blue Terrier
- King Charles Spaniel
- King Shepherd
- Kintamani
- Kishu
- Kombai
- Komondor
- Kooikerhondje
- Koolie
- Korea Jindo Dog
- Korean Mastiff
- Korthals Griffon
- Kraski Ovcar (Karst Shepherd Dog)
- Kritikos Ichnilatis (Cretan Hound)
- Kromfohrlander
- Kuchi
- Kunming Dog
- Kuvasz
- Kyi Leo

- Labradoodle (Mixed breed, primary breeds are Labrador Retriever and Poodle (Standard))
- Labrador Husky (Purebred, not mixed breed)
- Labrador Retriever
- Laekenois
- Lagotto Romagnolo
- Lakeland Terrier
- Lancashire Heeler
- Landseer (Continental-European type)
- Lapponian herder (Lapinporokoira)
- Large Munsterlander
- Latvian Hound
- Leonberger
- Leopard Cur
- Lhasa Apso
- Lithuanian Hound
- Llewellyn Setter
- Longdog
- Louisiana Catahoula Leopard Dog
- Lottatore Brindisino
- Löwchen
- Lucas Terrier
- Lurcher
- Magyar Agar
- Majestic Tree Hound
- Malinois
- Maltalier
- Mal-Shi
- Maltese
- Maltipoo
- Manchester Terrier

- Maremma Sheepdog
- Martin Mosa Mastiff
- Mastiff
- McNab
- Meliteo Kinidio
- Mexican Hairless Dog
- Middle Asian Owtcharka - see Central Asia Shepherd Dog
- Miniature Australian Shepherd
- Miniature Bull Terrier
- Miniature Fox Terrier
- Miniature Pinscher
- Miniature Schnauzer
- Miniature Siberian Husky
- Mioritic
- Mixed-breed dog
- Moscow Guard dog
- Moscow Watchdog (Moscovskaya Storozhevaya Sobaka)
- Mountain Burmese
- Mountain Cur
- Mongrel (Mixed breed)
- Mountain Feist
- Mucuchies (Venezuela)
- Mudi
- Mudhol Hound
- Munsterlander
- Mutt (Mixed breed)
- Native American Indian Dog
- Neapolitan Mastiff
- Nebolish Mastiff
- Neilmut (Neil Laing)
- Nenets Herding Laika

- Newfoundland
- New Guinea Singing Dog
- Norfolk Terrier
- Norrbottenspets
- Northeasterly Hauling Laika (Northeastern Sleigh Dog)
- Northern Inuit dog
- Norwegian Buhund
- Norwegian Elkhound
- Norwegian Lundehund
- Norwich Terrier
- Nova Scotia Duck-Tolling Retriever
- Old Danish Pointer
- Old English Sheepdog
- Old English Bulldog
- Old English Terrier
- Olde Englishe Bulldogge
- Österreichischer Kurzhaariger Pinscher
- Otterhound
- Otto
- Owczarek Podhalanski
- Pachon Navarro
- Panja, see American Mastiff
- Papillon
- Parson Russell Terrier
- Pashmi
- Pastor Garafiano
- Patterdale Terrier
- Pekingese
- Pembroke Welsh Corgi
- Perdiguero de Burgos
- Perro Cimarron

- Perro de Pastor Mallorquin
- Perro de Presa Canario
- Perro de Presa Mallorquin
- Perro de Toro
- Peruvian Hairless Dog (Perro Peruano sin Pelo)
- Peruvian Inca Orchid
- Petit Basset Griffon Vendeen
- Petit Berger
- Petit Bleu de Gascogne
- Petit Brabancon
- Petit Gascon Saintongeois
- Phalène (drop eared variety of the Papillon)
- Pharaoh Hound
- Phung San
- Picardy Shepherd
- Picardy Spaniel
- Pinscher
- Pit Bull
- Plott Hound
- Podenco Andaluz
- Podenco Canario
- Podenco Galego
- Podenco Ibicenco
- Pointer
- Poitevin
- Polish Scenthound (Gonczy Polski)
- Polish Greyhound
- Polish Sighthound
- Polish Hound (Polish Ogar)
- Polish Lowland Sheepdog (Polski Owczarek Nizinny or PON)
- Polish Tatra Sheepdog

- Pomeranian
- PON - see Polish Lowland Sheepdog
- Pont-Audemer Spaniel
- Poodle
    - Miniature Poodle
    - Standard Poodle
    - Toy Poodle
- Porcelaine
- Portuguese Podengo (Portuguese Podengo)
- Portuguese Pointer
- Portuguese Shepherd Dog
- Portuguese Water Dog
- Posavac Hound
- Prague Ratter
- Pudelpointer
- Pug
- Puggle (Mixed breed, Pug and Beagle)
- Pugnaces Britanniae
- Puli
- Pumi
- Pungsan (Poongsan)
- Pyrenean Mastiff
- Pyrenean Mountain Dog
- Pyrenean Shepherd (Pyrenees Sheepdog)
- Queensland Heeler
- Qimmiq
- Rafeiro do Alentejo
- Rajapalayam
- Rampur Greyhound
- Ratonero Bodeguero Andaluz
- Ratonero

- Rat Terrier
- Redbone Coonhound
- Red Setter
- Rhodesian Ridgeback
- Rottweiler
- Rough Collie
- Russian Black Terrier
- Russian Harlequin Hound (Russkaja Pegaja)
- Russian Hound
- Russian Setter
- Russian Spaniel
- Russian Toy Terrier
- Russian Tsvetnaya Bolonka
- Russo-European Laika (Russko-Evropeiskaia Laika)
- Russell Terrier
- Ryûkyû Inu
- Saarlooswolfhond
- Sabueso español
- Sage Koochee
- Sakhalin Husky
- Saluki
- Samoyed
- Sanshu
- Santal Hound
- Sapsali
- Šarplaninac
- Schapendoes
- Schillerstovare
- Schipperke
- Schnauzer - divided into Miniature Schnauzer, Standard Schnauzer, Giant Schnauzer,

- Schweizer Laufhund
- Schweizer Niederlaufhund
- Scottish Deerhound
- Scottish Terrier
- Sealyham Terrier
- Segugio Italiano
- Seppala Siberian Sleddog
- Serbian Hound
- Serbian Mountain Hound
- Serbian Tricolour Hound
- salvador rottweiller
- Shar Pei
- Shetland Sheepdog (Sheltie)
- Shi-lah
- Shiba Inu
- Shih Tzu
- Shikoku
- Shiloh Shepherd Dog
- Siberian Husky
- Silken Windhound
- Silky Terrier
- Sindh Mastiff
- Skye Terrier
- Sloughi
- Slovak Cuvac
- Slovakian Hound
- Slovensky Hrubosrsty Stavac (Ohar)
- Smalandsstovare
- Small Greek Domestic Dog
- Small Munsterlander
- Smithfield

- Smooth Collie
- Smooth Fox Terrier
- Snowfur Longsnout
- Soft-Coated Wheaten Terrier
- South Russian Ovtcharka
- Spanish Alano
- Spanish Greyhound
- Spanish Mastiff
- Spanish Pointer
- Spanish Water Dog
- Spinone Italiano
- Spitz
- Springer Spaniel
- St. Bernard
- Stabyhoun
- Staffordshire Bull Terrier
- Standard Schnauzer
- Stephens Stock (Stephens Cur)
- Styrian Coarse Haired Hound
- Sulimov dog (dog-jackal hybrid)
- Sussex Spaniel
- Swedish Elkhound
- Swedish Lapphund
- Swedish Vallhund
- Swiss Shorthaired Pinscher
- Tahltan Bear Dog
- Taigan
- Tainaker
- Taíno Dog - also *perro mudo* ("Mute Dog"); native name disputed. Extinct at least as a pure breed.
- Taiwan Dog - also Formosan

- Tamaskan dog - unrecognised breed
- Tasy
- Tatra Shepherd Dog
- Taylor Terrier
- Tenterfield Terrier
- Tervuren
- Thai Bangkaew Dog
- Thai Hairless Dog
- Thai Ridgeback
- Teddy Roosevelt Terrier
- Telomian
- Tibetan Kyi Apso
- Tibetan Lhasa Apso
- Tibetan Mastiff
- Tibetan Spaniel
- Tibetan Terrier
- Tornjak
- Tosa
- Toureg Sloughi
- Toy Bulldog
- Toy Fox Terrier
- Toy Manchester Terrier
- Toy Poodle
- Transylvanian Hound
- Treeing Cur
- Treeing Feist
- Treeing Tennessee Brindle
- Treeing Walker Coonhound
- Tsvetnaya Bolonka
- Turkish Kangal
- Tyrolean Hound

- Utonagan
- Valley Bulldog
- Vizsla
- Volpino Italiano
- Vorsteh
- White Whippet with brindle saddle and head
- Weimaraner
  - o Longhaired Weimaraner
  - o Smooth Haired Weimaraner
- Welsh Corgi
  - o Cardigan Welsh Corgi
  - o Pembroke Welsh Corgi
- Welsh Sheepdog (Welsh Collie)
- Welsh Springer Spaniel
- Welsh Terrier
- West Highland White Terrier
- West Siberian Laika (Zapadno-Sibirskaia Laika)
- Westphalian Dachsbracke
- Wire Haired Fox Terrier
- Wetterhoun
- Whippet
- White Shepherd Dog
- Winston Olde English Bulldogge
- Wire Fox Terrier
- Wirehaired Pointing Griffon
- Xoloitzcuintle
- Yorkillon - Half yorkshire terrier, half Papillon
- Yorkshire Terrier
- Yorkshire Terrier
- Yugoslavian Mountain Hound
- Yugoslavian Tricolour Hound
- Zapadno-Sibirskaia Laika

MIXED-BREED DOG

A mixed-breed dog is a dog that has characteristics of two or more types of breeds, or is a descendant of feral or pariah dog populations. The term "mutt" generally refers to a dog of unknown descent. Dogs interbreed freely, except where extreme variations in size exist, so mixed-breed dogs vary in size, shape, and color, making them hard to classify physically.

Terms for Mixed-Breed Dogs

There are a profusion of words and phrases used for dogs that are not purebred. The words cur, tyke, and mongrel are some used, but generally viewed as derogatory in North America. In the United Kingdom mongrel is the unique technical word for a mixed-breed dog, and is not a term of disparagement. North American owners generally prefer mix or mixed-breed. Mutt is also used (in the U.S.A and Canada), usually in an affectionate manner. In Hawaii, mixed breed dogs are referred to as poi dog, and in the Bahamas, they call them Pot Cakes (referring to the table-leftovers they are fed). Some American registries and dog clubs that accept mixed-breed dogs use the breed name All American. In South Africa, the tongue-in cheek expression pavement special is sometimes used as a description for a mixed-breed dog. Random-bred dog, mutt, and mongrel are often used for dogs who result from breeding without the supervision or planning of humans, especially after several generations, whereas *crossbreed* implies mixes of known breeds, sometimes deliberately mated. In Brazil and the Dominican Republic, the name for mixed-breed dogs is vira-lata (*vira*: to turn, to bring down; *lata*: tin can, trash can) because there are dogs without owners that feed on urban garbage on the streets, and often knock over trash cans to reach the food. Slang terms are also common. Heinz 57 is often used for dogs of uncertain ancestry, in a playful reference to the "57 Varieties" slogan of the H. J. Heinz Company. In some countries, bitsa (or bitzer) is common, meaning "bits o' this, bits o' that". A fice or feist is a small mixed-breed dog. In Newfoundland, a smaller mixed-breed dog is known as a cracky, hence the colloquial expression "saucy as a cracky" for someone with a sharp tongue. To complicate matters, many owners of crossbreed dogs identify them—often facetiously—by an invented breed name constructed from parts of their parents' breed names. These are known as portmanteau names. For example, a cross between a Pekingese and a Poodle is called a Peekapoo, possibly a play on *peek-a-boo*, along with the Goldendoodle, a cross between a poodle and a golden retriever. As another example, one of the UK's Queen Elizabeth's famous Corgis mated with her sister's Dachshund, and the resulting offspring were referred to as Dorgis.

Appearance

All possible body shapes, nose are closer to the genetic norm. Dogs that are descended from many generations of mixes are typically light brown or black and weigh about 18 kg (40 lb). They typically stand between 38 and 57 cm (15 and 23 inches) tall at the withers.

Determining Ancestry

Guessing a mixed-breed's ancestry is difficult for even knowledgeable dog observers, because mixed breeds have much more genetic variation than among purebreds. For example, two black mixed-breed dogs might each have recessive genes that produce a blond coat and, therefore, produce offspring looking unlike their parents. Starting in 2007, blood samples and cheek swabs have become available to the public to narrow down the ancestry of mixed-breed dogs. The companies claim their DNA-based diagnostic test that can genetically determine the breed composition of mixed breed dogs. These tests are still limited in scope because only a small number of the hundreds of dog breeds have been validated against the tests, and because the same breed in different geographical areas could have different genetic profiles. Also, the tests do not test for breed purity, but for genetic sequences that are common to certain breeds. With a mixed-breed dog, the test is not proof of pure-bred ancestry, but rather an indication that those dogs share common ancestry with certain purebreds. This might provide information to owners about breed-related health issues to be aware of.

Health

The theory of hybrid vigor suggests that as a group, dogs of varied ancestry will be healthier than their purebred counterparts. In purebred dogs, intentionally breeding dogs of very similar appearance over several generations produces animals that carry many of the same alleles, some of which are detrimental. This is especially true if the dogs are closely related. This inbreeding among purebreds has exposed various genetic health problems not readily apparent in less uniform populations. Mixed-breed dogs are more genetically diverse due to the more haphazard nature of their parents' mating. "Haphazard" is not the same as "random" to a geneticist. The offspring of such matings are less likely to express certain genetic disorders because there is a decreased chance that both parents carry the same detrimental recessive alleles. However, some deleterious recessives are common across many seemingly unrelated breeds, and therefore merely mixing breeds is no guarantee of genetic health. Crossing breeds to take advantage of the increased chance that a recessive detrimental allele will only be inherited from one parent, and therefore not expressed in the phenotype of the offspring, is only one strategy breeders can use to decrease the incidences of genetic faults. For example, large dogs such as the German Shepherd Dog often suffer from hip dysplasia, a multigene trait strongly affected by environment. Mating a German Shepherd, a breed known to have about a 20% incident of this disease (ref:OFA), with a dog of a different breed not known to suffer from it dysplasia like the Greyhound, reduces the likelihood that the cross-breed produced will suffer from hip dysplasia. Mate the same German Shepherd to an Otterhound (dysplasia incidence about 52%) or to a small dog such as a Pug (62%) or a Norfolk Terrier (37%), and the outcome might be considerably different. Breeding dogs that have been tested free of dysplasia offers a similar result, with the added benefit of producing offspring that are less likely to carry the defect, unexpressed. Knowing the disease incidence in the breed, and the genetic history of the individual, is ultimately important in dog breeding. Having the parental dogs genetically tested for defects known to be troublesome to their breed (or breeds, in the case of mixed parentage) will do as much, or more, than simply choosing dissimilar individuals with functional reproductive tracts. Genetic health must be approached from many angles. Some schemes are effective for the short-term, but disastrous in the long. Others are slow to take effect, but that effect is long-lasting. Mating two different purebreds to decrease the likelihood of inheriting a particular genetic disease, without consideration of all the other physical and mental attributes that comprise a suitable pet and companion, is as likely to fail as any other single-trait selection scheme. Overall, hybrid vigor is not associated with any of the traits in dogs that make them suitable companions. At the same time, there's nothing about a the genetic background of mixedbreed dog that makes it an inherently poorer companion than its purebred counterparts. Mixed breed dogs can make kind and caring companions and are often very good around children. There is no guarantee of good genetic health, or temperment, of any dog, purebred or otherwise, as not all damaging genes are recessive, and there are relatively few single-gene traits. Also, of course, purebred and mixed-breed dogs are equally susceptible to nongenetic ailments, such as rabies, distemper, injury, and infestation by parasites.

Types of Mixed Breeding

Recognized dog breeds are a result of human selection in that dogs were traditionally bred for specific functions. Most existing dog breeds began as mixed breeds, either by random occurrence or by deliberate crosses of existing breeds. Encouraging desirable traits and discouraging others, breeders sought to create their ideal appearance or behaviour, or both, for dogs, and, additionally, to ensure that the dogs could consistently produce offspring with the same appearance or behaviour. Mixing breeds can lead to desirable results, especially in the hands of an expert breeder. On the other hand, inexperienced crossbreeders can produce disastrous results. For example, the offspring of an obsessive Border Collie and an energetic, destructive Terrier could be dogs whose behaviour is so erratic that would make the dog a liability. Mixed-breed dogs can be divided roughly into three types:

(i) Crossbreed dogs, which are mixtures of two recognized breeds. Dogs that result from two different purebred parents are known as crossbreeds.

Some crossbreeds have traits that make them popular enough to be frequently bred deliberately, such as the Cockapoo—a cross between a Poodle and a Cocker Spaniel—and the Labradoodle, which crosses a Labrador Retriever with a Poodle. Other crossbreeds occur when breeders are hoping to create new breeds to add and reinforce characteristics from one breed into another breed. Most crossbreedings, however, occur accidentally.

(ii) Mixes that show characteristics of two breeds or more breeds. A mix might have some purebred ancestors, or might come from a long line of mixed-breeds. These dogs are usually identified by the breed they most resemble, such as a "Lab mix" or "Collie-Shepherd", even if their ancestry is unknown.

(iii) The generic pariah dog, or feral *Canis lupus familiaris*, where non-selective breeding has occurred over many generations. The term originally referred to the wild dogs of India, but now refers to dogs belonging to or descended from a population of wild or feral dogs. The Canaan Dog is an example of a recognized breed with pariah ancestry. Pariah dogs tend to be yellow to light brown and of medium height and weight. This may represent the appearance of the modern dog's ancestor. DNA analysis has shown pariah dogs to have a more ancient gene pool than modern breeds.

There is no scientific justification for the belief that a purebred bitch is in any way tainted after mating with a dog of another breed. Future matings with dogs of the same breed will produce purebred puppies.

Mixed Breeds in Dog Sports

Mixed-breed dogs can excel at dog sports, such as obedience, dog agility, flyball, and frisbee. Often, highly energetic mixed-breeds are left with shelters or rescue groups, where they are sought by owners with the caring, patience, and drive to train them for dog sports, turning unwanted dogs into healthy, mentally and physically stimulated award winners. Until the early 1980s, mixed-breed dogs were usually excluded from obedience competitions. However, starting with the American Mixed Breed Obedience Registry (AMBOR) and the Mixed Breed Dog Clubs of America (MBDCA), which created obedience venues in which mixed-breed dogs could compete, more opportunities have opened up for all dogs in all dog sports. Most dog agility and flyball organizations have always allowed mixed-breed dogs to compete. Today, mixed breeds have proved their worth in many performance sports. In conformation shows, where dogs' conformation is evaluated, mixed-breed dogs normally cannot compete. For purebred dogs, their physical characteristics are judged against a single breed standard. Mixed-breed dogs, however, are difficult to classify except according to height; there is tremendous variation in physical traits such as coat, skeletal structure, gait, ear set, eye shape and color, and so on. When conformation standards are applied to mixed-breed dogs, such as in events run by the MBDCA, the standards are usually general traits of health, soundness, symmetry, and personality. The Kennel Club (UK) operates a show called Scruffts (a name derived from its prestigious Crufts show) open only to mixed-breeds in which dogs are judged on character, health, and temperament. Some kennel clubs, whose purpose is to promote purebred dogs, still exclude mixed breeds from their performance events. The AKC and the FCI are two such prominent organizations. However, the AKC does allow mixed breed dogs to earn their Canine Good Citizen award.

Advantages and Drawbacks

The mature appearance and behaviour of purebred puppies is more predictable than that of mixed breeds, including cross-breeds. With purebred dogs, the genetic variations are well documented and a breeder has a fair estimation of what type of offspring a given pair will produce. Still, there is variation within breeds; for example, two champion sheep-herding Border Collies might produce offspring with no interest in sheep herding. Some trainers believe mixed-breeds exhibit higher average intelligence than purebreds, but others believe mixes are no more intelligent than purebreds. Both sets feature both slow learners and dogs with high learning capacity. For example, Benji, the hero in a series of films named for him, was a mixed-breed terrier. Many people enjoy owning mixed breeds, valuing their unique appearance and characteristics; while purebred dogs exhibit little variability of appearance within their breed, mixed-breed dogs exhibit often unique appearances. Although some dog owners prefer the status of owning a specific breed of dog or have a nostalgic attachment to a breed they wish to acquire, many others enjoy mixed-breed dogs that exhibit characteristics similar to their favourite breeds; in fact, with a mixed-breed, they can enjoy some aspects of appearance and personality of two favourite breeds with a single dog. There is usually an abundant supply of mixed-breed dogs wanting owners, available at negligible prices, while pedigreed dogs can cost hundreds or thousands of dollars and reputable breeders can be hard to find. Some owners value a dog's pedigree as a status symbol and, therefore, have no use for mixed-breed dogs; others particularly appreciate the physical or behavioural traits of certain breeds; still others ignore pedigree and, instead, value a dog's personality and health. Local animal shelters adopt out dogs of both purebred and mixed ancestry, emphasizing each dog's personality and suitability as a companion for each potential owner's lifestyle.

Famous Mutts

- Benji
- Muttley in Dastardly and Muttley in Their Flying Machines

Dog Types (that are not specific dog breeds)

- Crossbreeds (also Poodle hybrids)
- Bandogs
- Bichon
- Bird dogs
- Bull and Terrier
- Bulldogs
- Collies (Farm collie)
- Companion dogs
- Coonhounds
- Curs
- Fell Terrier
- Feist
- Fox Terriers
- Foxhounds
- Guard dogs
- Gun dogs
- Hounds
- Hunting dogs
- Lap dogs (also Toy dogs)
- Lurcher (also Longdog)
- Mixed-breed dogs
- Molossers
- Münsterländers
- Pariah dogs
- Pastoral dogs (related types Herding dogs, Sheep dogs, Drovers. Livestock guardian dogs)
- Pinschers
- Pit Bull
- Pointers
- Retrievers

- Sighthounds
- Scent hound
- Schnauzers
- Setters
- Sighthounds
- Sled dogs (and Eskimo dogs)
- Spaniels (and Cocker Spaniels)
- Spitz
- Terriers
- Turnspit dogs
- Water dogs
- Welsh Corgis
- Working dogs
- Working terrier.

BREED-SPECIFIC LEGISLATION

Breed-specific legislation (BSL), is any law, ordinance or policy which pertains to a specific dog breed or breeds, but does not affect any others. The term is most commonly used to refer to legal restrictions or prohibitions on the breeding and ownership of certain breeds.

Some examples of BSL:

- Australia:
    - Restrictions prohibiting the importation of the Pit Bull Terrier, American Pit Bull Terrier, the Japanese Tosa, Dogo Argentino, and the Brazilian Mastiff. There is also a more stringent ban in Sydney, restricting the sale, acquisition, breeding or giving away of any of the aforementioned breeds.
    - Proposed legislation in some Australian states that would prohibit the breeding of any breed of dog not recognized by the Australian National Kennel Council, or restrict or prohibit breeding certain breeds.
    - The requirement that Greyhounds wear muzzles in public in some Australian states.

- Britain:
  - Restrictions on or the prohibition of ownership of American Pit Bull Terriers.
  - The Dangerous Dogs Act legislatively drafted prohibitions for certain types of dogs; most notably pit-bull type dogs.
- Canada: Restrictions on the ownership and a prohibitation on the breeding of "pit bull type" dogs (i.e. Staffordshire Bull Terriers, American Staffordshire Terriers, American Pit Bull Terriers or any other dog with a substantially similar appearance) in Ontario.
- Germany: legislation banning the import of Pitbull Terriers, Bull Terriers, American Staffordshire Terriers, and Staffordshire Bull Terriers.
- Israel: legislation against the breeding of AmStaff, Bull Terrier, dogo argentino, Tosa, Staffordshire Bull Terrier, Pit Bull Terrier, fila brasileiro, and Rottweiler.
- USA:
  - Restrictions on or the prohibition of ownership of American Pit Bull Terriers in some municipalities.
  - Restrictions on the availability of homeowners insurance for owners of many breeds, including American Pit Bull Terriers, Rottweilers, Dobermanns, Akitas, in some areas .

Pros and Cons

Proponents of BSL usually cite the need to protect the public from dog breeds viewed to have inherent tendencies to aggressive behaviour. Many tend to believe that dangerous dogs must be certain breeds in order to make them this way. Some proponents believe that a pit bull has superior jaw strength, different muscular ability, and the like. However, the idea that a pit bull has superior jaw strength is a complete myth. Interestingly this myth was also supposedly true for Doberman Pinschers, but now it is no longer considered true for the Doberman. The musculature assertion is also false. Consider that while the American Pit Bull Terrier does well in weight pull competitions, it is by no means superior in that other breeds in the same weight classes have beaten these dogs.

Proponents point to specific studies done, particularly the CDC study (Center for Disease Control). However, it should be noted that the CDC itself and the authors of this study note that the study does not support Breed Specific Legislation nor that the article can be used to determine which dogs are inherently more dangerous. On the other hand, certain evidence shows that where BSL has been implemented, there is no proof that it actually produced the intended effect. Recently, laws are being pushed for implementing mandatory spay/neuter ONLY for those breeds proven to be dangerous, such as the law in California (Senate Bill 861, codified under the dangerous dog laws)and currently being challenged in Federal court.

Opponents believe that many of the policies created by BSL have been randomly or illogically developed, and are often capriciously or inconsistently enforced. For example, although "Pit Bulls" are primarily the focus of BSL, they are not a recognized breed; 'pit bull' is a term applied to several different kinds of terriers, and there is no consensus on what a "Pit Bull" actually is. The term "Pit-Bull-Type-Dog" has been used to describe over ten very different breeds, including Bulldogs, Boxers,Chow chows and Bullmastiffs. Even unrelated breeds such as the Labrador retriever and Jack Russel terrier can be mistaken for "pit bulls" by some individuals. Additionally, Rottweilers, though having shown a virtually equal propensity for dog attacks, are rarely included in dog bans or BSL. Doberman Pinschers, German Shepherd Dogs, Great Danes, Husky-type dogs, Alaskan Malamutes and Mastiff-Type Dogs also frequently injure people, but do not experience the same scrutiny as Bull Breeds. There is also the problem of mixed breed and cross breed dogs since it is often impossible to determine conclusively their breed(s).

Some believe an alternative to Breed Specific Legislation might be the consistent enforcement of existing dog laws. Others believe that offending dogs and people should be judged on a case by case basis, as dog attacks happen infrequently enough to warrant careful attention to each instance.

Additionally, there maybe some constitutional issues with Breed Specific Legislation. Since determining a dog's breed is extremely difficult in the cases of mixed breeds (which usually do not come with a pedigree of any sort) any such laws targeting breed may be considered vague from a constitutional stand point and violate the dog owner's right to due process. Further, any dog can be dangerous in the right situation. For example a 4 pound pomeranian killed an infant. As such, any law that targets one breed over another may violate the right to equal protection.

Recent Developments in USA

Some cities are removing breed specific ordinances to comply with the Sixth District ruling. Others are waiting to see how the Ohio Supreme Court will rule in the appeal of *City of* Toledo v. Tellings. *To see the Ohio Supreme Court docket in this case, search for case no. 2006-0690 at Ohio Supreme Court's Public Case Inquiry.*

Pending Ohio HB 533 would remove the language from ORC 955.11 automatically defining pit bulls as vicious dogs. Any action on this bill awaits the Ohio Supreme Court's decision in the Tellings case.

In the Ohio Supreme Court earlier this year (2007), the ASPCA (American Society for Prevention of Cruelty to Animals Society) filed an amicus brief in support of the Tellings decision, supra, in which it states:

"The trial court and the Sixth District agreed that the most current research indicates that dogs deemed to be pit bulls *are not more dangerous* than dogs of other breeds, so temperament cannot provide the rationale for the Ohio pit bull law and the other challenged statute, Toledo's ordinance prohibiting ownership of more than one adult dog deemed to be a pit bull (T.M.C. §505.14(a)).

Surely, then, these laws regulating pit bulls as vicious dogs must have a positive impact on public safety, their touted purpose. However, this, too, is not the case: while in 1996, 14.6% of animal control agencies reported local problems with dogfighting, by 2004, the number of agencies reporting local problems with dogfighting had skyrocketed to 29%.

Further, Lucas County's own data indicates a similar spiking in dog bite numbers (approximately 640 bites) in 2001 – more than a decade *after* the enactment of Ohio's pit bull law and the Toledo ordinance.Given the absence of evidence to support the notion that dogs deemed to be pit bulls are more dangerous than other dogs and should be regulated on this basis, as well as the failure of the Ohio and Toledo pit bull laws to address canine aggression in any meaningful way, these laws appear to have only two outcomes:

(1) the unavoidably *arbitrary enforcement* of irrational policy, including the unreasonably disparate treatment of different classes of persons, and (2) the consequent grievous *harm to property and* liberty that flows from such wholesale compromise of procedural and substantive due process rights."

(Amicus brief submitted on behalf of the Tellings case by ASPCA attorney Debora M. Bresch, Pro Hac Vice, *in the appeal of the Tellings case in the Ohio Supreme Court, November 28, 2006.)* The result in the Tellings case of Ohio shows the Ohio Supreme Court reversed the lower court's decision, which can be viewed by searching for case no. 2006-0690 at Ohio Supreme Court's Public Case Inquiry. Proponents have indicated they will submit an appeal to the Supreme Court of the United States, which is purely discretionary in review, and where few cases are chosen for hearing.

As for discussion regarding any anti-BSL groups, there are many of them. The Coalition of Human Advocates for K9's and Owners (CHAKO) filed a lawsuit against the City and County of San Francisco to challenge its Pit Bull ordinance. The American Dog Owners Association (ADOA) has also been involved with dog-related cases. The American Canine Foundation (ACF) has a federal case pending in both

California (Case No. C06 04713), Colorado (Case No. 061510), and ACF is giving assistance to federal cases being filed in both Washington (not assigned case number yet) and Arkansas (pending final draft) which deal with breed specific law; however you must be able to view the Federal PACER system to read the cases. See http://pacer.psc.uscourts.gov/ (There is a per page fee to view documents.)

For specific information regarding the political climate and breed specific legislation specifically, see http://www.dogpolitics.typepad.com, which gives up to date information on breed specific topics, groups,debates,proposed laws.

UNDERSTANDING THE BREEDING SEASON

The breeding season is the most suitable season, usually with favourable conditions and abundant food and water, for breeding among some wild animals and birds (wildlife). Species with a breeding season have naturally evolved to have sexual intercourse during a certain time of year in order to achieve the best reproductive success. Different species of wild animal and birds have different breeding seasons according to their particular requirements and food availability etc.

Communal Breeding

Many species breed in colonies or large communities which is known as communal breeding. It is common to see large congregations of these species in particular favourable locations in their breeding seasons. These breeding colonies and their location are generally protected by wildlife conservation laws to keep the species from going extinct. Some species have evolved for communal breeding in large breeding colonies and can not breed in smaller numbers or pairs alone.

These species can be threatened by imminent extinction if they are hunted on their breeding grounds or if their breeding colonies are destroyed. The Passenger pigeon is a famous example of probably the most numerous land bird on the American continent which had evolved for communal breeding that went extinct due to large scale hunting in its communal breeding grounds during the breeding season and its inability to breed in smaller numbers.

Closed Season in Hunting

Breeding season of wildlife is usually what is called the closed season when hunting is not permitted and is enforced by law for the conservation of the species and wildlife management, any hunting during closed season is punishable by law and termed as illegal hunting or poaching.

REFERENCES

Amer, P., and G. Fox. 1995. Imputing input characteristic values from optimal commercial breed or variety choice decisions: comment. *American Journal of Agricultural Economics* 77 (4): 1054-1058.

Anderson, S. 2003. Animal genetic resources and sustainable livelihoods. *Ecological Economics* 45 (3): 331-339.

Ayalew, W., J. M. King, E. Bruns, and B. Rischkowsky. 2003. Economic evaluation of smallholder subsistence livestock production: lessons from an Ethiopian goat development programme. *Ecological Economics* 45 (3): 473-485.

Campbell, B. M., D. Doré, M. Luckert, B. Mukamuri, and J. Gambiza. 2000. Economic comparisons of livestock production in communal grazing lands in Zimbabwe. *Ecological Economics* 33 (3): 413-438.

Cicia, G., E. D'Ercole, and D. Marino. 2003. Costs and benefits of preserving farm animal genetic resources from extinction: CVM and bio-economic model for valuing a conservation programme for the Italian Pentro horse. *Ecological Economics* 45 (3): 445-459.

CIP-UPWARD. 2003. *Conservation and sustainable use of agrobiodiversity: a source book.* Los Baños, Laguna, Philippines: CIP-UPWARD in collaboration with GTZ, IDRC, IPGRI and SEARICE.

Clark, C. W. 1973. Profit maximization and the extinction of animal species. *The Journal of Political Economy* 81 (4): 950-961.

Drucker, A. G. 2006. An application of the use of safe minimum standards in the conservation of livestock biodiversity. *Environment and Development Economics* 11 (1): 77-94.

Drucker, A. G., V. Gómez, N. Ferraes-Ehuan, O. Rubio, and S. Anderson. 1999. Comparative economic analysis of criollo, crossbreed and imported pigs in backyard production of Yucatan, Mexico. FMVZ-UADY Mimeo.

Drucker, A. G., and S. Anderson. 2004. Economic analysis of animal genetic resources and the use of rural appraisal methods: lessons from Southeast Mexico. *International Journal of Agricultural Sustainability* 2 (2): 77-97.

Drucker, A. G., E. Bergeron, U. Lemke, L. T. Thuy, and A. Valle Zrate. 2006. Identification and quantification of subsidies relevant to the production of local and imported pig breeds in Vietnam. *Tropical Animal Health and Production* 38 (4): 305-322.

Drucker, A. G., and S. Anderson. 2005. Putting the economic analysis of Animal Genetic Resources into practice. In *Participatory research and development for sustainable agriculture and natural resource management: a sourcebook. Volume 3: doing participatory research and development*, ed. J. Gonsalves, T. Becker, A. Braun, D. Campilan, H. De Chavez, E. Fajber, M. Kapiriri, J. Rivaca-Caminade, and R. Vernooy. Laguna, The Phillipines and Ottawa, Canada: International Potato Center-Users' Perspectives with Agricultural Research and Development and International Development Research Centre.

Drucker, A. G. 2001. The economic valuation of AnGR: importance, application and practice. Proceedings of a workshop on Community-based management of animal genetic resources: a tool for rural development and food security, May 7-11, in Mbabane, Swaziland. Rome, Italy and New York, U.S.A.: Food and Agriculture Organization of the United Nations and United Nations Development Programme.

Drucker, A. G., V. Gómez, and S. Anderson. 2001. The economic valuation of farm animal genetic resources: a survey of available methods. *Ecological Economics* 36 (1): 1-18.

Drucker, A. G. 2004. *The economics of farm animal genetic resource conservation and sustainable use: why is it important and what have we learned?* Commission on Genetic Resources for Food and Agriculture Background Study Paper 21.

Drucker, A. G., and R. Scarpa. 2003. Valuing animal genetic resources. *Ecological Economics* 45 (3).

Falconi, C. A., S. W. Omamo, G. d'Ieteren, and F. Iraqi. 2001. An ex ante economic and policy analysis of research on genetic resistance to livestock disease: trypanosomosis in Africa. *Agricultural Economics* 25 (2/3): 153-163.

Gandini, G. 1997. What economic value for local livestock breeds? Proceedings of the 48th Annual of the European Association of Animal Production. The Netherlands: Wageningen Academic Publishers.

Groen, A. F. 1988. Derivation of economic values in cattle breeding: a model at the farm level. *Agricultural Systems* 27: 195-213.

Hamath, A. S., M. D. Faminow, G. V. Johnson, and G. Crow. 1997. Estimating the values of cattle characteristics using an ordered probit model. *American Journal of Agricultural Economics* 79: 463-476.

ILRI (International Livestock Research Institute). 2000. *Report of the research planning workshop on "Economic valuation of animal genetic resources"*. Nairobi: International Livestock Research Institute.

Jabbar, M. A., and M. L. Diedhiou. 2003. Does breed matter to cattle farmers and buyers? evidence from West Africa. *Ecological Economics* 45 (3): 461-472.

Karugia, J. T., O. A. Mwai, R. Kaitho, A. Drucker, C. B. A. Wollny, and J. E. O. Rege. 2001. *Economic analysis of crossbreeding programmes in sub-Saharan Africa: a conceptual framework and Kenyan case study*. FEEM Working Paper 106.2001.

Kosgey, I. S., J. A. M. Van Arendonk, and R. L. Baker. 2004. Economic values for traits in breeding objectives for sheep in the tropics: impact of tangible and intangible benefits. *Livestock Production Science* 88: 143-160.

Krajcovic, V., E. Uhliarová, M. Michalec, and M. Zimková. 2003. Evaluation of some environmentally-friendly private livestock farms in Central Slovakia. In *Livestock farming systems in Central and Eastern Europe*, ed. A. Giban, and S. Mihina. Wageningen, The Netherlands: Wageningen Academic Publishers.

Kurian, V. 2001. Value-based crop-livestock production systems for the future in the semi-arid tropics. In *Future of agriculture in the semi-arid tropics*, ed. V. Kurien, M. C. S. Bantilan, P. Parthasarathy Rao, and R. Padmaja. Proceedings of an International Symposium on Future of Agriculture in Semi-Arid Tropics, November 14, 2000, in

ICRISAT, Patancheru, India. Patancheru, India: International Crops Research Institute for the Semi-Arid Tropics.

Ladd, G., and C. Gibson. 1978. Microeconomics of technical change: what's a better animal worth? *American Journal of Agricultural Economics* 60 (2): 236-240.

Melton, B. E., W. A. Colette, K. J. Smith, and R. L. Wilham. 1994. A time-dependent bioeconomic model of commercial beef breed choices. *Agricultural Systems* 45 (3): 331-347.

Melton, B. E., W. A. Colette, and R. L. Willham. 1994. Imputing input characteristic values from optimal commercial breed or variety choice decisions. *American Journal of Agricultural Economics* 76 (3): 478-491.

Melton, B. E., W. A. Colette, and R. L. Willham. 1995. Imputing input characteristic values from optimal commercial breed or variety choice decisions: reply. *American Journal of Agricultural Economics* 77 (4): 1059-1063.

Mendelsohn, R. 2003. The challenge of conserving indigenous domesticated animals. *Ecological Economics* 45 (3): 501-510.

Mitchell, G., C. Smith, M. Makower, and P. J. W. N. Bird. 1982. An economic appraisal of pig improvement in Great Britain 1. Genetic and production aspects. *Animal Science (Animal Production)* 35 (2): 215-224.

Mitchell, G., P. J. W. N. Bird, M. Makower, and C. Smith. 1983. An economic appraisal of pig improvement in Great Britain. *Oxford Development Studies (formerly Oxford Agrarian Studies)* (12): 48-62.

Mwacharo, J. M., and A. G. Drucker. 2005. Production objectives and management strategies of livestock keepers in South-East Kenya: implications for a breeding programme. *Tropical Animal Health and Production* 37 (8): 635-652.

Ouma, E., A. Abdulai, and A. G. Drucker. 2007. Measuring heterogeneous preferences for cattle traits amongst cattle keeping households in East Africa. *American Journal of Agricultural Economics* , forthcoming.

Ouma, E. A., A. O. Gideon, and S. J. Staal. 2004. The socio-economic dimensions of smallholder livestock management in Kenya and its effects on competitivenessof crop-livestock systems. Paper contributed to the NARO conference "Integrated agricultural research for development-achievements, lessons learnt and best practice," September 1-4, 2004, Kampala, Uganda.

Rege, J. E. O., and J. P. Gibson. 2003. Animal genetic resources and economic development: issues in relation to economic valuation. *Ecological Economics* 45 (3): 319-330.

Rege, J. E. O. 1999. Economic valuation of animal genetic resources. Proceedings of an FAO/ ILRI Workshop held at FAO Headquarters, March 15-17, 1999, in Rome, Italy. Nairobi, Kenya: International Livestock Research Institute.

Roosen, J., A. Fadlaoui, and M. Bertaglia. 2002. Economic evaluation and biodiversity conservation of animal genetic resources. University of Kiel, Germany.

Roosen, J., A. Fadlaoui, and M. Bertaglia. 2005. Economic evaluation for conservation of farm animal genetic resources. *Journal of Animal Breeding and Genetics* 122 (2005): 217-228.

Ruane, J. 1999. A critical review of the value of genetic distance studies in conservation of animal genetic resources. *Journal of Animal Breeding and Genetics* 116 (5): 317-323.

Ruane, J. 1999. Selecting breeds for conservation. In *Genebanks and the conservation of farm animal genetic resources*, ed. J. K. Oldenbroek. The Netherlands: IDO-DL Press.

Scarpa, R., A. G. Drucker, S. Anderson, N. Ferraes-Ehuan, V. Gómez, C. R. Risopatrón, and O. Rubio-Leonel. 2003. Valuing genetic resources in peasant economies: the case of 'hairless' creole pigs in Yucatan. *Ecological Economics* 45 (3): 427-443.

Scarpa, R., E. S. K. Ruto, P. Kristjanson, M. Radeny, A. G. Drucker, and J. E. O. Rege. 2003. Valuing indigenous cattle breeds in Kenya: an empirical comparison of stated and revealed preference value estimates. *Ecological Economics* 45 (3): 409-426.

Signorello, G., and G. Pappalardo. 2003. Domestic animal biodiversity conservation: a case study of rural development plans in the European Union. *Ecological Economics* 45 (3): 487-499.

Simianer, H., S. B. Marti, J. Gibson, O. Hanotte, and J. E. O. Rege. 2003. An approach to the optimal allocation of conservation funds to minimize loss of genetic diversity between livestock breeds. *Ecological Economics* 45 (3): 377-392.

Simianer, H. 2000. Valuation of indigenous farm animal populations and breeds in comparison with imported exotic breeds with a focus on sub-Saharan Africa.

Smith, C. 1984. Genetic aspects of conservation in farm livestock. *Livestock Production Science* 11: 37-48.

Smith, C., A. Gibson, and G. Mitchell. 1982. An economic appraisal of pig improvement in Great Britain 2. Factors affecting estimated benefits. *Animal Science (Animal Production)* 35 (2): 225-230.

Smith, C. 1984. *Estimated costs of genetic conservation of farm animals*. FAO Animal Production and Health Paper 44/1. Rome: Food and Agriculture Organization of the United Nations.

Smith, C. 1985. Scope for selecting many breeding stocks of possible economic value in the future. *Animal Science (Animal Production)* 41: 403-412.

Tano, K., M. Kamuanga, M. D. Faminow, and B. Swallow. 2003. Using conjoint analysis to estimate farmers' preferences for cattle traits in West Africa. *Ecological Economics* 45 (3): 393-407.

Tisdell, C. 2003. Socioeconomic causes of loss of animal genetic diversity: analysis and assessment. *Ecological Economics* 45 (3): 365-376.

Vaccaro, L. P. d. 1973. Some aspects of the performance of European pure bred and crossbred dairy cattle in the tropics: Part I: reproductive efficiency in females. *Animal breeding Abstracts* 41: 571-589.

Vaccaro, L. P. d. 1974. Some aspects of the performance of European pure bred and crossbred dairy cattle in the tropics: Part II: mortality and culling rates. *Animal breeding Abstracts* 42: 93-103.

Wickham, B., and G. Banos. 1998. Impact of international evaluations on dairy cattle breeding programmes. Proceedings of 6th World Congress on Genetics applied to livestock production. *Armidale* 23: 315-322.

Wurzinger, M., D. Ndumu, R. Baumung, A. G. Drucker, A. M. Okeyo, D. K. Semambo, N. Byamungu, and J. Sölkner. 2006. Comparison of production systems and selection criteria of Ankole cattle by breeders in Burundi, Rwanda, Tanzania and Uganda. *Tropical Animal Health and Production* 38 (7-8): 571-581.

Zander, K., and J. Mburu. 2004. *Compensating pastoralists for conserving animal genetic resources: the case of borana cattle in Ethiopia.* Bonn, Germany: Center for Development Research.

Zander, K. 2006. *Modelling the value of the borana cattle in Ethiopia - an approach to justify its conservation.* Bonn, Germany: Center for Development Research.

2

Animal Husbandry, Veterinary Science and Dairying in India: Analysis of Initiatives Taken at Centre and State Levels

Man has been domesticating animals either for cultivation and transport or for food. He has been able to rear good varieties of animals like sheep, birds, cattle,etc. Following are the importance of domestic animals:

(i) India is primarily an agricultural country. The villagers mainly depend on livestock like cows, buffaloes, goats, fows, horses etc. These animals are grouped into three types depending on their utility.

(ii) Milk-Yielding animals: The milk-yielding animals are cows, buffaloes, goats, etc. In our country the buffaloes form the major group of milk-yielding cattle, which form the basis of dairy industry. There were about 343 millions of cattle in our country during the year 1965-66. Out of these 176 millions were sheep, 64 millions were goats and the rest were others.

(iii) Animals from which meat is procured and egg laying animals: Meat for human consumption can be had from sheep, goat, fowls, ducks etc. Out of these the fowls and ducks lay eggs. The eggs and meat are rich in proteins and are the best nutrients.

(iv) Animals useful for agriculture and transport: Oxen are helpful in ploughing the land. Camels and horses are also used in ploughing . the horns and the skin of animals are very useful. The dung and urine of many animals are useful in increasing the fertility of the soil.

Animal Breeding: The animal breeders generally use following methods of breeding:

(i) Crossing the individual animals within the same breed.

(ii) Crossing a breed with the individual of another breed. In interbreeding, the animals chosen are those which have a good records or pedigree.

(iii) Line Breeding: The Line Breeding is the crossing among groups of animals with a gap of three or four generations. This yields better results than breading within the same generation.

(iv) Out Breeding: The animals belong to the same breed, the crossing will be between un-related groups.

(v) Mating outside the breed: This is of three types:- (a) cross between different species (b) cross between individuals of different breeds. (c) Cross between a true bread and a mixed breed. This also called Grading.

(vi) Grading: This method is used to increase the production of meat, milk, eggs etc.

Management of Cattle: Development pertaining to Animal Husbandry in India is hindered by the following reasons.

(i) Unsuitable climate and surroundings.

(ii) Low breeding capacity.

(iii) Insufficient nutritive feed.

Nutrition:- Animals should be fed with food rich in nutritive contents. The feed should contain carbohydrates, proteins, fats, minerals and vitamins. the feed is of two types are.

(a) Roughages:- It are nothing but the hay or the grassy material of paddy, wheat, maize etc. This is fibrous and nutritious.

(b) Concentrates:- It are the seeds of cotton, groundnuts, pulses, bran etc. The nutritive contents of concentrates are more than those of roughages.

Cattle Sheds: Cattle Sheds should be constructed to protect cattle from heat, cold, and rains. Cattle well maintained yield more milk. the cattle sheds should have plenty of ventilation and drainage. Clean water should be provided to the cattle.

Common Diseases: The diseases common to animals are categorized below:-

1. Viral Diseases: (a) Cows, Buffaloes, sheep, goats and fowls are infected with small-pox. (b)Viral skin infection in sheep and goats.

2. Bacterial and Fungal Diseases: (a) the fowls are infected with Tuberculosis.(b) Cholera among fowls (c) Infections causing abortion (d) Diarrohoea among cattle. (e) Necrosis disease affecting the hoofs and tail. (f)Infections of feet in sheep.
3. Liver not disease of sheep due to Liver fluke.

A HISTORICAL PERSPRECTIVE

Until 1892, there was no Department in the State dealing with matters concerning cattle welfare. There was, however, the Imperial Horse Breeding Department run by the Inspector-General of Civil Veterinary Department, Government of India with headquarters at Simla In1892, the Civil Veterinary Department was constituted in the State to supplement the operations of the Imperial Department and a Superintendent was appointed under the Director of Land Records and Agriculture with headquarters at Babugarh in District Meerut. In 1920, the department was taken away from the control of the Director of Land Records and Agriculture, and a Veterinary Advisor to Government was appointed to look after it. In 1929, his designation was changed to Director, Civil Veterinary Department. Animal Husbandry Department was created in January 1944. Advantage was taken to avail the services of Sir Frank Ware C.I.E., F.R.C.V.S. retired animal husbandry commissioner with the Government of India to form a new Department of Animal Husbandry in the United Province under his able guidance. Sir Frank Ware C.I.E., F.R.C.V.S. became the first Director of the Department. With the formation of this department the subjects which were being dealt earlier with either the Civil Veterinary Department or the Agriculture Department were transferred to the Animal Husbandry Department. The following subjects were dealt with the Animal Husbandry Department:

(i) Cattle Breeding and Cattle Farms.
(ii) Livestock Marketing.
(iii) Poultry Farms and Poultry Development.
(iv) Cattle shows and exhibitions.
(v) Sheep and Goat breeding.
(vi) Pig breeding.
(vii) Cattle census.
(viii) Ghee and Milk schemes.
(ix) Fisheries and milk development.
(x) Local purchase of the Army in connection with cattle, goat and sheep.

Soon after the creation of the department other posts were created for the development, in March 1944 to assist the Director of Animal Husbandry in the various branches of Animal Husbandry. In order to expand the activities of the Animal Husbandry Department through out the province the Civil Veterinary Department was amalgamated with the Animal Husbandry Department in 1945 whereby the entire field staff of the former department was utilized for the activities of the combined Animal Husbandry Department. Hitherto all the biological products required for the control of livestock diseases and epidemics were purchased from the IVRI, Izatnagar Bareilly. For the purpose of economy and facility it was decided to open a biological products selection at Lucknow to meet the requirements of biological products for the province. Therefore a post of officer incharge biological products section was created in 1945 having responsibility of biological product viz vaccines and sera manufacturing. Keeping in view the human resource development in the veterinary profession in the state in 1947-48 the U.P. College of Vety. Science and Animal Husbandry was established at Mathura. Which later came under C.S.A. University of Agriculture & Technology, Kanpur. Early in 1947 it was decided to form a separate Fisheries Department as a result of which the Fisheries was separated from the Department of Animal Husbandry in 1964. But Fisheries is dealt along with Animal Husbandry Department with the Secretary Animal Husbandry & Fisheries at Government Level. Latter on the Ghee demonstration scheme was also wounded up & Ghee grading stations along with the staff was transferred to the U.P. Cooperative Federation with effect from 1948. Presently the Department covers Livestock development, Poultry development, Disease control, Fodder development and other Animal Husbandry activities related to the socio-economic upliftment of rural masses and employment generations.

INDIA'S POSTION IN ANIMAL HUSBANDRY AND DAIRYING: AN INTRODUCTORY OVERVIEW

Cattle tending was one of the items of vartha and it was entrusted to a section of the people who thoroughly understood the business in ancient Insia. "When Prajapati created cattle, he made them over the Vaisya; and if a Vaisya is willing to keep them, it must not be kept by any other caste". Vaisyas were primarily agriculturists, formed a wealthy and respectable section of the community and produced fine breeds of cattle. There were experts and well-experienced cattle breeders in India who built fine and rich cattle herds—some of which existed in recent times in certain parts of India, such as Punjab, Bihar, Gujarat and Kathiawar. Agriculture, cattle-rearing, trade and commerce constituted the four-fold vartha or pursuits suitable for making fortune. Cattle-rearing has been noted in the Epics as important and universal and occupation as farming, it has already been mentioned in the beginning how agriculture and cattle farming flourished simultaneously. Contrary to the instruction of Manu that cattle rearing was the occupation of Vaisyas only, there are records in available literature that livestock and animal farming were not the business of a particular group or section of people. The kings themselves, the Ksatriyas, owned and reared the cattle and cattle-wealth was the mainstay of their household finances. The outstanding examples are the emperor of Kosala or of the prince of Kasi. Besides horses, elephants, cows, sheep, and goats they maintained buffaloes, camels, asses, mules swine and dogs for a variety of purpose (Arthasastra H. 29). In the Dhumakari Jataka (III. 410), the high-bred Brahmin is a goat keeper. "All the seventy families in Brahmana hamlet on the slopes of Grdhrakuta mountains near Rajagraha took to cattle breeding as the sole means of livelihood (Chinese Dharmapada by Beal)". The Setthis or merchants mentioned in Jataka (I. 388) were also keepers of cattle. The universality of cow-keeping and cattle trade is manifest in the common use of cows as a standard of value and medium of exchange in the transitional stage between barter and money transaction. Animal husbandry was the occupation, either sole or part, of most of the people. Economic condition of the farmer classes of people was mainly dependent upon agriculture. Some of them were very rich and possessed extensive farms. This is best illustrated by the example in the Dhaniya, the son of a setthi in Vedaka (as mentioned in Suttanipata by Paramatthajotika).

Dhaniya owned about 30,000 head of cattle, which 27,000 were milch cows, and luxuriant meadows for pasture. The Ganapati Mendaka enjoyed a bigger farm which had been managed by as many as 1,250 cow keepers. Every villager used to keep a few animals for draught purposes of to meet the supply to his own household. The village maintained on pay or on a share of produce, shepherds,who were entrusted with the work of taking the animals to the pasture ground in morning and bring them back in the evening. Cattle in India pose a difficult problem at a time when the economy is rapidly growing and agriculture is required to play a leading role.

Mixed farming in India enabled products of land to be fully utilised for both human food and animal feed. This situation is now slowly changing. Finer rice and wheat for which demand is rapidly rising are poor yielders of fodder. On the other hand, milk consumption is below the nutritional level and with increasing income and population the total demand for milk tends to rise. The religious traditions discourage cattle slaughter. Weeding of cattle is therefore done by nature through malnutrition which is indiscriminate in its impact. This keeps down quality without adequately controlling the number. Without mechanization and rural transportation, the farmer depends on bullocks and the use of cattle as a source of draught power and supply of organic manure still dominates. With expanding irrigation and double cropping, there is an increasing pressure of demand for cattle for which the present cattle population, despite its large number, is ill-equipped. A beginning has been made under the Five-Year Plans to improve systematically the quality of cattle with a long-term objective of increasing output of cattle products with proportionately less feeding costs. Livestock play and important role in the national economy of the country which is predominantly agricultural in character. India Accounts for nearly 10 per cent of the world livestock population and 25 per cent of cattle and buffalo. Although the animal and human population are almost equal the contribution of India's livestock to the national income of the country is not very significant. "Animal husbandry" accounts for only 14 per cent of the national income from the agriculture sector and its contribution to total national income does not exceed 7 per cent. Livestock products like hides and skins, fur skins, leather and leather products, dairy products and other crude animal materials are also an important source of foreign exchange in the country. The exports of these commodities during the past few years have steadily gone up from Rs. 40 crores during 1962-63 to Rs. 45.8 crores in 1965-66 and Rs. 88 crores in 1970-71. The major contributing item is "leather" which alone accounts for more than 50 percent of the total foreign exchange earnings from livestock products. Net foreign exchange earnings from the livestock sector are, however meagre as imports of livestock products have also gone up during the period under discussion from Rs. 12.8 crores to nearly Rs. 28 Crores. Dairy products, animal fats and oils are significant in the import list.

Such imports may be desirable in the context of improvement of Indian diet, at the same time they reflect the neglected development of the livestock industry. Along with the efforts to promote the export of livestock products, there would be need to reduce their imports not at the expense of consumption levels but through increased efficiency of the industry. From among the various species of livestock,cattle enjoy a pride of place not only because they account for roughly two-thirds of the total livestock population but also because they constitute and indispensable adjunct to agriculture and contribute in various ways to the prosperity of the farmer in particular and the country in general: They are other main source of supply of draught power for nearly all the agricultural operations, viz., ploughing fields, carrying manures to the fields, lifting water for irrigation, threshing the produce, transporting the produce from the fields to the threshing floors and thereafter to the market as well as driving sugarcane crushers and oil ghanies. It is estimated that the working bullocks put in about 12,000 million working hours for various agricultural operations. According to one estimate, their contribution of draught power in monetary terms is worth about Rs. 400 crores. A more scientific evaluation of the draught power would raise this figure much higher. Even after mechanization is introduced to the extent possible, cattle will continue to occupy an important position in so far as the requirements for farm operations are concerned. *The value figures given here include in addition to the value of cattle dung, the value of manures etc., from intestines, oesophagus, animal blood, edible offals, heads and legs, animal fat, useless meat, tail stumps, glands etc. The total quantity of dung and all these items would be, 5,025,000 tonnes. Cattle are a predominant source of all livestock products except meat, skins, wool and hair and poultry products. Their contribution in the total value of livestock products is of the order of 88 per cent. Cattle dung which is a very important source of soil nutrient, serves as a source of complementarity in the agricultural sector. Half of the cattle dung produced in the country is estimated to be burnt as fuel due to the scarcity of alternative source of fuel. With the increased use of chemical fertilizers the use of farm-yard manures will also have to be increased, to correct the imbalance caused in the soil due to the continued use of chemical fertilizers. Every effort has, therefore, got to be made to conserve and utilize dung for manurial purposes. According to rough estimates even under the present neglected state of affairs about two-thirds of the soil nutrients are being supplied by organic sources in which farm-yard manure figures prominently. Indian diets are miserably unbalanced in so far as the supply of proteins and vital vitamins are concerned. The cattle help not only to raise crops but also provide protective foods in the form of milk and milk products which are the main source of fats of high biological value. Nutritional experts have repeatedly stressed the importance of animal proteins in the diet of the people. The role of livestock in improving the living standards of the masses is thus significant.

Taking Stock of the Livestock Population

The total livestock population in India consisting of cattle, buffaloes and other small animals like goats, sheep, etc. was 343.8 million according to the Livestock Census for the year 1966. The bulk of the livestock population consists of cattle and buffaloes, which together accounted for 67 per cent of the total livestock. Numerically the livestock population in India, judged on the basis of the number of cattle and buffaloes only, is one of the largest in the world. According to 1951 data, India had about one-fifth of the world's total cattle population and one-half of the total buffalo population. Though cattle in India were then three times more than buffaloes, India was probably the world's most important buffalo-rearing country. Indian cattle and buffalo population accounted for over 68.5 per cent of the total number in the Asian continent. This inevitably puts a heavy pressure on India's limited land resources. It may be mentioned here that countries like Denmark have a larger density of cattle and buffalo per square kilometre than India and countries like New Zealand, Argentina and Australia have a larger ratio of cattle population to human population. Cattle, buffaloes, sheep and goats are the numerically important groups. The total livestock population increased by about 28.5 per cent during the 20 years. During the last five years the increase "is more spectacular in the case of work animals, *e.g.*, mules (41.6%), cames (13.9%), goats (16.1%). Male buffaloes have increased by 6.6 per cent while the female buffaloes have increased by 4.5 per cent only. The cattle population has increased only marginally. Male cattle are useful for work in the field or for transport while male buffaloes are used mainly for breeding, except in few states where they are used for work. Consequently we find that the number of male cattle is larger than that of cows, but in the case of buffaloes females outnumber males. Again, the ratio of young stock to cattle is relatively higher than that of young stock to total buffaloes.

Counting the Livestock in Rural and Urban Areas

Livestock population is proportionately much larger in rural areas, since the occupation of animal husbandry is largely combined with cultivation of land. However, owing to inadequate transport facilities for carrying fresh milk over long distances, a part of the livestock population is located in towns and cities. In the cities, horses are used for pulling carriages and bullocks for carts for local transport of goods. In 1956, of the total livestock of 306 millions in the country, 11.9 millions or about 4 per cent were in cities and towns, the rest were in rural areas. In both the cases, proportionately more female stock was found in urban areas. The proportion of sheep and goats in urban areas was as much as that of cattle. It was in the case of horses and ponies that a little larger proportion (8 per cent) was found in urban areas, as they carried the passenger traffic within cities. The animals in urban areas are fed on fodder and other feed brought from rural areas. Since this involves cost, most of the dry animals are maintained in rural areas. On the other hand, proportionately a much larger number of female animals in milk is found in urban areas. Uttar Pradesh with 14.5 per cent of cattle and 21.8 per cent of buffalo population is at the top of all the States so far as cattle population is concerned. With regard to cows and bullocks, Madhya Pradesh comes a close second with 14.3 per cent. Its position is fourth in respect of buffalo population. The second position with regard to buffalo goes to Andhra Pradesh. Maharashtra comes third with regard to the population of both cows and buffaloes. Rajasthan has 7.6 per cent of both cows and buffaloes. Punjab, while it possesses 8.5 per cent of buffaloes (males over three years being quite insignificant), has only 3.7 per cent of cows and bullocks. Between cows and bullocks the latter predominate. Variations in the number of cattle from state to state are due to climate, grazing facilities, and the area under fodder crops which determine to a great extent the quality of cattle and the carrying capacity of land. Agricultural development, introduction of irrigation reclamation of land, etc., affect adversely livestock as the land available for grazing is reduced and the cultivated area under fodder and feed crops does not go up correspondingly, resulting in the reduction of total livestock feed. Both from the viewpoint of quality and number, the states for livestock importance are Rajasthan, Punjab, Uttar Pradesh, Madhya Pradesh and Andhra Pradesh. As regards other animals, nearly 60 per cent of sheep are almost equally distributed in three States of Rajasthan, Andhra Pradesh and Tamil Nadu. A little more than 50 per cent of goats are found in Rajasthan, Maharashtra, Bihar and Uttar Pradesh, 65 per cent of horses and ponies are in Uttar Pradesh, Madhya Pradesh and Maharashtra. The mules are mainly in Uttar Pradesh and Punjab, camels in Rajasthan and Punjab and donkeys in Uttar Pradesh, Rajasthan and Maharashtra.

ANIMAL HUSBANDARY AND DAIRYING IN 4TH FIVE YEAR PLAN AS PER PLANNING COMMISSION

I. Animal Husbandary

The Third Plan and the subsequent Annual Plans attached considerable importance to animal husbandry. A new cattle breeding policy was evolved during this period. According to this policy, cross breeding would be undertaken in areas covered by Intensive Cattle Development (ICD) Projects and in key village blocks that lie in the milk sheds of existing and proposed dairies. Pure breeding would be confined to outstanding indigenous breeds in well-defined breeding tracts with a view to improving the quality of milch cattle. Simultaneous upgrading of indigenous cattle would be undertaken with recognised Indian breeds. Greater efforts would be made for the improvement of the productivity of buffaloes.

- The introduction of ICD Projects during the period 1961—69 represents a significant development. The programmes include improved methods of breeding, provision of feed and fodder and disease control. Earlier, the cattle development programmes taken up in small and scattered areas, could not make much impact on account of insufficient inputs, lack of tie-up with proper marketing and inadequate coverage of cattle population. The ICD Pro—ject was conceived to rectify these shortcomings.
- The Third Plan witnessed a notable breakthrough in poultry farming. The average egg production increased from 60 in 1960-61 to 80 in 1965-66. A large number of commercial poultry farms with 500 to 25,000 layers were set up in private sector in different parts of the country. Large private hatcheries were established as also poultry feed manufacturing units. As a result of these measures, egg production increased from about 2880 million in 19-61 to 5300 million in 1968-69. However the programme of poultry development continued to be adversely affected by shortage and high prices ot poultry feed. Another factor blocking the progress of poultry development was inadequate arrangements for marketing.
- Compared to eggs and poultry progress in other livestock products fell short of expectation. Wool production increased from 32.55 million kgs. in 1961 to 37.60 million kgs. (estimated) in 1969. The productio'Q of milk increased from 20 million tonnes in 1966-67 to 21.2 million tonnes in 1968-69, representing a growth rate of about 3 per cent per annum. This was much below the rate of growth in demand for milk.

Objectives and Targets

- The approach to livestock development in. the Fourth Plan is based on three major considerations. First, it is estimated that only about 12 pe< cent of the agricultural component of the Gross Domestic Product is accounted for by livestock production in India. The second consideration is nutritional. The following table indicates the present estimated availability of milk, meat, fish and eggs as against the level required by accepted nutritional standards:

TABLE 1

Availability and Requirements of Animals Proteins (in gms.)

Sl. No.	*Item*	*Level according to nutritional standards*	*Estimated availability*
(0)	(1)	(2)	(3)
1	Milk	284	105.02
2	Meat	28.4	
3	Fish	56.8	11.36
4	Eggs	28.4	

Compared with cereals, the demand for livestock products is more income elastic and it is likely to grow at a rate between 5.5 to 6.4 per cent per annum. The third major consideration; relates to the fact that animal husbandry offers considerable scope for the diversification of the economy of the small farmer and the landless labourer.

- In the light of these considerations, the Fourth Plan aims at increasing the supply of protective foods like milk, milk products, meat and eggs and at improving the output of certain animal products of commercial importance, such as wool, hides, skins, hair, bristles and bones. It is also one of the principal endeavours of the Plan to help ensure that animal husbandry programmes strengthen the economy of sub-marginal farmers and agricultural labour.
- In framing thp targets of livestock development, consideration has been given to the constraints that are still operating in this sector. The first major constraint concerns absence of a significant research break-through comparable to that in cereal crops. The second constraint concerns feed and fodder, while the third arises from the fact that a large percentage of the bovine population has to provide draught power for agriculture, thus leaving relatively limited scope for a milk-oriented cattle breeding policy.

There is also lack of sufficient integration between crop husbandry and livestock production with the result that these programmes do not adequately reinforce each other. Again, there is the limitation arising from an intense competition for land and water resources. An extremely small area -is devoted to fodder cultivation and pasturage. Certain animals, particularly sheep, are reared by nomadic groups moving up and down the mountain ranges in the Himalayas or from arid to semi-arid tracts in the plains. In these circumstances, a continuous and consistent programme of extension and improvement of quality of animals becomes doubly difficult. In the light of these constraints, the following production targets of selected animal products have been fixed:

TABLE 2

Targets of Production of Animal Products

Sl. No.	*Item*	*Unit*	*1965-66*	*1968-69 (estimated)*	*1973-74*
(0)	(1)	(2)	(3)	(4)	(5)
1	Milk	mill. tonnes	20	21.2	25.86
2	Wool	mill. kgs.	35.66	37.60	41.50
3	Eggs	mill. nos.	4100	5300	8000

- In the animal husbandry sector, the outlays included in the Fourth Plan are:—

TABLE 3

Outlay on Animal Husbandry

Sl. No.		*Outlay (Rs. Crores)*
(0)	(1)	(2)
1.	States	70.91
2.	Union territories	5.40
3.	Central sector	12.50
4.	Centrally sponsored schemes	5.25
5.	Total	94.06

The public sector outlays are proposed to be supplemented from various institutional sources. The Agricultural Refinance Corporation (ARC) has already finance several poultry and dairy development schemes. It is expected that, in the coming years, the role of ARC in this sector would be enlarged. Another significant source of finance particularly for purchase of milch animals is cooperative credit. The All-India Rural Credit Review Committee has recommended that such credit should also be available to persons who carry on animal husbandry activities without undertaking crop husbandry. These measures are expected to increase the flow of cooperative credit and, in particular make it accessible to small farmers and landless labour for their poultry and dairy activities.

Cattle Breeding Policy

The main features of recent policy are:

(i) Selective breeding in the breeding tracts of established or recognised milch, dual-purpose, or some important draught breeds of cattle.

(ii) Laying more emphasis on milk production in the breeding tracts of draught breeds or types of cattle and replacing the other existing draught breeds or types with dual-purpose breeds.

(iii) Grading up with recognised dual purpose or diary breeds in areas where cattle do not conform to any specific type of breed and are usually non-descript and of low productivity.

(iv) Cross breeding with exotic breeds in hilly areas and other places where there are facilities for the rearing and maintaining of high-yielding milch cattle and in urban areas and around industrial townships to ensure adequate supply of milk.

(v) Improvement of buffaloes by selective breeding in breeding tracts and grading up with recognised breeds in other areas where buffaloes have established themselves.

The main emphasis of the new cattle breeding policy is on cross breeding. The rate of progress in this respect will, however, depend upon the degree of the farmers' acceptance of cross-bred humpless animals as working stock. While a measure of progress has already been achieved in cross breeding, certain technical problems in regard to exotic inheritance are yet to be finally resolved, so as to achieve a measure of stability in the desired characteristics. Recently a set of detailed guidelines have been suggested by the Government of India to the State Directorates of Animal Husbandry. These guidelines relate to the area of operation where cross-breeding could be attempted most profitably, selection of one breed with a view to ensuring continuity and economy in breeding operations and precautions about the maintenance of cross-bred stock. It has also been suggested that while introduction of 50 per cent of exotic blood has proved to be satisfactory, the introduction of 75 per cent of exotic blood has generally tended to reduce body weight, viability and milk yield. In regard to cross-breeding with Jersey bulls, it has been considered advisable to fix the exotic nheri-tance at 5/8 level and inter-breed from stock having this exotic blood. The most crucial point on wnich the success of the cross-breeding prograJtnme depends is the quality of the cross-bred bulls used for breeding. It will be necessary to ensure that there is no uncontrolled cross-breeding and that action is taken within the frame-work of these guidelines.

Cattle Development Programmes

In addition to 31 ICD projects in existence, 15 large ICD projects will be set up in milk bhed areas of dairy plants with a minimum capacity of 20,000 litres. There will be 20 medium type ICD projects in the milk shed areas of dairy plants with a capacity of 15,000 litres. The key village scheme now operates in 490 blocks. Sixty new key village blocks will cover small dairies.

The availability of proven bulls is a pre-con-dition for i.he improvement of breeds. A beginning has been made with two progeny testing units. In tile Fourth Plan, a Centrally sponsored scheme will provide for progeny testing units at 10 State {aims. For cattle development, schemes in the Third Plan and the Annual Plans included breeding farms, bull rearing farms, goshala, development, control of wild and stray cattle and organization of mass castration. These pro-grammes will continue. Three central cattle breeding farms and eight bull reating farms will be set up. Sire evaluation cells will be established in each State.

Buffalo Development

The demand for the buffalo as a dairy animal lias increased in recent years on account of the high-yield and rich fat content of its milk. Of the eight well-defined breeds of buffalo, the Murrah breed, which is the most popular among the high-yielders, has adapted itself well all over the country. An important scheme, continued from the second Plan relates to the salvage of Murrah buffalo calves from milk colonies for distribution all over the country. In the Fourth Plan, an All-India Coordinated Research Project on buffaloes is envisaged. The objective is to improve the production potential of buffaloes through assessment of vital characters, selection for high economic value and development of breeds with the help of different systems of breeding-Research is in progress for overcoming the reproductive failures among buffalo cows during summer months. Some causes have already been identified. Further research work is contemplated in the Fourth Plan.

Sheep and Goat Development

The clip of Indian Sheep is generally of coarse quality and the bulk is classified as carpet wool. Wool and wool products such as carpets, blankets and druggets earn valuable foreign exchange. Considerable quantities of the indigenous wools are being utilised in the woollen manufacturing industries. There is also demand for raw wool in the export market. To improve the quality of wool from indigenous sheep, the development programme envisages cress-breeding of local sheep wuh exotic fine wool varieties as well as upgrading with some of the important local breeds. To produce quanity stud ranis of important indigenous breeds of sheep and exotic breeds, the programme envisages establishment of 8 large sheep breeding t'arms with a flock strength of 5000 or more sheep, expansion and reorganization of 15 State sheep breeding farms, establishment of 5 new sheep breeding farms and 50 sheep and wool extension centres besides the expansion of 80 centres established during the Third Plan. Import of fine wool breed of sheep and of mutton types is envisaged to popularise improved method cf sheep shearing, wool grading and marketing on the basis of quality. The programme will be taken up in 8 States. It is proposed to organise farms for Pashmina, Angora and dairy goats.

Poultry Development

With the growth of poultry as a commercial enterprise during the last decade, poultry fanning has become lucrative. The Agricultural Refinance Corporation has already provided finance for five poultry projects. A favourable atmosphere has been created for the growth of ancillary industries such as organised poultry feed, poultry equipment, and sales organizations for eggs and dressed birds. Propagation of stock with high feed conversion efficiency is important for bringing down costs. It is proposed to take up a coordinated poultry breeding programme at three Central and ten State farms to evolve superior lines and to cross them in various combinations with a view to explotiflg hybrid vigour. One hundred intensive egg and poultry production-cum-marketing centres will augment supplies.

Piggery Development

Development of piggery is becoming increasingly important. Pig-breeders are being supplied improved pigs and technical know-how in piggery development blocks. In some areas, supply of balanced feed for pigs has also been taken up. With a view to improving the economic condition of those who have adopted pig-rearing as a traditional occupation, it is proposed to supply them breeding Stock it subsidised rates. The becon factory at Harin^;iatta will be provided additional facilities. Work would be completed in remaining three becon factories and one pork processing plant. Four more pork processing plants are proposed to be set up in different States. To ensure regular supply of improved pigs, 10 piggery farms would be expard-cd and 25 new piggery development blocks would be set up.

Feed and Fodder Development

Increase in the productivity of livestock has been hampered by the shortage of feed and fodder. Only about 4.5 per cent of the cultivated area is under fodder. This is not capable of supp; rting more than a small fraction of the livestock population. Stress will be laid on developing feed and fodder resources under the ICD projects and key village blocks. For meeting emergent requirements it is proposed to set up 5 fodder banks in suitable areas where the available grass production will be harvested and conserved. It is also proposed to popularise silage and hay making by organizing demonstration on cultivators' holdings in the milk sheds of dairy projects. Seven regional forage demonstration stations will be set up. Foundation seeds will be multiplied at 20 seed farms.

Livestock Marketing

Marketing of livestock and livestock products has not developed to the same extent as that of other agricultural commodities. The marking activities have been confined to regulation of cattle markets and to some extent to the grading of live-sock products. Improvement in the conditions of marketing is an immediate need and regulation of markets would be an important Step in this context. It is proposed to establish a livestock marketing cell in the Directorate of Marketing and Inspection with the object of developing effective supervisory .and advisory control over the grading schemes for livestock products. In addition to this, other schemes proposed are classification of raw hides, improvement in the collection, preparation and grading of materials used for manufacture of animal casings -and management for grading of wool at producers' level.

With a view to creating conditions for hygienic production of meat, enabling proper ante-mortem and post-mortem examination, utilising valuable by-products and adopting humane methods-of slaughtering, a scheme for modernization of slaughter houses was taken up during the Third Plan. It did not make much progress. It is proposed to set up corporate bodies to develop a few large, medium and small slaughter houses and meat markets. Bacon factories and poultry dressing plans will be set up for processing piggery products and dressed birds. To ersure fuller utilization of fallen animal and slaughter house wastes for the production of stock feed, such as meat meal and bone meal, carcass utilization centres will be established in most of tlie States.

Animal Health

Maximum production can be ensured only when animals are healthy and protected against diseases and parasites. It is proposed to set up 200 new hospitals, 1000 veterinery dispensaries and 2000 stockman centres and to provide 60 mobile dispensaries. Five hundred existing dispensaries wi I be converted into hospitals and 60 clinical and investigation laboratories established. In addition to the continuance of the rinderpest eradication campaign in the southern States and follow-up programme in others, the immunization programme agiinst the disease will be intensified by establishing check-posts and creating an immune ?one to a de 3th of about 20 kilometers at ^nter-State borders. It is proposed to augment the production of tissue cu furs vaccines against rinderpest and foot and iw-'uth disease as cross bred animals are more sus-ccotible to these diseases than indigenous breeds. Tc prevent ingress of exotic diseases, an animal quarantine and certification service will be set up.

Research

The two central research institutes namely, In;iian Veterinary Research Institute (IVRI) Izat-nagar, I J.I'., and the Central Sheep and Wool Research Institute, Avikanagar, Rajasthan were transferred under the administrative control of the Indian Council of Agricultural Research in 1967. The existing research facilities at both the institutes are proposed to be strengthened and a new Plant Arimal Virus Research Institute would be set up. At IVRI, new divisions of Epidemiology, Veterinary Public Health, Experimental Medicine and Surgery and Livestock Products Technology are proposed to be set up. The work of these divisions will comprisc of invcstigations on the epidemiology . of various important diseases and their control measures, study of safety, efficacy, viability and applicability of newly developed medicines under con-trciied conditions and systematic research to evolve bCiter techniques of collection, processing, packaging and marketing of meat, meatmeal, bones, bone meal and animal casing.

'Since consumption of animal products has been steadily rising, various types of microbial food poisoning are likely to present problems. The two nuin aspects of research proposed concern the diseases communicable from animal to man and • vie "-versa and the problem of rendering animal products .safe for human consumption. For the con-t'rl of foot and mouth disease, a vaccine has al-ra; dy been developed. Further research on this di;;;ase wjll be intensified for developing a potent va-cine for pigs. Under a coordinated project, typing of foot and mouth disease viruses would bs undertaken. A number of coordinated multi-disciplinary research projects are proposed to be taken up for producing better productive strains of different species of livestock through adoption of scientific methods of breeding, feeding, management and disease centre]. At the Central Sheep and Wool Research Institute and at its two regional stations near Kulu and Kodaikanal, research will be undertaken on sheep breeding, management, health and wool processing technology.

II. Darying and Milk Supply

While in the Second Plan stress had been laid on establishment of colonies of milch cattle in metropolitan cities on the Aarey pattern, the Third Plan policy was to develop dairy projects with emphasis on milk production in rural areas linked with plants for marketing surplus milk in urban centres. It was envisaged that the supply and collection of milk would be undertaken by producers' cooperatives in rural areas and the processing and distribution of milk and manufacture of miik products would be organised through plants operated as far as possible on cooperative lines. During the period 1961-69, 22 liquid milk plants and 4 milk product factories had been commissioned and brought in operation. Besides, 25 liquid milk supply schemes and 8 milk product plants were under various stages of installation.

On the eve of the Fourth Plan, the total number of dairy plants in operation was 91 comprising 47 liquid milk plants, 7 milk product factories and 37 pilot milk schemes. Of these, 53 plants are in the public sector, while the rest in the cooperative sector. It is observed that most of them are operating at a loss on account of various reasons. In some cases, there was undue time-lag between the initiation and the commissioning of the project due to unavoidable circumstances and this added to the capital cost. Many plants have taken a long time to develop the operations to their full installed capacity. While some efforts to introduce balancing plant have been made, the inherent problem of fluctuation in milk supply between the flush and the lean season continues.

Objectives and Outlays

As some of the existing dairy projects are operating at a loss, one of the principal tasks in the Fourth Plan is to review the work of the projects and to take corrective measures. This will include changes in the milk pricing policy and introduction of modern management practices. The desirability of changing the management of public sector projects, from departmental of corporate form will have to be pursued. It will also be necessary to establish a direct link between the small producers and the public sector milk plants through cooperative organization. Dairy projects will need to be encouraged to take up extension work under their own auspices. At present, a number of dairy projects balance their operations by using imported milk powder. A phased programme is intended to be drawn up to increase production in the milk shed areas and gradually eliminate dependence on imported milk powder.

The organised sector of dairy industry will be extended to smaller towns with emphasis on milk production in tlie rural areas. Measures will be token to ensure that dairy projects are economically viable and, as far as possible, organised in the cooperative sector.

The financing of dairy development will be based on three principal sources, namely Plan outlays. institutional finance and counterpart funds generated by the sale of commodity gifts under the World Food Programme. As far as institutional sources are concerned, the Agricultural Refinance Corporation has already entered the field and financed one dairy project Further suitable schemes will have to be formulated for financing by ARC. As regards the commodity gifts from the World Food Programme, details are given later. Briefly, it is expected that funds of the order of nearly Rs. 95.40 crores will be generated and will form part of the Plan outlays. On this basis, the total outlay under the Plan will be of the order of Rs. 138.97 crores. The break-up will be as follows:

TABLE 4

Outlay on Dairying

Sl. No.		Outlay (Rs. Crores)
(0)	(1)	(2)
1.	States	39.77
2.	Union territories	1.95
3.	Central schemes	97.25*
4.	Centrally sponsored schemes	—
5.	Total	138.97

* Includes an outlay of Rupees 95.00 crores provided for Indian Dairy Corporation.

In the Fourth Plan, the first priority will be to complete the dairy schemes numbering 33 which spill over from the earlier period. In addition, organised dairy industry will be extended by taking up 24 new schemes in towns with a population of about 50,000. Further more, four milk product factories are proposed to be established. In addition, 64 rural dairy centres will be organised in areas with a population of less than 50,000 with a view to providing chilling and marketing facilities in isolated pockets of milk production.

The Government of India with the cooperation of the World Food Programme have formulated a project for stimulating milk marketing and dairy development. Under this project, the World Food Programme will supply free of cost during the next five years in a phased programme, 1.2 lakh tonnes of skimmed milk powder and 42,000 tonnes of butter oil at the international valuation of Rs. 41.90 crores which, when re-constituted into liquid milk by the four public sector dairies of Bombay, Calcutta, Delhi and Madras, will generate funds worth about Rs. 95.40 crores. These generated funds will be used for investment in increased milk processing facilities, improved breeding, feeding and management of milch animals. As a result of these steps, the production of milk in the milk shed areas of these plants will be stepped up so that by the time the supply of imported milk powder is tapered off, the level of supply from the dairies will be sustained through increased local milk production.

The project will also set up additional storage and transport facilities for balancing seasonal and regional variations in milk production. It is envisaged that about one lakh high-yielding milch animals along with their calves would be salv iged from the metropolitan cities. The project will further help in the economic growth and stability of the rural milk producers in 10 Stales and in the Union Territory of Delhi, which constitute the four dairy zones of milk supply for the four major dairy plants. Similar programmes will be under taken in smaller towns and their milk shed areas with a revolving fund which will be created in the course of implementation of the project, Fcr its efficient management, the Government of India will set up a new company which will work in collaboration with the National Dairy Development Board.

Milk Producers' Cooperatives

Except for a small percentage maintained in the urban areas, the rest of the mjilch animals are kept in villages often by small producers owning two or three animals. The hope of accomplishing general improvement in the, production and marketing of milk would depend substantially or; the extent of effectiveness of the organization of these scattered and small units of production. For this purpose, stress was laid in the Third Plan on the organization of a network of producers' cooperatives. Their total number is at present over 8,000. In some States, these cooperatives have had a fair amount of success. In the Fourth Plan, further efforts would be Made towards strengthening and development of milk producers' cooperatives in two directions. First, the pattern of organization of primary milk producers cooperatives would need modifica") tion. It is observed that the size of many of the existing societies is small, consequently the society is unable to undertake investment in special equipment for chilling. Apart from reorganization of pri-ma;y societies as viable units, with a minimum collection of 500 litres of milk per day on an average, it will be necessary to work towards a progressive cooperativisation of Government milk plants, so that the entire chain of operations from milk collection to transport, pasteurization and distribution gets integrated.

Research

The National Dairy Research Institute, Karnal was transferred to the Indian Council of Agricultural Research in October, 1967. The existing research facilities are proposed to be expanded with the addition of new divisions of dairy cattle genetics, physiology and nutrition and dairy economics. With the production of edible case in from skimmed milk through a simple process, not only has wasteful diversion of valuable milk protein been avoided, but the dairy industry has made a positive economic gain. Further investigations will be undertaken for the economic utilization of whey (from the manufacture of cheese) for the production of yeast protein which can be used directly as cattle feed or, after processing, as human food. Coordinated multi-disciplinary research projects are also proposed with the object of increasing the productivity of dairy cattle and improving the economics of milk production.

There has been considerable progress with mechanization in the fishing industry. About 5700 mechanised boats were brought into operation during 1961-69. A beginning was made in the development of fishing harbours. A programme of 16 small harbours was initiated during the Third Plan and the provision of landing and berthing facilities for mechanised boats has since been taken up at 30 other sites. A beginning has also been made in the construction of major fishing harbours,

ANNEXURE I

Physical Programmes—Levels Achieved and Anticipated

(Numbers)

Sl. No.	*Scheme*	*Levels achieved*			*fourth plan programme*	*level anticipated (1973-74)*
		1960-61	*1965-66*	*1968-69*		
(0)	*(1)*	*(2)*	*(3)*	*(4)*	*(5)*	*(6)*
1.	I.C.D. projects		19	31	35	66
2.	K. V. Blocks	407	498	490	60	550
3.	Cattle breeding farms	23	35	38	3	41
4.	Progeny testing units	—	2	2	10	12
5.	Fodder banks	1	4	4	5	9
6.	Sheep breeding farms	14	24	32	13	45
7.	Sheep and wool extension centres	305	461	503	50	553
8.	Sheep shearing, wool grading and marketing centres		2	6	5	11
9.	Intensive egg and poultry production-cum-marketing centres		77	92	100	192
10.	Poultry dressing plants		17	1	18	
11.	Bacon factories	2	7	8	4	12

1. Provisional and are subject to adjustment in consultation with State Governments.
2. Although 21 new centres were set up during 1966-69, 50 centres have been merged with I.C.D. projects or closed down.
3. Dhulia Fodder Bank was closed down.

ANNEXURE II

Physical Programmes—Levels Achieved and Anticipated

Sl. No.	scheme	Levels achieved 1960-61	1965-66	1968-69	fourth plan programme[1]	level anticipated (1973-74)
(0)	(1)	(2)	(3)	(4)	(5)	(6)
1.	Milk supply schemes	7	30	47	24 new 25 spillover	96
2.	Milk product factories		7	7	4 new 8 spillover	19
3.	Rural dairy centres				64	64

1. Provisional and are subject to adjustment in consultation with State Governments.

FUNCTION OF DEPARTMENT OF ANIMAL HUSBANDARY AND VETERINARY SCIENCE: A REFLECTION

The main functions of the Department of AH&VS are to provide Veterinary health care to the livestock and poultry population and augmenting the production of milk, meat & eggs through breed improvement, training and extension activities. The programmes of the department are implemented through the different plan schemes. The various activities of the Department as under:

Expansion of Animal Health Programme

The islands are free from many dreaded infections and contagious diseases of livestock such as Rabies, Anthrax, Rindpest and Foot & Mouth Disease. Govt. Of India and O.I.E Paris have also recognized the continuous disease free status of this territory by declaring it free from Rindepest from May 1994. The department provides veterinary health care through a network of Veterinary Hospitals, Veterinary Dispensaries & Sub Dispensaries spread over the entire territory which has increased from 57 institutions at the end of the 8thPlan period to 68 as on date. During the 9thplan a Veterinary Polyclinic and a Disease Diagnostic facility for Disease monitoring, forecasting and control were introduced. During the 10th Plan period a modern Veterinary Hospital Building has been inaugurated at Port Blair for providing specialized Veterinary treatment. The other important programmes under the scheme are:

- Expansion of Mobile Veterinary Dispensaries to remote areas.
- Strengthening and expanding the Veterinary health care facilities through
- Veterinary Institutions and maintenance of A&N Islands as Disease free zone.
- Providing regular Swine fever vaccination in the Nicobar District.

Expansion for Livestock Development Programme

Under this programme the local nondescript cattle are upgraded through artificial insemination using frozen semen of exotic bulls, to augment the milk production. During the 9th Plan a liquid Nitrogen Plant was established for ensuring uninterrupted supply of LN2 for maintaining cold chains. The Artificial Insemination which were only about 1200 Per annum in 1997 rose to 5500 per annum by 2001 and the number of crossbred claves born increased to 3000 per annum. During 2002-2003 the Artificial Insemination figures exceeded 6500. Milk production has increased from 17.65 thousand tons per annum in 1992 to 25.62 thousand tons in 2002-2003 and the average per capita consumption of milk per day is 167 gms per day. During the 10th plan new motivational programmes have been taken up such as calf rearing scheme, Elite cow insurance programme apart from inducting part-time volunteers for Artificial Insemination and castration in remote areas by providing them stipend and incentives.

Expansion of Animal Husbandry Programme

(a) Poultry Development Programme: Under this programme the department is providing practical training through the departmental poultry demonstration farms, free vaccination & health care to the poultry farmers, apart from supplying day old chicks of Broiler & larger varieties to the interested poultry farmers. By 2001 about 200 private commercial poultry farms have come up and the islands have become self sufficient in poultry meat production. The total egg production was 560.8 lakhs in 2002 and the per capita consumption per year came to 135 eggs . During the 10th plan the department has taken up a major programme of supplying and popularizing high yielding varieties of Backyard poultry egg and meat production especially in rural areas. Apart from this quail, turkey and ducks will also be popularized in these islands.

(b) Piggery Development Programme: Piggery is the most popular Animal Husbandry Venture in the tribal areas and the department is maintaining 5 pig demonstration farms to provide large white Yorkshire pigs for upgrading the indigenous pigs. A piggery farm has been established in the North Andaman area and with this pig keeping has become popular among the non-tribal population also. During the 10th plan the programme of supplying exotic pigs to farmers is being continued and a new scheme of insurance of pigs belonging to tribals has been taken up apart from providing practical training in pig keeping.

*(c) Goat Development Programme:*During the 9th Plan Malabari goats were popularized among farmers and tribals by supplying Malabari goats and bucks. Breeding facilities were provided through Malabari bucks maintained at the departmental institutions. The department plans to select evaluate and propagate for breeding of native terresa boats at the departmental farms.

Expansion of Fodder Development Programme

During the 9th plan the department had taken up the programme of developing fodder land in panchayats and fodder land has been developed in 39 panchayats. During the 10th plan, besides extending the programme to other panchayats, fodder seed production farm, developing grazing land, establishing Silvi Pastural System and fodder training cum extension programme as well as free supply of fodder seeds and cutting will be taken up.

Apart from the above programmes the department is also taking up extension and continuous training programmes in the different villages to popularize Animal Husbandry as a venture for augmenting milk, meat and egg production and to increase the rural employment potential in these islands.

ANIMAL HUSBANDRY AND STATE PLANNING COMMISSION: A CASE STUDY OF TAMIL NADU

Animal Husbandry plays a vital role in the development of livestock for increasing the production potential of milk, meat and egg which are required for improving the standard of nutrition for human beings apart from providing subsidiary income to them, particularly to the rural poor. It has grown into an industry with the application of advanced scientific methods. There is vast scope for further development since the demand for livestock products is growing day by day. The main objective is to increase the productivity of local livestock through genetic improvement by cross breeding and grading.

Tamil Nadu has a considerable livestock population. According to 1997 census, Tamil Nadu possesses 9.36 millions of cattle, 2.72 millions of buffaloes, 5.4 millions of sheep and 6.3 millions of goats. The poultry population is also considerable and Tamil Nadu is a major egg producing State in the country. With the increase in the individual animal yield, the percapita consumption of Animal Husbandry products is also on the rise. The estimated milk production in the State during 2001-2002 was 4990 thousand tonnes. The production of eggs and meat was 4,223 million nos. and 39 million kg respectively. The per capita availability of milk per day has increased from 166 grams during 1990-91 to 219 grams during 2001-02. During the same period the per capita availability of egg per annum has increased from 46 numbers to 68 numbers.

Veterinary Services and Animal Health

At present there are 1016 Veterinary institutions viz: 3 polyclinics, 25 Clinician centers, 139 Veterinary Hospitals and 849 veterinary dispensaries. There are 2123 subcentres functioning in the State manned by Veterinary Inspectors.

Production of Anti Rabies Caccine (BPL) at IVPM, Ranipet

To protect all dogs against the deadly disease rabies, the anti rabies vaccine is produced at the Institute of Veterinary Preventive Medicine, Ranipet. A sum of Rs.4.01 lakhs is expected to be spent during the year 2002-2003 to meet the recurring cost and a sum of Rs.4.01 lakhs has been proposed for 2003-2004.

Vaccination of Cattle and Buffaloes in Selected Areas

Foot and Mouth is a highly contagious disease, which causes great economic losses. To control the disease proper vaccination has to be undertaken well in advance. The expenditure is shared equally between the Centre and the State Governments at 25% each and the remaining 50% cost is collected from farmers. A sum of Rs.40 lakhs is likely to be spent during 2002-03.

Animal Disease Surveillance

This is a Centrally Sponsored Scheme equally shared by State and Central Governments. The department has been producing valuable animals through cross breeding programmes. As the cross breed animals are prone to suffer with diseases as compared to native stock, disease surveillance becomes very important. To monitor the disease in livestock and poultry, an Animal Disease Surveillance Unit is functioning in the Directorate of Veterinary Services. The district head quarters are connected through NICNET for effective disease surveillance. A sum of Rs.2.30 lakhs is expected to be spent during the year 2002-2003 and a sum of Rs.2.61 lakhs is proposed for 2003-2004. The expenditure is to be shared equally between State and Centre (50:50).

Creation of Disease Free Zone in Tamil Nadu

This is a Centrally Sponsored scheme shared equally between State and Central Governments. A Disease Free Zone is functioning in Kanyakumari District along with the adjoining two districts of Kerala State to eradicate major livestock diseases in a phased manner. This is expected to increase the export potential of livestock and livestock products from that area. An amount of Rs.13.75 lakhs is likely to be spent during 2002-03 and an amount of Rs.14.90 lakhs is proposed for 2003-04.

Canine Rabies Control

Rabies is a Zoonatic disease and the control of rabies therefore becomes a Veterinary public health problem. Hence the Canine Rabies Control programmes have been started with a view to protect the dogs against the deadly rabies disease. The scheme was started in Coimbatore, Trichy and Chennai under the Centrally Sponsored Scheme. An amount of Rs.26.36 lakhs is likely to be spent during 2002-03 and an amount of Rs.29.84 lakhs is proposed for 2003-04.

Livestock Protection Scheme

For the benefit of more than 2.40 crores of livestock in our State an innovative scheme viz., 'Livestock Protection Scheme' was first launched at Thirukalukundram, Kancheepuram district in January 2000, so that the health of the cattle could be maintained and their production potential increased and Veterinary Health Services and Breeding facilities provided at the door steps of farmers in the villages. A sum of Rs.214.14 lakhs is likely to be spent during the year 2002-2003. The same amount is proposed for the year 2003-2004.

Replication of DANIDA Project in Non-project Areas

During 2001-02, a sum of RS.67.20 lakhs had been provided to train the farmers in non-project areas in livestock farming and fodder development in selected 6 districts of 2096 Panchayat Unions. During 2002-2003,an amount of Rs.43.49 lakhs is likely spent. During 2003-2004, a provision of Rs.44.98 lakhs is made for this scheme.

Scheme for Sustainable Farming Integrated with Animal Husbandry Department

Animal Husbandry is an integral part of rural life and a key activity to provide sustainable growth in the rural economy. In order to give a new thrust to homestead farming that involves agriculture with animal husbandry in the coming year, a pilot scheme for homestead farms together with goat- rearing will be launched. An amount of Rs.100 lakhs is proposed for this scheme.

Cattle and Buffalo Development

Establishment of Frozen Semen Bank

For cross breeding of Cattle, exotic cattle breeds like Jersey, Friesian, and for up gradation of buffaloes Murrah are being used for the production of Frozen Semen. The Frozen Semen Bank at Eachenkottai was therefore started. It was expected to supply exotic semen straws throughout the State and also to produce Liquid Nitrogen for the storage of Frozen Semen Straws. An amount of Rs.18.76 lakhs is expected to be spent during the year during 2002-2003 towards recurring cost. A sum of Rs.22.30 lakhs is proposed for 2003-2004.

Livestock Development in Pudukottai District with DANIDA Assistance

The DANIDA assisted Livestock Development programme is being implemented in the districts of Pudukottai, Sivaganga, Virudhunagar and Tuticorin. The basic idea of the scheme is to intensively train farmers in project areas in livestock farming and fodder development so that they can get maximum return from livestock owned by them. This programme will provide much needed technical information to the local farmers. An amount of Rs.425.53 lakhs was provided for 2002-2003 for the implementation of the programme against which an amount of Rs.332.29 lakhs is expected to spend during 2002-2003. A sum of Rs.333.96 lakhs is proposed for the year 2003-2004.

Heifer Calf Rearing Project in Panchayat Unions

During 2001-2002, a sum of Rs.28.00 lakhs was sanctioned under Part II scheme for purchase of 800 graded buffalo heifer calves in 160 panchayat unions of 12 districts towards 50% subsidy to SC/STs and 33S! subsidy to others.

Strengthening the Stock of Murrah Breeding Bulls

During 2001-2002, a sum of Rs.6.00 lakhs had been sanctioned under Part-II scheme for purchase of Murrah bulls for the District Livestock Farm, Abishekapatti.

Poultry Development

Supply of Giriraja Chicks to Farmers in Southern Districts

During 2000-2001 a sum of Rs.30.65 lakhs had been provided to distribute chicks to the farmers in Southern Districts to increase their income.

Sheep and Wool Development

Intensive Sheep and Goat Development

With the closure of European Economic Committee aided Sheep Development scheme, Intensive Sheep and Goat Development work was undertaken in selected districts from the year 1998-99. During 2001-2002 a sum of Rs.45.00 lakhs had been sanctioned under Part-II scheme to distribute sheep in selected districts to the beneficiaries on a 50% subsidy for SC, STs and 33S! subsidy for others.

Fodder Development

Establishment of Fodder Bank

For the improvement of grazing facilities fodder production will be taken care of for the livestock. This scheme comes under Centrally Sponsored Scheme shared equally between State and Centre.

Supply of Enriched Paddy Straw to Agricultural Labourers and Farmers

By enriching the paddy straw it is possible to increase the nutritive value. Hence supply of enriched paddy straw will help the farmers to increase the production of their cattle. During the year 1997-98, Government provided a sum of Rs.7.5 lakhs for the benefit of 1500 beneficiaries.

Strengthening of Fodder Seed Production Unit

Providing bore well and open well strengthened the fodder seed production units functioning in the State at a cost of Rs.9.00 lakhs sanctioned under Part II for the year 1999-2000. To deepen the ponds in District Livestock Farm, Korukkai a sum of Rs.3.01 lakhs had been provided during 2001-2002.

Extension and Training

Strengthening of Training Centres

To carry out repairs and construction of buildings in order to conduct Livestock Inspector training course, a sum of Rs.7.00 lakhs was provided during the year 1999-2000.

Direction and Administration

Strengthening of Statistical Cell

A statistical cell has been established in the Directorate for conducting Sample Survey for estimating production of milk, meat and egg in the State. This is a Centrally Sponsored Scheme shared equally between State and Central Governments. During the year 2002-2003 a sum of Rs.16.83 lakhs is expected to be spent and a sum of Rs.18.21 lakhs is proposed for the year 2003-2004. The expenditure is to be shared equally between Centre and State.

Other Expenditure

Strengthening of Horse Breeding Unit

In hills and foot of the hills horses and ponies are in great demand. Hence a horse-breeding programme was started at the District Livestock Farm, Hosur during the year 1997-98. A sum of Rs.3.66 lakhs is likely to be spent during the year 2002-2003 for the feed cost of the animals and a sum of Rs.3.66 lakhs is proposed for the year 2003-2004.

Special Component Plan

Veterinary Dispensaries - SCP

The Veterinary Dispensaries opened in the colonies where people belonging to Scheduled Caste are living were brought under Special Component Plan during 1997-98. A sum of Rs.32.07 lakhs is likely to be spent during the year 2002-2003 and a sum of Rs.37.45 lakhs is proposed for the year 2003-2004.

Upgrading Veterinary Hospitals into Clinician Centres (SCP)

Veterinary Hospitals were upgraded as Clinician Centres to provide better treatment facilities in a phased manner. Recurring cost of Rs.3.71 lakhs is likely to be spent during the year 2002-2003. A sum of Rs. 4.17 lakhs is proposed for the year 2003-2004.

Mobile Veterinary Dispensaries at Block Level (SCP)

383 Mobile Veterinary Dispensaries were opened during the year 1994 and the Mobile Veterinary Dispensaries located in the Schedule Caste colony areas were brought under Special Component Scheme during 1997-98. A sum of Rs.3.70 lakhs is likely spent during the year 2002-03 as recurring cost and a sum of Rs.3.70 lakhs is proposed for the year 2003-2004.

Mobile Veterinary Units (SCP)

At present 54 Mobile Veterinary Units provide Health Cover to Livestock and Poultry in remote villages. The Mobile Veterinary Units located in Schedule Caste colonies were brought under Special Component Plan during 1997-98. A sum of Rs.10.73 lakhs is likely to be spent during the year 2002-2003 towards recurring cost and a sum of Rs.12.67 lakhs is proposed for the year 2003-2004.

Hill Area Development Programme

Lump Sum Provision for New Schemes Under Hill Area Development Programme

The Nilgiris District occupies a unique position as a hill station and hill terrain. For the upliftment of the people of this area, Hill Area Development Programme is being implemented as a Centrally Sponsored Programme from the year 1975 onwards. Beneficiary oriented schemes and infrastructure development form part of the programme. A sum of Rs.50.62 lakhs is likely to be spent during the year 2002-2003 and a sum of Rs.63.06 lakhs is proposed for the year 2003-2004.

Western Ghats Development Programme

Lump Sum Provision for New Schemes Under Western Ghats Development Programme

For the benefit of people in Western Ghats Development Districts, beneficiary oriented programmes and infrastructure development programmes have been taken up and Mobile Veterinary Units at Vettaikaranpudur and Uthamapalayam and Animal Disease Intelligence Unit at Dindugul and Farmers Training Centre at Abishekapatti have been established. A sum of Rs.47.74 lakhs is likely to be spent during the year 2002-2003 and a sum of Rs.47.76 lakhs is proposed for the year 2003-2004.

New Programmes for 2003-2004

During the year 2003-2004, the following new schemes are to be implemented:

1. Upgrading 10 Sub Centres into Veterinary Dispensaries - It is proposed to upgrade 10 sub centres at a cost of Rs.32.77 lakhs to provide Veterinary Health Care, based on demand and cattle population.
2. Distribution of Rams to Link Worker Couples - It is proposed to distribute rams to Link worker couples to eliminate inbreeding in sheep at a cost of Rs.18.14 lakhs to augment the meet production through scientific breeding, feeding and management practices and to generate additional income to the farmers through adoption of scientific management.
3. Providing Communication facilities to 15 Veterinary Institutions located at District Head quarters - It is proposed to provide the telephone facility to the veterinary institutions where these facilities are not available at a cost of Rs.0.75 lakhs for better supervision, direction, implementation of schemes, disease control, and management of disasters and natural calamities.
4. Establishment of Mobile Veterinary Unit at a cost of Rs.1.37 lakhs - It is proposed to provide veterinary and health care to animals in the rural areas at the doorsteps of the farmers and to extend all Veterinary assistance to the rural people who are unable to contact the nearest Veterinary Institutions at a total cost of Rs. 1.37 lakhs.
5. Establishment of Turkey Rearing Unit - Poultry farming has undergone a transformation from backyard units to present vibrant and dynamic commercial enterprises. Chick rearing has become more popular and awareness is not created in rearing turkey, Japanese quails, Guinea fowls, etc. Turkey rearing is one of the profitable backyard businesses, for which awareness must be created. An amount of Rs.4.24 lakhs is provided for this scheme.

The plan outlay for the year 2003-04 under various group heads are given below:

Sl.No.	Group Heads	Outlay for 2003-04 (Rs. in lakhs)
1.	Veterinary Services Animal Health	445.39
2.	Cattle and Buffalo Development	356.29
3.	Poultry Development	4.26
4.	Sheep & Wool Development	18.19
5.	Fodder and Feeds Development	0.03
6.	Extension and Training	0.01

7.	Direction and Administration	18.21
8.	Tribal Areas Sub-Plan	0.00
9.	Other Expenditure	3.70
10.	Special Component Plan	57.99
11.	Hill Area Development Programme	63.06
12.	Western Ghat Development Programme	47.76
	TOTAL - Animal Husbandry	1014.89

MODERN INDIA'S ANIMAL HUSBANDRY, DAIRY & FISHERIES SECTORS AT STATE LEVEL: A CASE STUDY OF RAJASTHAN

The animal husbandry sector comprises (i) Department of Animal Husbandry; (ii) Dairy Development; (iii) Fisheries and (iv) Veterinary Education & Research.

Agriculture plays an important role in Rajasthan economy and nearly one fourth of the total state income is generated by Agriculture and allied activities including Animal Husbandry. In Rajasthan, Animal Husbandry is not merely a subsidiary to Agriculture but it is a major economic activity specially in arid and semi-arid areas, thus providing the much needed insurance against prominently occurring scarcity conditions. Income from live stock accounts for 30 to 50% of the Rural households income, with wide variation in region and households.

Live Stock Sector tops in rural employment with 4.5 percent growth against 1.75 for all other Sectors and 1.1 for Agriculture. This Sector has also the highest potential for rural self-employment generation at the lowest possible investment per unit. Development of Livestock Sector therefore, is critical to rural prosperity.

The Animal husbandry sector is harbouring a fabulous livestock wealth having very significant role in providing subsidiary to major sources of income to the large numbers of cultivators, small farmers, marginal farmers, BPL families and agricultural labourers. Milk enterprise generates income on regular basis as against the crop enterprise, which is mostly seasonal and is more prone to droughts. Cattle are mainly looked after by the women folk. The provision of assured market for the milk leads to their increased participation and the availability of cash income encourages them to take up to social development programmes. In Rajasthan animal husbandry is major economic activity contributing 13 percent of the State's net domestic product. As against twenty five well defined breeds of cattle and seven buffaloes breeds in the country, the state is endowed with seven breeds with finest drought hardy milch breeds (Rathi, Gir and Tharparkar), dual purpose breeds (Kankrej and Haryana) and the famous draught breeds of Nagauri and Malvi.

Livestock production in general and cattle and buffaloes in particular is highly women oriented as it is labour intensive. Over 95% of households chore is related to the care and management of milch animals in livestock owing households are dealt by women and 60% of all labour engaged in rural livestock production are women. Live stock sector in Rajasthan is thus extremely livelihood intensive, closely interwoven into the social economic fabric of the rural society, making investments in development of this Sector the critical pathway for rural prosperity. The details of livestock population since 1951 are given below:

(No. in lakhs)

Item	*1951*	*1961*	*1972*	*1983*	*1992*	*1997*	*2003*
Cattle	107.82	131.36	124.70	135.04	116.66	121.41	108.54
Buffalo	30.45	40.19	45.92	60.43	77.75	97.70	104.14
Sheep	53.87	73.60	85.56	134.31	124.91	145.85	100.54
Goat	55.62	80.52	121.62	154.80	152.85	169.71	168.09
Camel	3.41	5.70	7.45	7.56	7.46	6.69	4.98
Others	3.99	3.72	3.53	4.36	4.82	5.19	5.07
Total	255.16	335.09	388.78	496.50	484.45	546.55	491.36

Source: Livestock Census.

The scarcity of fodder and low grass land productivity has forced the reduction in live stock population.

Present Position of Live Stock Population and Its Contribution to Development of Rajasthan

Rajasthan possess 11% of the total animal population of India that yield almost 9.16% of the total milk production, 30% of the Goat meat production and 39% of the total wool production and 35% of draught power. The animal husbandry is contributing about 13% of the states economy (GDP).

Development Constraints

- Low productivity and very large numbers are major development constraints faced the sector for all the species. Almost 60% of all cattle and about 80% buffaloes are non-descript and have very low milk and work output. Though the milk production is above national average but as compared to the growth potential it is much below the expectations.
- Compounding low productivity further, the important factor is the tiny stock holding and also shrinking of common grazing lands over a period of

time. This has widened the gap between demand and supply of fodder. Situation is further aggravated by the increasing production of unproductive cattle.

- All services in livestock sector like veterinary care, AI etc. are managed by the State Department of Animal Husbandry and by and large these services are free. This overwhelming presence of government in State of free services have compromised their quality and accountability and have crowded out possibilities for the emergence of free market for these services.
- The vast AI Centers now cover less than ten percent breedable female cattle and buffalo. Unselected bulls used for breeding leads to progenies with virtually no genetic progress from generation to generation, and the poor quality service with less than 30% conception rates together with the lack of coverage and zero genetic progress renders large investments in breeding of cattle and buffaloes infructuous.
- While reasonably effective extension network evolved in the crop production section nation wide, no such effort was made in the Live Stock Sector. Absence of well conceived extension support system in the live stock sector has undermined the pace of development in the sector in different plans. Therefore, accelerating livestock sector development in Rajasthan needs to be planned on priority.

Major Achievement during Tenth Five Year Plan Period

Some of the major achievements attained during Tenth Five Year Plan are as under:

- One new Polyclinic at Jhalawar, 11 new veterinary hospitals and 209 new sub centers were established in far-flung areas to provide animal husbandry services to the rural mass.
- 17th Livestock Census was conducted by the department and the Revenue Board with the help of the Department of Animal Husbandry.
- Breeding Policy for cattle and buffalo is being revised. The major stress is given on preservation and conservation of indigenous breeds and cross breeding of non-descript animal is being done only on demand.
- Sheep and Sheep Breeder's Insurance scheme were launched.
- Cow and Cattle Breeders' Insurance scheme were started.
- Rs. 199.50 lacs were provided to the Pathmera Goshala for the development and improvement of Kankrej breed.

- Total 113 Veterinary Officers were appointed during the 10th Five Year Plan.

Objectives and strategy during Eleventh Five Year Plan

The objectives for Animal Husbandry are:

- Prompt community participation with Public Private Partnership.
- Improvement of out reach services to increase livestock health & production.
- Increases in the income of the people engaged in animal husbandry.
- Promotion of Livestock industries and marketing in the State.
- Active participation of the local breeders.
- Shift from veterinary health care to animal husbandry practices and breed improvement.
- Enabling the small producer to participate in the process of Globalization, to gainfully participate in the process of growth and modernization of the livestock sector.

Strategy for Eleventh Five Year Plan

- Use of the national and global market pull to provide the energy and impetus for sectoral growth.
- Promote institutions and establish mechanism to ensure quality consciousness, encourage research and innovations, and enhance sector level efficiency in production, procurement, processing and marketing of all livestock products particularly through value addition.
- The state is endowed with vast livestock populations having wide genetic diversity suitable for milk, meat and fibre production. Therefore their is need for conservation of the valuable germ plasma for genetic improvement and efficient production.
- Research & development wing should made efforts for minimising the large and expanding gap between feed and fodder resource availability and demand.
- Breeding management through restructured AI programme comprising of input generation and delivery system.
- Camel is an essential & asset of the desert economy and is a very good source of draught power all over the State for short distance transport of human as well as goods. It also plays an important role in the economic development of weaker section of the society. Therefore, it is proposed to

encourage the breeding of quality animals.

- It is proposed to ensure active participation of non-governmental organizations in the livestock development programmes.
- For educating the animal owners regarding modern scientific methods of livestock management, extension activities needs to be strengthened.

Allocation for the Eleventh Five Year Plan 2007-12 & Annual Plan 2007-08

An amount of Rs. 17500.00 lacs has been proposed for Eleventh Five Year Plan period and Rs. 2069.90 lacs for the Annual Plan 2007-08.

The allocation is about 32.25% higher than the 10th plan allocations. The scheme wise details are as under:-

Training, Extension & Information System

In the year 1990-91 training cell was created to organize systematic training programmes. For providing training to the stockman government training schools at four places viz. Bassi (Jaipur), Jodhpur, Kota and Udaipur and at 39 places under private sector are functional. Human resource development and trainings are the essential component for the successful implementation of the various developmental programmes and to increase the productivity per unit time by optimum utilization of available resources and adoption of scientific methods of animal managements. Therefore, certain refresher training courses for the departmental para-veterinary staff would also be taken under this head. It is proposed to strengthen all departmental Stockman Training Schools with Photocopy machine, Audio-visual aids, digital camera, and Computer system with printer, furniture, along with furnishing of hostel and library facilities. For this purpose Rs. 5.00 lacs has been proposed for the Annual Plan 2007-08 and Rs. 30.00 lacs for the Eleventh Five Year Plan.

Rajasthan State Live Stock Management and Training Institute

The building and essential infrastructural facilities have been made available at Jaipur with world bank assistance under Agriculture Development Project for imparting regular training to the officers of Animal Husbandry Department. The building of institute and its hostel is completed. A provision of Rs. 10.00 lacs has been proposed for the Annual Plan 2007-08 and Rs. 75.00 lacs for the Eleventh Five Year Plan for modernization, furniture and fixture etc.

Establishment of Veterinary Polyclinics

There are at present 13 polyclinics in the state. It was envisaged earlier to establish polyclinics at all the district head quarters. At polyclinic specialized veterinary care is being provided under one roof. These Polyclinics are equipped with diagnostic aids like x-ray, clinical laboratory and there is facility to treat complex animal disease & infertility problems by the subject matter specialist. It is proposed to establish at least 4 new veterinary polyclinics in the districts of Bhilwara, Hanumangarah, Jhunjhunu and Tribal District of Dungarpur during the 11th five year plan period.

A provision of Rs. 30.00 lacs has been proposed for the Annual Plan 2007-08 and Rs. 300.00 lacs for the Eleventh Five Year Plan.

Disease Diagnostic Laboratories (DDL)

There are five Regional Disease Diagnostic Center (RDDC) in Rajasthan i.e. At Ajmer, Kota, Udaipur, Jodhpur and Bikaner; Jaipur has State level Disease Diagnostic Lab (SDDL). There is an urgent need of strengthening of Regional labs Regional Disease Diagnostic Center (RDDC) and facilities at state level labs State Disease Diagnostic Center (SDDC). During the Eleventh Five Year Plan, a Regional Disease Diagnostics Laboratories (RDDL) at Bharatpur is proposed to be set up. To provide essential chemicals, reagents and glassware for all the disease diagnostic labs and creation of new Diagnostic Lab at Bharatpur, a sum of Rs. 2.00 lacs are proposed for the year 2007-08 and Rs. 20.00 lacs for the Eleventh Five Year Plan.

Biological Products Unit (B.P.Lab.)

This institute is producing veterinary vaccines for the prevention of infections and contagious animal disease.

A provision of Rs. 0.50 lacs has been proposed for the Annual Plan 2007-08 and Rs. 4.00 lacs for the Eleventh Five Year Plan for strengthening & expansion of this unit.

Strengthening of Nutrition Laboratory

It is most essential to strengthening of the existing facilities of nutrition lab so that through advance technology, analysis of samples of feed and fodder can be done. For strengthening of the lab, Rs. 3.00 lacs have been proposed for Annual Plan 2007-08 and Rs. 20.00 lacs have been proposed for Eleventh Five Year Plan period.

Veterinary Hospital and Dispensary

The existing facilities of animal health care are inadequate. There is one veterinary center for each 14299 animals where as the recommendation of National Commission on Agriculture is one Veterinary Center for each 5000 animals. Thus the state is lagging far behind the NCA recommendation. During 10th five years plan period, 11 new veterinary hospitals and 209 sub centres were opened, to provide minimum facilities in the selected villages for rural growth center. Therefore it is necessary to increase and strengthen the existing veterinary health care facilities during 11th plan period to take care of precious livestock.

A provision of Rs. 822.50 lacs has been proposed for the Annual Plan 2007-08 and Rs. 8419.00 lacs for the Eleventh Five Year Plan Period. The scheme wise details are as under:-

Modernization and Renovation of Hospital Buildings

At present many hospital buildings are very old and in bad shape and need renovation.

A provision of Rs. 100.00 lacs has been proposed for the Annual Plan 2007-08 and Rs. 1000.00 lacs for the Eleventh Five Year Plan for the modernization and renovation of the existing hospital buildings

Strengthening of Veterinary Hospital

There are 1426 veterinary hospitals in the state. For furniture and equipments, a sum of Rs. 130.00 lacs has been proposed for Annual Plan 2007-08 and Rs. 1634.00 lacs for the Eleventh Five Year Plan period.

There is great demand from the public to establish Veterinary Hospitals in the rural areas of the state. To fulfill the gap and provide better veterinary services to livestock breeders all the veterinary dispensaries (285) would be up graded into veterinary hospitals in the 11th Plan.

A provision of Rs. 80.50 lacs has been proposed for the Annual Plan 2007-08 and Rs. 1425.00 lacs for the Eleventh Five Year Plan for the salary, equipments, and furniture in the up graded hospital.

Veterinary Extension Programme

Due to large distances and inadequate veterinary facilities, a large number of livestock breeders living in far-flung areas are unable to bring their animals to get the animal husbandry facilities. As a result most of the centers remain under utilized. To optimize the services of the existing staff and expand health coverage, it is proposed that each veterinary hospital would hold two days camps in the village of its jurisdiction every week, Therefore, about 50000 camps would be organized every year and will cover about 25000 revenue villages.

A provision of Rs. 230.00 lacs has been proposed for the Annual Plan 2007-08 and Rs. 1800.00 lacs for the Eleventh Five Year Plan.

Management Information & Evaluation System (MIS)

The limited success of the departmental schemes can be largely attributed to the lack of effective monitoring and feedback systems. All the schemes are monitored on the basis of quantitative data, and there is no means to judge the quality and direction of the programmes. It is proposed to get printed MIS formats for the progress of the veterinary institutions and district level offices.

A provision of Rs. 27.00 lacs has been proposed for the Annual Plan 2007-08 and Rs. 100.00 lacs for the Eleventh Five Year Plan.

Veterinary Services Through NGO Under Public Private Partnership Programme

The limited veterinary service of the department requires to be extended with the help of NGO under Private Public Partnership Programme. It is proposed to establish 300 veterinary centers in 3 districts. The NGO will run the centers and provide health and breeding services to the farmers.

A provision of Rs. 100.00 lacs has been proposed for the Annual Plan 2007-08 and Rs. 1200.00 lacs for the Eleventh Five Year Plan.

Livestock Breeders Training Camps

For the up-gradation of skill of livestock breeders, training camps are much more required. For this purpose two days and six days breeder training camps will be organized.

A sum of Rs. 100.00 lacs is proposed for the Annual Plan 2007-08 with a total of Rs. 900.00 lacs for the 11th Plan.

Expansion of Extension Units at Tehsil Level

Extension programmes have their own impact for the improvement of the skill of the livestock breeders for understanding modern techniques. Therefore, It is proposed to develop extension unit at the tehsil head quarter level hospitals.

A provision of Rs. 80.00 lacs has been proposed for the Annual Plan 2007-08 and Rs. 180.00 lacs for the Eleventh Five Year Plan.

Printing of Publicity Material

A provision of Rs. 20.00 lacs has been proposed for the Annual Plan 2007-08 and Rs. 150.00 lacs for the Eleventh Five Year Plan for the printing of posters, pamphlets, booklets and folders etc.

Strengthening of Extension Unit of the Directorate

For preparing of extension material and documentary films for all the extension units at district and tehsil level as well as training Centers, it is proposed to procure three CCD video camera, an attachment of high band video camera to convert into Beta cam, SLR digital camera with all attachments like Flash gun and lenses etc. colour Photocopy machines for the preparation of the exhibition material.

A provision of Rs. 10.00 lacs has been proposed for the Annual Plan 2007-08 and Rs. 30.00 lacs for the Eleventh Five Year Plan.

Private Veterinary Services

To fulfill the gap, the expansion of veterinary services would be taken up through involvement of community and NGO's. Under the scheme, if a Panchayat/NGO wants to open a veterinary hospital Government would provide salary of the veterinarian up to 5 years and equipment to the Panchayat/NGO.

A provision of Rs. 10.00 lacs has been proposed for the Annual Plan 2007-08 and Rs. 50.00 lacs for the Eleventh Five Year Plan.

Veterinary Council

State Veterinary Council has been constituted under the provisions of Indian Veterinary Council Act. It regulates the veterinary practices in the state and skill and knowledge up-gradation of veterinarians. It is a Centrally Sponsored Scheme on 50:50 basis.

A provision of Rs. 20.00 lacs has been proposed for the Annual Plan 2007-08 and Rs. 150.00 lacs for the Eleventh Five Year Plan for state matching share.

Livestock Development Programme

Dairy farming requires low investment and has the potential to create attractive livelihood opportunities for the economically challenged sections of rural Rajasthan, provided livestock can be genetically upgraded through systematic and scientific animal husbandry. For the improved and better services to livestock, it is desirable to test the semen before distribution to AI centers from the district supply unit for testing motility of semen to ensure quality. So, it is planned to procure 35 microscopes at a cost of Rs. 20000/- per microscope during 2007-2008 to increase milk yield.

A provision of Rs. 75.00 lacs has been proposed for the Annual Plan 2007-08 and Rs. 500.00 lacs for the Eleventh Five Year Plan for replacement of cryocan jars, AI equipments, input supply and feed supplement and medicines for the treatment of infertile animals.

Strengthening of Cattle Breeding Farm

Department has four cattle breeding farms at Nagaur, Dug (Jhalawar), Ramsar (Ajmer) and Kumher (Bharatpur). These farms are initially established for the rearing of elite breeding bulls. For breed improvement almost 50% success would depend on the provision of natural services through pedigree bulls because it is still the first choice among farmers. The department would procure bulls of Nagauri, Malvi, Gir and Murrah buffalo bulls from the breeding tract within the country and distribute on subsidized cost to Goshalas and farmers.

A provision of Rs. 20.90 lacs has been proposed for the Annual Plan 2007-08 and Rs. 200.00 lacs for the Eleventh Five Year Plan for bull distribution and establishment of farmers training centers at all four cattle breeding farms.

Goshala Development

For the development and monitoring of the Goshalas in the state a unit is working in directorate. Under this programme, a provision of Rs. 0.05 lacs has been proposed for the Annual Plan 2007-08 and Rs. 4.00 lacs for the Eleventh Five Year Plan to supervise and improvement of the Goshalas in the state.

Department is providing grant to the Rajasthan Go Seva Ayog. A provision of Rs. 25.00 lacs has been proposed for the Annual Plan 2007-08 and Rs. 150.00 lacs for the Eleventh Five Year Plan for the establishment and monitoring expenses of Ayog for 11th Plan Period.

Institutional Arrangement of Supplies (RLDB)

Rajasthan Livestock Development Board has been constituted in the 9th Five Year Plan for regular input supply to the departmental institutions and development. RLDB is getting funds from Government of India for the various livestock development programmes under NPCBB (National Project for Cattle and Buffalo Breeding). Department has been providing grant to RLDB to meet out salary expanses.

A provision of Rs. 30.00 lacs has been proposed for the Annual Plan 2007-08 and Rs. 200.00 lacs for the Eleventh Five Year Plan.

Poultry Development

The sector needs promotion and support through extension and input services besides technical guidance. Intensive Poultry Development Block and Poultry Training Center, Ajmer requires to be strengthened.

A provision of Rs. 6.25 lacs has been proposed for the Annual Plan 2007-08 and Rs. 39.00 lacs for the Eleventh Five Year Plan for Intensive Poultry Development Block and Poultry Training Center, Ajmer and Poultry Farm Jaipur respectively.

Equine Development

Marwari horse breed is well known worldwide and its main habitat is Rajasthan. It is proposed to establish few natural service centers in the state and pure bred stallion has to be provided for the improvement of the breed. It is also proposed to start a incentive programme to the breeders of Marwari horses. So a sum of Rs. 5.00 lacs and Rs. 30.00 lacs has been kept for the Annual Plan 2007-08 and 11th Plan.

Goat Development

As per 17th Livestock Census 2003, the goat population in Rajasthan is 168 lacs. For the development of goat in the state there is one goat farm, located at Ramsar. Buck distribution activity is the major breed improvement programme for the Sirohi goat. A sum of Rs. 10.00 lacs has been proposed for Annual Plan 2007-08 and Rs. 60.00 lacs for the 11th Plan respectively.

Livestock Census

18th livestock census will start in the year 2007 with the help of the Revenue Board. It is a central assisted programme with 100% central share. A provision of Rs. 0.75 lacs has been proposed for the Annual Plan 2007-08 and Rs. 4.00 lacs for the Eleventh Five Year Plan for office expenditure of the staff of department stationed at Revenue Board.

Direction and Administration

Department has vast area of institutions and offices including Joint Director offices at range level and at district level Dy. Director and Assistant Director level office. For the monitoring and supervision of different schemes and programmes vehicles are required. At present departmental vehicles are very old and requires major repairs. No. of offices have no vehicles. A provision of Rs. 85.00 lacs has been proposed for the Annual Plan 2007-08 and Rs. 500.00 lacs for the Eleventh Five Year Plan for purchase of new vehicles and other expenses.

Special Component Sub Plan (SCSP)

Under this head departmental schemes like calf rallies, buck distribution and purchase of medicines are undertaken. Following proposals are to be considered for the benefit of Schedule Caste families every year.

- *Buck & Ram Distribution:*

  Under this scheme 500 Sirohi bucks and 500 Rams will be purchased and distributed to the SC families on subsidized rate that is @ Rs. 400/- per Buck/Ram.

A provision of Rs. 30.00 lacs has been proposed for the Annual Plan 2007-08 and Rs. 200.00 lacs for the Eleventh Five Year Plan.

- *Bull Distribution:*

  Under this scheme 400 elite indigenous bulls (Gir, Nagauri, Tharparkar, Rathi, Kankrej and Murrah will be purchased and distributed to the SC families on subsidized rate that is @ Rs. 1000/- per buck.

  A provision of Rs. 20.00 lacs has been proposed for the Annual Plan 2007-08 and Rs. 125.00 lacs for the Eleventh Five Year Plan.

- Exemption of education fees for the students studying in Livestock Training School of the department. There are 4 training schools and total 64 students of the SC families will be benefited. Under this, total fees of the SC students would be exempted.

  A provision of Rs. 8.00 lacs has been proposed for the Annual Plan 2007-08 and Rs. 50.00 lacs for the Eleventh Five Year Plan.

- *Purchase and distribution of medicine:*

  For the treatment of animals of the SC families, a provision of Rs. 217.00 lacs has been proposed for the Annual Plan 2007-08 and Rs. 2125.00 lacs for the Eleventh Five Year Plan.

- *Farmers Training Programme (Two days):*

  Under this scheme 350 farmers training programmes will be conducted and a provision of Rs. 35.00 lacs would be required for Annual Plan 2007-08 and Rs. 175.00 lacs for 11th Plan.

- *Backyard Poultry Development Programme:*

  Under this scheme 4000 units of 20 chicks of low input chicks will be distributed free of cost and a sum of Rs. 65.00 lacs will be required during 2007-08 and Rs. 325.00 lacs during 11th Plan.

Tribal Sub Plan (TSP)

Under this head departmental schemes like calf rallies, buck distribution, backyard poultry development and trainings are undertaken. Following proposals are to be considered for the benefit of families belonging Schedule Tribes.

- *Buck & Ram Distribution:*

  Under this scheme 350 Sirohi bucks will be purchased and distributed to the ST families on subsidized rate that is @ Rs. 400/- per buck/ Ram.

  A provision of Rs. 30.00 lacs has been proposed for the Annual Plan 2007-08 and Rs. 200.00 lacs for the Eleventh Five Year Plan.

- *Bull Distribution:*

  Under this scheme 300 elite indigenous bulls (Gir, Nagauri, Tharparkar, Rathi, Kankrej and murrah will be purchased and distributed to the ST families on subsidized rate that is @ Rs. 1000/- per bull.

  A provision of Rs. 20.00 lacs has been proposed for the Annual Plan 2007-08 and Rs. 125.00 lacs for the Eleventh Five Year Plan.

- *Exemption of education fees*

  For the students studying in Livestock training school of the department. There are 4 training schools and total 48 students of the ST families will be benefited. Under this, total fees of the ST students would be exempted.

  A provision of Rs. 8.00 lacs has been proposed for the Annual Plan 2007-08 and Rs. 50.00 lacs for the Eleventh Five Year Plan.

- *Purchase and distribution of medicine*

  For the treatment of animals of the ST families. A provision of Rs. 40.00 lacs has been proposed for the Annual Plan 2007-08 and Rs. 600.00 lacs for the Eleventh Five Year Plan and also a provision of Rs. 730.00 lacs has been kept for the strengthening of polyclinic, DDL, LSD and school etc. for the institution of the tribal area for 11th Plan as well as Rs. 73.00 lacs for Annual Plan 2007-08.

- *Farmers Training Programme (Two days):*

  Under this scheme 350 farmers training programmes will be conducted and a provision of Rs. 35.00 lacs would be required for Annual Plan 2007-08 and Rs. 175.00 lacs for 11th Plan.

- *Backyard Poultry Development Programme:*

  Under this scheme 4000 unit of 20 chicks of low input chicks will be distributed free of cost and a sum of Rs. 65.00 lacs will be required for the year 2007-08 and 325.00 lacs for the 11th Five Year Plan Period 2007-2012.

Cattle Fair

The state is organizing ten state level cattle fairs at different areas. There is a wide scope to develop these fairs for the attraction of tourists as well to provide better market facilities to the farmers in the fairs. There is a need to provide adequate facilities for the incoming animals and animal owners on the fair ground. On these cattle fair grounds capital works for providing enough drinking water, animal shelters, fodder stores etc. are required to be developed. A provision of Rs. 75.00 lacs has been proposed for 11th Five year Plan. Rs. 10.00 lacs would be required for Annual Plan 2007-08.

Training/Seminars/Exhibitions

A provision of Rs. 2.00 lacs has been proposed for the participation of veterinarians in the trainings, seminars and exhibitions etc. for the Annual Plan 2007-08 and Rs 12.00 lacs for the 11th plan. It will include all the traveling expenses including registration fees and office expenses for the preparation of the events.

Construction Works

A provision of Rs. 50.00 lacs for the Annual Plan 2007-08 and Rs 250.00 lacs for 11th Plan has been kept for the repair and renovation works of the buildings of the departmental offices and veterinary institutions.

Assistance to States for Control of Animal Diseases (ASCAD)

Under this Centrally Sponsored Scheme (75:25), it is intended to fill up the critical gaps in terms of strengthening the laboratories and creating a disease management system, and to equip the personnel by providing them training on various aspects of disease diagnosis, control and management. Under the scheme, the biological product laboratory and the state as well as regional disease diagnostic laboratories are under the process of modernization/ strengthening.

Public awareness programme would also be taken up. During the plan period the B.P. Lab would be upgraded to ISO 2001 level and tissue culture lab will be fully established. Strengthening cold chain facilities through out state will also be taken up.

Under the scheme a provision of Rs. 900.00 lacs has been proposed for the 11th Plan, while Rs. 150.00 lacs is proposed for the Annual Plan 2007-08.

RAJASTHAN AGRICULTURE UNIVERSITY, BIKANER & MPUAT UDAIPUR

Research and Education

The Rajasthan Agriculture Unviversity Bikaner was established in 1987 to look after Agriculture Extension, Education and Research and to carry out production oriented agriculture research, rural mass education, adoption and propagation of new technologies in the field of agriculture including animal husbandry and allied services.

In the year 1999-2000 another Agriculture University was also establihed at Udaipur with a view to watch the interest of Agriculture Education and Research more effectively in the southern and eastern parts of the State. Now out of the 32 districts, 21 districts are served by RAU, Bikaner and 11 by Maharana Pratap University of Agriculture & Technology, Udaipur.

Following are the major activities related to Animal Husbandry which are under taken by these universities:

Veterinary Education and Research (Animal Husbandry)

Research

The responsibility of Animal Husbandry, Veterinary Education and Research and fisheries are entrusted to RAU, Bikaner & Maharana Pratap University of Agriculture & Technology, Udaipur in the State. A state livestock breeding strategy needs to be evolved to meet the requirement of milk, meat, egg and livestock products and to enhance the role of draught animals as a source of energy for farming operations and transport. Major thrust will be on genetic up-gradation of indigenous / native cattle and buffaloes using proven semen and high quality pedigreed bulls and by expanding artificial insemination network to provide services at the farmers' doorsteps. Following are the important thrust areas of research in the field of Veterinary & Animal Science, dairy technology and fisheries.

Livestock Research

There is need for research on conservation and improvement of indigenous germplasm of Tharparker and Rathi Cattle, Surti Buffaloes, Gir cows, Deogarhi and Parbatsari goats and magra, Chotla, Mawari, Sonadi sheep etc.

Dairy Technology Research

Research on food products - technology standardization, packaging, by-product utilization and development new foods.

Fisheries & Limnology Research

Studies on bio-diversity including survey of ichthyo fauna and other aquatic organisms contributing to the aquatic bio-diversity.

To work out nutritional requirement of local commercially important fishes for formulating ideal fish diet to promote intensive fish culture.

Education

In view of current needs of the State, syllabus has been re-oriented to include new disciplines viz. Veterinary Biochemistry, Livestock Product Technology and Epidemiology and Preventive Veterinary Medicine. In order to cater to the employer needs, new courses viz. Veterinary Ambulatory Clinic, Vety. Laboratory Diagnosis, Vety. Epidemiology and Computer Applications and Bio-statistics have been included. However, as per VCI norms, there are deficits in respect of equipments, and facilities in various departments of the College. One time financial assistance is required to make up the deficiencies of equipments and facilities. Significant infrastructure additions have been made under ADP, namely - Teaching Clinical Complex, Animal Biotechnology laboratory, Central Laboratory and Faculty House. These facilities need to be strengthened through equipment and work force. Other facilities requiring strengthening are Library, Internship Programme, Field Practical Mobility, Disease Investigation facility, PG Programmes, Internship stipend and PG Stipend.

A provision of Rs. 335.00 lakhs & Rs.325.00 lakhs has been proposed for Eleventh Plan period and Rs.34.00 lakhs & Rs.25.00 lakhs for Annual Plan 2007-08 for these Universities.

FISHERIES DEPARTMENT

Rajasthan possess a large area of inland water bodies which offer potential for development of both intensive and extensive system of culture based fisheries. From the available fresh water resources in the state 3.30 lakh ha. of inland water sheets in the form of reservoir (1.2 lakh ha.) tanks, and ponds (1.8 lakh ha.) and rivers (0.30 lakh ha.) have been identified for capture cum culture fishery management. Besides, there exists 0.04 lakh ha. brackish water bodies and perennial flowing system, 214 KM. Indira Gandhi Feeder Canal and about 500 KM under I.G.N.P. in north west Rajasthan.

Constraints in Increasing Production

- Uncertain and irregular monsoon.
- Draining/ pumping out of maximum stored water for Irrigation and Drinking purposes.
- Shortage of quality fish seed.
- Lack of traditional fishermen community.
- Lack of awareness among rural masses.
- Lack of technical know how in rural sector for fish culture.

Objectives during XIth Plan

The main objects of the fisheries and aquaculture development programme identified for Eleventh Five Year Plan are:

- Maximization of Biological Productivity with optimum utilization of available resources.
- Enhancement of the quality fish seed production.
- Generation of employment opportunities.

- Production of all commercially important fish species available in Rajasthan.
- Improvement in socio economical condition of Fishermen.
- Conservation of aquatic bio diversity and eco system.
- Intensive training and demonstration for propogation of advance technologies in aquaculture practices.
- Propagation of cage and pan fish culture.
- Establishment of fish marketing yard, cold chain and retail outlets.
- Promotion of nutritional fish feed production units.
- Strengthening of the fisheries research and education net work.
- Strengthening the data base information net work.

Financial Outlay and Programme during XI Plan

An amount of Rs. 615.00 lakhs has been proposed for the 11th Five Year Plan 2007-12 which includes Rs. 86.50 lakh for committed and Rs. 528.50 lakhs for new items. For the year 2007-08, an amount of Rs. 85.00 lakhs has been proposed which includes Rs. 27.00 lakhs for committed and Rs. 58.00 lakh for new items. Scheme wise details are as under:

Supervisory Staff- Direction and Administration

The total outlay proposed for Eleventh Five Year Plan is Rs. 60.00 lakh out of which Rs. 40.00 lakhs has been proposed for the creation of new office in Jaisalmer, Sikar & Dungarpur and Rs. 20.00 lakh has been proposed for the construction of office building at Alwar as new item and no provision has been proposed in the Annual Plan 2007-08.

Fish Seed Production

For the development of fishermen, the most important input component is the availability of quality fish seed. For renovation and extension fish farm with water supply system, an outlay of Rs. 113.70 lacs has been proposed for 11th Five Year Plan period and Rs. 27.70 lacs has been proposed for Annual Plan 2007-08. During 2007-08 Rs. 13.70 lakh are kept for the completion of boundary wall and water supply system at RPS., and Rs. 13.00 lakh is proposed for renovation of fish farm Bhimpur Banswara & Rs. 1.00 lakh for fish farm Jawai Pali.

Development of Inland Fisheries and Aquaculture

Under Centrally Sponsored Scheme, 15 FFDAs are functional in the state with the object:

(a) Development of water bodies through intensive fish culture particularly in rural area.

(b) Generation of rural employment potential.

(c) Additional source of income for local bodies.

During the 11th Five Year Plan period 4000 fish farmers are proposed to trained in aquaculture techniques for this purpose an outlay of Rs. 77.30 lacs has been proposed for 11th Five Year Plan period and Rs. 13.65 lacs has been proposed for Annual Plan 2007-08.

Fisheries Extension, Education and Training

For creating awareness for fish culture techniques and adoption of advance practices among progressive fish farmer, Rs. 19.50 lakh has been proposed for 11th Five Year Plan (2007-12), out of which Rs. 3.00 lakh for committed work and Rs. 16.50 lakh as new items out of which Rs. 2.00 lacs has been proposed for Annual Plan 2007-08.

Group Accident Insurance Scheme for Active Fishermen

Fish Farmers / Fishermen affiliated with FFDAs and Rajasthan Tribal Area Development Cooperative Federation are proposed to be insured through this scheme under CSS. The 50% of the premium is paid by Government of India directly to FISHCOPPED, and 50% has to be contributed by state government. A provision of Rs. 3.00 lakh has been proposed for 11th Five Year Plan and a sum of Rs.0.40 lakh for the year 2007-08 under group accident insurance scheme for active fishermen.

Fish Marketing

A total sum of Rs. 110.00 lakhs has been proposed for the 11th five year plan. Out of which Rs. 10.00 lakhs is proposed for the year 2007-08, for the state level fish marketing yard and purchase of mechanized boats.

Conservation & Biosphere Management

A total sum of Rs. 15.00 lakhs has been proposed for the 11th five year plan for this scheme & out of which Rs. 1.00 lakhs is proposed for the year 2007-08, for the restoration of indigenous fish species and establishment of ornamental fish hatchery.

National Scheme of Welfare of Fishermen

A total sum of Rs. 115.00 lakhs has been proposed for the 11th five year plan for this scheme & out of which Rs. 12.00 lakhs is being proposed for the year 2007-08, for the development of model fishermen village and saving cum relief programme.

Strengthening of Cooperative society and SHG

To increase the participation of local person in the development of reservoir fisheries on culture basis, it is proposed to strengthen the cooperative and self help groups. For this purpose an outlay of Rs. 7.00 lakh for the year 2007-08 and Rs. 35.00 lakhs for the Eleventh Five Year Plan 2007-12 has been proposed.

Diversification of Culture Practices

A total sum of Rs. 6.50 lakhs has been proposed for the 11th five year plan for this scheme & out of which Rs. 1.25 lakhs is proposed for the year 2007-08, for the artemia culture, breeding and culture of Cat / Air Breathing fishes.

Research and Development

A total sum of Rs. 60.00 lakhs has been proposed for the 11th five year plan for this scheme & out of which Rs. 10.00 lakhs is proposed for the year 2007-08, for the assistance to Fisheries College (MPUT) Udaipur, upgradation of Fisheries Survey and Investigation Unit, assistance to National Fisheries Board and Mobile Pathological Units.

DAIRY

In Dairy Sector, Rajasthan Co-operative Dairy Federation, without any financial assistance in the 10th Five Year Plan, kept continued its success journey with ever growing membership of 6.18 lakh fellow milk producers— with over 23% growth in improving the socio-economic status of rural masses. Attaining its 1st rank among the North Indian State Dairy Federations in milk procurement and marketing. RCDF recorded ever highest milk collection of 20.28 lakh Kg on a particular day in 2005 with average collection having gone to 14.67 lakh Kg per day. The Dairy Federation's economic policies resulted in increase of over 45% in milk pricing pay back to the producers. The demand for Saras milk scenario also grew breaking 40% increase milestone with added flavors of range of new milk products in all the trains, originating from Jaipur.

Apart from social benefit schemes- Saras Surakhsha Kavach and newly launched Saras Arogya Beema Yojna, Central Govt funded schemes SGSY and WDP (Women Development Programme) have contributed in the upliftment of the masses. Specially WDP has opened the doorway of Women Empowerment in the state- via 2,666 new Women Dairy Societies with additional 1.6 lakh fresh women membership.

During the Tenth Plan period with an outlay of Rs. 9.40 crore "Clean Milk Production Programme" has commenced under Centrally Sponsored Scheme with an objective of quality improvement thereby increasing returns and production enhancement. Projects have also been taken up under IDDP, another Centrally

Sponsored Scheme, with a total out-lay of Rs. 17.44 crore for development of infrastructure in the Distts. of Jhalawar, Chittorgarh, Baran, Rajsamand, Churu and Sriganganagar.

Rajasthan Co-operative Dairy Federation marching ahead has set its goal of achieving 36 lakh kg. average milk collection target by the end of 11th Five Year Plan by intensive coverage of milk-shed, cattle induction, breed improvement and productivity enhancement schemes.

A token provision of Rs. 0.05 lacs has been kept for Eleventh Five Year Plan period 2007-12.

MAJOR ACTIVITIES: RELATED TO ANIMAL HUSBANDRY: A CASE STUDY AN INDIA STATE UP

Animal Husbandry is an integral component of rural economy contributing substantially to the National G.D.P. to the tune of about 25% of the total agriculture sector. The state tops in the milk production in India . Of the total livestock population U.P. accounts for 24% and 15% of the buffaloe and cattle population respectively.

The Main Objectives of Animal Husbandry Related Activities include:

(a) To provide health cover and containment of diseases through prophylactic, diagnostic and curative service.

(b) To bring about qualitative and quantitative improvement in production potential of livestock through improved breeding practices.

(c) To provide self employment opportunities and subsidiary occupation, in order to ensure sustainable income to rural people.

(d) To encourage occupational diversification of people.

(e) To improve feed and fodder resources through biomass production and silvipasutre development.

Cattle and Buffalo Development

Department is being supported by World Bank through U.P. Diversified Agricultural Support Project for the improvement of Breeding services, conservation of indigenous breeds through establishment of Open Nucleus Breeding System and field milk recording system. The present breeding coverage of 23.6% is targeted to be increased to 40% by the end of IX Five Year Plan. For further information, contact Joint Director, I.C.D.P., Directorate of A.H. Lucknow.

Animal Health Care and Veterinary Services

Presently there is one veterinary hospital for every 21000 livestock population which is targeted to be expanded for 15000 livestock by the end of Ninth Five year Plan. There is one central disease diagnostic lab. at Directorate which fuctions as referral lab beside 12 regional labs for disease diagnostic facilities and containment of diseases in the state. Quality treatment services are beingprovided by nember of veterinary hospitals in the state.

State biological Product Institute is Producing 12 different types (5 Bacterial & 7 Viral) of vaccines for the effective control of diseases in the state. The modern techniques viz. (Fermentor,Deep Freeze drier) are being adopted in the routine production of the vaccine and existing labs are being modernized. In addition to this the department of Animal Husbandry is procuring nearly ten million doses of different types of vaccines (FMD, IBD, HS, Rabies, M.D.) from other sources. However we are able to cover only about 30% of the livestock population. Therefore, there is a substantial scope for vaccine production & procurement to cover whole of the livestock population. For further information contact Joint Director (Disease Control), Director of A.H. Lucknow .

Operation Zero Rinderpest

Through the activities of the department i.e. intensive immunization, monitoring and surveillance, Uttar Pradesh has achieved the status of Rinderpest disease fee state. Further with extensive surveillance and clinical search programme the state is heading towards Rinderpest infection free status. For f urther information contact Joint Director (Disease Control), Director of A.H. Lucknow.

Poultry Development

Uttar Pradesh is 9th largest producer of eggs in the country and has been able to achieve of late a growth rate of 20%. As a milestone achievement U.P. has become first state of country to give the status of industry to poultry production. For further information contact Joint Director, Poultry, Directorate of A.H. Lucknow.

Sheep and Goat Development

Sheep in U.P. are traditionally reared for wool and meat while goats are primarily for meat. U.P. has a unique resource of indigenous goat breeds like Jamunapari and Barbari while Jamunapari excell in large body size, dual purpose quality, hardiness and ideal for rough village condition, Barbari is known for small size, Stall fed and good quality of meat. The sheep breeds viz. Ramboullet and Marino are maintained in hilly areas with the objectives to improve local breeds for the production of fine apparel wool. Nali breed is being maintained in the eastern region for the improvement of local breed to enhance the production of coarse wool for carpet manufacturing. For further information contact Deputy Director, Small Animal, Directorate of A.H. Lucknow.

Pig Development

The cross breeding programmes has been taken up by using Middle White Yorkshire boars to improve the local stock. Mass awareness campaign and extensive training programmes has received major fillip during last five years and there has been a tremendous rise in piggery farms established by educated class of the society. The department has established one piggery development training center at C.D.F. Aligarh. 10 days training programme is being conducted by department free of cost. At this training center the training schedule from 4th to 14th every month is being followed. For application form and admission(High school is minimum qualification) to this course . There are only 30 seats on first come first serve basis. For further information contact Deputy Director, Small Animal, Directorate of A.H. Lucknow. In this institution free hostel facility is also available.

Feed, Fodder and Forage Development

During different plan periods, concerted efforts have been made to popularize nutritious fodder crops by giving various incentives to the farmers like subsidy in fodder demonstration & distribution of fodder seed minikits.For further information contact Fodder Development Officer at Directorate of Animal Husbandry,Lucknow .

State Epidemiological Unit

For surveillance & monitoring of disease there is extensive network of disease surveillance at division SEU at the apex. The epidemiological unit determines the disease endemic areas, control of outbreaks of animal diseases and its containment. Simultaneously it also helps in containment of zoonotic diseases of public health importance. For any out break the veterinary hospital at field level Chief Vety. Officer at district level, SEU, C.T.O. (EPD) & Joint Director (DC) can be contacted at Directorate level. Phone 329792

Self-Employment Schemes

Under Ambedkar Vishesh Rojgar Yojna the department has trained 2071 rural unemployed youths in A.I. Besides this rural unemployed youths are being encouraged to establish milch (Cow, Buff) animals, sheep, goat, pig, and poultry units for self employment and sustained income generation. Project Coordinator AVRY, Directorate of A.H. can be contacted for detail information.

SJRY (Swarn Jayanti Rojgar Yojna)

The objective of SGSY will be to bring the assisted poor families (Swarozgaris) above the poverty line in three years, by providing them income-generating assets through a mix of bank credit and government subsidy. It would mean ensuring that the family has a monthly net income of at least Rs. 2000. Subject to availability of funds, the effort will be to cover 30% of the poor families in each block during the next five years.

Quality will be the hallmark of SGSY, which has to be imaginatively used to bring people above the poverty line.

U.P. Diversified Agricultural Support Project

U.D.D.A.S.P. is a autonomous project by World Bank. In U.P. under Animal Husbandry sector it supports various activities of A.H. by financial & technical help.

U.P. Livestock Development Board

UPLDB comes in picture for rapid growth of livestock breeding and development in 1999. This board is an autonomous but subsidiary wing of A.H.D. under the financial assistance from central govt. for National Project of Cattle & Buffalo Breeding. The board is maintaining the infrastructure of production & supply of semen straw, LN2 and man power etc.

Marketing Development

There are several functions regarding the rate of livestock and livestock by products viz. to collect, compile and publish the average monthly , quarterly and annual reports regarding the rates of selected cities of milk and milk by products, meat and its by product, cattle seeds, cattle feed, green fodder etc. live livestock of various categories too. To maintain the byelaws of cooperative socities and help the desirable persons to form cooperative societies for pig, sheep, poultry and cattle and their by-products. For any other detail contact Livestock Marketing Officer in the marketing section of Directorate of Animal Husbandry, Lucknow.

Goshala

Under the U.P. Goshala Act 1964 till now 333 Goshalas has been registered. These Goshalas have been established by the public trusts for maintaining old, infirm, unproductive cows and its progeny. Under National Bull Production Programme these Goshalas has been used to produce good quality indigenous bulls and preservation & conservation of indigenous breeds. The central government as well as state government is also assisting these Goshals for cattle development. For detail contact to Registrar, U.P. Goshala cum Goshala Development Officer, Directorate of Animal Husbandry, Lucknow.

Goseva Ayog

Goseva Avog has been established in U.P. for the conservation of Cow breed as well as to promote Goshala etc. total ban on cow or its progeny slaughter and illegal transportation of cattle. For detail Chairman, Goseva Ayog, 1203 Mantri Awas, Dalibagh, Lucknow may be contacted.

Training Facilities

- One month Poultry Training — *Chakganjaria, Lucknow*
- Ten day Poultry Training — *Chakganjaria, Lucknow*
- Ten Days Piggery Training — *C.D.F. Aligarh*
- 2 month to 3 month inseminator training/private training.

Poultry Training

The training programme is now proposed to be classified in two parts:

(a) Training for the departmental staff.

(b) Training for the poultry farmers.

There is one month certificate course of poultry training for Livestock Extension Officer and private poultry breeders at Chakganjaria, Lucknow.

1. No fees charge for pultry training.
2. Hostel facilities are free.
3. After completing the one month course. Director of Animal Husbandry issue certificate.
4. For filling registration form:

   Need– (a) Photocopy of High School certificate.

   (b) Passport size photograph.
5. 20 seats for one batch.
6. 10 batches in a year.

10 days poultry training programme in I.P.D.P. (Intensive Poultry Development Programme) district for poultry breeders.

Under S.T.R.Y. – 10 days poultry training programme for poultry breeders.

Feed and Fodder

(a) Minikit fodder seed distribution from Vety. Hosp.

(b) Biomass & silvipasture development from Vety. Hosp.

(c) Urea-Mollasses treatment from Vety. Hosp.

Other Facilities on Nominal/Levy

- Health care & treatment.
- Large & small animal surgery & radiography.
- Artificial Insemination.
- Prophylactic Vaccination.
- Treatment of reproductive disorder.
- Castration.
- Mass Drenching for worm control.
- Cow & Buff. bulls, Ram, Buck, Boar etc. natural service to livestock in remote areas.

Incentives

- Cash prize to elite breeders in livestock & poultry shown for top quality breeds.
- Grant-in-aid to selected Goshalas.
- Rearing of indigenous male calves.
- Aid to sheep & poultry coop. societies.
- Natural breeding centers in Bundelkhand & hill area.

Levy and Service Charges

Description of Service	*Rates w.e.f. 1st April, 2000 (In Rupees)*
Artificial Insemination	
At head quarter	20.00
At door service	40.00
Natural Service	
Cow/Buffalo	30.00
Horse/Donkey	55.00
Goats	2.00
Pig	1.00
Castration (at head quarter)	

Large Animals	10.00
Small Animals	5.00
Castration (at door step)	
Large animals	15.00
Small animals	7.00
Treatment	
Large	10.00
Sheep/Goat/Pig	5.00
Dog/Cat	40.00
Vaccination	
H.S./B.Q.	2.00
R.P.	—
R.D.	.50
F.P. (Fowl Pox)	.50
Swine Fever	2.00
Enterotoxaemia	2.00
Sheep Pox	2.00
F.M.D.	2.00
Postmortem	
Large Animals	30.00
Small Animals	6.00
Poultry	2.50
Physical Checkup	
Large	15.00
Small	5.00
Animal Feed Analysis	-
Pathological Tests	
Blood	3.00
Stool	2.50
Urine	2.50

The Animal Husbandry carries out disease control, breed improvement programmes and fodder development. Simultaneously, it manages livestock farms consisting of cows, buffaloes, sheep, goats, pigs, rabbits and poultry. The department also deals with equine disease control programme and equine husbandry activities with emphasis on development of pack animals. Along these the department is also controlling zoonotic diseases which are hazardous to public health.

Objectives of the Department

- To provide health coverage to animals, containment of diseases by providing prophylactic vaccination, disease diagnostic and treatment services.

- To provide breeding facilities with advanced technologies and improved genetic stock.
- To evolve an effective and scientific breeding policy to develop animals for specific needs that is high milk yield, meat, egg, wool and draught power etc.
- To improve the feed and fodder resources through extension of area of fodder cultivation with improved fodder varieties , biomass production, silvipasture development and enrichment of cellulosic waste.
- To help in quality meat production based on international quality control measures and regulation of meat export.
- In order to maintain the genetic diversity, conservation of indigenous breeds of cows and buffaloes.
- To develop all livestock species and poultry with marketing back up and support services.

All the above Objective are Meant for Achieving

- Increase production of milk, meat, egg, wool and other livestock products through qualitative and quantitative improvement in production potential of livestock and through disease control activities.
- To provide self employment opportunity subsidiary occupation and income to larger section of population in the rural areas as well as to ensure proper availability of quality animal protein and nutritious diet.
- To encourage occupational diversification of people.
- To reduce the number of non-descript livestock population so as to reduce pressure on land and fodder resources.

CASE STUDY: ANIMAL HUSBANDRY AND VETERINARY SERVICES IN SIKKIM

Sikkim is a state of India, situated in the Eastern Himalayan Region and predominantly a land of hills, spread over an area of 7096 Sq.Km. between 27⁰ 00'46" and 28⁰ 07'48", North latitudes and 88⁰ 00'58" and 88⁰ 55'25" East Latitude is surrounded by high Himalayan ranges, having common border, Tibet in the North and North-East, Nepal in the West, Bhutan in the East and West Bengal in the South. Sikkim State being apart of inner mountain ranges of Himalayas is hilly having varied elevation ranging from 300 to 7000 meters. But the habitable areas are only up to the altitude of 2100 mtrs. Constituting only 20 % of the total area of the State. The highest portion of Sikkim lies in its Northwest direction. A large number of mountains having altitude of 7,000 meters stand herewith (i.e. Kanchenjanga – 3rd highest peak of the world).

Sikkim has a Himalayan or high mountain type of climate. Altitude is the main factor controlling the climate and weather condition of the State. Relief features such as High Mountain act as barriers for the movements of Monsoon wind. Low temperature, high rainfall on wind ward slopes, comparatively dry on the lee ward side and heavily precipitation in the form of snow at the mountain tops are the main features of the climate in Sikkim. Due to great variation in sharp edged mountains throughout the State, there is a large variation in the temperature in the State. The climatic conditions of the State vary greatly due to the wide fluctuations in elevation ranging from 800 to 20,000 ft.

Temperature varying from 17 °C to 27 °C maximum and 2 °C to 21 °C minimum annual, rainfall 162.5 Cms in South, 322.9 Cms in North, 358 Cms in East and 248.0 Cms in West.

As stated earlier, the terrain is hilly and there is very little flat area. In fact, there is not even a flat land of one Sq.Km. or a straight road of one Km. Length. The State is almost rectangle with 112.70 Km. Long and 64.4 Km. Wide. There is gradual elevation from south to the North, varying from 300 Mtrs. to 5500 Mtrs. But, the habitable area is normally upto 2100 mtrs. Which constitute about 20% of the total area of the States.

Amongst the various professions of Socio-economics importance, Animal Husbandry deserves a high priority as it plays an important role in the economic upliftment of the weaker section of the society engaged in Livestock rearing and processing of animal products. The Livestock wealth is the asset to the farmers and provides nutrition, draught power transportation employment and economic support.

Livestock Census

The first Livestock Census was carried out from December1919 to April 1920. It was under the British Indian Government that a systematic record of the periodical conduct of Livestock Census was initiated from this year. In this first Census only some provinces and 28 primary state constituting 29 % to that area participated. Subsequently percentage of participants increased and the Census has conducted simultaneously in a short period of time frame to ensure accurate and reliable results. The Livestock Census of 1951 was first official Census after attaining Independence with the initiation of the first five-year plan, the importance of Livestock Census is to provide various types of data ,it was decided to carry out the enumeration of animals with reference to a fixed data as adopted by the Human

Census. Unlike Human Census, which is conducted decinnially, the livestock census is conducted quinquennially. So livestock Census is conducted after every five years and carried out in the entire country at the same time.

Objectives of Livestock Census

To collect quantitative information and data regarding Livestock, Poultry, Fisheries and Agricultural implements. Essentially it provides Animal Husbandry Statistics to serve as a base for Planning and monitoring developments in the field of Animal Husbandry and Veterinary Sectors. Areas and aspects requiring priority can be ascertained for permanent upliftment of the countries economy. All related information like Veterinary Services back up, breeding programme requirements of feed and medicines etc .can be worked out for future plan strategy, therefore, it serves as a basis to assess the progress of different developmental programme in the Animal Husbandry sector for a given period.

Objectives of Sample Survey:

1. The aim of the survey is to estimate the production level of the major, livestock Product such as Milk, Meat and Eggs.
2. To study the utilization of Livestock Products and attendance practices of major Livestock at the State level.

Reference Date:- Initially the reference date for the conduct of 16th quenquennial Livestock Census was fixed on 15th October 1997 as conveyed by Ministry of Agriculture Department of Agriculture and Co-operation, Govt. of India for all States and UT's of India except Jammu and Kashmir.

Methodology

Livestock Census is a complete count of the Livestock, Poultry, Fisheries and also agriculture implements as per pre-defined reference point of time. As in population Census primary workers have to under take house to house enumeration and ascertain the number age, sex, breed etc. of Livestock and poultry possessed by every household or institution in rural or urban areas. In this State of Sikkim Primary workers engaged were Inspectors, Supervisors, Stockman of the department and locally available educated persons as enumerators temporarily. The 16th quenquennial Livestock Census was conducted as per schedule prescribed in the census calendar and also approved formats supplied by the Ministry of Agriculture, Govt. of India. (as per guide line issued by Govt. of India) 'ANNEXURE- I'.

To conduct the Census work smoothly, the State was divided four districts and the deputy Directors were entrusted direct responsibilities of conducting the Census in their respective districts. The Directors AH and the Directors/Vs were the chief of the Census operations In the State and Addl. Director /VS is appointed as a Nodel Officer for Livestock Census authority for Sikkim.

The Field work conducted by the Primary workers is checked by the Deputy Director, Sr. Veterinary officer of the District and Sub- division level. Even house to house survey work done by the enumerators were also checked and verified by the Sr. Official of the Head Office. Any draw back on the enumeration works had been rectified before sending it for compilation work. Annexure-II.

To carry out the Census on livestock and poultry including different breeds, sex and age in addition to it, Fisheries and Agricultural implements are also taken them in to account for Census, because of its co-related nature with the Livestock.

For easy assessment, the basic unit for Rural areas is the considered to the Revenue Block, which corresponds to a village. Besides, some areas like forest, tea Estate, Monasteries etc. have also covered under the rural area. Likewise, town of urban area is also incorporated in the Revenue Block like Gangtok Town under Revenue Block, Gangtok Pvt. Estate and so on. It is made just to simplify the easy Household enumeration work. Though- Revenue Block is the smallest unit and consists of several inhabited area which are commonly known as " Busty or Gaon".

Household is a group of persons who commonly live together and would take their meals from a common kitchen unless the exigencies of work prevented any of them form doing so. There may be household of persons related by blood or household of unrelated persons of having a mix of both. Example of correlated households are boarding houses, messes, hostels, residential hostels, rescues home, jails, ashram etc. these are called ' Institutional Household' These may be one member households, two member household or multi member households. For Census purposes each one of these type is regarded as " household". While conducting the livestock Census, above-mentioned methodology has been adopted to find out the exact livestock figure of the state. Unit for Census was – household and then Revenue Block, Sub-division level, District level and finally the state as a whole. Presentations of data are made in the table in the form of Departmental Institutional wise duly considering the Revenue Block. This Institution wise data is very convenient for the Department to assess requirement of Medicines, Vaccines, feed/fodder and also helps in the planning of the project.

ANNEXURE-I

Sixteenth Quinquennial of Live Stock, Form Equipment and Equipment of Fishery 16.10.97

CENSUS CALENDER

Operation Dates by which operations should be completed

1. Printing of forms 14th August 1997
2. Distribution of forms & stationary 21st August, 1997
3. Training 16th August-15th September, 1997
4. Preparation of lists of houses and 15th september-14th October, 1997 Households & enumeration
5. Final Check up (Date of Reference) 15th October, 1997
6. Forwarding Taluka Abstract Statements 13th November, 1997 (Provisional figures) to District Office
7. Forwading District Abstract Statements 13th December, 1997 (Provisional figures) to the Directorate Of EAS, Krishi Bhavan New Delhi and licve Stock Census Officer.
8. Compellation of final figures at Taluka 31st December Office
9. Compilation for final figure at District 31st January, 1998 Office
10. Compilation of final figure at State 31st March, 1988 Headquarters and forwarding the same To the Dte. Of E&S, New Delhi
11. Publication of provisional All- India total 28th February, 1998 And Live Stock Census items by the Directorate of EAS.
12. Publication of final results of Livestock 30th June 1998 Census by the Directorate of E&S New Delhi.

ANNEXURE-II

To conduct 16th quenquennial Livestock Census – State of Sikkim has constituted the following working group for smooth conduction of Livestock Census 1997.

1. Commissioner-cum-Secretary/A.H.&V.S-Chairman
2. Principal Director – Co-Chairman
3. Director/AH
4. Director/VS
5. Addl. Director/VS as a nodal Officer

6. Joint Director of the Districts
7. Deputy Director in-charge of the Districts – as a District Nodal Officer
8. Sr. Veterinary Officers
9. Veterinary Officer
10. Inspectors
11. Stockman
12. Enumerator

CONDUCT THE CENSUS

Some of the problems encountered during the Census Operation:

(a) Due to limited quota of fuel for vehicle the supervision was seriously hampered and could not be carried out as desired.

(b) The Departmental staff besides attending to their normal duty had to carry out the enumeration work, which affected the maximum workload.

(c) The Veterinary and Para Veterinary staff deployed for the Census was not enough which led to inordinate delay and had to engage extra enumerators on daily wages basis.

(d) Wrong entries and lesser numbers of household in some places were also noticed and necessary rectifications were made to minimize the errors.

(e) Certain constraints were felt in relation to some remote and high altitude areas where enumeration work is difficult and hazardous.

(f) Some of the items mentioned in the prescribed formats was not applicable to this State and caused confusion at the time of enumeration work.

(g) Financial Constraints – due to non-release of Central grant in time. The sanction of central grant had been released on installment basis, which has created a great complication for obtaining clearance from P&Dand Finance Department of the state.

GENERAL DESCRIPTION OF THE LIVESTOCK CONDITIONS IN THE STATE

Main Species and Breeds of Each Type of Livestock

The native cattle of Sikkim are Siri and Siritype but artificial insemination and placing of exotic breeding bulls in Veterinary dispensaries and Veterinary hospitals has influenced the native type into the crossbred. Nearly 50% of the population is crossbred the rest are Siri and non-descript type.

The majority of sheep belong to Banpala breed, which are heavy bodies with short tail. The local breeds of goat are of Bengal type and crossbred of Jammuna pari, Betal and Barabari.

The pigs found in Sikkim are largely crossbreeds with varying proportions of large white, Yorkshire and SaddleBack blood. Yorkshire variety is quite popular. Recently, the department introduced "Durac" breed of Pig from Royal Govt. of Bhutan.

The breed of Livestock and poultry found in Sikkim are:

Species Breed

Cattle Jersey, Jersey crossbred, Brown Swiss crossbred, Mixed-cross, Siri, Holstein.

Yak Nepalese breed, Tibetan breed

Goat Crosses of Jammuna pari, Bengal, Betal, Barabari with local.

Horses Hill Ponies

Sheep Banpala, Gharpala, Cross-breed, Merind, Cross breed Corrodale, Tibetian Sheep, Rambouilet Cross breed.

Mules Local

Pigs Durac, Saddleback, Large whiteYorkshire and their crosses and local (punri) landrace.

Fowls Black Austrolop and their crosses, white leg horn, Rode Island, indigenous KeystoneGolden.

Ducks Local, Khakhi Camble

Rabbit Angora – German, Russuan, British and German crosses.

Facilities for Artificial Insemination:

Artificial Insemination (A.I) was started in 1975 in the state. So far 26 centers using chilled semen were operating but the establishment of liquid Nitrogen Plant at Gangtok and Karfectar, A.I. with frozen semen has also started in 30 A.I. centers.

Facilities for grazing including restrictions by the state Govt.

There are extensive grassy meadows at a height varying from 14,000 to 20,000 ft. At lower altitudes the village herds are stall-fed. The forest herds are al grazed. Goacharan (Community forest) are planted with fodder trees, sub-tropical grasses, legumes species etc. In the forest area the animals are allowed only with proper plantation of high yielding varieties of fodder plants and feeding of tree leaves is very common. For this assistance is provided by the department.

Migration within and Outside the State

Migration of animals within the State is not restricted. Mainly sheep flocks and Yak have been found to be migratory.

Migration of animals from other State is not there. Animal intended for slaughter, however, get entry. These animals are permitted to enter after thorough check-up and vaccination in order to check the influx of diseases in the State.

With the sealing of borders with Tibet, migratory sheep do not find their way in the State now.

Housing Condition:

There are generally two types of housing for animal namely with the sea:

1. Permanent housing: These sheds are generally made up of permanent walls, cemented floors and roofs of G.C.I. sheets
2. Temporary sheds: These are made of bamboos and thatched roof. The flooring is made of stones. Bedding with grass is provided. Most of the sheds in the State are of this type.

DEVELOPMENTAL MEASURES

Cattle Development

The main objective is to increase the productivity of the local animals through crossbreeding so that it is economically beneficial to the farmers by providing production enhancement inputs. Programme under cattle development include setting up of A.I. centers with frozen semen, setting up heifers production and bull rearing farms and providing semen of highly pedigreed exotic breed.

1. Loans to marginal and progressive farmers provided through Banks and Subsidy to the tune of 25% of the cost in setting up cattle farms is provided by the department.
2. Vaccination programme against F.M.D., Anthrax H.S., B.Q., R.P., R.D. and other diseases at 100% subsidized rates as per the vaccination calendar set up by this department as a preventing measure. Anti Rabies vaccination is also administered to the pets at a nominal charges but post bite vaccination is done free of cost.

Sheep Development

(a) Sheep rearing is a traditional practice and demand for wool and meat is growing steadily in the state. Sheep farms are being strengthened by introduction of new exotic stock while preserving the indigenous breeds of Banpala and Tibetan Sheep.

(b) Subsidy to increase flock size as well as to provide better health and management is being extended by the department.

(c) Temporary facilities including transit camps are provided to migratory flocks. Health coverage and mass treatment against internal parasites is also provided by the department.

Goat Development

(a) The market for mutton in Sikkim is tremendous due to the presence of the Indian Army. As such farmers are keen to take up goat farming in a big way. The Department is improving the local breed by crossbreeding with Jammunapari and BetaL stock.

(b) Bucks are stationed at Veterinary Dispensaries and Panchayats for breeding purposes.

Piggery Development

(a) Farmers are encouraged to set up piggery units with loan. Subsidy is provided by the Department as incentive.

(b) Distribution of crossed piglets (2+1) under ST/SC programmes.

(c) Distribution of boars to Panchayats/blocks for breeding purposes under cross breeding programme.

Dairy Development

1. The district wise number of Dairy Plant and their capacities in the State may be given as under:-

Sl. No.	*District*	*Nos. of Dairy Plant / Capacity*	*Chilling Plant*
1.	East	1 10,000 L.P.D.	
2.	West	1 2,000 L.P.D.	
3.	South	1 5,000 L.P.D	
4.	North	2 5,000 L.P.D.& 1,030 L.P.D.	Chilling Plant at Kabi.

2. The total milk production in the state is estimated to 31.00 M.T. (thousand) per annum, average yield of milk/cow/Buffalo (Herd average) is

   Indigenous = 0.76 Ltrs.

   Crossbreed = 2,00 Ltrs.

   Buffalo -

3. Milk Products/anr

   (a) Ghee 5 M.T.

   (b) Butter 25 M.T

   (c) Curd

   (d) Khoa

4. Per Capita availability of milk is 150 – 180 M.T./day.

5. Schemes Of Dairy Development

At present two major schemes are in operation. These schemes are 100% C.S.S. and entitled as Integrated Dairy Development Programme both for Sikkim Milk Union Ltd. And for North Sikkim Dairy. The schemes started in the 8th five year Plan and will continue up to 9th five year Plan.

The marketing of milk and milk products in the state is being done through Sikkim Milk Union Ltd. The organization is autonomous and is being managed by its Board of Directors.

7. The activity under Dairy Development is to revitalize, assist the activities of present Sikkim Milk Union Ltd. to organize Dairying in North District. The main activity includes production enhancement, improve processing and marketing activities, training and extension, manpower Development, streamlining procurement activities.

FODDER AND GRAZING CONDITION

The Livestock rearing is an important aspect of each household total economy to supplement and compliments other components of farming. About 80-90 % of the nearly 1 lakh households are

Source of Grain / Dry fodder Availability Requirement	*(000 M.T)*	*(000 MT)*
Forest	250	305
Natural grass land	475	850

Agricultural Holding	920	1645
Fodder Cultivation	480	1770
Total	2125	4070

The details on fodder and grazing conditions are as under:

1. Types of Fodder Used

Various agricultural by product and crop residues are utilized by most households for feeding of cattle. Straw of paddy and ragi are used extensively but left over from maize crop is only used sparsely due to high fibre content. Amilisho (Thysanolaena agrostis) which is a versatile fodder and fodder for winter and early spring is cultivated in all parts of state. The Gautemala grass and Napier are also planted to meet the requirement of fodder grassess. The important fodder trees Nevaro (Ficus Hookeri) Gogun (Saurauvia nepalensis) Raikhanya (Ficus Benjarina) Kabra (Ficus intectoria) are also planted by the farmers. The fodder grasses like Oats, desmodium, signal grass Guinea grass are also under cultivation.

2. Animal Concentrated Feed

The concentrate feed for animals are being brought from outside the state (from Siliguri in West Bengal). Most of the ingredients are not available in the state to make concentrate feed.

3. Grazing Facilities

There is an extensive forest herding and intensive state feeding but free grazing of herds is common but decreasing in villages. Due to ban on entry of domestic animals in the forest, the farmers are facing problems for grazing of animals. Enough grazing land (Gaucharan area) are not available in the state.

4. State activities to improve the fodder and grazing condition.

Following activities are being taken up to increase the fodder production in the state.

1. Development of pasture and grazing ground.
2. Demonstrations in the farmers field by introduction extension fodder production scheme.
3. Introduction and implementation of extensive trials of temperate and tropical species of grasses and legumes.
4. Conservation of surplus fodder etc.
5. Consumption of Livestock feed and fodder.

On an average 2.5 quintal to 3 quintal dry and green fodder is being consumed by each animal per month. This fodder includes paddy straw, crop residues, green grasses etc. The animals are fed with concentrate ration as per the fixed schedule. More than 1000M.T. Of concentrates feed are fed to the animals in the state.

IMPORTANCE OF POULTRY HUSBANDRY IN THE STATE OF SIKKIM

Since 1995 the importance of poultry farming is increasing day by day. The main reason of development of this field all due to the following reasons.

1. Increased in population
2. Increased demands of eggs and chickens.
3. Increased of unemployment youth in the state.

There is availability of all the miscellaneous items required in the farm from the market. As there is no big farm in and around Gangtok, there are heavy demands of the products like table eggs and chickens. The poultry farmers are enthusiastic to increase quality and quantity of their products, but due to natural calamities which have no control. Under these farm, farmer cannot be able to fulfill their target. Government is also taking keen interest to uplift the poor poultry farmers of the state.

Under Jawahar Rojger Yojana, Industry department is disbursing loan amounting Rs.one lakhs per farmer under Government norms and condition. The Government is also giving subsidy too amounting Rs.7500/- per farmer. The department of Animal Husbandry is giving necessary guidelines and training. Side by side Industry department also imparting Training programme in every knock and corner of the state. Due to keen interest of the government now the public is very much aware of the importance of poultry farming.

ANNEXURE-III

Poultry farmers of the state are not willing to rear layer birds due to transportation problems especially during the month of May, June and July leading to heavy loss.

Method of keeping poultry: Deep litter system and Cage system.

No private and government hatcheries:

There is no hatchery in the private sector till the date, but the farmers are enthusiastic to set up their own hatchery to earn good sum of money by selling the chicks.

As far as government hatchery is concerned, there was a incubator of capacity 13000 eggs till 1992, but as approved by the government, the land handed over to the Health Department, Government of Sikkim to construct a refferal Hospital at Tadong near Gangtok

As far as Government hatchery is concerned Sikkim Hatchery is already installed at Tokal Bermoick which is at the distance of 20 Km from Gangtok.

This Poultry Farm is known as Sikkim Poultry Development Corporation, Govt. Undertaking, in which Sikkim Hatchery is also included to disburse day old chicks to the poultry farmers. Sikkim Poultry Dev. Corporation is set up in the state in collaboration of Venkateshwara Hatchery. Number of manufacturing poultry equipment- there is no firms manufacturing the equipment in the state till the date.

There was one feed mill at Ranipool namely: Sikkim Ani feed in the East district, which also closed since 1993 .

Production of eggs from exotic breed such as Key Stone Golden, Hisex-white and Hisex-brown was 70%

Percentage of Egg Consumed internally is 60-65%

Arrangement for marketing of eggs is by jeep, bus or van in the private sectors. In the government farms as there is a heavy demand of eggs and chicken the consumer itself drop in the farm to collect the products and there is no need of transportation.

Cost of Production

The state is playing key role in the field of Poultry Development to uplift unemployed youth and also to fulfill market demands. Government is disbursing loan amounting Rs. One lakhs to 5 lakhs to uplift the poor farmers. But simply Government cannot help the public; it is the duty of the public to impart Training Programme organized by the department. Secondly Technical know how is very important to run the farm smoothly. To render the Veterinary Aids in the village level, the department has established he network of Veterinary Stockman Centre in each block with target to establish such center at distance of 10Km. As a whole, the state is divided into four districts, Deputy Directors of the district is over all in charge of the district to implement the departmental schemes. Joint Directors of the concerned district will co-ordinate the work of district from Head Office Gangtok. Each district is again divided into Sub-division, where Sr. Veterinary Officer is the Sub-divisional head, who will control the farm, Veterinary Dispensaries and Stockman Centers. Sr. Veterinary Officers are supported by Veterinary Officers, Farm Manager, Inspectors, Supervisor and Stockman.

Every district Veterinary Hospital is provided with district laboratory, where routine works like blood analysis, urine test, stool examination etc. are done. Any complicated or unidentified sample is required to be sent to Disease Investigation Cell at Gangtok, which is headed by Deputy Director (D.I.C) and Joint Director (D.I.) at Head Office. DIC has to co-ordinate any Livestock outbreak in the district along with concerned Deputy Director/Sr.Veterinary Officer/Veterinary Officer of the district. They are also responsible for carry out the schedule vaccination programme. Rabies control, Deworming, Farm management, castration, A.I. work to attend outbreak cases and daily routine for treatment of animals in hospitals and also door step of Livestock owner in case of large animals

In addition to it, there are various centrally sponsored schemes viz. Rinderpest control, F.M.D.Animal Disease Surveillance and systematic control of Disease of National Importance, where priority are given for Rabies Control, Tuberculosis and Brucellosis. Emphasis is also given to control of Pullorum Disease of Poultry (Ref.Table No.I and II)

TABLE-I

Total Numbers of Veterinary Institution–Districtwise

Sl. No.	*Particulars*	*East District*	*West District*	*North District*	*South District*	*Total*
1.	State Veterinary Hospital	1	—	—	—	01
2.	District Veterinary Hospital	1	1	1	1	04
3.	Sub-Division Veterinary Hospital	2	2	2	1	07
4.	Veterinary Dispensary	8	5	4	6	23
5.	Stockman Center.	16	16	9	15	56
6.	(a) A .I. Centre	11	18	—	6	35
	(b) Bull Service Centre (Natural Service)	37	49	17	64	167
7.	Livestock Check Post	2	1	—	1	04
8.	D.I. Laboratory	1	1	1	1	04
9.	Cattle Breeding Form	1	1	—	1	03
10.	Demonstration Poultry Farm - Layer	1	1	1	1	04
11.	Piggery Farm	1	1	1	2	05
12.	Sheep & Goat Farm	—	2	—	1	03
13.	Rabbit Farm	—	—	1	—	01
14.	Vacational Training Center	1	—	—	—	01
15.	L_2N Plant	1	—	—	1	02

16.	Milk Processing Plant	1	—	—	1	02
17.	Milk Chilling Plant	—	1	—	—	01
18.	Milk Collection Centre	31	10	10	54	105
19.	M.P.C.S	31	10	10	54	105
20.	Cheese Plant	—	1	—	—	1

TABLE-II

Groupwise & Districtwise Incidence of Diseases

Sl. No.	*Groups of Disease*	*East District*	*North District*	*South District*	*West District*	*Total*
1.	Digestive disorder					
	(i) Anorexia	2145	175	1320	1016	4656
	(ii) Enterurtis/duarrhoea/dysentry	2920	146	1342	1054	5462
	(iii) Impaction/tympanny	723	36	673	219	1651
	Group total	5788	357	3335	2289	11769
2.	Respiratory disorders					
	(i) Cough/pneumonia	1249	80	1010	445	2784
3.	Metabolic diseases					
	(i) Milk fever	256	12	307	60	635
	(ii) Parturient paresis	70	0	91	04	165
	Group total	326	12	398	64	800
4.	Deficiency diseases					
	(i) Anemia	283	3	290	282	858
	(ii) Hypocalcimia	137	3	212	25	377
	Group total	420	6	502	307	1235
5.	Skin diseases					
	(i) Fungal (exema/ringworm)	389	40	211	102	742
	(ii) Parasitic (mange/ticks)	1361	26	541	303	2231
	(iii) Non specific	85`	0	147	93	325
	Group total	1835	66	899	498	3298
6.	Parasitic conditions					
	(i) Pyroplasmosis	41	0	34	2	77
	(ii) Ascariasis	596	10	234	372	1212
	(iii) Other round worms	3134	230	1192	1226	5782
	(iv) Tape worms	476	92	375	264	1207
	(v) Fasciola	552	0	294	733	1579
	Group total	4299	332	2129	2597	9857

7.	Gynaecological disorders					
(i)	Abortion	81	2	31	25	139
(ii)	Dystocia	327	19	238	110	694
(iii)	Retained placenta	518	23	332	176	1049
(iv)	Vaginal/Uterine prolapse	133	11	133	60	337
(v)	Metritis	43	0	4	10	57
(vi)	Vaginitis	26	0	4	10	40
(vii)	Anoestrus	288	1	63	103	455
(viii)	Cystic ovary	19	0	0	0	19
(ix)	Other reproductive disorders	71	1	5	85	
	Group total	1506	57	813	499	2875
8.	Poisoning					
(i)	Plant poisoning	174	6	59	100	339
(ii)	Other poisoning	46	1	42	17	106
	Group total	220	7	101	117	445
9.	Other diseases					
(i)	Mastitis	384	10	150	85	629
(ii)	Ephemeral fever	515	28	191	168	902
(iii)	Cancer/Tumer 16	0	16	3	35	
	Group total	915	38	357	256	1566
10.	Surgical condition					
(i)	Wound	107	14	103	68	292
(ii)	Fracture	954	96	580	283	1913
	Group total	1061	110	683	351	2205
11.	Infectious diseases					
(i)	Foot and mouth	409	3	375	13	800
(ii)	Black quarter	9	1	26	8	44
(iii)	Haemmorhagic septicaemia	65	1	179	52	297
	Group total	483	5	580	72	1140
12.	Haematuria	291	291	263	203	1048

LIVESTOCK PRODUCT- OTHER THAN MILK AND MILK PRODUCTS

(i) The main livestock products of the state other than milk and milk products are meat, skin, bone, manure, wool and fibbers from yaks.

(ii) Sikkim livestock and livestock products control Act –1985 regulates the entry of only healthy animals in the state. This act also empowers the department to regulate the selling price meat, its hygienic handling and curbson illegal slaughtering.

(iii) Slaughtering house registers are maintained in the registered slaughterhouse and the information compiled regularly.

(iv) There are four-slaughter house in the state and improved method is practiced in only the slaughterhouse.

(v) There are no arrangements for crushing bones and carcass utilization in the state.

(vi) Estimates of livestock products

(Meat in tons) Hide & skin Wool

1. Bull/Bullocks 1674 11,400 Nos. 600 MT
2. Buffalo 322 1,611 Nos. -
3. Goat 68.5 2,295 Nos. -
4. Chicken 140 MT. -
5. Pig 505. -

The state Government is continuously improving the local breed with the use of improved sires in respect of all the species. To improve the farm management practices, the farmers are given short courses in all the districts. Various types of fodder of improved variety are given to farmer's cultivation. Methods of preservation of fodder for lean season is also demonstrated to the of livestock owners. Farmers are also assisted in setting up small farms through NABARD.

Meat Production

- Bull/Bullocks 50-55% of live weight
  Buffalo -do-
  Goats -do-
  Pigs 60-70% -do-
  Chicken 70-75% in broilers and 180-220 eggs/year/hen

MEMORANDUM ON FISHERIES CONDITIONS IN SIKKIM STATE

I. Organization of Fisheries Services in the State

1. Fisheries Directorate is presently under the Secretary of Forest Department. The Directorate looks after following activities:-

(a) Carp culture: Under this programme the department has constructed demonstration and seed farms. Extension programme is done to encourage the people to adapt fish farming. Fish Farmers Development Agency has played an important role in this field.

(b) Trout Seed Production: Brown trout had been introduced in the state in 1954. The different resources of the state are stocked with fingerlings of it. These seeds are produced at different farms of the state.

(c) Mahseer Programme: Under this programme the seeds of Masher are reared in the farm and are released to different streams to raise their population.

(d) Conservation Programme: There are 42 Fishery guards and 10 Sub-Inspectors to look after the conservation programme. The licenses are issued to the Fisherman for fishing. Illegal fishing are checked like dynamiting, Poisoning, fishing without licenses etc.

(e) Research and Survey: The different resources are surveyed. Their physical and chemical conditions are checked. The final programmes are made after ascertaining all the measures.

(f) Education and Training: The staff is sent to different educational institutions for higher education in the field of fisheries. They are mostly sent to I.C.A.R. institutions. The department conducts the training programme to the farmers and taught about culture management aspects.

(g) The Sikkim Fisheries Act, 1980 has come in force in the state. It has brought streamlining in the activities of the department. This Act has been modified in Sikkim Fisheries Rule , 1990

I. Fisheries Statistics

1. Fisheries Directorate conducts the statistics programme. The staff of the Directorate does it. Fish catch statistics is done by the number of fisherman who is issued the licenses. Aquaculture production per unit of land is also done by the Directorate under Fish Farmer Development Agency Programme.

2. Every year the data on fish production, fish seed production, water area is done. The number of licenses and beneficiaries can be ascertained by the activities of the Directorate.

3. The fisheries products in Sikkim aren't preserved in any form. They are used fresh.

4. The resources of Sikkim are surveyed under Research and Survey Programme like the area of lake of high altitude and length of the rivers.

II. Fisheries Wealth of the State

Sikkim has 28 species as indigenous and many are introduced. The important indigenous species are Mahseer, Katli, Asla, Goonch, Gardi etc. The exotic species introduced in Sikkim are brown trout, common carp, grass carp, silver carp and Indian Major Carps:

1. Main species are breeding of each type of fish.
   (a) Mahseer (Tor putitora) is the most important species in spot fishery. They are reared in the farm. Breeding takes place during August and September. Being migratory they stay in the waters of Sikkim from March to October.
   (b) Brown Trout (Salmo trutta fario) has established in Sikkim. They are reared in farms and stripping is done. The seeds produced are stocked at different resources of Sikkim.
   (c) Common carp (Cyprinus carpio) is introduced in Sikkim. They breed naturally in farms. They are cultured by farmers.
2. State activities during the previous quinquennium to improve the fisheries wealth under the different schemes included in the five-year plan.
3. The main activity during 8th five-year Plan, the Fish Farmers Development Agency has covered 813 beneficiaries under extension programme. The Directorate has worked in maximizing benefits from the existing infrastructures.

III. Breeding Farm

1. Numbers of breeding farms, their locations and management and the experiment carried out.
   (a) Chinese Hatchery, Rothak, West Sikkim.
   (b) Soreng Farm, Soreng West Sikkim
   (c) Gyalshing Farm, Gyalshing West Sikkim
   (d) Mahseer Farm, Bagwa South Sikkim
   (e) Rorathang Farm, Rorathang East Sikkim
   (f) Pologround Farm, Gangtok East Sikkim
   (g) Kabi Farm, Kabi North Sikkim
   (h) Hee-Gyathang, Hee Gyathang North Sikkim

In the above farms the brooders are reared and breeding is done. In these farm common carp are bred. The seedlings are distributed among the fish farmers.

2. Types of Feed Used:

   The carps are fed with traditional feed i.e. Mustard oil cake and Rice bran in the ratio of 1:1. The grass carp are also fed with grass. Brown trout are fed with pellet type feed.

IV. Fisheries Products

The total production of state 140 tones of fish 60% of the product is from capture fisheries and 40% from culture fisheries.

V. Fisheries Research and Training

Research is looked after by Research Asstt. The work is mainly confined in finding the quality of water and soil of different Farm.

Under training programme the staff are sent to different institutions for higher education. The farmers are trained by the Directorate.

VI. Fisheries Trade

The products are sold locally. The important centers for landing of riverine catches are Legship, Jorethang, Melli, Rangpo and Singtam and Ranipool.

Due to the limitation in the production the products are consumed locally.

No work has been done in this field.

VII. Consumption of Fish and Fisheries Products

Due to scarcity of fish, it is imported from neighboring states. Mainly the carps are liked by the population of Sikkim

The people of Sikkim consume canned fish also.

VIII. Fishing Practices in the State

Fishing in Sikkim is done with cast-net and rod. Traditional practices are followed.

IX. Crafts and Tackles

No fishing crafts are used for fishing.

No work has been done in this field.

X. Preservation of Fish

1. The product is used fresh therefore the preservation is not done. At the landing centers there are cold-storage facilities in private sector.
2. There is not any proposal in both five-year Plans.

The fisherman community is constituted by the scheduled castes; Scheduled tribe and other Backward classes and represent one of the weakest sections in the society. A vast chunk of the fisherman population is illiterate and economically not sound. All together 25 fisherman pockets are scattered through the state along the river bank There are 650 fisherman both past time and occasional.

ANIMAL HUSBANDRY: A CASE STUDY OF PARVATHAPUR VILLAGE

Animal Husbanday is one of the main occupation of the village. This village has the largest number of buffaloes in the Ghatkesar Mandal. In this village there are more than 1600 buffaloes. It hold the record of the first place in the Ghatkesar Mandal for the largest number buffaloes in the whole mandal. Their is a veterinary hospital to meet the health requierments of the buffaloes. More than 10,000 liters of milk is supplied to the near by cities Hyderabad and Secunderabad areas.

There are 4 poultry farms in the village. Most of the production is supplied to the near by cities Hyderabad and Secunderab.

In Parvathapur the main crop is "Grass". Because of more number of buffaloes in the village the villagers are growing grass in their lands, to meet the feeding needs of the buffaloes in the village. More over the chemical water that flows in to the river musi that flows along the river is also a reason for growing grass other than any crop. And the farmers are also getting good income of Rs.1200 to 1300 every month.

REFERENCES

Anand, K.J.S. & McGrath, P.J. (Eds) (1993) *Pain in neonates.* Publ: Elsevier Science, Amsterdam.

Back, W. & Clayton, H. (2000) *Equine Locomotion*. Publ: W.B. Sanders, Harcourt Health Sciences, London.

Flecknell, P. & Waterman-Pearson, A.(2000) *Pain Management in Animals*. Publ: W.B.Sanders, Harcourt Health Sciences, London.

Hellebrekers, L.J. (2000) *Animal Pain. A practical-oriented approach to an effective pain control in animals.* Publ: van der Wees Uitgevery, Utrecht, The Netherlands.

McGrath, P.A. (1990) *Pain in Children: Nature Assessment and Treatment*. Publ: Guildford Press, New York.

Sanford, J.; Ewbank, R.; Molony, V.; Tavernor, W.D.; Uvarov, O. (1986) *Guidelines for the recognition and assessment of pain in animals. Veterinary Record* 118, 334-338.

Schmidt, R.F. & Thews, G. (Eds) (1989) *Human Physiology*. Publ: Springer-Verlag, Berlin, New York.

Wall, P.D. & Melzack, R. (Eds) (1994) *Textbook of Pain*. Publ: Churchill Livingstone, Edinburgh.

International Association for the Study of Pain. http://www.iasp-pain.org/pubsopen.html

IASP (1991) *Core curriculum for professional education in pain*. Ed. H.L. Fields. Publ: IASP Publications, Seattle, USA.

Journal References

Abbott, F. V., Franklin, K. B., and Connell, B. (1986). The stress of a novel environment reduces formalin pain: possible role of serotonin. *European Journal of Pharmacology* 126, 141-144.

Abbott, F. V., Franklin, K. B., and Westbrook, R. F. (1995). The formalin test: scoring properties of the first and second phases of the pain response in rats. *Pain* 60, 91-102.

Abbott, F. V., Ocvirk, R., Najafee, R., and Franklin, K. B. (1999). Improving the efficiency of the formalin test. *Pain* 83, 561-569.

Bateson, P. (1991). Assessment of pain in animals. *Animal Behaviour*. 42, 827-839.

Berriatua, E., French, N. P., Broster, C. E., Morgan, K. L., and Wall, R. (2001). Effect of infestation with Psoroptes ovis on the nocturnal rubbing and lying behaviour of housed sheep. *Applied Animal Behaviour Science* 71, 43-55.

Coderre, T. J., Fundytus, M. E., McKenna, J. E., Dalal, S., and Melzack, R. (1993). The formalin test: a validation of the weighted-scores method of behavioural pain rating. *Pain* 54, 43-50.

Cook, C. J. (2002). Rapid noninvasive measurement of hormones in transdermal exudate and saliva. *Physiology & Behaviour* 75, 169-181.

Corke, M. J. and Broom, D. M. (1999). The behaviour of sheep with sheep scab, Psoroptes ovis infestation. *Veterinary Parasitology*. 83, 291-300.

Dolan, S. K. and Nolan, A. M. (2000). Behavioural evidence supporting a differential role for spinal group I and II metabotropic glutamate receptors in inflammatory hyperalgesia in sheep. *Neuropharmacology* 39, 1132-1138.

Dubuisson, D. and Dennis, S. G. (1977). The formalin test: a quantitative study of the analgesic effects of morphine, meperidine, and brain stem stimulation in rats and cats. *Pain* 4, 161-174.

Earley, B and Crowe, M. A. (2002). Effects of ketoprofen alone or in combination with local anesthetic during the castration of bull calves on plasma cortisol, immunological, and inflammatory responses. *Journal of Animal Science* 80, 1044-1052.

Eckersall, P. D. (2000). Recent advances and future prospects for the use of acute phase proteins as markers of disease in animals. *Revue de Medecine Veterinaire*. 151, 577-584.

Eckersall, P. D., Young, F. J., McComb, C., Hogarth, CJ, Safi, S, Weber, A, McDonald, T, Nolan, A. M., and Fitzpatrick, J. L. (2001). Acute phase proteins in serum and milk from dairy cows with clinical mastitis. *Veterinary Record* 148, 35-41.

Eshraghi, HR, Zeitlin, IJ, Fitzpatrick, J. L., Ternent, H, and Logue, D. N. (1999). The release of bradykinin in bovine mastitis. *Life Sciences* 64, 1675-1687.

Faulkner, P. M. and Weary, D. M. (2000). Reducing pain after dehorning in dairy calves. *Journal of Dairy Science* 83, 2037-2041.

Firth, A. M. and Haldane, S. L. (1999). Development of a scale to evaluate postoperative pain in dogs. *JAVMA* 214, 651-659.

Fisher, A. D., Crowe, M. A., Nuallain, E. M. O., Monaghan, M. L., Prendiville, D. J., OKiely, P., and Enright, W. J. (1997). Effects of suppressing cortisol following castration of bull calves on adrenocorticotropic hormone, in vitro interferon- gamma production, leukocytes, acute-phase proteins, growth, and feed intake. *Journal of Animal Science* 75, 1899-1908.

Fisher, A. D., Knight, T. W., Cosgrove, G. P., Death, A. F., Anderson, C. B., Duganzich, D. M., and Matthews, L. R. (2001). Effects of surgical or banding castration on stress responses and behaviour of bulls. *Australian Veterinary Journal.* 79, 279-284.

Fitzpatrick, J. L., Nolan, A. M., Scott, E. M., Harkins, LS., and Barrett, D. C. (2002). Observers perception of pain in cattle. *Cattle Practice.* 10, 209-212.

Graf, B. and Senn, M. (1999). Behavioural and physiological responses of calves to dehorning by heat cauterization with or without local anaesthesia. *Applied Animal Behaviour Science* 62, 153-171.

Graham, M.J., Kent, J.E. and Molony, V. (1997) Effects of four analgesic treatments on the behavioural and cortisol re-sponses of 3-week old lambs to tail docking. *British Veterinary Journal* 153, 87-97.

Grant, C. (2002) The safety and efficacy of intramuscular Xylazine for pain relief in sheep and lambs. Faculty of Health Sciences, Adelaide University, South Australia. MSc Thesis.

Grant, C. and Upton, R. N. (2001). The anti-nociceptive efficacy of low dose intramuscular xylazine in lambs. *Research in Veterinary Science* 70, 47-50.

Gruys, E., Obwolo, MJ, and Toussaint, MJM (1994). Diagnostic significance of the major acute phase proteins in veterinary clinical chemistry: a review. *Veterinary Bulletin* 64, 1009-1018.

Harkins, LS., Fitzpatrick, J. L., Nolan, A. M., Barrett, D. C., and Scott, E. M. Perception of pain in Sheep. *Sheep Veterinary Society* 57, 17-20. 2002.

Hay, M., Vulin, A., Genin, S., Sales, P. and Prunier, A (2003). Assessment of pain induced by castration in piglets: behavioural and physiological responses over the 5 subsequent days. *Applied Animal Behaviour Science* 82, 201-218.

Holton, L., Reid, J., Scott, E. M., Pawson, P., and Nolan, A. (2001). Development of a behaviour-based scale to measure acute pain in dogs. *Veterinary Record* 148, 525-531.

Holton, L. L., Scott, E. M., Nolan, A., Reid, J., Welsh, E., and Flaherty, D. (1998). Comparison of three methods used for assessment of pain in dogs. *Journal of the American Veterinary Medical Association.* 212, 61-67.

Hosie, B. D., Carruthers, J., and Sheppard, B. W. (1996). Bloodless castration of lambs: results of a questionnaire. *British Veterinary Journal.* 152, 47-55.

International Association for the Study of Pain (IASP) (1979) Pain terms: a list with definitions and notes on usage. *Pain* 6, 249-252.

Jackson, R E., Molony V., and Kent, J. E. (1999) Behavioural effects of chronic neuropathic pain from the tail in lambs. p144. *Proceedings of the 10th World Congress on Pain.* Vienna, Austria, August 1999.

Kania, B. F., Zaremba-Rutkowska, M., and Romanowicz, K. (1999). Experimental intestinal stress induced by duodenal distention in sheep. *Journal of Animal & Feed Sciences.* 8, 233-245.

Kent, J. E. and Goodall, J. (1991) Assessment of an immunotur-bidimetric method for measuring equine serum haptoglobin concen-tration. Equine veterinary J. 23 59-66.

Kent, J.E. (1977) The effect of road transportation on young cattle. University of Liverpool. MSc Thesis.

Kent, J. E. (1992) Acute phase proteins: their use in veterinary diagnosis. Guest Editorial for British Veterinary J. 148 279-281.

Kent, J.E., Molony, V. and Robertson, I.S. (1993) Changes in plasma cortisol concentration in lambs of three ages after three methods of castration and tail docking. *Research in Veterinary Sciences* 55 246-251.

Kent, J.E., Molony, V. and Robertson, I.S. (1995) Comparison of Burdizzo and rubber ring methods of castration and tail dock-ing of lambs. *Veterinary Record* 136 192-196.

Kent, J. E., Molony, V., and Graham, M. J. (1998). Comparison of methods for the reduction of acute pain produced by rubber ring castration or tail docking of week-old lambs. *The Veterinary Journal.* 155, 39-51. (

Kent, J. E., Molony, V., Jackson, R. E., and Hosie, B. D. (1999). Chronic inflammatory responses of lambs to rubber ring castration: are there any effects of age or size of lamb at treatment. *Occasional Publication - British Society of Animal Science.* 23, 160-162.

Kent, J. E., Jackson, R. E, Molony, V., and Hosie, B. D. (2000). Effects of acute pain reduction methods on the chronic inflammatory lesions and behaviour of lambs castrated and tail docked with rubber rings at less than two days of age. *The Veterinary Journal* 160, 33-41. (Abstract)

Kent, J. E., Molony V., and Graham, M. J. (2001a). The effect of different bloodless castrators and different tail docking methods on the response of lambs to the combined Burdizzo rubber ring method of castration. *The Veterinary Journal* 162, 250-254.

Kent, J. E., Meikle, L., Molony V., and McKendrick, I. J. (2001b) Qualitative versus quantitative assessment of an acute pain in lambs. Sheep Veterinary Society 25, 65-66.

Kent, J. E., Thrusfield, M. V., Molony, V., Hosie, B. D., and Sheppard, B. W. (2004). A randomised, controlled field trial of two new techniques for castration and tail docking of lambs less than two days of age. *Veterinary Record* 154, 193-200..

Lester, S. J., Mellor, D. J., and Ward, R. N. (1991a). Effects of repeated handling on the cortisol responses of young lambs castrated and tailed surgically. *New Zealand Veterinary Journal.* 39, 147-149.

Lester, S. J., Mellor, D. J., Ward, R. N., and Holmes, R. J. (1991b). Cortisol responses of young lambs to castration and tailing using different methods. *New Zealand Veterinary Journal.* 39, 134-138.

Lester, S. J., Mellor, D. J., Holmes, R. J., Ward, R. N., and Stafford, K. J. (1996). Behavioural and cortisol responses of lambs to castration and tailing using different methods. *New Zealand Veterinary Journal.* 44, 45-54.

Ley, S. J., Livingston, A., and Waterman, A. E. (1991). Effects of chronic lameness on the concentrations of cortisol, prolactin and vasopressin in the plasma of sheep. *Veterinary Record.* 129, 45-47.

Ley, S. J., Waterman, A. E., Livingston, A., and Parkinson, T. J. (1994). Effect of chronic pain associated with lameness on plasma cortisol concentrations in sheep: a field study. *Research in Veterinary Science.* 57, 332-335.

Liles, J. H. and Flecknell, P. A. (1993a). The effects of surgical stimulus on the rat and the influence of analgesic treatment [see comments]. *British Veterinary Journal* 149, 515-525.

Liles, J. H. and Flecknell, P. A. (1993b). The influence of buprenorphine or bupivacaine on the post-operative effects of laparotomy and bile-duct ligation in rats. *Laboratory Animals* 27, 374-380.

Manson, F. J. and Leaver, J. D. (1988). The influence of concentrate amount on locomotion and clinical lameness in dairy cattle. *Animal Production* 47, 185-190.

Mellor, D. J. and Murray, L. (1989a). Changes in the cortisol responses of lambs to tail docking, castration and ACTH injection during the first seven days after birth. *Research in Veterinary Science.* 46, 392-395.

Mellor, D. J. and Murray, L. (1989b). Effects of tail docking and castration on behaviour and plasma cortisol concentrations in young lambs. *Research in Veterinary Science.* 46, 387-391.

Mellor, D. J., Stafford, K. J., Todd, K. S., Jr., Lowe, TE, Gregory, N. G, Bruce, R. A., and Ward, R. N. (2002). A comparison of catecholamine and cortisol responses of young lambs and calves to painful husbandry procedures. *Australian Veterinary Journal* 80, 228-233.

Melzack, R. (1975). The Mcgill pain questionaire:major properties and scoring methods. *Pain* 1, 277-299.

Molony, V. (1997) Comments on Anand and Craig (Letters to the Editor). *Pain* 70, 293.

Molony, V., Kent, J.E. and Robertson, I.S. (1993a) Behavioural responses of lambs of three ages in the first three hours after three methods of castration and tail docking. *Research in Veterinary Science* 55 236-245.

Molony, V., Kent, J. E., Fleetwood-Walker, S. M., Munro, F., and Parker, R. M. C. (1993b) Effects of Xylazine and L659874 on behaviour of lambs after tail docking. p80. Proceedings of the 7th IASP World Congress on pain, Paris.

Molony, V., Kent, J.E. and Robertson, I.S. (1995) Assessment of acute and chronic pain after different methods of castration of calves. *Applied Animal Behavioural Sciences* 46 33-48.

Molony, V. and Kent, J.E. (1997) Assessment of acute pain in farm animals using behavioural and physiological measurements. Journal of Animal Science 75 266-272. Written paper of presentation given by V. Molony at 87th Annual meeting of the American Society of Animal Science, Florida, USA, July 1995.

Molony, V., Kent, J.E., Hosie, B. and Graham, M.J. (1997) Reduction in pain suffered by lambs at castration. *British Veterinary Journal* 153 205-213.

Molony, V., Kent, J. E., and McKendrick, I. J. (2002). Validation of a method for the assessment of an acute pain in lambs. *Applied Animal Behaviour Science* 76, 215-238.

Mostl, E., Maggs, J. L., Schrotter, G., Besenfelder, U., and Palme, R. (2002). Measurement of cortisol metabolites in faeces of ruminants. *Veterinary Research Communications.* 26, 127-139.

Noonan, G. J., Rand, J. S., Priest, J., Ainscow, J., and Blackshaw, J. K. (1994). Behavioural observations of piglets undergoing tail docking, teeth clipping and ear notching. *Applied Animal Behaviour Science.* 39, 203-213.

O'Callaghan, K (2002). Lameness and associated pain in cattle - challenging traditional perceptions. *In Practice* April, 212-219.

Otto, K. A. and Short, C. E (1998). Pharmaceutical control of pain in large animals. *Applied Animal Behaviour Science* 59, 157-169.

Owens, M. E. (1984). Pain in Infancy: Conceptual and methodological issues. *Pain* 20, 213-230.

Peers, A., Mellor, D. J., Wintour, E. M., and Dodic, M (2002). Blood pressure, heart rate, hormonal and other acute responses to rubber ring castration and tail docking of lambs. *New Zealand Veterinary Journal* 50, 56-62.

Pepys, M.B., Baltz, M.L., Tennent, G.A., Kent, J., Ousey, J and Rossdale, P.D. (1989) Serum amyloid A protein (SAA) in horses: objective measurement of the acute phase response. Equine Veterinary J. 21 106-109.

Petrie, N. J., Mellor, D. J., Stafford, K. J., Bruce, R. A., and Ward, R. N. (1996). Cortisol responses of calves to two methods of disbudding used with or without local anaesthetic. *New Zealand Veterinary Journal* 44, 9-14.

Pollard, J. C., Littlejohn, R. P., Johnstone, P, Laas, FJ, Corson, ID, and Suttie, J. M. (1992). Behavioural and heart rate responses to velvet antler removal in red deer. *New Zealand Veterinary Journal* 40, 56-61.

Price, J and Nolan, A. M. (2001). Analgesia of newborn lambs before castration and tail docking with rubber rings. *Veterinary Record* 149, 321-324.

Pritchett, LC, Ulibarri, C, Roberts, MC, Schneider, RK, and Sellon, DC (2003). Identification of potential physiological and behavioural indicators of postoperative pain in horses after exploratory celiotomy for colic. *Applied Animal Behaviour Science* 80, 31-43.

Robertson, I.S., Kent, J.E. and Molony, V. (1994) Effect of different methods of castration on behaviour and plasma cortisol in calves of three ages. *Research in Veterinary Sciences* 56, 8-17.

Ross, M.R. and Dyson, S.J. (2003) Diagnosis and Management of Lameness in Horses. Publ. Sanders, Elsevier Science, USA.

Roughan, J. V. and Flecknell, P. A. (2001). Behavioural effects of laparotomy and analgesic effects of ketoprofen and carprofen in rats. *Pain* 90, 65-74.

Sanford, J.; Ewbank, R.; Molony, V.; Tavernor, W.D.; Uvarov, O. (1986) *Guidelines for the recognition and assessment of pain in animals. Veterinary Record* 118, 334-338.

Schatz, S and Palme, R. (2001). Measurement of faecal cortisol metabolites in cats and dogs: A non-invasive method for evaluating adrenocortical function. *Veterinary Research Communications.* 25, 271-287.

Scott, P. R., Dun, K., Penny, C. D., Strachan, W. D., and Keeling, N. (1996). Field assessment of lamb behaviour after xylazine hydrochloride epidural injection for castration using rubber rings. *Agri-Practice.* 17, 19-22.

Sprecher, D. J., Hostetler, D. E., and Kaneene, J. B. (1997). A lameness scoring system that uses posture and gait to predict dairy cattle reproductive performance. *Theriogenology.* 47, 1179-1187.

Stafford, K. J., Mellor, D. J., Todd, S. E., Gregory, N. G., Bruce, R. A., and Ward, R. N. (2002). Effects of local anaesthesia or local anaesthesia and nonsteroidal anti-inflammatory drug on cortisol responses of calves to castration by five different methods. *Research in Veterinary Science* 73, 61-70.

Stafford, K. J., Mellor, D. J., Todd, S. E., Ward, R. N. and McMeekan, C.M. (2003) The effect of different combinations of lignocaine, ketoprofen, xylazine and tolazoline on the acute cortisol response to dehorning in calves. *New Zealand Veterinary Journal* 51, 219-226. (Abstract)

Sutherland, M. A., Mellor, D. J., Stafford, K. J., Gregory, N. G., Bruce, R. A., and Ward, R. N. (2002a). Modification of cortisol responses to dehorning in calves using 5-hour local anaesthetic regimen plus phenylbutazone, ketoprofen or adrenocorticotropic hormone injected prior to dehorning. *Research in Veterinary Science* 73, 115-123.

Sutherland, M. A., Mellor, D. J., Stafford, K. J., Gregory, N. G., Bruce, R. A., and Ward, R. N. (2002b). Effect of local anaesthetic combined with wound cauterization on the cortisol response to dehorning in calves. *Australian Veterinary Journal* 80, 165-167.

Sylvester, S. P., Stafford, K. J., Mellor, D. J., Bruce, R. A., and Ward, R. N. (1998). Acute cortisol responses of calves to four methods of dehorning by amputation. *Australian Veterinary Journal.* 76, 123-126.

Tarbourton, IS, Bray, AR, and Wilson, JA (2002). Incidence and perceptions of cryptorchid lambs in 2000. *Proceedings of the New Zealand Society of Animal Production* 62, 334-336.

Thornton, P. D. and Waterman-Pearson, A. E. (1997). Castration in young lambs produces changes in mechanical nociceptive threshold responses and behaviour as assessed by a dynamic and interactive visual analogue scale. *Journal of Veterinary Anaesthesia.* 24, 41.

Thornton, P. D. and Waterman-Pearson, A. E. (2002). Behavioural responses to castration in lambs. *Animal Welfare* 11, 203-212.

Wallace MS; et al. (1990) Gastric ulceration in the dog secondary to the use of non-steroid anti-inflammatory drugs. Journal of American Animal Hospital Association 26, 467-472.

Weary, D. M., Braithwaite, L. A., and Fraser, D. (1998). Vocal response to pain in piglets. *Applied Animal Behaviour Science* 56, 161-172.

Wells, S. J., Trent, A. M., Marsh, W. E., and Robinson, R. A. (1993). Prevalence and severity of lameness in lactating dairy cows in a sample of Minnesota and Winsconsin herds. *Journal American Veterinary Medicine Association* 202, 78-82.

Welsh, E. M., Gettinby, G., and Nolan, A. M. (1993). Comparison of a visual analogue scale and a numerical rating scale for assessment of lameness, using sheep as a model. *American Journal of Veterinary Research.* 54, 976-983.

Welsh, E. M. and Nolan, A. M. (1995). Effect of flunixin meglumine on the thresholds to mechanical stimulation in healthy and lame sheep. *Research in Veterinary Science.* 58, 61-66.

Welsh, E. M. and Nolan, A. M. (1995). The effects of abdominal surgery on thresholds to thermal and mechanical stimulation in sheep. *Pain* 60, 159-166.

Wemelsfelder, F., Hunter, E. A. Mendl, M. T. Lawrence, A. B (2000)The spontaneous qualitative assessment of behavioural expressions in pigs: first explorations of a novel methodology for integrative animal welfare measurement. *Applied Animal Behaviour Science.* 67, 193-215

Wesselmann, U. and Lai, J (1997) Mechanisms of referred visceral pain: uterine inflammation in the adult virgin rat results in neurogenic plasma extravasation in the skin. *Pain* 73, 309-317

White, R. G., DeShazer, J. A., Tressler, C. J., Borcher, G. M., Davey, S., Waninge, A., Parkhurst, A. M., Milanuk, M. J., and Clemens, E. T. (1995). Vocalization and physiological response of pigs during castration with or without a local anesthetic. *Journal of Animal Science.* 73, 381-386.

Woodbury, M. R., Caulkett, N. A., Baumann, D., and Read, M. R. (2001). Comparison of analgesic techniques for antler removal in wapiti. *Canadian Veterinary Journal-Revue Veterinaire Canadienne* 42, 929-935.

Zimmermann, M. (1986) Behavioural investigations of pain in animals. In: Assessing pain in farm animals. Eds. I.J.H.Duncan and Molony V. pp16-27. Luxembourg, Commission of the European Communities.

3

Global Dairy Science and Technology: An Introductory Overview

INTRODUCTION

This illustration is about the study of milk and milk-derived food products from a food science perspective. It focuses on the biological, chemical, physical, and microbiological aspects of milk itself, and on the technological (processing) aspects of the transformation of milk into its various consumer products, including beverages, fermented products, concentrated and dried products, butter and ice cream. Milk is as ancient as mankind itself, as it is the substance created to feed the mammalian infant. All species of mammals, from man to whales, produce milk for this purpose. Many centuries ago, perhaps as early as 6000-8000 BC, ancient man learned to domesticate species of animals for the provision of milk to be consumed by them. These included cows (genus Bos), buffaloes, sheep, goats, and camels, all of which are still used in various parts of the world for the production of milk for human consumption. Fermented products such as cheeses were discovered by accident, but their history has also been documented for many centuries, as has the production of concentrated milks, butter, and even ice cream. Technological advances have only come about very recently in the history of milk consumption, and our generations will be the ones credited for having turned milk processing from an art to a science. The availability and distribution of milk and milk products today in the modern world is a blend of the centuries old knowledge of traditional milk products with the application of modern science and technology. The role of milk in the traditional diet has varied greatly in different regions of the world. The tropical countries have not been traditional milk consumers, whereas the more northern regions of the world, Europe (especially Scandinavia) and North America, have traditionally consumed far more milk and milk products in their diet. In tropical countries where high temperatures and lack of refrigeration has led to the inability to produce and store fresh milk, milk has traditionally been preserved through means other than refrigeration, including immediate consumption of warm milk after milking, by boiling milk, or by conversion into more stable products such as fermented milks.

World-wide Milk Consumption and Production

The total milk consumption (as fluid milk and processed products) per person varies widely from highs in Europe and North America to lows in Asia. However, as the various regions of the world become more integrated through travel and migration, these trends are changing, a factor which needs to be considered by product developers and marketers of milk and milk products in various countries of the world.

Even within regions such as Europe, the custom of milk consumption has varied greatly. Consider for example the high consumption of fluid milk in countries like Finland, Norway and Sweden compared to France and Italy where cheeses have tended to dominate milk consumption. When you also consider the climates of these regions, it would appear that the culture of producing more stable products (cheese) in hotter climates as a means of preservation is evident. Table 1 illustrates milk per capita consumption information from various countries of the world. Table 2 shows the quantity of raw milk produced around the world.

TABLE 1

Per Capita Consumption of Milk and Milk Products in Various Countries, 2006 Data

Country	*Liquid Milk Drinks (Litres)*	*Cheeses (kg)*	*Butter (kg)*
Finland	183.9	19.1	5.3
Sweden	145.5	18.5	1.0
Ireland	129.8	10.5	2.9
Netherlands	122.9	20.4	3.3
Norway	116.7	16.0	4.3
Spain (2005)	119.1	9.6	1.0
Switzerland	112.5	22.2	5.6
United Kingdom (2005)	111.2	12.2	3.7
Australia (2005)	106.3	11.7	3.7
Canada (2005)	94.7	12.2	3.3
European Union (25 countries)	92.6	18.4	4.2
Germany	92.3	22.4	6.4
France	92.2	23.9	7.3
New Zealand (2005)	90.0	7.1	6.3
United States	83.9	16.0	2.1

Austria	80.2	18.8	4.3
Greece	69.0	28.9	0.7
Argentina (2005)	65.8	10.7	0.7
Italy	57.3	23.7	2.8
Mexico	40.7	2.1	N/A
China (2005)	8.8	N/A	N/A

Source: International Dairy Federation, Bulletin 423/2007.

TABLE 2

Cow Milk Production ('000 Tonnes) in Selected Countries in the World (2006)

United States	82,462
India	39,759
China	31,934
Russia	31,100
Germany	27,955
Brazil	25,750
France	24,195
New Zealand	15,000
United Kingdom	14,359
Ukraine	13,287
Poland	11,970
Italy	11,186
Netherlands	10,995
Mexico	10,352
Argentina	10,250
Turkey	10,000
Australia	9550
Canada	7854

Source: International Dairy Federation, Bulletin 423/2007.

Milk Composition

The role of milk in nature is to nourish and provide immunological protection for the mammalian young. Milk and honey are the only articles of diet whose sole function in nature is food. It is not surprising, therefore, that the nutritional value of milk is high.

TABLE 3

Composition of Milk from Different Mammalian Species (Per 100 g Fresh Milk)

	Protein (g)	*Fat (g)*	*Carbohydrate (g)*	*Energy (kcal)*
Cow	3.2	3.7	4.6	66
Human	1.1	4.2	7.0	72
Water Buffalo	4.1	9.0	4.8	118
Goat	2.9	3.8	4.7	67
Donkey	1.9	0.6	6.1	38
Elephant	4.0	5.0	5.3	85
Monkey, rhesus	1.6	4.0	7.0	73
Mouse	9.0	13.1	3.0	171
Whale	10.9	42.3	1.3	443
Seal	10.2	49.4	0.1	502

Source: Webb, B.H., A.H. Johnson and J.A. Alford. 1974. Fundamentals of Dairy Chemistry. Second Ed. AVI Publishing Co., Westport, CT., Chap. 1.

TABLE 4

Gross Composition of Milk of Various Breeds, g/100g

	Body Wt. (kg)	*Milk Yield (kg)*	*Fat (%)*	*Protein (%)*	*Lactose (%)*	*Ash (%)*	*Total Solids (%)*
Holstein	640	7360	3.54	3.29	4.68	0.72	12.16
Brown Swiss	640	6100	3.99	3.64	4.94	0.74	13.08
Ayrshire	520	5760	3.95	3.48	4.60	0.72	12.77
Guernsey	500	5270	4.72	3.75	4.71	0.76	14.04
Jersey	430	5060	5.13	3.98	4.83	0.77	14.42
Shorthorn	530	5370	4.00	3.32	4.89	0.73	12.9

Holstein: 12.16% T.S. x 7360 kg/lactation = 895 kg of total solids produced/lactation (140% of her body wt.!)

Jersey: 14.42% T.S. x 5060 kg/lactation = 730 kg of total solids produced/lactation (170% of her body wt.!)

MODERN DAY DAIRY FARMING

Dairy farmers work hard every day to bring you fresh, great tasting, wholesome milk products. Almost all dairies are family-owned, and as active members of their communities, farm families take pride in maintaining natural resources. That means preserving the land where they live and work, protecting the air and water they share with neighbors, and providing the best care for their cows—the lifeblood of their business.

TOWARDS UNDERSTANDING SCIENCE OF MODIFIED ATMOSPHERE PACKING

Many foods spoil rapidly in air due to moisture loss or uptake, reaction with oxygen and the growth of aerobic micro-organisms i.e., bacteria and moulds. Microbial growth results in changes in texture, colour, flavour and nutritional value of the food. These changes can render food unpalatable and potentially unsafe for human consumption. Storage of foods in a modified gaseous atmosphere can maintain quality and extend product shelf life by slowing chemical and biochemical deteriorative reactions and by slowing or in some instances preventing the growth of spoilage organisms.

Modified atmosphere packaging (MAP) is defined as "the packaging of a perishable product in an atmosphere which has been modified so that its composition is other than that of air" (Hintlian and Hotchkiss, 1986). Whereas controlled atmosphere storage involves maintaining a fixed concentration of gases surrounding the product by careful monitoring and addition of gases, the gaseous composition of fresh MAP foods is constantly changing due to chemical reactions and microbial activity. Gas exchange between the pack headspace and the external environment may also occur because of permeation across the package material. Packing foods in a modified atmosphere can offer extended shelf life and improved product presentation in a convenient container, making the product more attractive to the retail customer. However, MAP cannot improve the quality of a poor quality food product. It is therefore essential that the food be of the highest quality prior to packing in order to optimise the benefits of modifying the pack atmosphere. Good hygiene practices and temperature control throughout the chill-chain for perishable products are required to maintain the quality benefits and extended shelf life of MAP foods.

Gases Used in Modified Atmosphere Packaging

The three main gases used in modified atmosphere packaging are O2, CO2 and N2. The choice of gas is very dependent upon the food product being packed. Used singly or in combination, these gases are commonly used to balance safe shelf life extension with optimal organoleptic properties of the food. Noble or 'inert' gases such as argon are in commercial use for products such as coffee and snack products; however, the literature on their application and benefits is limited. Experimental use of carbon monoxide (CO) and sulphur dioxide (SO2) has also been reported.

Carbon Dioxide

Carbon dioxide is a colourless gas with a slight pungent odour at very high concentrations. It is an asphyxiant and slightly corrosive in the presence of moisture. CO2 dissolves readily in water (1.57 g/kg @ at 100 kPa, 20° C) to produce carbonic acid (H_2CO_3) that increases the acidity of the solution and reduces the pH. This gas is also soluble in lipids and some other organic compounds. The solubility of CO_2 increases with decreasing temperature. For this reason, the antimicrobial activity of CO_2 is markedly greater at temperatures below 10° C than at 15° C or higher. This has significant implications for MAP of foods. The high solubility of CO_2 can result in pack collapse due the reduction of headspace volume. In some MAP applications, pack collapse is favoured, for example in flow wrapped cheese for retail sale.

Oxygen

Oxygen is a colourless, odourless gas that is highly reactive and supports combustion. It has a low solubility in water (0.040 g/kg at 100 kPa, 20° C). Oxygen promotes several types of deteriorative reactions in foods including fat oxidation, browning reactions and pigment oxidation. Most of the common spoilage bacteria and fungi require oxygen for growth. Therefore, to increase shelf life of foods the pack atmosphere should contain a low concentration of residual oxygen. It should be noted that in some foods a low concentrations of oxygen can result in quality and safety problems (for example unfavourable colour changes in red meat pigments, senescence in fruit and vegetables, growth of food poisoning bacteria) and this must be taken into account when selecting the gaseous composition for a packaged food.

Nitrogen

Nitrogen is a relatively un-reactive gas with no odour, taste, or colour. It has a lower density than air, non-flammable and has a low solubility in water (0.018 g/kg at 100 kPa, 20° C) and other food constituents. Nitrogen does not support the growth of aerobic microbes and therefore inhibits the growth aerobic spoilage but does not prevent the growth of anaerobic bacteria. The low solubility of nitrogen in foods can be used to prevent pack collapse by including sufficient N_2 in the gas mix to balance the volume decrease due to CO_2 going into solution.

Carbon Monoxide

Carbon monoxide is a colourless, tasteless and odourless gas that is highly reactive and very flammable. It has a low solubility in water but is relatively soluble in some organic solvents. CO has been studied in the MAP of meat and has been licensed for use in the USA to prevent browning in packed lettuce. Commercial application has been limited because of its toxicity and the formation of potentially explosive mixtures with air.

Noble Gases

The noble gases are a family of elements characterised by their lack of reactivity and include helium (He), argon (Ar), xenon (Xe) and neon (Ne). These gases are being used in a number of food applications now e.g. potato-based snack products. While from a scientific perspective, it is difficult to see how the use of noble gases would offer any preservation advantages compared with N_2 they are being used.

Dairy Products

MAP has the potential to increase the shelf life of a number of dairy products. These include fat-filled milk powders, cheeses and fat spreads. In general these products spoil due to the development of oxidative rancidity in the case of powders and or the growth of micro-organisms, particularly yeasts and moulds, in the case of cheese. Whole milk powder is particularly susceptible to the development of off-flavours due to fat oxidation. Commercially the air is removed under vacuum and replaced with N_2 or N_2/CO_2 mixes and the powder is hermetically sealed in metal cans. Due to the spray drying process air tends to be absorbed inside the powder particles and will diffuse into the container over a period of 10 days or so. This typically will raise the residual O_2 content to 1%-5% or higher. Because some markets require product with low levels of residual O_2(<1%) some manufacturers re-pack the cans after 10 days storage. Use of O_2 scavenging may also be useful. English territorial cheeses e.g. Cheddar have traditionally been vacuum packed. Increasingly MAP is being used with high CO_2 concentration CO_2/N_2 gas mixes. This has the advantage of obtaining a low residual O_2 content and a tight pack due to the CO_2 going into solution. It is important to balance this process using the correct N_2 level in the gas mix so as to avoid excessive pressure on the pack seal. Use of N_2 / CO_2 atmospheres has significant potential for extending the shelf life of cottage cheese. The latter is a high moisture, low fat product that is susceptible to a number of spoilage organisms including Pseudomonas spp. Use of gas mixtures containing CO_2 balanced with N_2 can increase the shelf life significantly.

Raw Red Meat

Microbial growth and oxidation of the red oxymyoglobin pigment are the main spoilage mechanisms that limit the shelf life of raw red meats. The packaging technologist has to maintain the desirable red colour of the oxymyoglobin pigment, by having an appropriate O_2- concentration in the pack atmosphere, and at the same time minimise the growth of aerobic micro-organisms. Aerobic spoilage bacteria, such as Pseudomonas species normally constitute the major flora on red meats. Since these bacteria are inhibited by CO_2 it is possible to achieve both red colour stability and microbial inhibition by using gas mixtures containing CO_2 and O_2. These mixtures can extend the chilled shelf-life of red meats from 2-4 days to 5-8 days. For information on the gas compositions and the recommended gas/product ratio please see Food Packaging Technology. The maintenance of recommended chilled temperatures and good hygiene and handling practices throughout the butchery,

MAP, distribution and retailing chain is of critical importance in ensuring both the safety and extended shelf-life of red meat products. Because raw red meats are cooked before consumption the risk of food poisoning can be greatly reduced by proper cooking.

Raw Poultry

Microbial growth, particularly of Pseudomonas and Achromobacter species, is the major factor limiting the shelf life of raw poultry. These Gram-negative aerobic spoilage bacteria are effectively inhibited by CO_2. Consequently the inclusion of CO_2 in MAP at a concentration in excess of 20% can significantly extend the shelf-life of raw poultry products. Since poultry meat provides a good medium for the growth of pathogenic micro-organisms, including some that are not inhibited by CO_2, it is critical that recommended chilled temperatures, good hygiene and handling practices throughout the supply chain are adhered to and that products are properly cooked prior to consumption. Early research into gas mixes for MAP of poultry meat reported discolouration of the meat at carbon dioxide concentrations higher than 25%. This research is at variance with the lack of problems reported from the commercial use of relatively high levels of CO_2 with meat products, up to 100% with some products. It would appear that the problems that have been occasionally encountered with high levels of CO_2 may simply be due to high residual levels of O_2 refer to Food Packaging Technology for further information. Research into the optimal gas composition and package type and size should be conducted for individual food products.The ratio of headspace pack volume to food product volume is also important as are the types and thickness of the package material and the package design. Shelf life evaluations must reflect the conditions from manufacture to consumption of the product. It may also be necessary to consider the effect of pack opening on the subsequent shelf life of the product.

Cooked, Cured and Processed Meat Products

The principal spoilage mechanisms that limit the shelf life of cooked, cured and processed meat products are microbial growth, colour change, and oxidative rancidity. For cooked meat products, the heating process should kill vegetative bacterial cells, inactivate degradative enzymes, and fix the colour. Consequently, spoilage of cooked meat products is primarily due to post-process contamination by micro-organisms as a result of poor hygiene and handling practices. The colour of cooked meats is susceptible to oxidation and it is important have only low levels of residual O_2 in packs. MAP using CO_2/N_2 mixes along with a recommended gas/product ratio are given in and if used will maximise shelf life and inhibit the development of oxidative off flavours and rancidity. Processed meat products such as sausages, frankfurters and beef burgers generally contain sodium metabisulphite, which is an effective preservative against a wide range of spoilage micro-organisms and pathogens.

Cooked, cured and processed meat products containing high levels of unsaturated fat are liable to be spoiled by oxidative rancidity, but MAP with CO_2/N_2 mixtures is effective at inhibiting this undesirable reaction. Potential food poisoning hazards are primarily due microbial contamination or growth resulting from post-cooking, curing or processing contamination. These can be minimised by using recommended chilled temperatures, good hygiene and handling practices. The low aw and addition of nitrite in cooked, cured and processed meat products inhibits many food poisoning bacteria, particularly Cl. botulinum. This inhibition may be compromised in products formulated with lower concentrations of chemical preservatives than those used in traditional foods. The potential effects of any changes in product formulation on growth and survival of pathogens should always be considered. Cooked meats stored without any added preservatives will be at risk from growth of Cl. botulinum under anaerobic MAP conditions particularly when held at elevated storage temperatures.

Fish and Fish Products

There has been a very significant increase in the sale of MAP fish products in Europe and particularly in the UK. Nevertheless packaging technologists should be aware of a major concern limiting the development of MAP for this product group, namely Cl. botulinum. There is also debate about the cost-benefits of MAP since in some applications only relatively small increases in safe shelf life have been reported. Spoilage of fish results in the production of low molecular weight volatile compounds therefore packaging technologists need to consider the odour barrier properties of packaging films and select appropriate high barrier materials for packaging strong flavoured fresh, smoked and brined fish and fish products. Spoilage of fish and shellfish results from changes caused by three major mechanisms (i) the breakdown of tissue by the fish's own enzymes (autolysis of cells), (ii) growth of micro-organisms, and (iii) oxidative reactions. MAP can be used to control mechanisms (ii) and (iii) but has no direct effect on autolysis. Generally the major spoilage bacteria found on processed fish are aerobes including Pseudomonas, Moraxella, Acinetobacter, Flavobacterium and Cytophaga species. There are several micro-organisms that are of particular importance when dealing with MAP fish products, these include Cl. botulinum. Use of CO_2 can effectively inhibit the growth of some of these species. The aerobic spoilage organisms tend to be replaced by slower growing and, less odour producing, bacteria particularly lactic acid bacteria such as lactobacilli during storage. Because fish and shellfish contain much lower concentrations of myoglobin, the oxidation status of this pigment is less important than in other meats. Because of the high moisture content and the lipid content of some species N_2 is used to prevent pack collapse.One of the concerns about MAP of fish is that removal of O_2 and its replacement by N_2 or N_2/CO_2 results in anaerobic conditions that are conducive to the growth of protease-negative strains of Cl. botulinum. Because these bacteria can grow at temperatures as low as 3oC and do not significantly alter the sensory properties of the fish, there is the potential for food poisoning that can lead to fatalities. While there is no evidence that CO2 promotes growth of psychotropic strains of Cl. botulinum there is, as discussed previously, some concerns about CO2 promoting the germination of spores of this organism. Considerable research has been undertaken to assess, and to control the risks associated with the growth of Cl. botulinum in MAP of fish and other products. The Advisory Committee on the Microbiological Safety of Food (ACMSF) (Anon, 1992) have recommended controlling factors that should be used singly or in combination to prevent the growth of, and toxin production in, prepared chilled food by psychotropic Cl. botulinum. Some fish processors include O2 in their MAP to further reduce the risk of growth of clostridia. Since botulinum toxin is relatively heat sensitive, correct cooking of seafood should eliminate any problem with preformed toxin.

Modified Atmosphere Packaging of Fruits and Vegetables

Consumers now expect 'fresh' fruit and vegetable produce throughout the year. MAP has the potential to extend the safe shelf life of many fruits and vegetables. Packaging fresh and unprocessed fruit and vegetables poses many challenges for packaging technologists. Unlike other chilled perishable foods, fresh produce continues to respire after harvesting. The products of aerobic respiration include CO2 and water vapour. In addition, respiring fruits and vegetables produce C2H4 that promotes ripening and softening of tissues. The latter if not controlled will limit shelf life. Respiration is affected by the intrinsic properties of fresh produce as well as various extrinsic factors including ambient temperature. It is accepted that the potential shelf life of packed produce is inversely proportional to respiration rate. Respiration rate increases by a factor of 3-4 for every 10oC increase in temperature. Hence the goal of modified atmosphere packaging for fruits and vegetables is to reduce respiration to extend shelf life while maintaining quality. Respiration can be reduced by lowering temperature, lowering the O2 concentration, increasing the CO2 concentration and by the combined use of O2 depletion and CO2 enhancement of pack atmospheres. If the O2 concentration is reduced beyond a critical concentration, this is dependant on the species and cultivar, then anaerobic respiration will be initiated. Anaerobic respiration, or anaerobiosis, is usually associated with undesirable odours and flavours and a marked deterioration in product quality. While increasing the CO2 concentration will also inhibit respiration, high concentrations may cause damage in some species and cultivars.The use of low concentrations of O2 and elevated levels of CO2 can have a synergistic effect on slowing down respiration, and indirectly, ripening. While the mechanisms whereby MAP can extend the shelf life of fresh produce are not fully understood it is known that the low O2/high CO2 conditions reduce the conversion of chlorophyll to pheopytin, decrease the sensitivity of plant tissue to C2H4, inhibit the synthesis of carotenoids, reduce oxidative browning and discolouration and inhibit the growth of micro-organisms. These mechanisms are all temperature dependant. Packaging technologists should be aware of several major pathogens as far as MAP fresh produce is concerned, in particular L. monocytogenes and Cl. botulinum. As previously discussed L. monocytogenes can grow under reduced O2 levels and is not markedly inhibited by CO2. This combined with its ability to grow at temperatures close to 0oC helps to explain the concern.The use of MAP atmospheres containing low concentrations of O2 and elevated CO2 concentrations may permit the growth of psychotropic protease-negative strains of Cl. botulinum. However, provided packs are stored at 3oC or below for not more than 10 days there is unlikely to be a problem with clostridia. Temperature control is critical since temperature abuse could lead to pack contents becoming toxic. The environment in which fruits and vegetables are grown may harbour pathogens including Salmonella species, enterotoxegenic E. coli and viruses. While these micro-organisms may not grow in MAP packs, particularly if the storage temperature is maintained around 3o C, they may survive throughout storage and could cause food poisoning through cross contamination in the home or due to the consumption for raw or under processed product. Hygienic preparation, sanitation in chilled-chlorinated water, rinsing and dewatering prior to MAP are now considered essential treatments to fruits and vegetables prior to packaging to ensure low microbial counts and assure safety.

Since there is a risk of anaerobic pathogens, such as Cl. botulinum growing in MAP packs, a minimum level of O2 (e.g. 2-3%) is usually recommended to ensure that potentially hazardous conditions are not created.

MODERN AGRICULTURE AND DAIRY FARMING

Until about four decades ago, crop yields in agricultural systems depended on internal resources, recycling of organic matter, built-in biological control mechanisms and rainfall patterns. Agricultural yields were modest, but stable. Production was safeguarded by growing more than one crop or variety in space and time in a field as insurance against pest outbreaks or severe weather. Inputs of nitrogen were gained by rotating major field crops with legumes. In turn rotations suppressed insects, weeds and diseases by effectively breaking the life cycles of these pests. A typical corn belt farmer grew corn rotated with several crops including soybeans, and small grain production was intrinsic to maintain livestock. Most of the labor was done by the family with occasional hired help and no specialized equipment or services were purchased from off-farm sources. In these type of farming systems the link between agriculture and ecology was quite strong and signs of environmental degradation were seldom evident.

But as agricultural modernization progressed, the ecology-farming linkage was often broken as ecological principles were ignored and/or overridden. In fact, several agricultural scientists have arrived at a general consensus that modern agriculture confronts an environmental crisis. A growing number of people have become concerned about the long-term sustainability of existing food production systems. Evidence has accumulated showing that whereas the present capital- and technology-intensive farming systems have been extremely productive and competitive, they also bring a variety of economic, environmental and social problems.

Evidence also shows that the very nature of the agricultural structure and prevailing policies have led to this environmental crisis by favouring large farm size, specialized production, crop monocultures and mechanization. Today as more and more farmers are integrated into international economies, imperatives to diversity disappear and monocultures are rewarded by economies of scale. In turn, lack of rotations and diversification take away key self-regulating mechanisms, turning monocultures into highly vulnerable agroecosystems dependent on high chemical inputs.

Expansion of Monocultures

Today monocultures have increased dramatically worldwide, mainly through the geographical expansion of land devoted to single crops and year-to-year production of the same crop species on the same land. Available data indicate that the amount of crop diversity per unit of arable land has decreased and that croplands have shown a tendency toward concentration. There are political and economic forces influencing the trend to devote large areas to monoculture, and in fact such systems are rewarded by economies of scale and contribute significantly to the ability of national agricultures to serve international markets. The technologies allowing the shift toward monoculture were mechanization, the improvement of crop varieties, and the development of agrochemicals to fertilize crops and control weeds and pests. Government commodity policies these past several decades encouraged the acceptance and utilization of these technologies. As a result, farms today are fewer, larger, more specialized and more capital intensive. At the regional level, increases in monoculture farming meant that the whole agricultural support infrastructure (i.e. research, extension, suppliers, storage, transport, markets, etc.) has become more specialized. From an ecological perspective, the regional consequences of monoculture specialization are many-fold:

a. Most large-scale agricultural systems exhibit a poorly structured assemblage of farm components, with almost no linkages or complementary relationships between crop enterprises and among soils, crops and animals.

b. Cycles of nutrients, energy, water and wastes have become more open, rather than closed as in a natural ecosystem. Despite the substantial amount of crop residues and manure produced in farms, it is becoming increasingly difficult to recycle nutrients, even within agricultural systems. Animal wastes cannot economically be returned to the land in a nutrient-recycling process because production systems are geographically remote from other systems which would complete the cycle. In many areas, agricultural waste has become a liability rather than a resource. Recycling of nutrients from urban centers back to the fields is similarly difficult.

c. Part of the instability and susceptibility to pests of agroecosystems can be linked to the adoption of vast crop monocultures, which have concentrated resources for specialist crop herbivores and have increased the areas available for immigration of pests. This simplification has also reduced environmental opportunities for natural enemies. Consequently, pest outbreaks often occur when large numbers of immigrant pests, inhibited populations of beneficial insects, favourable weather and vulnerable crop stages happen simultaneously.

d. As specific crops are expanded beyond their "natural" ranges or favourable regions to areas of high pest potential, or with limited water, or low-fertility soils, intensified chemical controls are required to overcome such limiting factors. The assumption is that the human intervention and level of energy inputs that allow these expansions can be sustained indefinitely.

e. Commercial farmers witness a constant parade of new crop varieties as varietal replacement due to biotic stresses and market changes has accelerated to unprecedented levels. A cultivar with improved disease or insect resistance makes a debut, performs well for a few years (typically 5-9 years) and is then succeeded by another variety when yields begin to slip, productivity is threatened, or a more promising cultivar becomes available. A variety's trajectory is characterized by a take-off phase when it is adopted by farmers, a middle stage when the planted area stabilizes and finally a retraction of its acreage. Thus, stability in modern agriculture hinges on a continuous supply of new cultivars rather than a patchwork quilt of many different varieties planted on the same farm.

f. The need to subsidize monocultures requires increases in the use of pesticides and fertilizers, but the efficiency of use of applied inputs is decreasing and crop yields in most key crops are leveling off. In some places, yields are actually in decline. There are different opinions as to the underlying causes of this phenomenon. Some believe that yields are leveling off because the maximum yield potential of current varieties is being approached, and therefore genetic engineering must be applied to the task of redesigning crop. Agroecologists, on the other hand, believe that the leveling off is because of the steady erosion of the productive base of agriculture through unsustainable practices.

Impacts of Climate Change

The risks associated with climate change lie in the interaction of several systems with many variables that must be collectively considered. Agriculture (including crop agriculture, animal husbandry, forestry and fisheries) can be defined as one of the systems, and climate the other. If these systems are treated independently, this would lead to an approach which is too fragmentary. The issue is more global. It is now held as likely that human activities can affect climate, one of the components of the environment. Climate in turn affects agriculture, the source of all food consumed by human beings and domestic animals. It must be further considered that not only climate may be changing, but that human societies and agriculture develop trends and constraints of their own which climate change impact studies must take into consideration.

An expert meeting held at FAO Headquarters in Rome from 7 to 10 December 1993 considered the direct effects of changing hydrological, pedological and plant physiological processes on agricultural production and concentrated on mechanisms. Firstly, sustainability of agricultural and rural development. How will the links between environmental resources and demography be affected in the coming 50 years? Will it be possible, at the same time, to increase food production without irremediably losing environmental resources like soils or biodiversity? Secondly, improved food security and nutrition, two members of a spiral which also includes rural poverty and demographic pressure. What are the prospects of breaking the vicious poverty circle in many developing countries under changing climate conditions?

World Agricultural Context

In general, global food production has been growing faster than human population. There are, however, marked disparities between continents. For example, in Africa local production of cereals cannot keep pace with population increase, and production of root and tuber crops is growing faster than that of the nutritionally more valuable cereals.

In contrast, arable land growth lags behind population growth, which indicates some intensification of production. In Asia, the upper limit of available land has been reached in several countries, resulting in very high cropping intensities and a dominant role for irrigation. In Latin America, the increase in arable land is achieved only at a high ecological cost (especially deforestation) which may have direct relevance to climate change.

According to a recent FAO prospective study covering the years 1988/1990 to 2010 (*World Agriculture: Towards 2010 or AT-2010;* FAO, 1995a), 'the rate of growth of agricultural land will be further reduced during the next two decades. The pressures on *fresh* water *resources,* however, will be considerable, as will be those on the environment arising from the intensification of land use'.

A significant reduction in world population growth rates is foreseen from 1.8% (1980/1990) to 1.4% (2000/2010), roughly equivalent to an increase of the population doubling time from 40 to 50 years. The projected continuation of the high population growth in Africa can be related to slow economic development during the coming two decades, since in general a reduction of poverty precedes reduction of population growth.

AT-2010 also lists the following significant trends for the near future: (i) world production of cereals will continue to grow, but not in per capita terms; (ii) export cereals will undergo a modest growth in demand; (iii) the livestock sector in developing countries will continue to grow; (iv) root crops, tubers and plantains will retain their importance; (v) oil crops will undergo rapid growth in developing countries; and, most importantly, (vi) many developing countries will become net agricultural importers.

As indicated above for population increase, large differences exist in the rate of agricultural development among the developing continents, while differences among the developed countries tend to level out.

It is also likely that hunger and under-nutrition, which currently affects 800 million people (about 20% of the 4 000 million inhabitants of developing countries), will be reduced, but large pockets of malnutrition will persist. Pressure on environmental resources will continue to build up.

Climate constitutes a complex of inter-related variables. On average, through a set of regulatory mechanisms, a smooth change in one variable triggers smooth changes in most others. With the exception of possible qualitative and abrupt variations, which will be mentioned below, such inter-relations are independent of atmospheric carbon dioxide (CO_2). The latter and other greenhouse gases play a part largely through their effect on the radiation balance of the atmosphere. There is only a weak link between such factors as cloudiness and wind. Temperature, evaporation and rain are strongly correlated, which illustrates the likely intensification of the hydrological cycle. Combined with the projected pressure on land and water use, competition for land and water will certainly become a key social and political issue. Climate variability is likely to increase under global warming (Katz and Brown, 1992), both in absolute and in relative terms. This is linked with thresholds which affect the occurrence of many meteorological phenomena. For instance, tropical cyclones are 'fed' by water vapour evaporating from oceans at a temperature above 26 or 27°C. Therefore, higher average sea surface temperatures are bound to result in a higher frequency of tropical cyclones. The rate of change itself is extremely important. For example, recent work (Lehman, 1993; Paillard and Labeyrie, 1994; Rahmsdorf, 1994; Holmes, 1995) on the saw-tooth temperature changes in the past as observed in the Arctic, raise concern that changes may occur abruptly, with average temperatures changing by 10 or 12°C in just a matter of decades. The mechanism of such changes is not clear as yet, but seems to involve the mechanical stability of ice sheets, and sudden changes in the hydrological cycle and the Atlantic conveyor current. Such changes would, of course, be associated with dramatic changes in the distribution and quantities of ocean products, and cause havoc to established national fishery activities. They would also make adaptation to climate change, together with most agricultural planning, extremely difficult.

In addition to water vapour, important greenhouse gases are carbon dioxide (CO_2), methane (CH_4), nitrous oxide (N_2O), tropospheric ozone (O_3) and chlorofluorocarbons (CFCs). The basic characteristics of the first three gases. The degree to which these greenhouse gases stem from agricultural sources is also given. The exact ratio between these land-use related emissions and those from natural ecosystems (swamps, tundra), fossil fuels (coal, oil, gas) and geological sources (volcanoes) are a bone of contention among industrialized, oil producing and developing countries. Moral rights and duties in relation to the international flow of development and conservation funds are involved. Therefore, hetter estimates based on exact measurements of the net greenhouse gas emissions from agricultural practices in developing countries are urgently required to assess responsibilities properly. The required reductions of emissions to achieve stabilization of atmospheric concentrations of current levels are believed to be >60% for CO_2 15-20% for CH_4 and 70-80% for N_2O. This implies the need for up-to-date and complete information on land cover and land uses per country: not only the kind of crops grown but also their intensity, their rotation and the amount of inputs (energy, fertilizers).

The current trends of some of the agricultural sectors directly associated with greenhouse gas emission sources are listed. The loss of 'natural land', including tropical forests to agriculture, grazing, logging and urbanization, may continue, though at a slower pace (*AT-2010*). Many earlier estimates are now regarded with suspicion, as the borderline between crop agriculture, forest and cattle agriculture appears to be a fuzzy one. Finally, both recent trends of fertilizer use and *AT-2010* projections indicate major increases in consumption. Even with appropriate measures to optimize fertilizer use, it is likely that N_2O losses from fertilizers will continue to increase.

In qualitative terms, many indirect effects of climate change on agriculture can be conjectured. Most of them are estimated to be negative and they catch most of the attention of the media. These effects include:

- the overall predictability of weather and climate would decrease, making the day-to-day and medium-term planning of farm operations more difficult;
- loss of biodiversity from some of the most fragile environments, such as tropical forests and mangroves;
- sea-level rise (40 cm in the coming 100 years) would submerge some valuable coastal agricultural land;
- the incidence of diseases and pests, especially alien ones, could increase;
- present (agro) ecological zones could shift in some cases over hundreds of kilometres horizontally, and hundreds of metres altitudinally, with the hazard that some plants, especially trees, and animal species cannot follow in time, and that farming systems cannot adjust themselves in time;
- higher temperatures would allow seasonally longer plant growth and crop growing in cool and mountainous areas, allowing in some cases increased cropping and production. In contrast, in already warm areas climate change can cause reduced productivity;
- the current imbalance of food production between cool and temperate regions and tropical and subtropical regions could worsen.

The greenhouse gases CH_4, N_2O and chlorofluorocarbons (CFCs) have no known direct effects on plant physiological processes. They only change global temperature and are therefore not discussed further. Instead, concentration should be on the effects of increased CO_2 tropospheric O_3, increased UV-B through depleted stratospheric ozone, increased temperatures and the associated intensification of the hydrological cycle.

CO_2 is an essential plant 'nutrient', in addition to light, suitable temperature, water and chemical elements such as N, P and K, and it is currently in short supply. Higher concentrations of atmospheric CO_2 due to increased use of fossil fuels, deforestation

A total of 10 to 20% of the approximate doubling of crop productivity over the past 100 years could be due to this effect (Tans *et al.*, 1990) and forest growth or regrowth may have been stimulated as well. Further productivity increases may occur in the coming century, in the order of 30% or more where plant nutrients and moisture are adequate. Higher CO_2 values would also mitigate the plant growth damage caused by pollutants such as NO_x and SO_2 because of smaller stomatal openings. Higher percentages of starch in grasses improves their feeding quality, implying less need for feed mixes when silaging.

With increased atmospheric CO_2 the consumptive use of water becomes more efficient because of reduced transpiration. This is induced by a contraction of plant stomata and/or a decrease in the number of stomata per unit leaf area. This restricts the escape of water vapour from the leaf more than it restricts photosynthesis. With the same amount of available water, there could be more leaf area and biomass production by crops and natural vegetation. Plants could survive in areas hitherto too dry for their growth.

Increased ultraviolet radiation (UV-B, between 280 and 320 nanometres), due to depletion of the stratospheric ozone layer, mainly in the Antarctic region, may negatively affect terrestrial and aquatic photosynthesis and animal health. Over the last decade, a decrease of stratospheric ozone was observed at all latitudes (about 10% in winter, 0% during summer and intermediate values during spring and autumn). However, the 'Biological Action Factor' of UV-B can vary over several orders of magnitude with even slight changes in the amount and wavelength of UV-B. The subject is treated in detail elsewhere in this book. In particular it can be noted that:

- there are damaging effects of increasing UV-B on crops, animals and plankton growth. It has been reported that UV-B affects the ability of plankton organisms to control their vertical movements and to adjust to light levels;
- reductions in yield of up to 10% have been measured at experimentally very high UV-B values, and would be particularly effective in plants where the CO_2 fertilization effect is strongest. On the other hand, UV-B increase could increase the amount of plant internal compounds that act against pests.

Tropospheric ozone originates about half from photochemical reactions involving nitrogen oxides (NO_x), methane or carbon monoxide, and half by downward movement of stratospheric ozone. High ozone concentrations have toxic effects on both plant and animal life (German Bundestag, 1991; MacKenzie and El-Ashry, 1988; UNEP, 1993). It is likely that ozone, in conjunction with other photo-oxidants, is contributing towards the 'new type of forest damage' observed in Europe and the United States. In the tropics, tropospheric ozone concentrations are generally lower than at northern mid-latitudes. However, this does not apply to periods when biomass burning releases precursor substances for the photochemical formation of ozone.

Temperature

In general, higher temperatures are associated with higher radiation and higher water use. It is relatively difficult to separate the physiological effects (at the level of plants and plant organs) of temperatures from the ecological ones (at the level of the field or of the region). There are both positive and negative impacts at the two levels, and only crop- and site-specific simulation can assess the global 'net' effect of temperature increases. It is generally agreed that:

- rising temperatures — now estimated to be 0.2°C per decade, or 1 °C by 2040 (Mitchell *et al.*, 1995) with smallest increases in the tropics (IPCC, 1992) — would diminish the yields of some crops, especially if night temperatures are increased (the temperature increase since the mid-1940s is mainly due to increasing night-time temperatures, while CO_2-induced warming would result in an almost equally large rise in minimum and maximum temperatures (Kukla and Karl, 1993);
- higher temperatures could have a positive effect on growth of plants of the CAM type. They would also strengthen the CO_2 fertilization effect and the CO_2 anti-transpirant effect of C_3 and C_4 plants unless plants get overheated;
- higher night temperature may increase dark respiration of plants, diminishing net biomass production;
- higher cold-season temperatures may lead to earlier ripening of annual crops, diminishing yield per crop, but would allow locally for the growth of more crops per year due to lengthening of the growing season. Winter kill of pests is likely to be reduced at high latitudes, resulting in greater crop losses and higher need for pest control;
- higher temperatures will allow for more plant growth at high latitudes and altitudes.

Combined Effects and Some Uncertainties

The changes in CO_2 tropospheric ozone and increased UV-B do not necessarily occur simultaneously: CO_2 increase is worldwide, but with a strong seasonality in middle and higher latitudes; significant increase of UV-B is largely limited to subpolar regions (and mainly during the northern hemisphere winter months); high near-surface O_3 levels are restricted to the neighbourhood of major cities, airports, etc. (Seitz, 1994)

The Hydrological Cycle and Soils

Even a slight increase in surface temperatures will affect evaporation, atmospheric moisture and precipitation. While it is generally agreed that rainfall will increase (by an estimated 10 to 15%), two aspects have to be elucidated: how will rainfall intensities be affected, and what are the details of spatial changes. This is still largely a matter of discussion among experts. Based on palaeoclimatic analogies, certain authors predict more favourable rainfall conditions in the present-day Sahel (Petit-Maire, 1992). If the increase in precipitation should be associated with increased rainfall intensities, then the quality and quantity of soil and water resources would decline, for instance through increased runoff and erosion, increased land degradation processes, and a higher frequency of floods and possibly droughts. However:

- the extra precipitation on land, if indeed including present subhumid to semi-arid areas, will increase plant growth in these areas, leading to an improved protection of the land surface and increased rainfed agricultural production; in already humid areas the extra rainfall may, however, impair adequate crop drying and storage;
- the extra precipitation predicted to occur in some regions provides possibilities for off-site extra storage in rivers, lakes and artificial reservoirs (on-farm or at subcatchment level) for the benefit of improved rural water supply and expanded or more intensive irrigated agriculture and inland fisheries:
- the effects on water resources and water apportioning of international river and lake basins can be very substantial, with political overtones.

The greatest risks are often estimated to be associated with increased soil loss through erosion. Soils, as a medium for plant growth, would be affected in several other ways:

- increased temperatures may lead to more decomposition of soil organic matter;
- increased plant growth due to the CO_2 fertilization effect may cause other plant nutrients such as N and P to become in short supply; however, CO_2 increase would stimulate mycorrhizal activity (making soil phosphorus more easily available), and also biological nitrogen fixation (whether or not symbiotic). Through increased root growth there would be extra weathering of the substratum, hence a fresh supply of potassium and micronutrients;
- the CO_2 fertilization effect would produce more litter of higher C/N ratio, hence more organic matter for incorporation into the soil as humus; litter with high C/N decomposes slowly and this can act as a negative feedback on nutrient availability;

- the 'CO_2 anti-transpirant' effect would stimulate plant growth in dryland areas, and more soil protection against erosion and lower topsoil temperatures, leading to an 'anti-desertification effect'.

Global climate change, if it occurs, will definitely affect agriculture. Most mechanisms, and two-way interactions between agriculture and climate, are known, even if not always well understood. It is evident that the relationship between climate change and agriculture is still very much a matter of conjecture with many uncertainties; it remains largely a conundrum. Major uncertainties affect both the Global Circulation Models (GCMs) and the response of agriculture, as illustrated by differences among models, especially as regards effects at the national and subregional levels. In addition, many of the models do not take into consideration CO_2 fertilization and improved water-use efficiency, the effect of cloud cover (on both climate and photosynthesis), or the transient nature of climate change. It is also worth remembering that enormous knowledge gaps still affect the carbon cycle (with a missing sink of about 2 Gt of carbon), the factors behind the recent near-stabilization of the atmospheric methane concentrations or the unexplained reduced rate of CO_2 increase in recent years, the effect of volcanic eruptions (such as the recent Pinatubo eruption), the effect of any increased cloudiness, etc.

Designing Sustainable Farming Systems

The concept of sustainable agriculture is a relatively recent response to the decline in the quality of the natural resource base associated with modern agriculture (McIsaac and Edwards 1994). Today, the question of agricultural production has evolved from a purely technical one to a more complex one characterized by social, cultural, political and economic dimensions. The concept of sustainability although controversial and diffuse due to existing conflicting definitions and interpretations of its meaning, is useful because it captures a set of concerns about agriculture which is conceived as the result of the co-evolution of socioeconomic and natural systems (Reijntjes et al. 1992). A wider understanding of the agricultural context requires the study between agriculture, the global environment and social systems given that agricultural development results from the complex interaction of a multitude of factors. It is through this deeper understanding of the ecology of agricultural systems that doors will open to new management options more in tune with the objectives of a truly sustainable agriculture.

The sustainability concept has prompted much discussion and has promoted the need to propose major adjustments in conventional agriculture to make it more environmentally, socially and economically viable and compatible. Several possible solutions to the environmental problems created by capital and technology intensive farming systems have been proposed and research is currently in progress to evaluate alternative systems (Gliessman 1998). the main focus lies on the reduction or elimination of agrochemical inputs through changes in management to assure adequate plant nutrition and plant protection through organic nutrient sources and integrated pest management, respectively. Although hundreds of more environmentally prone research projects and technological development attempts have taken place, and many lessons have been learned, the thrust is still highly technological, emphasizing the suppression of limiting factors or the symptoms that mask an ill producing agroecosystem. The prevalent philosophy is that pests, nutrient deficiencies or other factors are the cause of low productivity, as opposed to the view that pests or nutrients only become limiting if conditions in the agroecosystem are not in equilibrium (Carrol et al. 1990). For this reason, there still prevails a narrow view that specific causes affect productivity, and overcoming the limiting factor via new technologies, continues to be the main goal. This view has diverted agriculturists from realizing that limiting factors only represent symptoms of a more systemic disease inherent to unbalances within the agroecosystem and from an appreciation of the context and complexity of agroecological processes thus underestimating the root causes of agricultural limitations (Altieri et al. 1993). On the other hand, the science of agroecology, which is defined as the application of ecological concepts and principles to the design and management of sustainable agroecosystems, provides a framework to assess the complexity of agroecosystems (Altieri 1995). The idea of agroecology is to go beyond the use of alternative practices and to develop agroecosystems with the minimal dependence on high agrochemical and energy inputs, emphasizing complex agricultural systems in which ecological interactions and synergisms between biological components provide the mechanisms for the systems to sponsor their own soil fertility, productivity and crop protection (Altieri and Rosset 1995).

In the search to reinstate more ecological rationale into agricultural production, scientists and developers have disregarded a key point in the development of a more self-sufficient and sustaining agriculture: a deep understanding of the nature of agroecosystems and the principles by which they function. Given this limitation, agroecology has emerged as the discipline that provides the basic ecological principles for how to study, design and manage agroecosystems that are both productive and natural resource conserving, and that are also culturally sensitive, socially just and economically viable (Altieri 1995). Agroecology goes beyond a one-dimensional view of agroecosystems — their genetics, agronomy, edaphology, and so on,- to embrace an understanding of ecological and social levels of co-evolution, structure and function. Instead of focusing on one particular component of the agroecosystem, agroecology emphasizes the interrelatedness of all agroecosystem components and the complex dynamics of ecological processes (Vandermeer 1995). Agroecosystems are communities of plants and animals interacting with their physical and chemical environments that have been modified by people to produce food, fibre, fuel and other products for human consumption and processing. Agroecology is the holitstic study of agroecosystems, including all environmental and human elements. It focuses on the form, dynamics and functions of their interrelationships and the processes in which they are involved. An area used for agricultural production, e.g. a field, is seen as a complex system in which ecological processes found under natural conditions also occur, e.g. nutrient cycling, predator/prey interactions, competition, symbiosis and successional changes. Implicit in agroecological research is the idea that, by understanding these ecological relationships and processes, agroecosystems can be manipulated to improve production and to produce more sustainably, with fewer negative environmental or social impacts and fewer external inputs (Altieri 1995). The design of such systems is based on the application of the following ecological principles (Reinjntjes et al. 1992):

1. Enhance recycling of biomass and optimizing nutrient availability and balancing nutrient flow.
2. Securing favourable soil conditions for plant growth, particularly by managing organic matter and enhancing soil biotic activity.
3. Minimizing losses due to flows of solar radiation, air and water by way of microclimate management, water harvesting and soil management through increased soil cover.
4. Species and genetic diversification of the agroecosystem in time and space.
5. Enhance beneficial biological interactions and synergisms among agrobiodiversity components thus resulting in the promotion of key ecological processes and services.

These principles can be applied by way of various techniques and strategies. Each of these will have different effects on productivity, stability and resiliency within the farm system, depending on the local opportunities, resource constraints and, in most cases, on the market. The ultimate goal of agroecological design is to integrate components so that overall biological efficiency is improved, biodiversity is preserved, and the agroecosystem productivity and its self-sustaining capacity is maintained. The goal is to design a quilt of agroecosystems within a landscape unit, each mimicking the structure and function of natural ecosystems.

From a management perspective, the agroecological objective is to provide a balanced environments, sustained yields, biologically mediated soil fertility and natural pest regulation through the design of diversified agroecosystems and the use of low-input technologies (Gleissman 1998). Agroecologists are now recognizing that intercropping, agroforestry and other diversification methods mimic natural ecological processes, and that the sustainability of complex agroecosystems lies in the ecological models they follow. By designing farming systems that mimic nature, optimal use can be made of sunlight, soil nutrients and rainfall (Pretty 1994). Agroecological management must lead management to optimal recycling of nutrients and organic matter turnover, closed energy flows, water and soil conservation and balance pest-natural enemy populations. The strategy exploits the complementarities and synergisms that result from the various combinations of crops, tree and animals in spatial and temporal arrangements (Altieri 1994). In essence, the optimal behaviour of agroecosystems depends on the level of interactions between the various biotic and abiotic components. By assembling a functional biodiversity it is possible to initiate synergisms which subsidize agroecosystem processes by providing ecological services such as the activation of soil biology, the recycling of nutrients, the enhancement of beneficial arthropods and antagonists, and so on (Altieri and Nicholls 1999). Today there is a diverse selection of practices and technologies available, and which vary in effectiveness as well as in strategic value. Key practices are those of a preventative nature and which act by reinforcing the "immunity" of the agroecosystem through a series of mechanisms. Various strategies to restore agricultural diversity in time and space include crop rotations, cover crops, intercropping, crop/livestock mixtures, and so on, which exhibit the following ecological features:

1. Crop Rotations. *Temporal diversity incorporated into cropping systems, providing crop nutrients and breaking the life cycles of several insect pests, diseases, and weed life cycles (Sumner 1982).*

2. *Polycultures*. Complex cropping systems in which tow or more crop species are planted within sufficient spatial proximity to result in competition or complementation, thus enhancing yields (Francis 1986, Vandermeer 1989).

3. *Agroforestry Systems*. An agricultural system where trees are grown together with annual crops and/or animals, resulting in enhanced complementary relations between components increasing multiple use of the agroecosystem (Nair 1982).

4. *Cover Crops*. The use of pure or mixed stands of legumes or other annual plant species under fruit trees for the purpose of improving soil fertility, enhancing biological control of pests, and modifying the orchard microclimate (Finch and Sharp 1976).

5. Animal integration in agroecosystems aids in achieving high biomass output and optimal recycling (Pearson and Ison 1987).

All of the above diversified forms of agroecosystems share in common the following features (Altieri and Rosset 1995):

a. Maintain vegetative cover as an effective soil and water conserving measure, met through the use of no-till practices, mulch farming, and use of cover crops and other appropriate methods.

b. Provide a regular supply of organic matter through the addition of organic matter (manure, compost, and promotion of soil biotic activity).

c. Enhance nutrient recycling mechanisms through the use of livestock systems based on legumes, etc.

d. Promote pest regulation through enhanced activity of biological control agents achieved by introducing and/or conserving natural enemies and antagonists.

Research on diversified cropping systems underscores the great importance of diversity in an agricultural setting (Francis 1986, Vandermeer 1989, Altieri 1995). Diversity is of value in agroecosystems for a variety of reasons (Altieri 1994, Gliessman 1998):

- As diversity increases, so do opportunities for coexistence and beneficial interactions between species that can enhance agroecosystem sustainability.
- Greater diversity often allows better resource-use efficiency in an agroecosystem. There is better system-level adaptation to habitat heterogeneity, leading to complementarity in crop species needs, diversification of niches, overlap of species niches, and partitioning of resources.
- Ecosystems in which plant species are intermingled possess an associated resistance to herbivores as in diverse systems there is a greater abundance and diversity of natural enemies of pest insects keeping in check the populations of individual herbivore species.
- A diverse crop assemblage can create a diversity of microclimates within the cropping system that can be occupied by a range of noncrop organisms — including beneficial predators, parasites, pollinators, soil fauna and antagonists — that are of importance for the entire system.
- Diversity in the agricultural landscape can contribute to the conservation of biodiversity in surrounding natural ecosystems.
- Diversity in the soil performs a variety of ecological services such as nutrient recycling and detoxification of noxious chemicals and regulation of plant growth.
- Diversity reduces risk for farmers, especially in marginal areas with more

unpredictable environmental conditions. If one crop does not do well, income from others can compensate.

Sustainable Agroecosystems

Most people involved in the promotion of sustainable agriculture aim at creating a form of agriculture that maintains productivity in the long term by (Pretty 1994, Vandermeer 1995):

- optimizing the use of locally available resources by combining the different components of the farm system, i.e. plants, animals, soil, water, climate and people, so that they complement each other and have the greatest possible synergetic effects;
- reducing the use of off-farm, external and non-renewable inputs with the greatest potential to damage the environment or harm the health of farmers and consumers, and a more targeted use of the remaining inputs used with a view to minimizing variable costs;
- relying mainly on resources within the agroecosystem by replacing external inputs with nutrient cycling, better conservation, and an expanded use of local resources;
- improving the match between cropping patterns and the productive potential and environmental constraints of climate and landscape to ensure long-term sustainability of current production levels;
- working to value and conserve biological diversity, both in the wild and in domesticated landscapes, and making optimal use of the biological and genetic potential of plant and animal species; and
- taking full advantage of local knowledge and practices, including innovative approaches not yet fully understood by scientists although widely adopted by farmers.

Agroecology provides the knowledge and methodology necessary for developing an agriculture that is on the on e hand environmentally sound and on the other hand highly productive, socially equitable and economically viable. Through the application of agroecological principles, the basic challenge for sustainable agriculture to make better use of internal resources can be easily achieved by minimizing the external inputs used, and preferably by regenerating internal resources more effectively through diversification strategies that enhance synergisms among key components of the agroecosystem. The ultimate goal of agroecological design is to integrate components so that overall biological efficiency is improved, biodiversity is preserved, and the agroecosystem productivity and its self-regulating capacity is maintained.

The goal is to design an agroecosystem that mimics the structure and function of local natural ecosystems; that is, a system with high species diversity and a biologically active soil, one that promotes natural pest control, nutrient recycling and high soil cover to prevent resource losses. Agroecology provides guidelines to develop diversified agroecosystems that take advantage of the effects of the integration of plant and animal biodiversity such integration enhances complex interactions and synergisms and optimizes ecosystem functions and processes, such as biotic regulation of harmful organisms, nutrient recycling, and biomass production and accumulation, thus allowing agroecosystems to sponsor their own functioning. The end result of agroecological design is improved economic and ecological sustainability of the agroecosystem, with the proposed management systems specifically in tune with the local resource base and operational framework of existing environmental and socioeconomic conditions. In an agroecological strategy, management components are directed to highlight the conservation and enhancement of local agricultural resources (germplasm, soil, beneficial fauna, plant biodiversity, etc.) by emphasizing a development methodology that encourages farmer participation, use of traditional knowledge, and adaptation of farm enterprises that fit local needs and socioeconomic and biophysical conditions.

Ecological Processes to Optimize in Agroecosystems

- Strengthen the immune system (proper functioning of natural pest control)
- Decrease toxicity through elimination of agrochemicals
- Optimize metabolic function (organic matter decomposition and nutrient cycling)
- Balance regulatory systems (nutrient cycles, water balance, energy flow, population regulation, etc.)
- Enhance conservation and regeneration of soil-water resources and biodiversity
- Increase and sustain long-term productivity

Mechanisms to Improve Agroecosystem Immunity

- Increase of plant species and genetic diversity in time and space.
- Enhancement of functional biodiversity (natural enemies, antagonists etc.)
- Enhancement of soil organic matter and biological activity
- Increase of soil cover and crop competitive ability
- Elimination of toxic inputs and residues

Environmental Problems

The specialization of production units has led to the image that agriculture is a modern miracle of food production. Evidence indicates, however, that excessive reliance on monoculture farming and agroindustrial inputs, such as capital-intensive technology, pesticides, and chemical fertilizers, has negatively impacted the environment and rural society. Most agriculturalists had assumed that the agroecosystem/natural ecosystem dichotomy need not lead to undesirable consequences, yet, unfortunately, a number of "ecological diseases" have been associated with the intensification of food production. They may be grouped into two categories: diseases of the ecotope, which include erosion, loss of soil fertility, depletion of nutrient reserves, salinization and alkalinization, pollution of water systems, loss of fertile croplands to urban development, and diseases of the biocoenosis, which include loss of crop, wild plant, and animal genetic resources, elimination of natural enemies, pest resurgence and genetic resistance to pesticides, chemical contamination, and destruction of natural control mechanisms. Under conditions of intensive management, treatment of such "diseases" requires an increase in the external costs to the extent that, in some agricultural systems, the amount of energy invested to produce a desired yield surpasses the energy harvested.

The loss of yields due to pests in many crops (reaching about 20-30% in most crops), despite the substantial increase in the use of pesticides (about 500 million kg of active ingredient worldwide) is a symptom of the environmental crisis affecting agriculture. It is well known that cultivated plants grown in genetically homogenous monocultures do not possess the necessary ecological defense mechanisms to tolerate the impact of outbreaking pest populations. Modern agriculturists have selected crops for high yields and high palatability, making them more susceptible to pests by sacrificing natural resistance for productivity. On the other hand, modern agricultural practices negatively affect pest natural enemies, which in turn do not find the necessary environmental resources and opportunities in monocultures to effectively and biologically suppress pests.

Due to this lack of natural controls, an investment of about 40 billion dollars in pesticide control is incurred yearly by US farmers, which is estimated to save approximately $16 billion in US crops. However, the indirect costs of pesticide use to the environment and public health have to be balanced against these benefits. Based on the available data, the environmental (impacts on wildlife, pollinators, natural enemies, fisheries, water and development of resistance) and social costs (human poisonings and illnesses) of pesticide use reach about $8 billion each year. What is worrisome is that pesticide use is on the rise. Data from California shows that from 1941 to 1995 pesticide use increased from 161 to 212 million pounds of active ingredient. These increases were not due to increases in planted acreage, as statewide crop acreage remained constant during this period. Crops such as strawberries and grapes account for much of this increased use, which includes toxic pesticides, many of which are linked to cancers.

Fertilizers, on the other hand, have been praised as being highly associated with the temporary increase in food production observed in many countries. National average rates of nitrate applied to most arable lands fluctuate between 120-550 kg N/ha. But the bountiful harvests created at least in part through the use of chemical fertilizers, have associated, and often hidden, costs. A primary reason why chemical fertilizers pollute the environment is due to wasteful application and the fact that crops use them inefficiently. The fertilizer that is not recovered by the crop ends up in the environment, mostly in surface water or in ground water. Nitrate contamination of aquifers is widespread and in dangerously high levels in many rural regions of the world. In the US, it is estimated that more than 25% of the drinking water wells contain nitrate levels above the 45 parts per million safety standard. Such nitrate levels are hazardous to human health and studies have linked nitrate uptake to methaemoglobinemia in children and to gastric, bladder and oesophageal cancers in adults.

Fertilizer nutrients that enter surface waters (rivers, lakes, bays, etc.) can promote eutrophication, characterized initially by a population explosion of photosynthetic algae. Algal blooms turn the water bright green, prevent light from penetrating beneath surface layers, and therefore killing plants living on the bottom. Such dead vegetation serve as food for other aquatic microorganisms which soon deplete water of its oxygen, inhibiting the decomposition of organic residues, which accumulate on the bottom. Eventually, such nutrient enrichment of freshwater ecosystems leads to the destruction of all animal life in the water systems. In the US it is estimated that about 50-70% of all nutrients that reach surface waters is derived from fertilizers.

Chemical fertilizers can also become air pollutants, and have recently been implicated in the destruction of the ozone layer and in global warming. Their excessive use has also been linked to the acidification/salinization of soils and to a higher incidence of insect pests and diseases through mediation of negative nutritional changes in crop plants.

It is clear then that the first wave of environmental problems is deeply rooted in the prevalent socioeconomic system which promotes monocultures and the use of high input technologies and agricultural practices that lead to natural resource degradation. Such degradation is not only an ecological process, but also a social and political-economic process. This is why the problem of agricultural production cannot be regarded only as a technological one, but while agreeing that productivity issues represent part of the problem, attention to social, cultural and economic issues that account for the crisis is crucial. This is particularly true today where the economic and political domination of the rural development agenda by agribusiness has thrived at the expense of the interests of consumers, farmworkers, small family farms, wildlife, the environment, and rural communities.

Despite that awareness of the impacts of modern technologies on the environment increased, as we traced pesticides in food chains and crop nutrients in streams and aquifiers, there are those that confronted to the challenges of the XXI century still argue for further intensification to meet the requirements of agricultural production. It is in this context that supporters of "status-quo agriculture" celebrate the emergence of biotechnology as the latest magic bullet that will revolutionize agriculture with products based on natures' own methods, making farming more environmentally friendly and more profitable for the farmer. Although clearly certain forms of non-transformational biotechnology hold promise for an improved agriculture, given its present orientation and control by multinational corporations, it holds more promise for environmental harm, for the further industrialization of agriculture and for the intrusion of private interests too far into public interest sector research.

What is ironic is the fact that the biorevolution is being brought forward by the same interests (Monsanto, Novartis, DuPont, etc.) that promoted the first wave of agrochemically-based agriculture, but this time, by equipping each crop with new "insecticidal genes", they are promising the world safer pesticides, reduction on chemically intensive farming and a more sustainable agriculture. However, as long as transgenic crops follow closely the pesticide paradigm, such biotechnological products will do nothing but reinforce the pesticide treadmill in agroecosystems, thus legitimizing the concerns that many scientists have expressed regarding the possible environmental risks of genetically engineered organisms. So far, field research as well as predictions based on ecological theory, indicate that among the major environmental risks associated with the release of genetically engineered crops can be summarized as follows:

- The trends set forth by corporations is to create broad international markets for a single product, thus creating the conditions for genetic uniformity in rural landscapes. History has repeatedly shown that a huge area planted to a single cultivar is very vulnerable to a new matching strain of a pathogen or pest;
- The spread of transgenic crops threatens crop genetic diversity by simplifying cropping systems and promoting genetic erosion;
- There is potential for the unintended transfer to plant relatives of the "transgenes" and the unpredictable ecological effects. The transfer of

genes from herbicide resistant crops (HRCs) to wild or semidomesticated relatives can lead to the creation of super weeds;

- Most probably insect pests will quickly develop resistance to crops with Bt toxin. Several Lepidoptera species have been reported to develop resistance to Bt toxin in both field and laboratory tests, suggesting that major resistance problems are likely to develop in Bt crops which through the continuous expression of the toxin create a strong selection pressure;
- Massive use of Bt toxin in crops can unleash potential negative interactions affecting ecological processes and non-target organisms. Evidence from studies conducted in Scotland suggest that aphids were capable of sequestering the toxin from Bt crops and transferring it to its coccinellid predators, in turn affecting reproduction and longevity of the beneficial beetles;
- Bt toxins can also be incorporated into the soil through leaf materials and litter, where they may persist for 2-3 months, resisting degradation by binding to soil clay particles while maintaining toxic activity, in turn negatively affecting invertebrates and nutrient cycling;
- A potential risk of transgenic plants expressing viral sequences derives from the possibility of new viral genotypes being generated by recombination between the genomic RNA of infecting viruses and RNA transcribed from the transgene;
- Another important environmental concern associated with the large scale cultivation of virus-resistant transgenic crops relates to the possible transfer of virus-derived transgenes into wild relatives through pollen flow.

Although there are many unanswered questions regarding the impact of the release of transgenic plants and micro-organisms into the environment, it is expected that biotechnology will exacerbate the problems of conventional agriculture and by promoting monocultures will also undermine ecological methods of farming such as rotations and polycultures. Because transgenic crops developed for pest control emphasize the use of a single control mechanism, which has proven to fail over and over again with insects, pathogens and weeds, transgenic crops are likely to increase the use of pesticides and to accelerate the evolution of "super weeds" and resistant insect pest strains. These possibilities are worrisome, especially when considering that during the period 1986-1997, approximately 25,000 transgenic crop field trials were conducted worldwide on more than 60 crops with 10 traits in 45 countries. By 1997 the global area devoted to transgenic crops reached 12.8 million hectares. Seventy-two percent of all transgenic crop field trials were conducted in the USA and Canada, although some were also conducted in descending order in Europe, Latin America and Asia. In most countries biosafety standards to monitor such releases are absent or are inadequate to predict ecological risks. In the industrialized countries from 1986-1992, 57% of all field trials to test transgenic crops involved herbicide tolerance pioneered by 27 corporations including the world's eight largest pesticide companies. As Roundup and other broad spectrum herbicides are increasingly deployed into croplands, the options for farmers for a diversified agriculture will be even more limited.

Reduction and, especially, elimination of agrochemical require major changes in management to assure adequate plant nutrients and to control crop pests. As it was done a few decades ago, alternative sources of nutrients to maintain soil fertility include manures, sewage sludge and other organic wastes, and legumes in cropping sequences. Rotation benefits are due to biologically fixed nitrogen and from the interruption of weed, disease and insect cycles. A livestock enterprise may be integrated with grain cropping to provide animal manures and to utilize better the forages produced. Maximum benefits of pasture integration can be realized when livestock, crops, animals and other farm resources are assembled in mixed and rotational designs to optimize production efficiency, nutrient cycling and crop protection.

In orchards and vineyards, the use of cover crops improve soil fertility, soil structure and water penetration, prevent soil erosion, modify the microclimate and reduce weed competition. Entomological studies conducted in orchards with ground cover vegetation indicate that these systems exhibit lower incidence of insect pests than clean cultivated orchards. This is due to a higher abundance and efficiency of predators and parasitoids enhanced by the rich floral undergrowth.

Increasingly, researchers are showing that it is possible to provide a balanced environment, sustained yields, biologically mediated soil fertility and natural pest regulation through the design of diversified agroecosystems and the use of low-input technologies. Many alternative cropping systems have been tested, such as double cropping, strip cropping, cover cropping and intercropping, and more importantly concrete examples from real farmers show that such systems lead to optimal recycling of nutrients and organic matter turnover, closed energy flows, water and soil conservation and balanced pest-natural enemy populations. Such diversified farming exploit the complementarities that result from the various combinations of crops, trees and animals in spatial and temporal arrangements.

In essence, the optimal behaviour of agroecosystems depends on the level of interactions between the various biotic and abiotic components. By assembling a functional biodiversity it is possible to initiate synergisms which subsidize agroecosystem processes by providing ecological services such as the activation of soil biology, the recycling of nutrients, the enhancement of beneficial arthropods and antagonists, and so on. Today there is a diverse selection of practices and technologies available, and which vary in effectiveness as well as in strategic value.

Ecological Impact

We might call this the Age of the Recognition of Limits. A generation ago, the energy reserves that fueled the extractive economy were seen to be the primary limits affecting human progress in every endeavor, including agriculture. American scientists and policymakers alike were comfortable in their belief that these acknowledged finite reserves were large enough to buy the time necessary to bring the peaceful atom on line. It isn't that renewable resources were ignored. Americans did worry about the consequences of soil erosion and water pollution, of course, but there was a nearly full confidence in the technical fix for problems in the renewable sector of the economy. For agriculture, this meant that terraces, grass waterways, and proper incentives would take care of erosion. Simple solutions like indoor plumbing and chlorine in our drinking water would handle the water problem. And as for this nonsense about pesticides, Rachel Carson was just a hysterical extremist who was not an expert. The attitude about commercial fertilizers was that "nitrogen is nitrogen" and "plants don't care whether it comes from an anhydrous ammonia tank or from the nodule of a legume." And besides that, "We must feed the world." We had a sense of the heroic urge in us. It was also an era in which "hard-headed realism" was a favourite expression. But as author Wendell Berry has said, "A hard-headed realist is usually a person who uses a lot less information than is available." We have more information now than a generation ago, and the public is better informed. The "problem of agriculture" is as old as agriculture itself, and although the core of the problem has always been soil erosion, new problems have been added. The high-energy epoch that fueled industry, which in turn made the industrialization of agriculture possible, simply exacerbated the old problems in agriculture and added more besides. And so we are dealing with an old subject with new data and the need for a fresh approach in our search for solutions. We will consider these problems one at a time even though they are inextricably intertwined.

Raindrops bombarding bare soil result in the oldest and still most serious problem of agriculture. The long history of soil erosion and its impact on civilization is one of devastation. Eroded fields record our failure as land stewards. W. C. Lowdermilk (1953) described several civilizations that collapsed because of erosion. At the end of his historical sweep of failed civilizations, Lowdermilk warned that unless we want to experience a similar fate, we had better heed these lessons of the past and safeguard our soils. Soil erosion came to the nation's attention dramatically during the dust-bowl 1930s. Dust storms that blackened the sky were harder to ignore than the more severe problem of sheet and rill erosion due to rain. The Soil

Conservation Service (SCS) was formed during this period under the energetic and imaginative direction of Hugh Hammond Bennett. The SCS encouraged planting shelterbelts of trees and grassed waterways and using contour ploughing and terracing. But by the mid-1970s soil erosion once again became a national concern. Farmers were encouraged both by the export potential and by Secretary of Agriculture Earl Butz to plant from fencerow to fencerow (Weasel, 1983), bringing 24 million additional hectares (60 million acres) into production (Crosson and Stout, 1983; Batie, 1983). By and large, the extra land was more prone to erosion. Larger machinery was purchased and pulled into the fields, and suddenly the terraces on sloping land and the windbreaks were nuisances (Weasel, 1983). Soil conservation meant increased costs to farmers, who were already burdened by the additional investments. The possibility of higher profits overshadowed any long-term benefits of conservation.

In 1977 the SCS, in an attempt to understand the amount of this precious resource being lost, began a new survey called the National Resources Inventory (NRI), published in 1981. The public was given a comprehensive review of sheet, rill, and wind erosion in the United States over a cropland base of 167 million hectares (413 million acres) (Crosson and Stout, 1983; Batie, 1983). An estimated total of 5.8 billion metric tons (6.4 billion tons) of topsoil was washed or blown away in the United States (USDA, 1980; Crosson and Stout, 1983). Some 2 billion metric tons of soil was lost from American cropland alone, an average of 15.4 metric tons/ha (6.8 tons/acre) (Crosson and Stout, 1983). Sheet and rill erosion were responsible for 1.7 billion metric tons being lost, while wind swept away 816 million metric tons (Crosson and Stout, 1983). The average "official" tolerable losses of 11.3 metric tons/ha (5 tons/acre) for deep soils and of 2.3 metric tons/ha (1 ton/acre) for shallow soils were violated (Crosson and Stout 1983; Batie, 1983).

Since erosion is not evenly distributed, in some areas the erosion rate greatly exceeded 15.4 metric tons/ha (6.8 tons/acre) In the Palouse Hills region of southeastern Washington and southwestern Idaho, where dryland farming is devoted to wheat, barley, beans, and lentils, there are numerous slopes ranging in steepness from 15 to 25 percent; here the soil erosion due to snow and rain was 113 to 227 metric tons/ha (50 to 100 tons/acre) (USDA, 1980). In southeastern Idaho, where hard red wheat is planted on slopes as steep as 35 percent, the annual erosion rate was 36 metric tons/ha (16 tons/acre) (Batie, 1983). Steep but fertile ground in southern Mississippi experienced a loss of 45 metric tons/ha (20 tons/acre) or more (USDA, 1980). Corn-belt losses up to 45 metric tons/ha were common (Batie, 1983).

Although the NRI was the most extensive survey yet produced, it did not give a complete picture of the amount of soil loss. The NRI relied on the universal soil loss equation (USLE) and the wind erosion equation (WEE) for estimating total soil loss (Batie, 1983). The USLE can estimate sheet and rill erosion, but it will not determine the amount of soil lost in gullies (Crosson and Stout, 1983). Wind erosion was determined for only 10 states in the 1977 survey (Crosson and Stout)

The more comprehensive NRI in 1982 documented an increase in the nat~on's cropland base from 167 million to 170 million hectares (413 million to 421 million acres) during the 5 years between surveys (Lee, 1984). The acreage in the land-capability classes and subclasses most subject to erosion increased disproportionately. The size of the highly erodable land-capability classes of V, VI, VII, and VIII increased 9 percent, while cropland in the high-quality classes I, II, and III increased only 1.2 percent (Lee, 1984). The sum of wind and water erosion averaged more than 18 metric tons/ha (8 tons/acre), or 3.1 billion metric tons (3.4 billion tons) on cropland. Forty-four percent of the cropland lost soil above the tolerance level (Lee, 1984). Nearly a fifth of the acreage was devoted to corn, a crop linked to high erosion rates (Bills and Heimlich, 1984). Global losses are also high. Lester Brown (1981) reports that around 21 billion metric tons of soil is being lost worldwide. Even India and China, presently self-sufficient in food production, are losing soil beyond replacement levels. In India an estimated 6 billion metric tons of soil is lost annually from their croplands, representing a 4.7 billion metric tons/yr loss beyond the tolerance level (Brown, 1984). Erosion is not always obvious, especially sheet erosion. Three steps characterize the erosion process: Soil particles are detached, then displaced by wind and rain, and finally deposited. Raindrops striking the bare soil can throw soil particles several centimeters. Sheet erosion, or overland flow, occurs when the water removes a relatively uniform thickness of soil. Rill erosion is easier to detect, since we can readily see the small channels, or rills, where water concentrates. The erosive power of the rill variety is responsible for a majority of our soil losses. (All from Morgan, 1979.) Wind erosion also separates soil particles. Fine particles less than 0.2 mm (0.01 in) in diameter are transported high in the air over long distances (Lyres and Tatarko, 1986). Larger particles roll along the ground, become abraded, and add to the fine particles that are removed by the wind (Lyres and Tatarko, 1986; Morgan, 1979). Soil structure is damaged and soil texture altered as the coarser particles are left behind and organic matter decreases (Lyres and Tartako, 1986).

With such a wholesale altering of its basic structure, texture, and organic matter, the deposited soil no longer has the potential to produce its former yields (Rosenberry et al., 1980; Meilke and Schepers, 1986; National Soil Erosion-Soil Productivity Research Planning Committee, 1981). As the soil erosion continues, the subsoil becomes part of the tillage layer, and the fertilizer inputs and power requirements increase to the point where it is no longer feasible to cultivate the land (Rosenberry et al., 1983). The problems of soil erosion are not limited to the farm. Soil from eroded cropland is the largest contributor to nonpoint pollution in the United States (Meyers et al., 1985). Sediment makes up the greatest volume by weight of materials transported in streams and rivers, and other pollutants (such as pesticides and fertilizers) can be carried along with it, either adsorbed on the soil particles or in solution (Clark, 1985). These pollutants adversely affect aquatic ecosystems either by directly affecting fish and other aquatic life or by altering spawning areas and food sources (Clark, 1985; Meyers et al., 1985). In addition, sediment accumulation is reducing the capacity of waterstorage facilities; approximately 1727 m3 (1.4 million acre-feet) of reservoir and lake capacity is filled each year with sediment (Clark, 1985). Dredging is a costly way of maintaining this capacity, and alternative reservoir sites are becoming difficult to find (Poster, 1985). The consequence is that municipal water sources are affected. The costs to clean the water are substantially higher since cleaning devices (sedimentation basins, filters, and chemical coagulants) need to be installed (Clark, 1985). The USDA has promoted no-till and minimum-tillage practices as erosion control measures. In 1982, 4.7 million hectares (11.6 million acres) of farmland were in no-till, and the USDA has predicted that 85 percent of all cultivated cropland will be in some form of minimum tillage by the year 2000 (Hinkle, 1983). Unfortunately, no-till practices rely heavily on herbicides.

Some promising attempts to reduce the amount of soil erosion are the two conservation provisions of the 1985 Farm Bill (Wenzel, 1986). Under the first, the SCS will identify highly erodable land. Farmers who convert any highly erodable land to cropland will be denied such benefits as price supports and crop insurance (Wenzel, 1986), possibly keeping 13 million hectares (32 million acres) from being converted to cropland. The second provision, the Conservation Reserve, pays farmers to take highly erodable land out of production; it caused 3.6 million hectares (9 million acres) to be taken out of production in 1986 (American Farmland Trust, 1986). By 1986, Congress had a target of 18.2 million hectares (45 million acres) (American Farmland Trust, 1986), but the number is likely to go higher. Future efforts at soil conservation should concentrate on developing farming practices that preserve and enhance the soil. Farmers should be subsidized to initiate soil-conserving practices, such as terracing and contour plowing. Also, we should look to natural ecosystems to provide our best example, such that soil surfaces are always protected by a canopy. This is difficult to obtain in an annual cropping system where the soil is tilled every year. Reduced-tillage practices have helped conserve soil, but the use of herbicides with this practice makes it undesirable in a sustainable agriculture. Perhaps in our search for new and more sustainable agricultural methods we will find other ways to conserve soil. Planting perennial cropping systems on sloping ground may allow continued use of this fragile land, or we may have to convert it to grassland and limit our cropping systems to flat bottomlands.

Irrigation

Adequate rainfall is never guaranteed for the dryland farmer in arid and semiarid regions, and thus irrigation is essential for reliable pro auction. Irrigation ensures sufficient water when needed and also allows farmers to expand their acreage of suitable cropland. In fact, we rely heavily on crops from irrigated lands, with fully one-third of the world's harvest coming from that 17 percent of cropland that is under irrigation (Poster, 1985). Unfortunately, current irrigation practices severely damage the cropland and the aquatic systems from which the water is withdrawn and partially returns. Currently, the 255 million irrigated hectares (631 million acres) in the world use 70 percent of the total world water consumption (Poster, 1985). In the United States more than 20 million irrigated hectares (49 million acres), concentrated in the 18 western states and in the southeast, use 83 percent of the total water consumed (Fredrick and Hanson, 1982; USDA, 1985). Very few conservation efforts are currently devoted to water, probably because it has been regarded as a limitless resource. Farmers worldwide face the problem of limited supplies and the resultant competition over what is left. But even in the face of decreased supplies, the number of irrigated acres continues to expand. In the 1970s an additional 5.1 million hectares (12.5 million acres) per year were irrigated (Poster, 1985). Water from underground aquifers supplies 40 percent of our irrigation needs in the United States (Pimentel et al., 1982). Groundwater provides an excellent high-quality source. In the United States the storage capacity of aquifers greatly exceeds that of rivers, lakes, and reservoirs (U.S. Geological Survey, 1984). But supplies are not limitless; aquifers recharge slowly from water percolating through the soil, and water in some deep aquifers can be considered a nonrenewable resource. The amount of water that is pumped from the aquifers exceeds the amount replenished in many areas of the world. In the United States approximately one-fourth of the groundwater extracted exceeds the replenishment (Stokes, 1983). As water tables fall, springs dry up and streams experience reduced flow (Poster, 1985; Pimentel et al., 1982).

The Ogallala aquifer is a good example of rapid aquifer depletion. This aquifer lies under parts of seven states: Texas, New Mexico, Colorado, Kansas, Nebraska, Wyoming, and South Dakota. It supports one-fifth of the U.S. irrigated cropland (Poster, 1985). The export push in the 1970s encouraged farmers to install extensive irrigation systems (Weasel, 1983). Farmers irrigated 3.2 million hectares (8 million acres) in 1978, compared to 0.85 million hectares (2.1 million acres) in the 1940s (Poster, 1985). A center-point irrigation system (the type now prevalent in this area) can use 3030 L (800 gal) of water per minute (Adams, 1981). Five hundred cubic kilometers (405.4 million acre-feet) of water has been withdrawn in the last four decades, and some hydrologists predict that given the current rate of drawndown, the aquifer will fail to yield in another 40 years (Ferguson, 1983). The exhaustion of the Ogallala aquifer is showing its effects; land in Texas, New Mexico, and Kansas is being taken out of production because of diminishing well yields and rising pumping costs (Ferguson, 1983). A total of 6 million hectares (15 million acres) in 11 states may be taken out of production because of water shortages; Texas alone may lose 1.2 million hectares (3 million acres) by the year 2000 (Stokes, 1983). Declining water levels not only affect farmers. In the southeast, several cities depend on groundwater to meet their water needs. Tucson, Arizona, is a city completely dependent on an aquifer to meet its needs, yet only 35 percent of the water withdrawn is replaced by recharge (Stokes, 1983). Exhausted water resources will effectively curtail the expansion of these cities. Land subsidence, the dropping of the land surface, is another major problem when aquifers are overdrawn (Bouwer, 1981). Numerous incidences of land subsidence have been recorded in California (Carbognin, 1985). Heavy pumping of the groundwater began after World War II and has continued to the present (Carbognin, 1985). In the 51 years between 1925 and 1977, 20 billion cubic meters (26.2 billion cubic yards) of land has dropped due to the extraction of 70 million cubic meters 57,000 acre-feet of groundwater (Carbognin, 1985). Groundwater withdrawal in Mexico City has caused 9 m (30 ft) of land subsidence, creating problems with their drainage system and with construction (Carbognin, 1985). In addition to subsidence, the depletion of aquifers along the coastlines of California, Georgia, and Florida has caused saltwater intrusions (Pimentel et al., 1982).

Depletion of streams and rivers is also a serious problem. River ecosystems require minimum levels of instream flow. Aquatic and riparian habitats are severely affected or destroyed when rivers are overdrawn (U.S. Water Resources Council, 1978). In the l8 western states most dependent on irrigation, 70 percent of the instream flow is often depleted (Fredrick and Hanson, 1982). For example, in the lower Colorado River more than 80 percent of the instream flow is depleted (Pimentel et al., 1982). River runoff may not be adequate for optimal fish and wildlife habitats in rivers of 14 western and southwestern states by the end of this century (Stokes, 1983). To protect these natural habitats, irrigation would have to be reduced to one-fifth of what it is now (Stokes, 1983).

Beyond depleting water resources, current irrigation systems can destroy cropland and reduce water quality. Waterlogging, salinization, and alkalinization have damaged about half of the world's irrigated lands (CEQ, 1980). Although this is only a small percentage of the world's total cropland, the losses are significant when the higher yields obtained with irrigation are considered. The soil and groundwater in arid zones typically have a high salt content, and natural drainage is usually poor (Kovda, 1980). Irrigation water drains slowly from fields, and as the water table rises, the soils become waterlogged in the crop root zone (Kovda, 1980). Continued application of irrigation water and seepage from unlined canals cause the salty groundwater to reach the surface (Kovda, 1980). As the water evaporates from the soil, a salty white crust forms on the surface, leaving the land unfit for cultivation unless expensive reclamation procedures are used. These problems are present wherever large acreages of cropland are irrigated (Kovda, 1977). Pakistan has enormous problems with salinization; 9.1 million of its 14.6 million irrigated hectares (22.4 million of 36 million acres) have some form of salinity problem (CEQ, 1980). In the San Joaquin valley about 162 million hectares (400,000 acres) are affected by high brackish water tables. Unless adequate drainage is installed, approximately 13 percent of this acreage will become unproductive (Poster, 1985). Cropland salinity also creates water-quality problems in arid and semiarid river basins. Less than 50 percent of the water applied for irrigation is used by plants; the rest eventually returns to rivers and streams (Poster, 1985). The quality of the return flow is seriously degraded by salts, sediments, fertilizers, and pesticides. In addition, groundwater sources can become contaminated by the deep percolation of degraded irrigation water (El-Ashry et al., 1985). As a result, municipal and industrial water sources become polluted and users must pay higher water-treatment costs.

Applying large amounts of water in arid regions can also alter the ecosystem; in addition, such changes in farming practices as plowing, fertilizing, and monocropping (which usually accompany irrigation) also affect the ecosystem. Plant diseases, insects, and weeds, previously unable to withstand the dry conditions, can become serious pests (Ghabbour, 1977). The soil fauna, usually sparse in desert climates and adapted to drought conditions, can undergo drastic changes with irrigation; certain groups of soil fauna become dominant, destroying the previous balance. For example, earthworms, collemba, mites, and nematodes will invade newly irrigated soils at the expense of the original fauna unable to adapt to the new moisture levels (Ghabbour, 1977).

Human diseases have also increased with the widespread use of irrigation, especially water-transmissible diseases such as malaria and schistosomiasis (Obeng, 1977; Farid, 1977). The aquatic snail that transmits schistosomiasis inhabits sluggish waters common in irrigation projects, and schistosomiasis is now estimated to affect 200 million people in Asia, Africa, the Caribbean, and South America (Obeng, 1977). The disease has reached its current levels with the construction of irrigation projects. In the Gezira project in the Sudan, for example, the incidence of the disease increased almost 50 percent in a 20-year period after the expansion of the irrigation project in 1950 (Obeng, 1977). Although irrigation in arid areas cannot be totally abandoned, we need to realize that water is a limited resource and that the arid ecosystems are fragile. Upgrading irrigation systems will ameliorate some of the problems outlined above. Instead of financing massive water-diversion projects, improving irrigation efficiencies can free water for other demands. Lining canals, reusing water runoff, and increasing the efficiency of flood and furrow systems can save enormous quantities of water. Unfortunately, the incentive to save water is lacking in the United States since farmers pay only one-fifth of the cost of the water they use (Ferguson, 1983). Limited supplies and rational costs will force changes in the allocation and use of water. In addition, the type and extent of agriculture practiced in arid regions should be reexamined. Agricultural systems transferred from humid areas are not suitable. Investigating the unique qualities of arid ecosystems will help develop more appropriate agricultural practices.

Genetic Diversity

As modern agriculture converts an ever-increasing portion of the earth's land surface to monoculture, the genetic and ecological diversity of the planet erodes. Both the conversion of diverse natural ecosystems to new agricultural lands and the narrowing of the genetic diversity of crops contribute to this erosion. An estimated 852 million hectares (2.1 billion acres) worldwide have been converted to regular cropping or grazing in the past 120 years through deforestation, draining wetlands (or converting to rice), irrigation, and conversion to grazing (Richards, 1984; Salick and Merrick, Chap. 19 of this book). Although in North America from 1920 to 1978 the reversion of cultivated land to wild land about equaled the conversion to new agricultural land, elsewhere conversion has far exceeded reversion. In Africa 90.5 million hectares (223.6 million acres) were converted during this same period, and worldwide net conversion (conversion minus reversion) equaled about 420 million hectares (1.04 billion acres) (Richards, 1984). Current world population trends and the unequal distribution of power and resources will keep the pressure for continued conversion high (Wilson, 1985). In the United States the increased demand for exported food in the 1970s caused the conversion rate to climb dramatically; an estimated 24 million hectares (59 million acres) were converted to agriculture between 1972 and 1980 (Richards, 1984). Inevitably, some ecologically sensitive areas, such as the vital breeding grounds for wetland birds and waterfowl known as the "prairie potholes" in North Dakota, were victims of this agricultural expansion.

In the tropics, rates of land conversion to agriculture remain high (Richards, 1984), and there the impact on global species diversity is the greatest. The ongoing destruction of tropical forests is estimated to cause thousands of species extinctions annually (Wilson, 1985). Scientists estimate that 3 million species of plants (two-thirds of all species on earth) reside in the tropics, and about half of these are confined to tropical forests (U.S. Department of State, 1981; White, 1983). At the current rates of deforestation, one-fourth to one-third of all modern species will be gone in the next 30 to 40 years (U.S. Department of State, 1981; Batten, 1983). The pressure on tropical forests is twofold. On the one hand, the need for foreign exchange, coupled with the industrial nations' voracious consumption of forest products, produces lucrative economic incentives for logging operations (World Commission on Environment and Development, 1987; Nations and Komer, 1983). Historically, economic factors have been the prime cause of most deforestation

(Richards, 1984). On the other hand, inequitable land distribution has created a rapidly growing population of landless peasants. In Latin America a mere 7 percent of the landowners control 93 percent of the arable land. From Brazil and Colombia to Kenya, Thailand, Indonesia, and Madagascar the story is similar. Logging roads provide access to unclaimed lands, and once the valuable timber species have been logged out, the landless poor are encouraged to move in and clear the land. In the process, indigenous peoples are likely to be displaced, thus increasing the pool of landless poor. Typically, the cleared land will yield for only 5 to 7 years of intensive cropping, and then the peasants must look for a new plot. Often the newly cleared land is taken over by large landowners to raise export crops, such as beef in Latin America, at pitifully low productivity rates (World Commission on Environment and Development, 1987; Nations and Komer, 1983). Grazing cattle in cleared rain forest in Chiapas, Mexico, yields a mere 10 kg/ha (8.9 lb/acre) of meat per year. In sharp contrast, the traditional methods of the indigenous Maya yield 5896 kg/ha (5263 lb/acre) of shelled corn plus 4535 kg/ha (4049 lb/acre) of roots and vegetables per year for 5 to 7 years, then additional yields of citrus, rubber, cacao, avocado, and papaya for the following 5 to 10 years while the plots are allowed to regrow (Iltis, 1983). To remove the pressure on tropical forests, a complex set of changes must occur, ranging from (1) economic disincentives to the industrial nations' importation of tropical wood to (2) extensive agrarian reforms in the developing nations. These reforms include land redistribution that returns productive land to displaced peasants, the development and adoption of ecologically sound farming practices, and the goal of self-sufficiency in order to reduce the need for foreign exchange.

Modern agriculture's impact on the genetic diversity of crop plants is also profound. Recent gains in yield have been achieved by increased dependence on very few genetically narrow cultivars of only a handful of major food plants (Committee on Germplasm Resources, 1978; N. J. Brown, 1981; Mayer, 1981; Wilson, 1983; Duvick, 1984; Kannenberg, 1984). Only 10 to 20 crops provide 80 to 90 percent of the world's calories (N. J. Brown, 1981; Mayer, 1981)—or as W. L. Brown (1981) states, 15 species of cultivated plants "literally stand between man and starvation." As of 1970, in the United States 56 percent of the soybean crop, 71 percent of the hybrid corn crop, and 41 percent of wheat acreage were occupied by only six cultivars each (Duvick, 1984). For red winter wheat, only two varieties made up 75 percent of the wheat land, with a single variety covering more than 50 percent of the acreage. Although the dominance of the top six cultivars was reduced somewhat by 1980—to 42 percent of soybeans, 43 percent of corn, and 38 percent of wheat (Duvick, 1984)—still, the genetic base of the varieties is limited (Kannenberg, 1984). The major cultivars of red winter wheat trace back to only two original cultivars: Turkey and Marquis. The hundreds of corn hybrids grown in the United States and Canada are largely based on about 12 inbred lines that originated from a few open-pollinated varieties of a single race (out of some 200 known races of corn). The U.S. soybean varieties were derived originally from six plants from Asia (N.J. Brown, 1981). Throughout the world the attraction of high yields from new strains has caused farmers to abandon the more diverse, locally adapted varieties, or landraces (Committee on Germplasm Resources, 1978; Pluckiest et al., 1983; Van Sloten, 1984). For example, F1 hybrids have virtually replaced landraces of cultivated Brassica spp. in Japan, Korea, North and South America, and Northern Europe. China, the last holdout for genetic diversity of this group, is now losing landraces rapidly (Van Sloten, 1984). In the past 40 years, 95 percent of Greek wheat landraces have been lost because of the introduction of highyielding varieties (Pluckiest et al., 1983). Similarly, nearly all sorghum landraces disappeared from South Africa following the introduction of high-yielding hybrids from Texas (Pluckiest et al., 1983). Landraces of white potatoes, sugarcane, and tomatoes are already mostly lost (Committee on Germplasm Resources, 1978). The problem is compounded by governmental policy where official seed agencies emphasize uniformity and purity of seeds, as in the United States and Canada (Committee on Genetic Vulnerability of Major Crops, 1972). Solutions to this threat will need to be multifaceted.

Reduced genetic diversity in crops results in a loss of flexibility to meet future breeding challenges. Ironically, the landraces in jeopardy are essential to the maintenance of the high-yielding, genetically narrow cultivars that replace them (Committee on Germplasm Resources, 1978; Bertrand, 1981; Pluckiest et al., 1983; Smartt, 1984). Landraces are the breeders' sources of resistance to diseases and pests and of adaptation to climatic extremes and poor soils (Brady, 1981; Pluckiest et al., 1983; Kannenberg, 1984). For example, the source of powdery mildew resistance in California melons was a wild melon from India. Resistance to ripe rot in North American pepper plants came from a Peruvian species. High-yielding hybrid cucumber seed was obtained from a Korean strain (Bertrand, 1981). Wild genotypes of rice have helped to boost rice yields by an estimated $1.5 billion per year (Myers, 1983). Several recent trends further increase the need for new sources of genetic diversity: Wide use of uniform varieties in monoculture has increased vulnerability to biological and physical stresses; expansion of agriculture into marginal lands requires special adaptation to climatic extremes, poor soil, and low water quality; consumer demands for appearance, flavor, and quality require continual innovation, as do such changes in farm technology as no-till farming, multiple cropping, and biological farming (Pimentel, 1977; Bertrand, 1981). Without the landraces and wild species to draw on in the future, breeders will reach the limits set by the variation available in the cultivars and have nowhere to turn for future innovation (Kannenberg, 1984).

Crop breeders have begun to respond to this dangerous loss of genetic diversity. Reduction in dependence on the top six cultivars of wheat, soybeans, and corn between 1970 and 1980 in the United States is one example (Duvick, 1984). Corn breeding has shifted most dramatically toward genetically broader varieties (Kannenberg, 1984). Pedigree breeding (where the best cultivars of each generation are the parents for the next generation), which results in a very narrow genetic base, is less popular now than it once was. By 1975 synthetic populations (intercrosses of several elite inbred lines), crosses of exotic and adapted lines, and population-based recurrent selection programmes were producing 49 percent of the new lines of corn. Kannenberg (1984) advocates adoption of an even richer genetic foundation, which he achieves by a hierarchical open-ended system in which new genotypes are continually introduced into the breeding programme. Widespread recognition of the severity of the genetic erosion of crop species came in the 1960s, and by the early 1970s this recognition had turned to alarm among crop breeders (Williams, 1984). The Consultative Group on International Agricultural Research (CGIAR) was formed in 1971 to study the problem, and in 1973 the CGIAR created the International Board for Plant Genetic Resources (IBPGR). This board undertook the major tasks of collecting, organizing, and preserving seed sources representing as much of the worldwide diversity of crop and related species as possible. The funding for this programme is international, and the seeds and information gathered are meant to be freely available to all. Although this programme achieved a great deal in its first 10 years, especially in the collection of lines of the major grain crops, in 1984 the job was still far from complete (Committee on Germplasm Resources, 1978; Williams, 1984). Systematic ecogeographic searches in remote areas are still needed, with special attention to disease and pest resistance. Closely related wild species and most tropical crops have been largely neglected. Effective methods for preserving vegetatively propagated crops and those with short-lived seeds have yet to be found. Even though many seed sources are now safely preserved, most are still essentially unavailable to breeders because they have not been evaluated or characterized. This huge job will require major effort and funding. In many cases, in situ conservation of wild crop relatives and landraces is necessary to effectively preserve their genetic diversity, particularly for perennial vegetatively propagated crops (Ingram and Williams, 1984). The most direct way to accomplish the in situ preservation of genetic diversity may be to preserve indigenous subsistence agriculture. Where local varieties are found to be nutritionally superior, ecologically sound, and more reliable in unpredictable growing conditions, these systems should be maintained rather than replaced with modern, genetically limited varieties and farming techniques..

Chemical Contamination

"In nearly all respects agriculture became an industry, sharing with the traditional manufacturing industries the problems of waste byproducts disposal." Young (1983) is referring in this quote to the changes in modern agriculture that have occurred over the past 45 years since the introduction of cheap inorganic nitrogen fertilizers. Six kilograms (13 lb) of pesticides, 63.5 kg (140 lb) of actual nitrogen, 19 kg (42 lb) of phosphate—these were the average chemical inputs per irrigable acre in California in 1980. In 1 year, California alone uses 55 million kilograms (121 million pounds) of restricted-use pesticides (Afford and Ferguson, 1982). Along with the promise of high production that these figures imply is the potential environmental pollution that these massive chemical inputs can cause.

Pollution of Groundwater and Surface Water

Nitrate is currently recognized as the most serious agricultural chemical pollutant (Alfoldi, 1983). As nitrate shows up in drinking water, where it can cause methemoglobinemia in infants, attention is increasingly focused on nitrates in the groundwater source. Agriculture contributes to groundwater nitrate through the leaching of nitrogen fertilizer, animal wastes from feed lots, and organic matter from plowing under grasslands and crop residues (Guenzi, 1974; Aldwell et al., 1983; Alfoldi, 1983; Young 1983). Studies in Illinois from 1969 to 1972 showed that 73 percent of farm wells in areas with shallow aquifers had nitrate levels above the critical level considered safe to drink (Singh and Sekhon, 1978/1979). Reports of elevated nitrate levels in groundwater from England, Israel, Germany, France, and the Netherlands surfaced in the 1970s, and at a recent symposium other European countries, as well as Nigeria and India, joined the growing list (Young, 1983).

Where water tables are shallow, nitrate can reach the water table quickly (Zaporozec, 1983). A number of studies have shown correlations between nitrogen fertilizer use and polluted groundwater (Singh and Sekhon, 1978/1979; Zaporozec, 1983; McWilliams, 1984). In addition, irrigation on porous, shallow soils exacerbates the problem of nitrate leaching. An experiment conducted in Minnesota on sandy glacial outwash soils above a 4.5-m (14.8-ft) water table clearly demonstrated an increase in nitrate in the groundwater after onetime applications of nitrogen fertilizer (Gerwing et al., 1979). Deeper water tables can also become contaminated by nitrate, but fewer cases are known. In a 1966 survey of thousands of rural Illinois wells, only 1.4 percent of the water samples from deep wells [deeper than 30 m (98 It)] exceeded the Environmental Protection Agency (EPA) limit of 10 mg/L (38 mg/gal), as compared to 23 percent of the samples from shallow wells. But in 1982 in rural Rock County, Wisconsin, 10 percent of the water samples from deep wells [deeper than 46 m (151 ft)] exceeded the limit (Zaporozec, 1983). In England, high nitrate concentrations were found to a depth of 35 m (115 ft) below the minimum water table (Young, 1983). The movement of nitrate into deep groundwaters occurs so slowly in relation to horizontal groundwater movement that it is difficult to prove a direct association with fertilizer applications (Singh and Sekhon, 1978/1979); even so, nitrate levels are usually correlated with overlying land use. One study found the highest nitrate concentrations below cattle feedlots and irrigated alfalfa, followed by fallow wheat and then by virgin grasslands, which had the lowest concentrations

(Singh and Sekhon, 1978/1979). Decontamination will take place as slowly as contamination. Once below the root zone, nitrate is highly stable (Singh and Sekhon, 1978/1979) and can percolate slowly toward deep aquifers for years. Even if all fertilizer application were suspended immediately in areas where deep aquifers are beginning to show nitrate pollution, the inputs of past years would still slowly percolate toward the aquifers, and nitrate input to the groundwater would continue for many years.

Deep aquifers are fairly well protected from pesticide contamination (Alfoldi, 1983), since all but the most persistent pesticides degrade before they reach a deep aquifer. Shallow aquifers, however, are more commonly contaminated with pesticides. In the shallow aquifers of Wisconsin's Central Sands region, the pesticide aldicarb has reached the groundwater in amounts above the EPA limit (McWilliams, 1984). In California, agricultural use of DBCP, a highly leachable nematicide used on a variety of crops, contaminated wells until its use was suspended in 1977 (FAO, 1979; Alford and Ferguson, 1982; Lecos, 1984).

Surface waters receive agricultural chemicals by runoff from field applications and feedlots and from the dumping of excess chemicals (Afford and Ferguson, 1982). Phosphorus from fertilizer applications binds to soil particles, whereby it can be exported in erosive runoff. Soluble organic forms in animal wastes or plant residues can also move into surface waters. Phosphorus is often the limiting nutrient in aquatic systems and is thus a major contributor to eutrophication (Oglesby, 1971). Nitrate, which is highly soluble and easily transported in runoff, also contributes to eutrophication (Oglesby, 1971). In combination, these two fertilizers have contributed much to the degradation of streams and other bodies of water in agricultural regions.

Pesticide runoff is normally low, with losses of 0.5 percent or less (Wauchope, 1978), giving concentrations of 1 rLg/L (3.8 g/gal) or less (Guenzi, 1974). Under the worst conditions—a major rainfall or heavy irrigation within 3 days of application, the most mobile pesticides, sloping ground, and fine-textured soils—the losses may reach up to 15 percent; however, under these conditions the pesticide is likely to be in low concentration (Wauchope, 1978). Higher concentrations of pesticides in runoff occur when the runoff volume is low; even so, the concentrations are usually below the acute toxicity levels for aquatic organisms (Guenzi, 1974). Ironically, contaminated runoff is less likely to poison aquatic life than to damage other crops. Irrigation water contaminated with herbicides in one field can reduce crop productivity in another field. Picloram concentrations as low as 0.1 ppm can reduce sugar beet yield significantly (Guenzi, 1974).

The dangers of persistent pesticides in aquatic ecosystems, especially the organochlorines (e.g., DDT), have been known for many years. These chemicals, although not highly toxic at low concentrations, stay in the bodies of consumers and thus accumulate up the food chain (A. W. A. Brown, 1978; Guenzi, 1974). Top predators (such as fish-eating birds) consume toxic dosages with disastrous results. Agricultural use of toxaphene from 1957 to 1961 in the Klamath basin of northern California killed white pelicans and other fish-eating birds in wetland refuges (A. W. A. Brown, 1978). As these consequences became known, and as many organochlorine compounds came to be banned in the richer countries, they began to be replaced with shortlived, highly toxic organophosphates (such as parathion). Although toxic in minute quantities, the short life of such chemicals makes them unlikely to accumulate in bodies of water, and it also makes them difficult to study (Alfoldi, 1983; Guenzi, 1974; Wauchope, 1978). So far, the presence of nonpersistent pesticides has been documented, but no evidence of permanent impact in aquatic systems has been found (Wauchope, 1978). Studies set up to determine the impact of these pesticides have been limited to the direct application of pesticides to streams at higher concentrations than those found in agricultural runoff. These studies show that aquatic organisms vary widely in their sensitivity to different pesticides. Studies of organophosphates have shown that stone flies are sensitive to malathion, but less so to trichlorfon. In one study, malathion caused only a temporary depression of stream insect fauna, whereas fenitrothion caused extensive kills of stream insects (A. W. A. Brown, 1978). Species-specific responses to particular chemicals have been used to infer possible damage from natural agricultural runoff. A stream community in North Carolina that was affected by agricultural runoff showed a pattern of species reduction that matched the pattern caused by the direct application of carbaryl to streams in other studies—a correspondence that led Lenat (1984) to infer that this pesticide may have been present in the runoff. So far, studies of the effects of long-term, low-dose exposure of aquatic systems to nonpersistent pesticides are rare to nonexistent. Such studies should be pursued in order to ensure that we avoid serious environmental consequences.

Residues on Food

Pesticides also pose hazards by direct contamination of foodstuffs. To what extent is our food contaminated with pesticide residues, and how much of a hazard is this?

The long-lived pesticides, such as DDT, other organochlorines, and parathion, are most likely to leave persistent residues (Bull, 1982). Even though most persistent pesticides are banned in the United States and other wealthy countries, they are still persistent where previously used, are still used elsewhere, and show up on foods. Others, such as organophosphates, tend to break down so rapidly that they are unlikely to show up on food unless applied to crops very close to harvest time, but they are often more acutely toxic. Approved pesticides leave little residue on crops when used according to directions. National and international standards of acceptable residue levels are based on approved usage and maximum acceptable daily intake (MADI) levels; thus, theoretically, foods should be safe. However, it is not practical to monitor all crops and shipments, residues are not always detected even when tests are done, and pesticides are not always used according to directions; thus, dangerous residues do make their way to the foods that people eat.

In the United States, public concern centers on the residues on imported foods (Lecos, 1984). Although the Food and Drug Administration (FDA) maintains an active monitoring programme that makes it advantageous for growers to comply with U.S. standards, only a small percentage of shipments can actually be tested. According to FDA figures, some 2000 food shipments crossing the Mexican border are tested annually (Thompson, 1984). This is indeed a small percentage of the Mexican import traffic, which exceeds 900 truckloads per day during the peak of the January-to-April shipping season. In 1986 the FDA checked nearly 8000 samples, less than 1 percent of the imported shipments of fruits and vegetables. Rather than hold perishable produce at border checkpoints for the average 18 days it takes for the test results, shipments are routinely allowed to pass (GAO, 1986b). Positive tests, indicating dangerous pesticide residues, result only in the stoppage of future shipments or in recalls, if the food can be traced. From 1979 to 1985 an average of 6 percent of the shipments sampled were found to have illegal residues—a rate about twice the average rate of contamination of domestic foods. Online little more than half (55 percent) of the illegal shipments were prevented from reaching consumers, and in only a few cases were damages collected from offending shippers (GAO, 1986a; Bull, 1982).

How often pesticide residues actually turn up on the table in the United States is largely unknown. The FDA's total-diet studies, conducted once a year in four regions of the country, reveal that pesticides do show up frequently but are "consistently" below the limits set by the Food and Agriculture Organization (FAO) and the World Health Organization (WHO) (Farley, 1988). DDE, a breakdown product of DOT, was still found in about 40 percent of the food sampled in 1983. This chemical, along with other organochlorines, has declined steadily in food since the 1970s. At the same time, organophosphates have increased, corresponding to their increased usage. Malathion was found in about 20 percent of the food sampled in 1983. Increases in pesticide residues found in potatoes and peanuts in the early 1980s stemmed from the greater use of sprout suppressants and soil fungicides, respectively (Penner, 1984). The EDB scare of the early 1980s is an example of a case where residues went undetected for a long time. EDB (ethylene dibromide) had been used since 1948 as a soil fumigant and as a postharvest fumigant on citrus and grains. Although a suspected carcinogen, EDB was thought to leave no residue, but the sensitive tests developed in the 1970s found that it did leave a residue. In 1983 it was found in cake mixes, cereals, and other foods from grains, and it was found in groundwater in several states where it had been used as a soil fumigant.

Public reaction was so strong that in September 1983 the EPA issued an emergency ban on the use of EDB as a soil fumigant. Bans on its use on grains and citrus followed in early 1984 ("Pesticide Is Banned," 1983; "New Moves," 1984; "EPA to Limit," 1984; Thompson, 1984). Although our food supply may now be nearly free of this dangerous chemical, over 30 years of use left a legacy of contaminated water and untold increases in cancer risk for the generation of consumers and farmers raised with it ("EDB's Long-Lasting Legacy," 1985; Barles and Kotas, 1987).

In third-world countries, pesticides are commonly misused because information on their proper use is not generally available. Their labeling is poor and often not in the native language, extension advisory services are usually unavailable, and proper usage outside of the temperate zone may be unknown, even by the manufacturers (Bull, 1982). In one market in India, a 7-year study showed that organochlorine residues were above the tolerance limits in more than 35 percent of the food. Forty percent of the samples of horticultural crops in Kenya contained excessive residues. A 1978 study. turned up rice samples from Sri Lanka that contained excessive residues of dieldrin and heptachlor. While exported foods are usually monitored carefully so that they will pass the inspections that the richer importing nations impose, locally marketed foods are usually not monitored, yet these are most likely to have residues of the short-lived, highly toxic organophosphates because of the short time between harvesting and marketing.

The responsibility for correcting this problem seems to lie with the richer nations of the world, which have profited by the export of pesticides and the technology of modern agriculture at the expense of the citizenry of third-world nations.

Pesticide Resistance

In addition to directly poisoning our environment and our food, pesticides pose a serious threat to our food-production system itself. From one viewpoint, pesticides are the wonder chemicals that have increased food production by 20 percent since 1940 by reducing pest damage. Yet over the same period they have also created at least 261 strains of insect species, 67 strains of plant pathogens, 2 strains of nematodes, and 4 (by some counts 19) strains of weeds that they cannot kill (Conway, 1982). While insecticide use increased tenfold since the 1940s, crop losses to insects doubled (Pimentel et al., 1983). The key to this paradox is the strong selection for resistance that pesticides exert on their target pests. Pesticides never kill 100 percent of a pest population, and the survivors tend to have a lower susceptibility to that particular chemical. With every repeated application of the same pesticide, these naturally resistant individuals make up a higher percentage of the population, until a highly resistant strain of pest evolves. When the conditions are right—the pesticide kills a large percentage of the pest population, the pest completes several life cycles per year, and little immigration from untreated populations occurs—resistance can develop very rapidly.

Resistance to one pesticide often confers resistance, or faster development of resistance, to a whole family of related pesticides (Bull, 1982). Alternating different pesticides or applying a mixture of chemicals can sometimes delay the development of resistance, but it can also promote the development of superresistant pests. Superpests, resistant to multiple pesticides, have already developed and threaten a number of crops throughout the world. So far, widespread famine has not resulted from superpests on food crops, but regional crises have occurred and some larger problems loom ahead. In North and Central America, cotton production has the worst record of superresistant pests. The cotton bollworm and budworm are both resistant to organochlorines, organophosphates, and carbamates and are becoming resistant to the new pyrethroids (FAO, 1979; Conway, 1982). Worldwide, 33 species of cotton pests were resistant to pesticides as of 1982, an increase from 20 in 1965. Rice, the staple food of a large proportion of the world population, now has 14 pest species resistant to one or more insecticides. Without pest control, experts estimate that 50 percent of the rice crop in Japan, India, and the Philippines would be lost to pests (Conway, 1982). The diamondback moth, which feeds on brassicaceous crops, has developed multiple resistance to at least 11 insecticides, including some from all the major groups, including pyrethroids (Bull, 1982). This pest now threatens the production of these crops in southeast Asia (FAG, 1979; Conway, 1982). Farmers there may spray three times a week and use 11 different insecticides in the course of a season, often spraying less than a week before harvest and in higher-than-recommended doses (Bull, 1982).

Pesticide resistance is considered inevitable, with no type of chemical immune (A. W. A. Brown, 1978); formerly effective pesticides cannot be successfully reintroduced (Bull, 1982). In order to avoid major pest-caused crop failures in the near future, we cannot simply rely on chemists developing new chemicals. Rotation of pesticides may delay the development of resistance in some cases, but these cases will require careful, regionally specific tailoring, and the research has yet to be done (Conway, 1982). Under normal field conditions, tolerating low kills slows the rate of resistance development because of the lower selection pressure exerted. Any increasing migration of pests from untreated areas into treated areas also slows the evolution of resistance (Conway, 1982). Both lower kills and higher migration probably mean that farmers would have to tolerate greater crop damage by pests. Integrated pest management is one approach to a longer-term solution, but it still may be vulnerable to the development of resistance. The development of pest-resistant crop strains, a major thrust of crop breeding programmes, suffers much the same malady as the pesticide solution; it becomes a race between the breeder and the pest as to which will develop resistance first.

Rather than looking at ways to kill pests without exerting selection pressure for resistance, we might do better to turn the question around and ask, How can we successfully do agriculture without having to kill pests? Perhaps the cropping systems themselves can be altered to minimize pest population buildups in the first place. Traditional crop rotations and mixed-cropping systems hold potential for pest control by suppression rather than by killing (Pimentel, 1977). Development of new perennial crops, drawing on the greater natural resistance of perennials, holds great promise (Jackson, 1987).

Despite the above chronicle of ecological disasters that attend it, agriculture is still potentially a renewable enterprise. In a certain sense, nothing is new here. In every century for 10,000 years on almost every agricultural acre in the world, ecological capital has slipped seaward from its wilderness-built home or been degraded by toxic materials. Still, on a global scale, agriculture is seen as potentially renewable and therefore as fundamentally different from the industrial sector of society. It is only in the last 50 years, with the expansion of industry and the chemicalization of agriculture, that the inherently extractive economy has acted as though the renewable resources that support agriculture are fair targets for exploitation in industrial terms. That is what makes the modern era different. That is what makes the current agricultural economy more brittle than almost any agricultural economy in history. It is hard for us to see this, perhaps, because it is hard for us to imagine our energy and aquifer mines lying hollow. It is hard for us to believe that our well water and air, our pleasingly packaged food, and our perfect produce could contain invisible poisons. It is hard for us to grasp the value of genetic diversity or to foresee the consequences of rampant extinctions. It is hard for us to believe that we may one day come to the end of our magical ability to produce everhigher-yielding crops. It is hard for us to comprehend the total loss of our vast tropical forests or to anticipate the climatic changes when the earth's belly is belted in barren, sterile soils rather than the green, moist vegetation. It is hard for us to imagine a world where ancient salts sterilize the land and young chemical pesticides and fertilizers lie below our agricultural surfaces like demons in quiet prisons of degraded soil.

The outlook is not entirely bleak, for solutions to all these problems lie in lessons we have learned and can still learn from nature. If we turn our attention away from the extractive industrial model and begin to focus on nature's models of productive ecosystems as our guide for agricultural systems, we may yet see truly sustainable agriculture emerging. It isn't that nature learns faster than humans. It is just that she has been at it longer.

Barriers for the Implementation of Alternatives

The agroecological approach seeks the diversification and revitalization of medium size and small farms and the reshaping of the entire agricultural policy and food system in ways that are economically viable to farmers and consumers. In fact, throughout the world there are hundreds of movements that are pursuing a change toward ecologically sensitive farming systems from a variety of perspectives. Some emphasize the production of organic products for lucrative markets, others land stewardship, while others the empowerment of peasant communities. In general, however, the goals are usually the same: to secure food self-sufficiency, to preserve the natural resource base, and to ensure social equity and economic viability.

What happens is that some well-intentioned groups suffer from "technological determinism". and emphasize as a key strategy only the development and dissemination of low-input or appropriate technologies as if these technologies in themselves have the capability of initiating beneficial social changes. The organic farming school that emphasizes input substitution (i.e. a toxic chemical substituted by a biological insecticide) but leaving the monoculture structure untouched, epitomizes those groups that have a relatively benign view of capitalist agriculture. Such perspective has unfortunately prevented many groups from understanding the structural roots of environmental degradation linked to monoculture farming.

This narrow acceptance of the present structure of agriculture as a given condition restricts the real possibility of implementing alternatives that challenge such a structure. Thus, options for a diversified agriculture are inhibited among other factors by the present trends in farm size and mechanization. Implementation of such mixed agriculture would only be possible as part of a broader programme that includes, among other strategies, land reform and redesign of farm machinery adapted to polycultures. Merely introducing alternative agricultural designs will do little to change the underlying forces that led to monoculture production, farm size expansion, and mechanization in the first place.

Similarly, obstacles to changing cropping systems has been created by the government commodity programmes in place these last several decades. In essence, these programmes have rewarded those who maintained monocultures on their base feed grain acres by assuring these producers a particular price for their product. Those who failed to plant the allotted acreage of corn and other price-supported crops lost one deficit hectrage from their base. Consequently this created a competitive disadvantage for those who used a crop rotation. Such a disadvantage, of course, exacerbated economic hardship for many producers. Obviously many policy changes are necessary in order to create an economic scenario favourable to alternative cropping practices.

On the other hand, the large influence of multinational companies in promoting sales of agrochemicals cannot be ignored as a barrier to sustainable farming. Most MNCs have taken advantage of existing policies that promote the enhanced participation of the private sector in technology development and delivery, positioning themselves in a powerful position to scale up promotion and marketing of pesticides. Realistically then the future of agriculture will be determined by power relations, and there is no reason why farmers and the public in general, if sufficiently empowered, could not influence the direction of agriculture along sustainability goals.

Clearly the nature of modern agricultural structure and contemporary policies have decidedly influenced the context of agricultural technology and production, which in turn has led to environmental problems of a first and second order. In fact, given the realities of the dominant economic milieu, policies discourage resource-conserving practices and in many cases such practices are not privately profitable for farmers. So the expectation that a set of policy changes could be implemented for a renaissance of diversified or small scale farms may be unrealistic, because it negates the existence of scale in agriculture and ignores the political power of agribusiness corporations and current trends set forth by globalization. A more radical transformation of agriculture is needed, one guided by the notion that ecological change in agriculture cannot be promoted without comparable changes in the social, political, cultural and economic arenas that also conform agriculture. In other words, change toward a more socially just, economically viable, and environmentally sound agriculture should be the result of social movements in the rural sector in alliance with urban organizations. This is especially relevant in the case of the new biorevolution, where concerted action is needed so that biotechnology companies feel the impact of environmental, farm labor, animal rights and consumers lobbies, pressuring them to re-orienting their work for the overall benefit of society and nature.

REFERENCES

Collier RJ, Miller MA, McLaughlin CL, Johnson HD, Baile CA (2008). "Effects of recombinant bovine somatotropin (rbST) and season on plasma and milk insulin-like growth factors I (IGF-I) and II (IGF-II) in lactating dairy cows". *Domest. Anim. Endocrinol.*. doi:10.1016/j.domaniend.2008.01.003. PMID 18325721.

Mullan, W.M.A. (2002). Science and technology of modified atmosphere packaging. [On-line] UK: Available: file:///D:/dairy-w/Dairy-27.htm. Accessed: 16 May 2008.

North, R (2007-01-10). Safeway & Chipotle Chains Dropping Milk & Dairy Derived from Monsanto's Bovine Growth Hormone. Oregon Physicians for Social Responsibility. Retrieved on 2008-01-29.

Production Responses to Bovine Somatotropin in Northeast Dairy Herds - Bauman et al. 82 (12): 2564 - Journal of Dairy Science

4

Plans and Projects Related to Intensive Cattle Development in India: Analyzing Schemes and Evaluating Processes

NEEDS OF INTENSIVE CATTLE DEVELOPMENT IN INDIA: SCHEMES & PROCESSES

IN SUM THE need for improving the quality of cattle and raising their productivity, particularly with regard to milk, has been fully recognised and this has been receiving increasing attention with the initiation of the First Plan. In percentage terms India has about 20 per cent and 54 per cent of the world's cattle and buffaloes populations respectively and together they form 23 per cent of the world's bovine population. In spite of such an impressive bovine population, India has been able to produce only 6.56 per cent of the world's milk production. This has been mainly due to low average yields of cows and buffaloes in the country. The average annual milk yield of cows and buffaloes in India has been estimated at only 382 lbs. and 1117 lbs. respectively. With such low yield it is no wonder that we have a very unsatisfactory level of milk consumption in the country. There are 25 well-defined breeds of cattle and eight breeds of buffaloes in the country, but vast majority of these cattle and buffaloes are non-descript and of poor productivity. It is estimated that 75 per cent of Indian bovine population falls under this category.

The Case of Key Village Scheme: An Analysis

Certain research projects aimed at cattle development were taken up from the thirties. Herd books were opened for eight improved breeds of cattle, namely,

Haryana, Kankrej, Gir, Tharparkar, Red Sindhi, Sahiwal, Kangayam and Ongole and the Murrah breed of buffalo. Certain measures for encouraging selective breeding for improving the milk yielding capacity and relating or improving draught quality of cattle were laid down. The first organised attempt to develop village cattle on an effective scale was initiated with the introduction of key village scheme during the First Plan. This scheme was introduced in the light distribution of a few premium bulls for upgrading village cattle and the production of a limited number of breeding stock in a few Government farms was not adequate for bringing about any substantial improvement in the quality of the stock. They key village scheme, therefore, embraced all aspects of cattle improvement, *viz.* controlled breeding, improved feeding, disease and sexual health control, better management and marketing and adoption of improved animal husbandry practices through proper extension methods. Since the First Plan, a number of key village blocks have been established in the different States and the total breedable female bovine population covered by these centres was spread over 529 blocks at the end of 1972. In concept of the key village scheme is indeed sound and this scheme has helped greatly in developing pockets of good quality cattle in different parts of the country. It has also given a sound approach for taking up programmes of cattle development in an integrated manner, but considering the vast magnitude of the problem this scheme has not been able to make the necessary impact for immediately making available the increased quantities of milk that are required to meet the demand in several places. It has been estimated that this scheme is at present handling only about three per cent of the breedable cow and she-buffalo population of the country. Moreover, these centres are generally dispersed all over the country, and in many places all the items of activities envisaged in the scheme have not been taken up. Each key village blocks is thus only a tiny area of well-organised activity surrounded all round by vast areas where indiscriminate breeding is continuing. As a result the effect of the good work done in the key villages gets obliterated by the uncontrolled breeding going on all round. It is desirable that the key village programme should be continued and wherever necessary further extended and improved so that the general programme of cattle development is continued without break. It is at the same time very essential to close down those blocks which have been chronically lagging behind. In addition to the key village programme, it is essential to take up some large Intensive Cattle Development Projects on the lines of the package programme for agricultural production. These could be special projects for creating the necessary impact and confidence in cattle industry. These projects may be located both in the breeding tracts of the indigenous breeds of cattle and buffaloes which we would like to develop in the future and also in the milk shed areas of the large diary projects (even though they may be largely non-descript cattle) to enable the diary plants to procure milk to their installed capacities economically. In the former areas the policy of breeding should be selective. These areas should developed for the supply of quality breeding stock to other parts of the country to provide the necessary incentive to the producers. These areas will have to be linked up with the milk supply scheme but where this is not possible, dairy products in factories will have to be provided for the utilization of milk. In the later areas, cross-breeding with suitable exotic breeds and/or up grading with improved indigenous breeds may be practised. In determine the breeding policy for the various localities recommendations of the Working Group on Breeding Policy set up by the Ministry of Food and Agriculture and the Committee on Cross-breeding set up by the Central Council of Gosamvardhana will give helpful guidance. Under each project envisaged to be taken up during the Fourth Plan, it is proposed to cover one lakh cows and she-buffaloes of breedable age so that it will have a significant impact on milk production in the area. The exact size of these projects will have to be determined after a proper survey of the areas where these centres are to be located. In this connection it should be borne in mind, as already experienced in the case of key village scheme, that good cattle cannot be raised everywhere and for development of cattle on right lines for improving their productivity expeditiously, certain natural conditions as well as incentives are necessary. It would, therefore, be desirable that these centres are located in places which will provide conditions conducive to good cattle raising. In the location of these centres this should be primary consideration. With a view to achieving concrete results within the shortest possible time, concentrated efforts will have to be made in all aspects of cattle development which have a breaking on increasing the productivity of the animals. The programme popularly known as ICDP, was started in 1964-65 and there were 52 such projects in operation in 1971-72. The following programme of work is being taken up at each of the centres.

(a) *Survey.* A village to village survey of the area selected for the location of the project is undertaken and completed in a short time to ascertain the initial animal husbandry conditions, level of milk production, marketing and utilization of milk, feeds and fodder resources etc. Initial data would be useful for periodical evaluation of the progress under the project. The survey also introduces the staff to all the livestock owners in his area and gives an opportunity of acquainting them with the broad outline of the programme envisaged. When the survey has been completed a summary should be made which could be used all the year round for reference. The survey may be repeated preferably in the beginning of each year subsequently. Through this preliminary survey it will be ensured that the sites chosen for the location of projects offer suitable conditions for raising good cattle and are conducive for animal production.

(b) *Breeding:* As controlled breeding alone would ensure progressive improvement in the genetic make-up of cattle for efficient production, adequate breeding facilities are provided for all the breedable bovine population both by artificial insemination and natural services, A cattle development officer is provided for the whole project who will be assisted by the officer incharge of the breeding operations in the four regions into which the project is divided for operational convenience. Each veterinary officer is incharge of the artificial insemination centre in his region and organises the breeding operation for the 25,000 breedable cows and buffaloes.

(c) *Castration of scrub bulls:* For the successful operation of the project which envisages progressive breeding control it is very helpful if the Livestock Improvement. Act is enforced in the project area. Complete-removal of all scrub bulls from the area is necessary to ensure full breeding control. For this purpose a mass castration campaign is organised throughout the area initially by the project authorities as soon as the preliminary survey is completed. A mass castration team is then located in each Block to undertake castration of all scrub bulls in a planned and systematic manner.

(d) *Veterinary aid and disease control:* During the preliminary survey information is gathered as to the contagious diseases which are prevalent in the area and necessary measures are immediately initiated for periodic prophylactic vaccination of livestock against these diseases. It would be advantageous if the Control of Contageous Disease Act is enforced in the project area. Arrangements are made at each stockmen centre for the provision of veterinary aid. For this purpose the stockmen would be provided with required veterinary equipment and medicines to enable him to undertake this work.

(e) *Registration and milk recording:* With a view to ensuring proper breeding control and locating high yielding stock and utilising their progeny for breeding purposes for ensuring progressive improvement in the quality of stock registration and milk recording of milch bovines and their progeny is carried out. Every effort is made by the staff to get the subsidised calves registered in the Central Hard Books. The regular milk recording work in the respective stockmen units is carried out by the milk recorders.

(f) *Introducing high-yielding milch cattle:* In order to encourage cattle owners to maintain high yielding milch animals they are advanced loans on usual terms for the purchase of good quality milch cattle of a sufficiently high productivity. The owners are required to maintain them properly and supply the milk to the local cooperative for the milk collection centres. The recovery of the loan is made in convenient instalments through deductions from the milk collection area of a large dairy project.

(g) *Subsidies and production incentives:* In order to foster a spirit of healthy competition among the cattle owners and their 'cooperatives for breeding better-cattle and increasing milk production etc. production incentives and prizes for cattle owners are provided for maintaining highest milk yield in animals and also to cooperatives for organizing and stepping up the collection of milk to make its collection and handling economically. Subsidised rearing of selected progeny both young bull and calves on the lives of the calf subsidy under the key village scheme is also provided.

(h) *Feeds and fodder development:* An adequate supply of feeds and fodder is essential to support production of increased quantities of milk in the intensive cattle-development areas. In fact, feeding constitutes the major portion of the expenditure on maintenance of cattle and, therefore, augmenting the available feeds and fodder resources locally to the extent possible is expected to be more economical than their procurement from other areas. Thus increased cultivation of high yielding and more nutritious feeds and fodder crops particularly crops like the hybrid maize and jowar, perennial crops like hybrid Napier grass para-grass and leguminous crops like berseem, lucerne, cow-pea, gram, guar, phillipeacra etc. could help in reducing the cost of maintaining cattle for milk production. This is achieved partly by increasing the area under fodder cultivation and partly by increasing per care yields through the cultivation of high yielding varieties of fodder crops and adoption of better cultural, manurial and irrigation practices. Emphasis is also to be laid on the cultivation of quick growing fodder crops as 'catch crops' thereby increasing fodder resources without effecting the existing crop patterns.

It would be equally necessary to popularise the conservation of green fodder during the flush season for the requirements of summer season when green fodder is generally in short supply and preventing the wastage of fodder by popularising the use of chaff-cutters in areas where these are not yet popular. In order to supplement the local resources, it is necessary to augment the supply of feeds and fodders by bulk purchase at economic rates from surplus areas and arrange for their supply to cattle owners through their cooperatives on no profit no-loss basis.

- *Demonstrating cultivation of fodder crops:* In a programme of popularising cultivation of fodder crops and adoption of improved cultural, manurial and irrigation practices it would be necessary to demonstrate these on the cultivator's own field. Extensive demonstrations are, therefore, organised on small plots on the cultivators fields with regard to the cultivation of different fodder crops. In each intensive cattle development project area 400 demonstrations per year are taken up. In organizing these demonstrations, seeds, fertilizers and technical assistance is provided free to the cultivator.

- *Distribution of seeds and planting material:* It is necessary to arrange for the timely supply of seeds and planting material of recommended varieties of fodder crops at subsidised rates for an initial period till the cultivation of these crops becomes a regular practice with the farmers. For this purpose the project authorities make necessary arrangements for the requirements of seeds and planting material from the fodder seeds farms and distribute the same at subsidised rates.
- *Irrigation facilities:* Adequate irrigation facilities are one of the essential requirements for extending cultivation of fodder crops, increasing the yields of green fodder and cultivating yearly summer fodder crops, necessary for maintaining milk production during the lean period. In areas where irrigation facilities are not readily available it would be necessary to assist the farmers liberally by way of loans and subsidies for wells and pumping-sets as provided for under the minor irrigation schemes.
- *Silage making:* Adequate supply of nutritious green fodder is essential for maintaining optimum milk production. To provide succulent fodder to cattle during summer months when green fodder is generally scanty, it is necessary to popularise conservation of green fodder during the flush seasons as silage. Silage making is, therefore, popularised by subsiding the construction of medium size silopits.
- *Popularising the use of chaff-cutter:* In a number of areas chaffing of fodder, particularly the thick stemmed fodder crops like maize and jowar, is not yet quite popular and this results in substantial wastage of fodder due to trampling the soiling. The use of chaff-cutters of improved types of choppers is being popularised in areas where this is commonly not used, by supplying chaff-cutters to the cattle owners at subsidised rates.
- *Dairy Extension:* With a view to making the farmer interested in better breeding, feeding and management of his cattle, it is necessary to provide him with a remunerative market for milk. For this purpose, these projects have been linked up with the milk supply schemes wherever possible and with the milk product factories in other cases. It is also necessary to support this programme by initiating dairy extension activities. Milk producers are organised into a co-operative society and they are given financial assistance for obtaining necessary dairy equipment. Under this programme loan are provided for the purchase of cattle feeds and fodder and milch stock. A suitable number of milk chilling stations is provided to

facilitate the transport of milk from interior areas to the main dairy plant. At the project level, a dairy extension officer is incharge of this work.

INTENSIVE CALLTE DEVELOPMENT PROJECT: A CASE STUDY OF THANJAVUR

Livestock growth in the district has shown a marginal increase over the decade. Animal Husbandry which allied activity of Agriculture cannot grow as fast as agriculture since its breeding programme is a slow process the district has more indigenous cattle than any special breed of cattle. We depend on cattle purchases outside the district. Thanjavur poled cattle are distinguished by dehorned and possession of clipped ears. The main stream of the District is fed by Cauvery river, when the river is dry flock owners of the sheep from Ramnad and Southern Districts are coming with their migratory stock for pasturing temporarily. The Cattle Breeding and Fodder Development has been replaced for Intensive Cattle Development Project which functions from Thanjavur. Under Cattle Breeding and Fodder Development a Semen Bank is established at Ammapettai about 19 Kms., from Thanjavur from which Liquid Nitrogen and Frozen Semen straws are being supplied to various institutions and Veterinary Sub-centres of the Animal Husbandry Department in this district. Since, January, 2000, the Hon'ble Chief Minister's Special Animal Husbandry camps are being held in various Panchayat Union Villages where Livestock breeders are not having access to Veterinary aid, in which Veterinary aids, Artificial Insemination (at free of cost) Disease investigation works are being carried out. Demonstration of Urea enrichment of paddy straw Audio Visuals are exhibited in the special camps by Cattle Breeding and Fodder Development. Cattle owners in remote villages are benefitted by the various department activities of this district.

1997- Cattle Census (Thanjavur District)

Cattle		*Male*	*Female*	*Total*
1.	Exotic	247	69	316
2.	Cross breed	11688	41626	53314
3	Non-descript	31553	43123	74676
4.	Indegeneous	105241	142581	247822
	Total Cattle	148729	227399	376128
Buffaloes				
1.	Murrah	78	631	709
2.	Graded	1206	6385	7591
3.	Non-descript	9133	41440	50573
	Total Buffaloes	10417	48456	58873

Ovines			
1. Sheeps	**11384**	**15759**	**27143**
2. Goats	**65906**	**164564**	**230470**
Total	**77290**	**180323**	**257613**
Horse & Ponies	**28**	**37**	**65**
Donkeys	**46**	**54**	**100**
Pigs-crossbreed	**45**	**135**	**180**
Non descript	**1360**	**1734**	**3094**
Total 1479	**1960**	**3439**	
Dogs			
Domestic	**46642**	**23034**	**69676**
Others	**2522**	**1853**	**4375**
Total	**49164**	**24887**	**74051**
Rabbits	**31**	**38**	**69**
Poultry			
Non-descript	**201728**	**449604**	**651332**
Exotic	**15256**	**15624**	**30880**
Total	**216984**	**465228**	**682212**
Ducks	**1736**	**382**	**2118**
Turkey, Guinea, Fowl, Quail	**—**	**—**	**62**

The district has two livestock Farms. One Exotic cattle Breeding Farm at Eachenkottai, in Orathanad Taluk, another Progeny Testing Scheme (Buffaloe) District Livestock Farm, Orathanad. There are 2 Poultry Extension Centres one at Orathanad and another in Pattukkottai. The Animal Husbandry Department of this District looks after the welfare of the livestock through 2 clinician Centres, 6 Veterinary Hospitals, 45 Veterinary Dispensaries, 73 Sub-Centres, 14 Extension Veterinary Dispensaries, 3 Mobile Veterinary Dispensaries and 16 visiting Sub-centres in the district. Work done particulars of the department during 1999-2000 as follows:

1. **No. of Artificial Inseminations done : 1,78,514**
2. **No. of Calves Born: 57,382**
3. **No. of Mass contact Programmes conducted : 881**
4. **No. of Vaccinations done : 9,82,964**

5.	No. of Castrations done	:	38,061
6.	No. of cases treated	:	5,97,423

TOWARDS UNDERSTANDING MANAGED INTENSIVE GRAZING

Management Intensive Grazing (MIG) is the practice of using rotational grazing and careful, usually daily, management to get optimal production. The technique is applied with herds of sheep, cattle, and occasionally other animals. The term "MIG" or "MiG" was popularized by writers and graziers Jim Gerrish and Allan Nation. One hallmark of MIG systems is rotational grazing, that is, the practice of dividing up available pasture into multiple smaller areas, called paddocks, and then moving the animals from one paddock to the next after a number of days. However, in some instances continuous grazing is an accepted strategy under MIG. The grazier manages the grazing by determining the number, size, and layout of the paddocks, when to move animals from one paddock to the next, and when to cut hay or provide supplemental feed. Also, the grazier can choose to add or remove animals from the herd to match the herd size to the available pasture. The decisions are based on estimates of the amount of forage in each paddock, soil conditions, present and forecast weather conditions, season of the year, and condition of the animals. Some MIG operations make objective measurements of forage condition using devices that measure the height of the sward. Others rely more upon personal observation and assessment. One of the key concepts in MIG is the grazing wedge, which is the range of sward heights where the forage grows most rapidly. The monthly magazine The Stockman Grass Farmer is a leading forum of MIG ideas. Graze is a primary source of information on dairy grazing and grazing in the northern U.S.

Comparisons with Traditional Grazing and Cattle Ranching

For farmers and ranchers with cattle in open fields, there is a tendency for the animals to beat down and trample the plants across a wide area. The animals also typically congregate in one area such as around a water tank, feeding wagon, and often in riparian areas where degradation of banks can have negative impacts on wildlife. This repeated trampling of the same areas over and over destroys plant life faster than it can recover. Eventually sections of the field become a permanent swath of exposed soil. When it rains this turns into muck a foot deep, which in turn covers the animals and makes maintaining sanitary conditions difficult. These exposed tracts of land often serve as seed beds for invasive species of weeds. The main idea of the paddock is the concept of rest. Rather than the same large areas being repeatedly trampled, the animals are instead forced to only occupy just a small area of the total field inside the paddock. By keeping the animals in this one small area, the trampled and grazed plants in other previously occupied parts of the field are given time to recover and re-establish themselves. Additionally, constantly moving the animals every few days between paddocks prevents animal wastes from building up to extreme levels in small areas. It also permits time for the wastes to naturally break down so that there is minimal odor from a field of paddocks, as opposed to a feedlot that is constantly trampled into a wet smelly mixture of mud, manure, and urine.

REFERENCES

Ashley, S.D., Holden, S.J. & Bazeley, P.B.S. 1996. *The changing role of veterinary services: a report of a survey of chief veterinary officers' opinions.* Somerset, UK, Livestock in Development.

Baker, J.L. 1995. *Profile of veterinary services in New Zealand.* Rome, FAO.

Cheneau, Y. 1984. Towards new structures for the development of animal husbandry in Africa south of the Sahara. *Rev. Sci. Tech. Off. Int. Epiz.*, 3(3): 621-627.

CTA (Technical Centre for Agriculture and Rural Cooperation). 1985b. *Proceedings of the Seminar on Primary Animal Health Care in Africa.* 25-28 September 1985, Balantyre, Malawi.

FAO. 1991. *Guidelines for strengthening animal health services in developing countries.* Rome.

FAO. 1994. *Expert Consultation on Application of effective herd health and production programmes to increase livestock productivity in developing countries.* 30 March-1 April 1993, Rome.

FAO. 1996. *World livestock production systems - current status, issues and trends.* FAO Animal Production and Health Paper No. 127. Rome.

Fassi-Fehri, B.E.C. & Bakkoury, M. 1995. *Profile of the veterinary services in Morocco.* Rome, FAO.

Gatongi, P.M., Scott, M.E., Ranjan, S., Gathuma, J.M., Munyna, W.K., Cheruiyot, H. & Prichard, R.K. 1993. Effects of three nematode anthelmintic treatment regimes on flock performance of sheep and goats under extensive management in semi-arid Kenya, *Vet. Parasitology*, 68: 323-336.

McInerney, J.P., Howe, K.S. & Schepers, J.A. 1992. A framework for the economical analysis of disease in farm livestock. *Prev. Vet. Med.*, 13: 137-154.

Swallow, B.M., Mulatu, W. & Leak, S.G.A. 1995. Potential demand for a mixed public-private animal health input: evaluation of a pour-on insecticide for controlling tsetse-transmitted trypanosomiasis in Ethiopia. *Prev. Vet. Med.*, 24: 265-275.

Till, J.E. 1995. Building credibility in public studies. *Am. Sci.*, 83(Sept-Oct): 468-473.

Umali, D., Feder, G. & de Haan, C. 1992. *The balance between public and private sector activities in the delivery of livestock services.* World Bank Discussion Papers No. 63. Washington, DC, World Bank.